TEXTS AND READINGS IN MATHEMATICS **64**

Combinatorial Techniques

Combinatorial Techniques

Sharad S. Sane
Indian Institute of Technology
Mumbai

HINDUSTAN
BOOK AGENCY

Published by

Hindustan Book Agency (India)
P 19 Green Park Extension
New Delhi 110 016
India

Email: info@hindbook.com
www.hindbook.com

Paper cover Edition, 2016

ISBN 978-93-802508-4-7

To my wife Suneeta

Contents

Preface

The idea of writing a text book on combinatorics has been on my mind for a very long time. The question was that of judgment in deciding which topics are to be considered basic and must be included as also the organization of the material. I hope that the reader will find both the content and the organization of material in this book sufficiently interesting and coherent.

Whether combinatorics consists of merely techniques (read, a slightly derogatory, tricks) or also encompasses a sequence of theorems that can properly be perceived to be purporting some deep theory is not clear. The truth lies somewhere in between. To a section of the mathematical community, combinatorics is nothing more than a pastime of solving puzzles. Cleverness is an acknowledged essential ingredient of creating all mathematics. That appears to be more so in case of combinatorics where cleverness is seemingly the only tool required to achieve the goal. However, "Mathematics is not cleverness" declares the objecting section of the mathematical community. When a mathematician, working in some other area of mathematics uses the expression, "that is combinatorial", it sometimes means clever but at times also means that it amounts to messy computations devoid of any theory and an ennui inducing exercise. It is the author's intention to show to the reader that combinatorial techniques can be well studied and that these techniques, are far more systematic and sophisticated than the puzzles and tricks that they engulf and encompass.

In terms of the material presented here, I have loosely followed the contents of the texts by Brualdi, Liu, Krishnamurty and Cohen [13, 36, 35, 18]. In terms of style as well as contents, I am highly impressed by the book on combinatorics by van Lint and Wilson [57]. A large part of modern combinatorics seems to have its origin in the gambling problems of the last few centuries European society, particularly in the work of Laplace and de Moivre. Questions in probability, therefore, form a right setting for combinatorial problems, both in terms of understanding and historical perspective. An algebraization of the discipline was obtained in the concept of generating functions championed by Euler, the founder of modern combinatorics. This paved a way for the systematic use of algebra in combinatorics. To many people, algebra itself is a discrete discipline. Algebra has played such a major role in modern combinatorics that the present state of knowledge and direction in combinatorics make it appear as if it were all the time a subtopic of algebra. This, to me, appears a serious shortcoming on the pedagogical aspects of the combinatorial discipline. It is not my intention to suggest here that the role of algebra has been overestimated. On the contrary, algebra makes things very systematic and smooth. My assertion mainly pertains to the pedagogy of combinatorics, where, I believe that overuse of algebraic language pushes the reader into a jungle of symbols and the sophisticated write-up drives him away rather than inviting him to the plain cleverness of combinatorics. This also amounts to losing some historical perspective. A lack of knowledge of the framework of discrete probability on the part of a mathematician and more importantly on the part of a combinatorialist is a serious shortcoming. I have been driven by the article of Mumford [43] where a strong appeal in favour of stochasticness is made. Wherever possible, the book will also try to connect the material under consideration with other areas of mathematics particularly number theory, analysis and topology. The rigid distinction between pure and applied (or applicable) mathematics as well as the distinction between discrete and continuous mathematics is fast vanishing and in this regard, I am highly impressed by Concrete Mathematics of Knuth, Graham Knuth and Patashnik [27].

I have been associated with the Mathematics Olympiad activity in India for more than two decades. That has certainly influenced my choice of problems and exercises, some of which are borrowed from the Olympiad contests including the International Mathematical Olympiad, the IMO. However, the major contribution to the contents of this book has come from teaching the combinatorics course at the University of Mumbai besides at the University of Florida, Central Michigan University and the Michigan Technological University. Besides the books mentioned in the earlier paragraphs, I have been impressed by the article on Polya theory by de Bruijn, [4] and the essays in [33].

Every chapter discusses a famous and important result in Combinatorics which demonstrates the tremendous power of some of the very simple ideas. Organization of chapters in the book is as follows. The first six chapters could form a semester course at an undergraduate level. These include the basics of counting parameters (Chapter 1), listing combinatorial objects (Chapter 2), combinatorics of permutations (Chapter 3) and the basic inclusion exclusion principle (Chapter 4). Chapter 5 and Chapter 6 deal with probability and random variables respectively. Chapters 7, 8 and 9 have material for the Olympiad level audience and contain a large number of exercises. Many situations of the occurrence of parity arguments in combinatorics are discussed in Chapter 7. This chapter also includes the Gauss quadratic reciprocity law. Chapter 8 is on pigeonhole principle and after discussing Ramsey theory, this chapter also includes various Erdös-Szekeres theorems that use pigeonhole principle in some form. Chapter 9 deals with geometric results that have combinatorial flavour and includes the Euler equation, classification of regular polytopes, tilings and Sperner's result on triangulations. The first nine chapters can form a semester course at a slightly advanced level. Chapter 10 deals with Stirling and Catalan numbers. Chapter 11 is on recurrence relations and Chapter 12 deals with generating functions. Besides standard material on generating functions, this chapter discusses at some length the coin exchange theorem and the Dirichlet generating functions. I have separated the partition theory of integers by making it an independent Chapter 13. This chapter ends with the Euler pentagonal theorem and the material here should be useful to people interested in combinatorial number theory. Chapter 14 can be viewed as a forerunner to Chapter 15 but it can also be of independent interest since it shows the use of group action as a major tool in finite group theory. This chapter includes the class equation, Sylow theorems, automorphisms of symmetric groups and the classification of finite subgroups of the orthogonal groups in 2 and 3 dimensions. Chapter 15 deals with Polya's theory of enumeration where a large number of examples are discussed and it also includes de Bruijn's generalization of Polya theory. Chapter 16 deals with the systems of distinct representatives and includes the Birkhoff von Neumann theorem on doubly stochastic matrices and Dilworth's theorem on posets. Finally, I would like to emphasize that wherever possible, I have tried to draw a comparison between the combinatorics of infinite and finite. This is particularly visible in Chapters 7 and 9 where an idea of infinite version of Ramsey results and the Euclidean Ramsey theory respectively are discussed as well in the last chapter 16 where the Rado selection principle is discussed.

An elementary first level course could include the first eight chapters in that order. Chapters 3 through 11 are more suited for Mathematical Olympiad students. A course with emphasis on generating functions and recurrence relations should follow Chapters 3, 4, 10, 11, 12 and 13. Advanced Chapters in the text are Chapters 12 through 16. The author believes that the strength of the text also lies in the very large number of exercises at the end of each chapter. The exercises are at various levels of difficulty and range from very simple to more advanced such as Euler convergence (Exercise 6.31) and Conway's soldiers in the desert (Exercise 11.45).

I trust that the book will prove useful and interesting to a wide range of mathematical and non-mathematical community.

Sharad S. Sane

Acknowledgements

I thank all the referees of the manuscript of the text for their careful reading and detailed comments that led to considerable improvement in the final form of the book. It is a pleasure to thank a number of senior colleagues in the Mathematics Department of Mumbai University. They include M.G.Nadkarni, Nirmala Limaye, Prafullata Chawathe and Anjana Wirmani-Prasad. Special thanks to R.C.Cowsik. Discussing mathematics with all these people has been a pleasure. I also wish to thank Rajendra Pawale, Santosh Shende and Anand Kumtha, who made careful reading of parts of the manuscript of the book. University Mathematics Department library was the main source of material and I have extensively used it in the book writing project. I wish to thank colleagues C.R. Pranesachar and B.J.Venkatachala of the Mathematics Olympiad cell at the Indian Institute of Science, Bangalore for helpful discussions.

Some friends have been instrumental in pushing me to complete the book writing project and I wish to thank them. They include Ravindra Bapat of the Indian Statistical Institute, Delhi, S.A. Katre of Pune University and Sham Navathe of Computer Science Department at Georgia Tech. Sham Navathe had warned me decades ago that ninety percent of text book writing projects that authors have on their minds never see the completion. Special thanks are due to Jet Wimp for making a copy of his review [62] in the Mathematical Intelligencer (excerpts used in Chapter 8) available to me.

My father was an embodiment of hard work and perseverance and my mother that of cleverness; both would have liked to see the completion of the book writing project.

Sharad S. Sane

Chapter 1

Basic counting

1.1 Introduction

A large part of combinatorics is concerned with counting. As such this is not a very difficult activity, at least in principle. However, lack of clarity can very much make the counting obscure. The material covered in this chapter forms a basis for all the other chapters in this book because it sets the basic counting parameters required throughout the book. We begin by introducing the following elementary principles.
A man wants to travel from place A to place B either by a bus or by a train. He knows that there are five different buses he can choose from and three different trains that he take to go from A to B. Obviously then, there are eight different ways in which he can travel. If there are three different sized apples and two different sized mangoes on a table, then the number of ways of picking up one fruit among these is $3+2 = 5$. We have:
Addition Principle: If the first box contains m objects and the second box contains n objects, then the number of ways of choosing one object from either of the two boxes is $m+n$.
In a purely set-theoretic language, the addition principle tells us about the order of the union of two disjoint sets X and Y, if we know the number of objects in each of them. The main point to note here is that the sets must be disjoint. For example, The order of the set of numbers less than 30 that are either prime or perfect squares is obtained by finding the possibilities of the occurrences of the two events separately and then adding them and hence the answer is $10+5 = 15$. However if we wish to find the order of the set of numbers below 30 that are divisible by 2 or 3, the required number is not $14+9$.

Multiplication Principle: If the first box contains m objects and the second box contains n objects, then the number of ways of choosing a pair of objects, the first from the first box and the second from the second box is mn.

While the addition principle gives union of two disjoint sets, the multiplication principle gives the (order of) the Cartesian product of two sets. For example, suppose that in order to go from A to B, one must pass through C. If there are five ways of going from A to C and three ways of going from C to B, then there are fifteen ways of going from A to B.

There is a slightly more profound but equally basic technique that is applied in combinatorics. Suppose in a classroom where the students occupy seats on the benches arranged in m rows and n columns, not all the seats on all the benches are filled. If you wish to count the total number of students in the class, then there are two ways of doing it. You could fix a row and carefully count the number of students in that row. Do this for each row and then sum over all the rows. You could then ask your friend to do the same thing fixing a column first and then summing over all the columns (do not forget to count yourself if you are occupying a seat on some bench!). The two numbers must be equal. This elementary observation leads to:

Two-Way Counting : Let S be a subset of the Cartesian product of two sets A and B. Let, for a in A, R_a denote the subset $\{y \in B : (a, y) \in S\}$. Similarly, for b in B, let C_b denote the subset $\{x \in A : (x, b) \in S\}$. Then

$$\sum_{a \in A} |R_a| = \sum_{b \in B} |C_b|$$

All throughout this book, we denote the *order or cardinality* of a set T by $|T|$. Since most of the sets we deal with are *finite* the word *size* will also be used to denote this number. Also, the Cartesian product of two sets X and Y (denoted by $X \times Y$) consists of all the pairs (x, y) for which x is in X and y is in Y.The innocuous technique of two-way counting has a large number of applications in branches of Combinatorics such as *graph theory and design theory*. We illustrate this with an application.

Example 1.1.1. A graph G is a pair (V, E) where V is the set of vertices and E the set of edges. An edge is an unordered pair of vertices. We denote an edge e by the pair (xy) or (yx) with the understanding that the edge e is an edge between the two vertices x and y. In that case, we say that e is incident with x (and also with y). An edge of the form (xx) is called a loop . If we have two edges e and e' such that both are equal to (xy), then e, e' are called double edges (or multiple edges, in general) between x and y. For x in V, $d(x)$, the degree of the vertex , is simply the number of edges (xy) with y in V. Here, we count a loop (xx) twice and hence it contributes two to the degree of x. In the first part in Figure 1.1, we have a loop at x, two loops at y and a double edge between x and z. A graph G is called a *simple graph* if it has no loops or multiple edges. The second part in Figure 1.1 is a simple graph. In the first graph, the degrees of x, y, z, w are $6, 5, 2, 1$ respectively while the second graph has 5 vertices of degree 2 each and one vertex with degree 0.

If we sum over all the degrees in the graph G, then the result must be an even number since this simply counts each edge two times. We thus can not have a graph on 7 vertices with each vertex of degree 3.

We use the term an n-set to mean a set of order n. Likewise a k-subset will mean a subset of order k (similar term also applies to a superset). A *k-permutation* of a set S is an *ordered* k-tuple of elements of S. Thus a k-permutation is a sequence $(x_1, x_2, \ldots, x_k)$ where the k elements are all different (and come from S) and with an understanding that $(x_2, x_1, \ldots, x_k)$ is not the same as $(x_1, x_2, \ldots, x_k)$. How many k-permutations does an n-set have? The first element can be chosen in n ways since any of the n elements can be the first element. Having chosen the first element, the second can be chosen in $n-1$ ways (out of the remaining $n-1$ elements). Proceeding in this manner, we have the last, i.e., the k-th element chosen in $n - (k-1)$ ways.

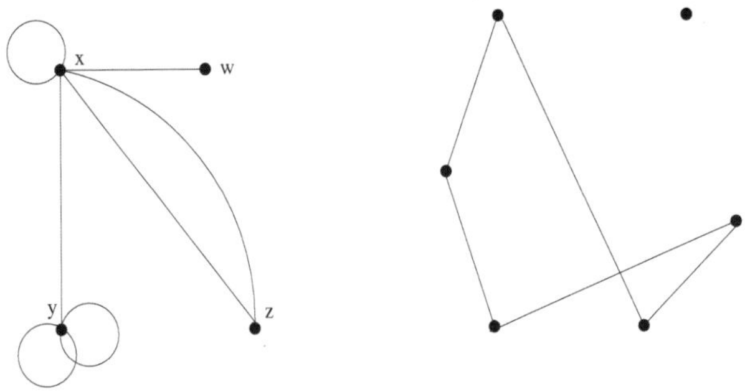

Figure 1.1: A graph and a simple graph

Lemma 1.1.2. *The number of k-permutations of an n-set is given by:*

$$P(n,k) \;=\; n(n-1)\cdots(n-k+1)$$

A slight trouble with the argument is what happens if k exceeds n. The answer then should be zero, which is what it obtains in the formula given in Lemma 1.1.2. The most important case (in Lemma 1.1.2) occurs here when k equals n. In that case we call an n-permutation of an n-set simply a permutation (of an n-set S) and $P(n,n)$ is denoted by $n!$ (pronounced n factorial; the exclamation is probably due to the fact that $n!$ is very large even for moderate values of n) which is equal to the product of all the integers from 1 to n.

Example 1.1.3. If a class has 15 students and if 5 prizes (first, second, third, fourth and fifth) are to be given to some 5 students among them, then the number of ways of doing this is $P(15,5)$ which equals $3,603,600$. However, if the class has 10 boys and 5 girls, then the number of ways in which three prizes are to be separately given to boys and two prizes separately given to girls is equal to $P(10,3) \times P(5,2) \;=\; 14,400$ by using the multiplication principle.

Notice that a permutation is an ordered selection of objects (from a set). In a similar manner, an *unordered* selection of k objects from a set of n objects is called a k-*combination.* We denote the number of k-combinations of an n-set by $C(n,k)$. If n is a positive integer, then $C(n,0)$ equals 1 and $C(n,k)$ is 0 for all $k \;>\; n$. Since a set, by definition, is an unordered collection, a k-combination of a set is simply a k-subset of the given n-set.

Example 1.1.4. Consider the problem in 1.1.3 and suppose that we are merely interested in selecting three boys and two girls to distribute the prizes (without ranking them). Then the required number is $C(10,3) \times C(5,2) = 1,200$.

Lemma 1.1.5. *Let n be a positive integer and let k be a non-negative integer. Then*

$$C(n,k) = \frac{P(n,k)}{P(k,k)} = \frac{n(n-1)\cdots(n-k+1)}{k!}$$

Proof Make a two-way counting. How many k-permutations does an n-set have? We first select a k-subset in $C(n,k)$ ways and then order (permute) the elements of the chosen k-subset in $P(k,k)$ ways. But this number is just $P(n,k)$ which obtains the required formula using Lemma 1.1.2. Also, the statement trivially holds if $k > n$. □

A large number of problems of (elementary) combinatorics are solved by determining whether the required answer is a combination or a permutation. Since $n! = n \times (n-1) \times \cdots \times 2 \times 1$, we write $[n]_k$ to denote $P(n,k) = n(n-1)\cdots(n-k+1)$ and call it the *falling factorial*. By convention, $0! = 1$. We then have:

$$C(n,k) = \frac{[n]_k}{k!} = \frac{[n]_k(n-k)!}{k!(n-k)!} = \frac{n!}{k!(n-k)!}$$

Observe the arithmetical symmetry between k and $n-k$ in the above formula. This, of course, is a reflection of the fact that $C(n,k)$ and $C(n,n-k)$ are the same numbers:

Lemma 1.1.6. *If k is a non-negative integer and n a positive integer such that $k \leq n$, then $C(n,k) = C(n,n-k)$.*

Proof Let S be an n-set. In choosing a k-subset A of S, we are also rejecting an $(n-k)$-subset of S, namely the $(n-k)$-subset B which is the complement of A in S. It is like this. Any time you select k elements of S, ask your friend to select the remaining $n-k$ elements of S. The process is reversible. If your friend picks up $n-k$ elements, then you pick up the remaining k elements. This sets up a bijection (one-to one correspondence) between the set of k-subsets (that you select) and the set of $n-k$-subsets (that your friend selects). Hence we have $C(n,k) = C(n,n-k)$. □

1.2 Bijections

If we have an alphabet consisting of only three letters a, b and c and we wish to form words of length 5 using the three letters of our alphabet, then the number of ways of doing this is $3 \times 3 \times 3 \times 3 \times 3 = 3^5 = 243$. Similarly the number of binary sequences (i.e., sequences of 0's and 1's) of length n is 2^n. Interestingly, this also indirectly counts the total number of subsets of a set of order n. Let $T = \{1,2,\ldots,n\}$. *Throughout this book, we use the symbol $[n]$ to mean the set* $\{1,2,\ldots,n\}$. For each subset X of T associate a binary sequence b_X of length n where b_X has 1 in the position i if i is in X and has 0 in the i-th position if i is not in X. For example, if $n = 5$, and if X, Y, Z are respectively the empty set, the set $\{2,3,5\}$, and the set $\{1,2,3,4,5\}$, then the corresponding sequences b_X, b_Y and b_Z are 00000, 01101 and 11111 respectively.

This procedure obtains a pairing of the members of the set of all the subsets of T with the members of the set consisting of all the binary sequences of length n. Since the latter set has size 2^n, we readily have:

Corollary 1.2.1. *The total number of subsets of a set of order n is 2^n.*

As the discussion preceding Corollary 1.2.1 shows, more is true than what is stated in Corollary 1.2.1. Namely, we actually have an explicit bijection between the two sets: The set of all the subsets of an n-set and the set of all the binary sequences of length n. The proof technique used here is called *proof by bijection.* This is among the most powerful and elementary techniques of combinatorics and a substantial part of combinatorial activity today is devoted to finding ingenious bijections to count the number of objects under certain stipulations. We recall that a bijection is an injective (one-to-one) and surjective (onto) function. An interested reader is encouraged to look up the book by Stanton and White [50], where a large number of results are proved using bijections. One important point to observe here is that the *proofs that use bijection are genuinely combinatorial in nature.* An example of this is furnished in the proof of Lemma 1.1.6. We content ourselves with giving some obvious bijections.

Theorem 1.2.2. *Let m and n be positive integers. Then there exists a bijection between any two of the following sets.*

(a) The set of all the functions from an n-set to an m-set.

(b) The set of words of length n on an alphabet of m letters.

(c) The set of n-tuples (sequences) with entries from an m-set.

(d) The set consisting of all the ways of distributing n distinct objects into m distinct boxes (or cells).

In each case, the cardinality of the set in question is m^n.

Proof Let $D = \{1, 2, \ldots, n\}$ be the n-set and let $R = \{r_1, r_2, \ldots, r_m\}$. Given any function f from D to R, we write down the sequence $(f(1), f(2), \ldots, f(n))$ of length n with entries from R. This process is reversible and clearly sets a bijection between the set in (a) with those in (b) or (c). For (d), let the m distinct boxes be denoted by $B_1, B_2, \ldots, B_m$ and let the n distinct objects be denoted by the elements of D. Given a function f from D to R, put object i in the box B_j if and only if $f(i) = r_j$. This obtains the required bijection. □

Recall that an injective function is a function f for which $f(a) = f(b)$ implies $a = b$.

Theorem 1.2.3. *Let m and n be positive integers. Then there exists a bijection between any two of the following sets.*

(a) The set of all the injective functions from an n-set to an m-set.

(b) The set of words of length n on an alphabet of m letters with the condition that the word consists of distinct letters.

(c) The set of n-tuples (sequences) with distinct entries from an m-set.

(d) The set consisting of all the ways of distributing n distinct objects into m distinct boxes (or cells) with the condition that no box holds more than one object.

(e) The set of all the n-permutations of an m-set.

In each case, the cardinality of the set in question is $P(m,n) = [m]_n$.

Proof Let $D = \{d_1, d_2, \ldots, d_n\}$ and let R be the set of integers $1, 2, \ldots, m$. Any n-permutation of R say, $(a_1, a_2, \ldots, a_n)$ may be viewed as an injective function from D to R which sends d_1 to a_1, d_2 to a_2, $\cdots$, d_n to a_n (note that a_i's are distinct and are ordered). This sets up a bijection between the sets in (e) and (a). For (b), we may think of an alphabet precisely consisting of the elements of R. Since (b) stipulates that words must consist of distinct letters, we have a bijection between the set in (a) and the set in (b). The bijection between (c) and (a) is similar and is left to the reader. For (d), treat the elements of D as the objects, and let $B_1, B_2, \ldots, B_m$ be the set of m (distinct) boxes. Given any permutation, i.e., a function say $(a_1, a_2, \ldots, a_n)$ as above, put the object d_i in the k-th box B_k iff $a_i = k$. *All throughout this book, the term iff is used to mean "if and only if"*. It is easy to check that this sets a bijection between the set in (d) and the one in (a) (or (e)). □

Definition 1.2.4. For a real number α and any non-negative integer k,

$$\binom{\alpha}{k} = \frac{\alpha(\alpha-1)\cdots(\alpha-k+1)}{k!}$$

Pronounce $\binom{\alpha}{k}$ as α choose k. By convention, we let $\binom{\alpha}{0}$ equal 1. Thus, if α is a positive integer, then $\binom{\alpha}{k}$ is same as $C(\alpha, k)$. As an example,

$$\binom{-1/2}{3} = \frac{-1/2 \times -3/2 \times -5/2}{6} = -\frac{3 \times 5}{3 \times 2^4}$$

equals $\frac{-5}{16}$. For reasons that will become clear after we prove the following theorem (Theorem 1.2.5), *the numbers $\binom{\alpha}{k}$ are called binomial coefficients. The two-line notation given in Definition 1.2.4 that we follow throughout this book is the modern notation and the reader is strongly urged to follow it.* At school level, one encounters expressions such as $(x+y)^2 = x^2+2xy+y^2$ and $(x+y)^3 = x^3+3x^2y+3xy^2+y^3$. Observe that the right hand sides of these equations are homogeneous polynomials in x and y, i.e., polynomials in which in each summand, the exponents of x and y add to a fixed number. This fixed number is called the degree of the homogeneous polynomial. We have the following theorem which dates back to at least the medieval times and was known to various civilizations.

Theorem 1.2.5. *(The Binomial Theorem) Let n be a non-negative integer. Then:*

$$(x+y)^n = \binom{n}{0}x^n + \binom{n}{1}x^{n-1}y + \binom{n}{2}x^{n-2}y^2 + \cdots + + \binom{n}{n}y^n$$

Proof We first apply the multiplication principle. Indeed, $(x + y)^n$ is simply $(x + y)(x + y) \cdots (x + y)$, n times. Multiplying out all the terms amounts to doing the following. There are n boxes with each box containing both x and y. We choose x's from some boxes and y's from the remaining. If we choose x's from $n - k$ boxes and y's from the remaining k boxes, then that obtains a summand $x^{n-k}y^k$. Thus we already know that the resulting polynomial is homogeneous of degree n. It just remains to find the coefficient of the term $x^{n-k}y^k$. Obviously, we have to choose $n - k$ boxes for x (and hence the remaining k boxes for y). Since there are n boxes in all, the coefficient of $x^{n-k}y^k$ must be $\binom{n}{k}$. This completes the proof of the binomial theorem. □

For a brief history of the binomial theorem, refer to Knuth [34]. In the statement of the binomial theorem (Theorem 1.2.5), we did not specify as to what x and y are. In some sense, the reason for choosing to do so is that *it does not matter*. In other words given the usual laws of addition and multiplication (of natural numbers), the binomial theorem (Theorem 1.2.5) is an always true statement. This is expressed in the mathematical parlance by saying that the binomial theorem is a 'formal identity' (or a combinatorial identity). This, in particular, means that the statement of binomial theorem is true if we let the variables (x and y) to be any real or complex numbers. But still more important is the fact that the theorem is a formal identity. To be mathematically more precise, the binomial theorem holds over any commutative ring with identity. All through combinatorics, we shall have occasions to meet many such formal identities, a combinatorial theory of which will be reasonably formulated and formalized in Chapter 12 on generating functions.An uninteresting proof of the binomial theorem (Theorem 1.2.5)involves induction on n via the use of Pascal identity.

Theorem 1.2.6. *(Pascal identity) Let n and k be positive integers. Then*

$$\binom{n}{k} = \binom{n-1}{k} + \binom{n-1}{k-1}$$

Proof Let X be an n-set and fix an element x of X. and let Y denote the set $X - \{x\}$. For any k-combination (i.e. a k-subset) A of X, either x is in A or x is not in A. In the first case if B is the set $A - \{x\}$, then B is a $(k - 1)$-subset of Y and hence can be chosen in $\binom{n-1}{k-1}$ ways, while in the second case, A is itself a k-subset of Y and hence can be chosen in $\binom{n-1}{k}$ ways. The proof is complete by invoking the addition principle. □

It is also easy to see that Pascal's identity (Theorem 1.2.6) follows from the binomial theorem (Theorem 1.2.5) (if we have proved the latter without using the former as we did). Pascal's identity gives rise to the famous Pascal triangle, initial portion of which is drawn below. Each entry is obtained from the two entries directly above it as given Pascal's identity. This obtains all the binomial coefficients $\binom{n}{k}$, where n runs from 1 to 6 and the horizontal lines correspond to a fixed value of n. Note that Pascal triangle is an infinite triangle; only a finite portion of this triangle (from 1 to 6) is shown in

Figure 1.2. A more familiar form of the binomial theorem (Theorem 1.2.5) is:

$$(1+x)^n = \sum_{k=0}^{n} \binom{n}{k} x^k$$

which is obtained by letting $y = 1$ in the binomial theorem. By making the substitution $x = 1$ in the above expression, we obtain:

$$\sum_{k=0}^{n} \binom{n}{k} = 2^n$$

Stated in other words, the above identity tells us that a set of order n has 2^n subsets in all. We have already seen this in Corollary 1.2.1. Again, the substitution $x = -1$ yields

$$\sum_{k} \binom{n}{2k} = \sum_{k} \binom{n}{2k+1}$$

Thus, in any set the number of subsets of odd order is the same as the number of subsets of even order.

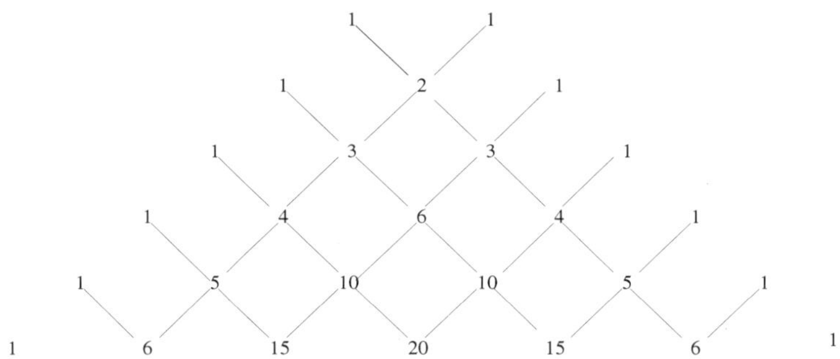

Figure 1.2: Pascal Triangle

A large number of identities involving binomial coefficients are actually proved either using a known combinatorial identity or a known polynomial expression such as the binomial theorem (Theorem 1.2.5) and manipulating it. For example, to prove that

$$\binom{n}{1} - 2\binom{n}{2} + 3\binom{n}{3} - 4\binom{n}{4} + \cdots + (-1)^{n-1} n \binom{n}{n} = 0$$

we observe that each summand on the left hand side (ignoring the sign) has the form

$$\begin{aligned} k\binom{n}{k} &= k\frac{n}{k}\binom{n-1}{k-1} \\ &= n\binom{n-1}{k-1} \end{aligned}$$

Hence the left hand side reduces to the alternating sum

$$n \times \sum_{i=0}^{n-1} (-1)^i \binom{n-1}{i} = 0$$

using the binomial theorem (Theorem 1.2.5). Similarly, to find $n \sum \frac{1}{k+1}\binom{n}{k}$, we take the familiar form of the binomial theorem and integrate both sides (as polynomials in x) w.r.t. x from 0 to 1.

Example 1.2.7. A binary block code C of length n is a set of binary sequences of length n. For two words $x = x_1 x_2 \cdots x_n$ and $y = y_1 y_2 \cdots y_n$ we define the *Hamming distance (or distance)* between x and y to be the number of i's where x_i and y_i differ. For example, let $x = 1000110$ and $y = 1101001$. Then the Hamming distance between x and y which we write as $d(x, y)$ is 5. The distance d is a metric on the set of all the binary words of length n. For a word x, how many words are exactly at distance k from x? Since any such word differs from x at exactly k places, and since the length of a word is n, there are precisely $\binom{n}{k}$ words that are at distance k from x. The codewords are transmitted over a communication channel (which is like a telephone line). Since the channel is noisy (prone to make mistakes), the word that is received at the other end of the channel may not be the same as the transmitted word x. The person (or a device, called the decoder) at other side of the channel has, however, a list of all the codewords, i.e., the list of all the words in C. He tries to match the received word with a codeword that is nearest to it. If y is the received word and z is the nearest codeword (word in the code C), then the decoder interprets this as "z must have been sent". It then follows that if it is known beforehand that the channel makes no more than r errors (i.e. at most r of the n positions can have 0's and 1's interchanged by the channel) and if any word is at distance less than or equal to r from at most one codeword, then we will be able to correct all the errors and recover the codeword that was sent. This is called the *nearest neighbor decoding.* (In the above case, $d(x, y) \leq r$, and hence, if z and x are different, then $d(y, z) \geq r + 1$, so that the decoder will recover x and not z). For this to happen, we must have the following condition satisfied: The distance between any two codewords must be at least $2r + 1$. If we draw balls of radius r around each codeword, then no word should belong to two such balls. We call a code C with words of length n an (n, r)-code if any two balls of radius r drawn with centers at two codewords are disjoint from each other, i.e., contain no word in common. For a good code, we should have r as large as possible and also $|C|$ as large as possible since having more codewords amounts to being able to send more information. The following gives an upper bound (called the *Hamming bound*) on the number of codewords in an (n, r)-code C.

$$|C| \leq \frac{2^n}{\binom{n}{0} + \binom{n}{1} + \cdots + \binom{n}{r}}$$

For a proof, observe that there are 2^n words in all (some of which are codewords). If x is a codeword, then a ball of radius r drawn around x contains no codeword other than x. Since the number of words at distance at most r from x is given by the denominator of the right hand side of the above expression and since the total number of words is 2^n, a two-way counting produces the desired result.

Definition 1.2.8. Let r be a real number. *By* $\lfloor r \rfloor$, *we mean the largest integer less than or equal to* r *and by* $\lceil r \rceil$, *we mean the smallest integer greater than or equal to* r.

Thus, $\lfloor \pi \rfloor = 3$ while $\lceil \pi \rceil = 4$ and $\lfloor 7 \rfloor = \lceil 7 \rceil = 7$.

Theorem 1.2.9. *Fix a positive integer* n. *Then the number of odd binomial coefficients among the* $n+1$ *numbers:* $\binom{n}{0}, \binom{n}{1}, \ldots, \binom{n}{n-1}, \binom{n}{n}$ *is a power of* 2.

Though a shorter proof of this theorem will be given as an exercise in Chapter 12 (Exercise 12.55), we prefer to give the following elementary and elegant proof, which we break into seven parts.

(a) Let p be a prime number. We say that p divides an integer u exactly m times if p^m divides u but p^{m+1} does not. Let n be natural number and let p a prime. Suppose p divides $n!$ exactly m times. Then

$$m = \left\lfloor \frac{n}{p} \right\rfloor + \left\lfloor \frac{n}{p^2} \right\rfloor + \left\lfloor \frac{n}{p^3} \right\rfloor + \cdots$$

To prove this, write down $n!$ as the product $1 \times 2 \times \cdots \times (n-1) \times n$. In that expression, the multiples of p are $p, 2p, \ldots$ and hence $\left\lfloor \frac{n}{p} \right\rfloor$ in number. The terms such as p^2 should actually be counted two times and they have already been counted once. Again the multiples of p^2 are $p^2, 2p^2, \ldots$ and hence $\left\lfloor \frac{n}{p^2} \right\rfloor$ in number. For these numbers that have been already counted once, we need to add one for each such and hence we obtain the second summand. Then on to multiples of p^3 and continuing in this fashion, we obtain the required result.

(b) Given a non-negative integer n, an expression of the form $a_r a_{r-1} \cdots a_1 a_0$ is called the binary representation of n if

$$n = \sum_{i=0}^{i=r} a_i 2^i$$

and provided that each a_i is 0 or 1. For the sake of uniqueness, We also stipulate that a_r is not 0. It is then easily seen (exercise) that the binary expansion of any non-negative integer is uniquely determined by that integer.

(c) Let a, b, c be non-negative integers and let u be a positive integer. Suppose $a = b + c$. What can we say about $\left\lfloor \frac{a}{u} \right\rfloor$, $\left\lfloor \frac{b}{u} \right\rfloor$ and $\left\lfloor \frac{c}{u} \right\rfloor$? We leave it to the reader to check that

$$\left\lfloor \frac{a}{u} \right\rfloor = \left\lfloor \frac{b}{u} \right\rfloor + \left\lfloor \frac{c}{u} \right\rfloor$$

except when the remainders obtained on dividing b by u and c by u add to a number greater than or equal to u in which case the L.H.S. exceeds the R.H.S. by 1.

(d) Turning back to the binary representation in (b), if n has binary representation $a_r a_{r-1} \cdots a_1 a_0$,then $\lfloor \frac{n}{2^j} \rfloor$ has binary representation $a_r a_{r-1} \cdots a_j$ (check this).

(e) Now let n be a positive integer and let m and k be non-negative integers with $n = m + k$. Since $\binom{n}{k} = \frac{n!}{k!m!}$, the number $\binom{n}{k}$ will be odd (in view of (a) and (c)) if and only if

$$\left\lfloor \frac{n}{2^j} \right\rfloor = \left\lfloor \frac{k}{2^j} \right\rfloor + \left\lfloor \frac{m}{2^j} \right\rfloor$$

for all j. For example, with $j = 0$, if k and m are both odd, then $\binom{n}{k}$ will have a power of 2 surviving and hence will not be an odd integer.

(f) If we now let $a_r a_{r-1} \cdots a_1 a_0$, $b_r b_{r-1} \cdots b_1 b_0$ and $c_r c_{r-1} \cdots c_1 c_0$ to be respectively the binary representations of n, k, m (where, we may add zeros to the left of the representation so as to make them all of the same length) then using (d) and (e) the binomial coefficient $\binom{n}{k}$ will be odd if and only if in writing

$$a_r a_{r-1} \cdots a_j = b_r b_{r-1} \cdots b_j + c_r c_{r-1} \cdots c_j$$

with binary (i.e. modulo 2) addition, there is no carry at any stage of addition. (For example, in the usual 10-based system, a child with no knowledge of 'carry', will make the correct addition of 23 and 45 but will not be able to add 47 and 38 correctly!)

(g) Let the binary representations of n and k be as given above. Given n, if we have to choose k so that the binary coefficient $\binom{n}{k}$ is odd, then we must have a "no carry" situation at each level of addition in (f). Hence if a_i is 0, then b_i must be 0 (and hence c_i is 0), while if a_i is 1, then b_i can be either 1 or 0 (and correspondingly c_i will be 0 or 1 respectively to make the correct binary addition). In any case, each such i gives two choices for b_i. Using the multiplication principle, the number of odd binomial coefficients is a multiple of some 1's and some 2's and is therefore a power of 2. □

For the sake of completeness, we give the following general form of the binomial theorem called Newton's binomial theorem. This can be proved using Taylor's theorem (in fact, it is just a power series expansion with a suitable radius of convergence).

Theorem 1.2.10. *Let α be real number and let x be a real number with absolute value less than 1. Then*

$$(1 + x)^\alpha = \sum_{k=0}^{\infty} \binom{\alpha}{k} x^k$$

1.3 Counting objects with repetitions

In the example of apples and mangoes given earlier (at the beginning of the present chapter), we had three apples (of different sizes) and two mangoes (of different sizes).

If the three apples were all identical and the two mangoes can not be distinguished from each other, and if we still have to pick up one fruit among these, in how many ways can that be done? The answer is not 5 but is 2 since picking up either of the two mangoes or any one of the three apples makes no difference. In mathematical terms, we say that we are counting objects with repetition or objects are drawn with replacement or we are dealing not with a set of elements but a multi-set. A multi-set $M = \{a_1.x_1, a_2.x_2, \ldots, a_n.x_n\}$ is a collection of distinct objects $x_1, x_2, \ldots, x_n$ with the object x_i occurring a_i times, $i = 1, 2, \ldots, n$. Here a_i's are all non-negative integers and we say that a_i is the multiplicity of the object x_i. The set $S = \{x_1, x_2, \ldots, x_n\}$ will be called the underlying set of the multi-set M. We also allow the possibility of some or all a_i's equal to ∞. This is to be understood as 'the object is available in unlimited supply'. We talk of permutations and combinations of a multi-set in the same way as that of a set. Initially, it might appear that the formulas for permutations and combinations of a multi-set might be as easy as those for sets and might even have straightforward relationships of the kind given in the two Lemmas 1.1.2 and 1.1.5. It is not that simple. For example, if $M = \{4.x_1,\ 2.x_2,\ \infty.x_3\}$ then to pick up 10 objects of M, we need to consider different cases: we must pick up j copies of x_2, where j equals $0, 1$ or 2. The number of 10-combinations of M is then 15. As another example, the number of r-permutations of the multi-set $M = \{\infty.x\}$ is just one for any value of r.

Theorem 1.3.1. *Let M be a multi-set consisting of r distinct objects, each with infinite multiplicity. Then the total number of d-permutations of M is r^d.*

Proof This simply amounts to counting the number of sequences of length d on an alphabet that consists of r distinct letters. We never run short of any letter since it is allowed to repeat any number of times. Hence by Theorem 1.2.2 (c), the required number is r^d. □

In defining the binomial coefficient $\binom{n}{k}$, we actually count the number of ways of putting n distinct objects into two distinct boxes labeled B_1 and B_2 such that the first box holds k objects and the second box holds the remaining $n-k$ objects. Generalizing this situation prompts us to make the following definition of multinomial coefficient: The number of ways of putting n distinct objects in r distinct boxes $B_1, B_2, \ldots, B_r$ such that the i-th box B_i holds n_i objects is called a multinomial coefficient and is denoted by $\binom{n}{n_1, n_2, \ldots, n_r}$. Necessarily then, $n_1 + n_2 + \cdots + n_r = n$. Thus $\binom{n}{k} = \binom{n}{k,\ n-k}$.

Theorem 1.3.2. *Let S be an n-set and suppose the n objects in S are to be put in r distinct boxes $B_1, B_2, \cdots, B_r$ such that the i-th box B_i contains n_i objects with $n_1 + n_2 + \cdots + n_r = n$. Then the number of ways of doing this is equal to*

$$\binom{n}{n_1, n_2, \ldots, n_r} = \frac{n!}{n_1!n_2!\cdots n_r!}$$

Proof Though a direct proof can be given, we prefer to make an induction on r, purely for pedagogical reasons. For $r = 2$, Lemma 1.1.5 and the discussion following it show that the statement of the theorem is true. Let $r \geq 3$. Out of n objects, we may first choose n_1 objects to put in the first box B_1 in $C(n, n_1) = \binom{n}{n_1}$ ways and then try to put the remaining $n - n_1$ objects in the other $r - 1$ boxes. Having performed the first task, that can be done in $\binom{n-n_1}{n_2, n_3, \ldots, n_r}$ ways. Now using induction on r,

$$\binom{n - n_1}{n_2, n_3, \ldots, n_r} = \frac{(n - n_1)!}{n_2! \cdots n_r!}$$

Hence the required number equals

$$\frac{n!}{n_1!(n - n_1)!} \times \frac{(n - n_1)!}{n_2! \cdots n_r!} = \frac{n!}{n_1! n_2! \cdots n_r!}$$

□

Corollary 1.3.3. *Let M be a multi-set consisting of r distinct objects $x_1, x_2, \ldots, x_r$ such that the i-th object x_i has multiplicity n_i. Let $n = n_1 + n_2 + \cdots + n_r$. Then the total number of n-permutations of M is*

$$\binom{n}{n_1, n_2, \ldots, n_r} = \frac{n!}{n_1! n_2! \cdots n_r!}$$

Proof We set up a bijection between the required set of all the ways of putting the elements of S in r boxes and all the n-permutations of the multi-set

$$M = \{n_1.x_1, n_2.x_2, \ldots, n_r.x_r\}$$

First number the elements of S from 1 to n. If the element i is put in the box B_j, then make an n-permutation in which the i-th place is occupied by x_j. Conversely, given an n-permutation of M, if we find the i-th place occupied by x_j then put the element i in the box B_j. Hence the result is proved using bijection and Theorem 1.3.2. □

It is also possible to prove Corollary 1.3.3 without using Theorem 1.3.2.

Theorem 1.3.4. *(The multinomial theorem) Let n be a non-negative integer. Then:*

$$(x_1 + x_2 + \cdots + x_r)^n = \sum_{n_1+n_2+\cdots+n_r = n} \binom{n}{n_1, n_2, \ldots, n_r} x_1^{n_1} x_2^{n_2} \cdots x_r^{n_r}$$

where the sum is taken over all the non-negative integers n_i that satisfy the given stipulation.

Proof We sketch three different proofs.
First Proof Make induction on r using the fact that the assertion holds for $r = 2$, which is the binomial theorem (Theorem 1.2.5).
Second Proof Make induction on n and use Pascal type identities.
Third (direct) Proof Expand directly imitating the proof of the binomial theorem (Theorem 1.2.5) and use Theorem 1.3.2. □

Theorem 1.3.5. *Let M be a multi-set with r distinct objects $x_1, x_2, \ldots, x_r$ each with infinite multiplicity. Let $\overline{C}(r,k)$ denote the number of k-combinations of M. Then $\overline{C}(r,k) = \binom{k+r-1}{k}$.*

Proof Every k-combination is uniquely determined by a sequence $b_1, b_2, \ldots, b_r$ where b_i's are all non-negative integers and $b_1 + b_2 + \cdots + b_r = k$. For each k-combination of M make a binary sequence of length $k + r - 1$ as follows. At the beginning, write b_1 zeros and follow this by a 1, then write b_2 zeros and then write a 1 and so on. There will be a final 1 separating the b_{r-1} zeros and the last b_r zeros. Thus the binary sequence will consist of exactly $b_1 + b_2 + \cdots + b_r = k$ zeros and $r - 1$ ones. Conversely given a binary sequence of length $k + r - 1$ consisting of k zeros and $r - 1$ ones, we read the number of zeros to the left of the first one and call it b_1, then the number of zeros between the first one and the second and call it b_2 and so on. Finally the number of zeros to the right of the last one is b_r. For example, with $r = 4$ and $k = 6$, the sequence $2, 1, 1, 2$ gives the binary sequence 001010100 while the binary sequence 100011000 must have come from $(0, 3, 0, 3)$. Since the number of binary sequences of length $k + r - 1$ with exactly k zeros equals $C(k + r - 1.k)$, the result is proved using bijection. □

Let S be an r-set. We call S an ordered set if there exists an order $<$ on the elements of S which is a total order (or a chain). That is, the elements of S can be written in the form $c_1, c_2 \ldots, c_r$ where $c_1 < c_2 < \cdots < c_r$. A sequence $(x_1, x_2, \ldots, x_k)$ with entries from the set S is said to be a monotonically increasing sequence if $x_1 \leq x_2 \leq \cdots \leq x_k$. We then have

Theorem 1.3.6. *The following sets are in bijective correspondence.*

(a) The set of all increasing sequences of length k on an ordered set with r elements.

(b) The set of all the ways of putting k identical objects into r distinct boxes.

(c) The set of all the k-combinations of a multi-set with r distinct elements.

In all the three cases, the cardinality of the set is $C(k + r - 1, k)$.

While we leave the proof of the above theorem to the reader, to conclude this section, we also note some interesting connections.

$$\overline{C}(r,k) = \binom{k+r-1}{k} = (-1)^k \binom{-r}{k}$$

Now let $[r]^k$ denote the product $r(r+1)\cdots(r+k-1)$. This called the *rising factorial*. We then have

$$\overline{C}(r,k) = \frac{[r]^k}{k!}$$

Theorem 1.3.7. *The number of ways of putting k identical objects into r distinct boxes with each box containing at least one object is $\binom{k-1}{r-1}$.*

Proof Follow the proof of Theorem 1.3.5. We have to find the totality of binary sequences of length $k + r - 1$ with no two 1's adjacent and the sequences do not begin or end with a 1. Given such a binary sequence remove one zero between every two adjacent 1's and also one zero each from the left and the right ends. This leaves us with a binary sequence of length $k-1$ with exactly $r-1$ ones. The process is reversible. Since the number of binary sequences of length $k-1$ with $r-1$ ones is $\binom{k-1}{r-1}$, the assertion is proved. □

Example 1.3.8. *An application to Statistical Mechanics:* In Statistical Mechanics, one encounters the situation of putting k particles into r distinct energy levels. The particles can thus be considered as objects and the different energy levels as distinct boxes or cells. Three different situations (statistics) are obtained by making three different assumptions. These are

(a) Maxwell-Boltzman: Here the particles are all distinct and any number of particles can be put in any of the r boxes. The number of possibilities (as given by Theorem 1.3.1) is r^k.

(b) Bose-Einstein: Here the particles are all identical and any number of particles of particles can be put in any of the r boxes. The number of possibilities (as given by Theorem 1.3.6) is $\binom{k+r-1}{k}$.

(c) Fermi-Dirac: Here the particles are all identical but no box can hold more than one particle. The number of possibilities is $\binom{r}{k}$.

1.4 Two-way counting revisited: the de Bruijn-Erdös Theorem

In this last part of the discussion on basic counting techniques, we give a somewhat sophisticated and deep application of two-way counting. The result known as the de Bruijn-Erdös theorem was proved by the authors using repeated applications of two-way counting. This theorem first appears in the literature in 1948. However, Erdös knew its proof ten years prior to its appearance in print. But he did not publish it at that time because "It was considered relatively less important to do mathematics of that sort!" All the combinatorial proofs of the de Bruijn-Erdös theorem tend to be somewhat messy. A short proof (due to Conway) given in van Lint and Wilson [57] is discussed here.

A (finite) incidence structure $\mathbf{I}$ is a pair $\mathbf{I} = (\mathbf{P}, \mathbf{L})$ where $\mathbf{P}$ is (finite) set called the set of points and $\mathbf{L}$ is a set of subsets of $\mathbf{P}$. Each member of $\mathbf{L}$ is called a line. A *linear space* is an incidence structure in which every pair of points is contained in a unique line (thus *no* two lines intersect in two or more points). To avoid obvious trivialities, we stipulate that every line has size at least two but no line contains all the points (in that case there will be no other line). Letting the number of points to be v and the number of lines to be b, what relationship do these two integers have in general?

There are two special linear spaces of interest. A linear space is called a *near pencil* if some line has size $v-1$ (and hence necessarily other lines have size two each). In this case, there is just one more point outside the line of size $v-1$ and this point is on $v-1$ lines each of size two. We thus have $v = b$. Clearly a near pencil can be constructed for all the values of $v \geq 3$. A more special linear space is what is called a *projective plane*. Here all the lines have the same size $n+1$ for some $n \geq 2$ and further any two lines intersect each other. This object is called a projective plane of order n. Figure 1.3 shows a projective plane of order two and a near-pencil with $v = 6$.

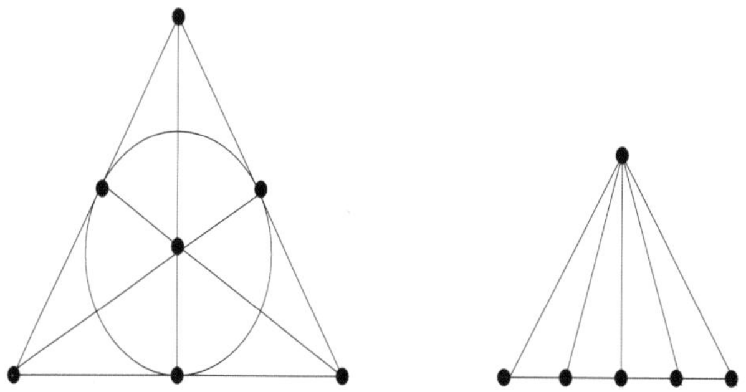

Figure 1.3: Fano Plane and Near-Pencil

Answering the question posed in the previous paragraph, we have the following theorem of de Bruijn and Erdös:

Theorem 1.4.1. *In a linear space* **I**, *we must have* $v \leq b$ *with equality iff the linear space is a near pencil or a projective plane.*

Proof All the combinatorial proofs to the above theorem depend on one or more applications of two-way counting. We first make some elementary observations. For a point x of the linear space **I** under consideration, let r_x denote the number of lines of **I** that contain x and for a line L of **I**, let k_L denote the number of points contained in L. We proceed through the following claims.

(a) If $x \notin L$, then $r_x \geq k_L$ with equality iff each line containing x meets L.

Proof For each y on L we must have a line containing both x and y and no two such lines can be identical for then the line L will intersect some line on x in two points. Further equality can hold iff there is no line containing x which is disjoint from L.

(b) Every two lines meet each other (i.e. are not disjoint) if for all $x \notin L, r_x = k_L$. This is obvious from claim (a).

(c) We have

$$\sum_{y \in \mathbf{P}} r_y = \sum_{L \in \mathbf{L}} k_L$$

Proof Make a two-way counting on the set S consisting of all the pairs (y, L) where y is a point on the line L.

At this point, we assume that $b \leq v$ *and then prove that this leads to* $b = v$ *and* $\mathbf{I}$ *must be a near-pencil or a projective plane.*

(d) Let L and M be two lines such that $L \cup M = \mathbf{P}$. Then we have

$$L \cap M \neq \emptyset; \;\; b = v$$

and **I** is indeed a near pencil.

Proof If $L \cap M \neq \emptyset$ then let $\{p\} = L \cap M$. It then follows that L and M are the only lines containing p and hence every other line Z must intersect each one of L and M in a single point. In particular, we must have $|Z| = 2$ for such a line Z. Let $\alpha = |L| \geq |M| = \beta \geq 2$. We see that if $L \cap M = \emptyset$, then $v = \alpha + \beta$ and $b = 2 + \alpha\beta$ while if $L \cap M \neq \emptyset$, then $v = \alpha + \beta - 1$ and $b = 2 + (\alpha - 1)(\beta - 1)$. In the former case, we get (using $b \leq v$), $(\alpha - 1)(\beta - 1) + 1 \leq 0$ which is absurd and in the latter case, we get $(\alpha - 2)(\beta - 2) \leq 0$ showing that $\beta = 2$ and hence $\alpha = v - 1$ showing that we have a near pencil as desired.

(e) Let **I** be not a near pencil. Let every two lines have a non-empty intersection. Then **I** is a projective plane.

Proof Let L and M be two lines. Using claim (d), we see that there is a point $x \notin L \cup M$. Then the hypothesis implies that $k_L = r_x = k_M$ showing that all the lines have the same number of points $k = n + 1$ where $n \geq 2$. Then all the points x also satisfy $r_x = n + 1$ and we have a projective plane as desired.

We now finish the proof of the theorem using the following clever argument of Conway as quoted in the book [57] by van Lint and Wilson. Using (b) and (d), it will suffice to show that $r_x = k_L$ for all the pairs (x, L) with x not on L. To that end, fix such a pair (x, L) with x not on L. Then (a) shows that $r_x \geq k_L$ and hence because of the assumption $vr_x \geq bk_L$, i.e. $-vr_x \leq -bk_L$, i.e. $v(b - r_x) \leq b(v - k_L)$. Thus

$$\frac{1}{v(b - r_x)} \geq \frac{1}{b(v - k_L)}$$

Then sum both sides of the inequality over all the elements of the set T consisting of all the pairs (x, L) with x not on the line L. Consider the L.H.S. of the above inequality. Fix a point x and sum the expression over all the lines not containing x. These are $b - r_x$ in number, and then summing over all the points x must sum the L.H.S. to 1. Use two-way counting to change the order of summation and sum the expression on the R.H.S. over all the elements of T to obtain 1 again. We thus have:

$$1 = \sum_{x \in \mathbf{P}} \sum_{x \notin L} \frac{1}{v(b - r_x)} \geq \sum_{L \in \mathbf{L}} \sum_{x \notin L} \frac{1}{b(v - k_L)} = 1$$

Therefore equality must hold everywhere and $r_x = k_L$ for all the pairs x, L with x not all L and we are done. □

A slightly different proof of Theorem 1.4.1 will be given in Exercise 16.19.

1.5 Exercises for Chapter 1

1.1 The card game of bridge is played among four players, called north, south, east and west occupying the (corresponding) four sides of a table. The game uses all the 52 cards in a deck and the cards are dealt with each player receiving 13 cards. What is the number of ways in which cards can be dealt?

1.2 How many distinct positive divisors do each of the following numbers have? (a)$5^7 \cdot 7^5 \cdot 11^3 \cdot 17$ (b) $2,926,125$.

1.3 A club consisting of 10 men and 12 women has to select a committee of four people with at least one man and one woman. In how many ways can this be done? If among the four committee members, one is the chairperson, one convener, one secretary and one treasurer then in how many ways can the four office bearers of the committee be selected ?

1.4 Show that the number of ways of seating n people on a round table is $(n-1)!$ Any such arrangement is called a circular permutation. (Hint: Show that there is an n to 1 correspondence between the set of (linear) permutations and the set of circular permutations).

1.5 There are 16 books on a shelf. In how many ways can six of these books be selected if the selection is not to include two neighboring books?

1.6 Let $m \geq 2$ be a fixed natural number. Show that every non-negative integer n has a unique m-ary expansion

$$n = \sum_{i=0}^{i=r} a_i m^i.$$

(Hint: You may use induction on n)

1.7 Prove that:

$$1(1!) + 2(2!) + \cdots + n(n!) = (n+1)! - 1.$$

Use this to prove that every non-negative integer n can be uniquely expressed as:

$$n = a_1(1!) + a_2(2!) + a_3(3!) + \cdots$$

where $0 \leq a_i \leq i$ for all i.

1.8 There are n^2 balls of n different colors with n balls of each color. All the balls have the same size. Show that the number of permutations of this set of n^2 balls is equal to

$$\frac{(n^2)!}{(n!)^n}.$$

By permutations of the colors, show further that $(n^2)!$ is divisible by $(n!)^{n+1}$.

1.9 Show that there is no graph on 11 vertices with 6 vertices each of degree 6 and 5 vertices each of degree 5.

1.10 Consider the non-empty subsets of the set $[n]$.

(a) Prove that the number of subsets in which the greatest element is j is equal to 2^{j-1}.

(b) Use (a) to prove the identity:

$$1 + 2 + 2^2 + \cdots + 2^m = 2^{m+1} - 1.$$

1.11 Prove that $\binom{2n}{n}$ is always even.

1.12 Prove that no four consecutive binomial coefficients $\binom{n}{r}$, $\binom{n}{r+1}$, $\binom{n}{r+2}$, $\binom{n}{r+3}$ can be in arithmetic progression.

1.13 Show that the number of ways of solving the equation $x_1 + x_2 + \cdots + x_r = k$ in positive integers is equal to $\binom{k-1}{r-1}$.

1.14 Show that the number of r-subsets of $[n]$ that do not contain a pair of consecutive integers is $\binom{n-r+1}{r}$.

1.15 Determine the number of r-combinations of the following multi-set

$$\{1.a_1,\ \infty.a_2,\ \ldots,\ \infty.a_k\ \}$$

More generally, find a formula that determines the number of r-combinations of a multi-set with repetition numbers either 1 or ∞.

1.16 Let n and k denote positive integers with n greater than k. For a k-subset A of $[n]$. let $w(A)$ denote the sum of all the elements of A. Find the $\sum_A w(A)$ where the sum is taken over all the k-subsets.

1.17 Prove that $1! + 2! + \cdots + n!$ is not a perfect square if $n > 3$. (Hint: what are the squares modulo 5?)

1.18 Show that if n identical dice are rolled, there are $\binom{n+5}{5}$ distinct possibilities.

1.19 Show by a combinatorial argument that $\binom{n}{k} = \frac{n}{k}\binom{n-1}{k-1}$. Then show that $\binom{n}{k} = \frac{n}{n-k}\binom{n-1}{k}$.

1.20 What is the coefficient of x^{20} in the binomial expansions of $(2x^3 - \frac{3}{x})^8$?

1.21 Use the binomial theorem to prove Pascal's identity. Then prove the binomial theorem (Theorem 1.2.5) assuming Pascal's identity (Theorem 1.2.6).

1.22 Prove the following combinatorial identities:

$$\sum k\binom{n}{k} = n2^{n-1} \tag{1.1}$$

$$\sum k(k-1)\binom{n}{k} = n(n-1)2^{n-2} \tag{1.2}$$

$$\sum (2k+1)\binom{n}{k} = (n+1)2^n \tag{1.3}$$

$$\sum \frac{1}{k+1}\binom{n}{k} = \frac{1}{n+1}(2^{n+1}-1) \tag{1.4}$$

$$\sum (-1)^k \frac{1}{k+1}\binom{n}{k} = \frac{1}{n+1} \tag{1.5}$$

$$\sum \frac{(-1)^{k-1}}{k}\binom{n}{k} = 1 + \frac{1}{2} + \cdots + \frac{1}{n} \tag{1.6}$$

$$\sum \binom{n}{k}^2 = \binom{2n}{n} \tag{1.7}$$

$$\sum \frac{(2n)!}{(k!)^2((n-k)!)^2} = \binom{2n}{n}^2 \tag{1.8}$$

$$\sum_{k=0}^{n}\sum_{r=0}^{n-k}\binom{n}{k}\binom{n-k}{r} = 3^n \tag{1.9}$$

$$\binom{n}{r}\binom{r}{k} = \binom{n}{k}\binom{n-k}{r-k} \tag{1.10}$$

$$\binom{m}{0}\binom{n}{0} + \binom{m}{1}\binom{n}{1} + \binom{m}{2}\binom{n}{2} + \cdots + \binom{m}{n}\binom{n}{n} = \binom{m+n}{n} \tag{1.11}$$

$$4\sum_k \binom{n}{4k} = 2^n + 2^{\frac{n}{2}+1} cos\frac{\pi n}{4} \tag{1.12}$$

$$\sum_{k=0}^{n} \binom{n}{3k} = \frac{1}{3}(2^n + 2(-1)^n) \tag{1.13}$$

$$\binom{n}{0} + 2\binom{n}{1} + \binom{n}{2} + 2\binom{n}{3} + \binom{n}{4} + 2\binom{n}{5} + \cdots = 3\times 2^{n-1} \tag{1.14}$$

$$n(n+1)2^{n-2} = \sum_{k=1}^{n} k^2\binom{n}{k} \tag{1.15}$$

$$2^n = \sum_{k=0}^{n}(-1)^k\binom{n}{k}3^{n-k} \tag{1.16}$$

1.23 How many rational terms are there in the following binomial expansions?

$$(\sqrt{2}+\sqrt[6]{3})^{20}$$

$$(\sqrt{3}+\sqrt[4]{5})^{50}$$

$$(\sqrt[3]{6}+\sqrt[4]{2})^{100}$$

$$(\sqrt[3]{12}+\sqrt[6]{3})^{30}$$

1.24 Prove Corollary 1.3.3 without using Theorem 1.3.2.

1.25 The (integer) lattice in the real plane is the set of points (i, j) where both i and j are integers. The lattice points are the elements of the (integer) lattice. A lattice path is a path that begins and ends at a lattice point and moves horizontally or vertically in full units. In other words, a lattice path moves along the straight lines $x = m$ or $y = n$ where m and n are integers and changes direction only at the lattice points. This is somewhat like arrangement of roads in town-planned city where the lattice points represent the intersections of the roads. A man wants to travel from $(0, 0)$ to (m, n) where m and n are positive integers along a lattice path with shortest distance. Show that the total number of lattice paths of shortest distance is equal to $\binom{m+n}{n}$.

1.26 In Exercise 1.25, let k and r be positive integers such that $k \leq m$ and $r \leq n$. Find the number of shortest distance lattice paths from $(0, 0)$ to (m, n) that avoid the lattice point (k, r).

1.27 Suppose we have a set of m identical marbles of blue color and k identical red marbles. Show that if no two red marbles are adjacent then the number of ways of putting all the marbles on a line is $\binom{m+1}{k}$.

1.28 Show that if a and b are positive integers such that $a > b$, then for any prime p,

$$\binom{pa}{pb} \equiv \binom{a}{b} (mod\ p).$$

1.29 A set of 3 girls and 5 boys is to be divided into two (disjoint) teams consisting of 4 players each. If each team must contain at least one girl, in how many ways can this be done?

1.30 There are nine rooks, five of which are identical red and four of which are identical blue. In how many ways can they be placed on a 10×10 chessboard so that no two rooks attack each other?

1.31 How many integer solutions are there to the equation $x_1 + x_2 + x_3 = 0$ with each $x_i \geq -5$?

1.32 Evaluate the sum:

$$\binom{n}{0} + 2\binom{n}{1} + 2^2\binom{n}{2} + \cdots + 2^n\binom{n}{n}$$

1.33 What is the largest power of 10 in 50!?

1.34 In how many ways can 12 people be seated at a round table if a certain pair of people refuse to sit next to one another?

1.35 How many integers between 1,000 and 9,999 have distinct digits? Of these, how many are even numbers? How many consist entirely of odd digits?

1.36 In how many different orders can the seventeen letters

$$x, x, y, y, y, y, y, y, z, z, z, z, z, z, z, z, z$$

be written?

1.37 What is the coefficient of x^4 in the expansion of $(1 + x + x^2)(1 + x)^5$?

1.38 What is the coefficient of x^{12} in the expansion of $(1 + x + x^{-1})(1 + x)^{26}$?

1.39 By using the multinomial expansion $(1+1+\cdots+1)^p$, show that $n^p \equiv n(mod\, p)$.

1.40 We are given 8 rooks, 5 of which are red and are identical and 3 of which are blue and identical.

(a) In how many ways can the 8 rooks be placed on an 8×8 chessboard so that no rook attacks another?

(b) In how many ways can the 8 rooks be placed on an 12×12 chessboard so that no rook attacks another?

1.41 Find the number of integer solutions to the following equations satisfying the given constraints:

(a) $x + y + z = 7,\ x, y, z \geq 0$

(b) $x + y + z + w = 14,\ x, y, z, w > 0$

(c) $x + y + z = 9,\ x > 0\ ,\ y > 2\ ,\ z > 1$

(d) $x + y + z = 10,\ x > -1\ ,\ y > -2\ ,\ z > -3$

(e) $x + y - z = 7,\ x > 0\ ,\ y > 2\ ,\ z < 2.$

(f) $2x - 2y + 2z = 10,\ x \geq 1\ , y \leq 1,\ z \geq 0$

(g) $2y + z = 17, y \geq 1\ ,\ z \geq 1$

(h) $x + y + z + w = 18; x \geq 1,\ y \geq 1,\ z \geq 2\ ,\ w \geq 2$

1.42 A particle, starting at the origin, can move one unit at a time either in the positive x-direction or the positive y-direction in the plane. Show that the number of distinct paths the particle can take from the origin to the point $(n - r, r)$ is $\binom{n}{r}$.

1.43 An ice cream shop has n different flavors of ice cream available in unlimited supply. If a choice of k ice creams is to be made in such a way that each flavor must be represented, in how many ways can this be done?

1.44 There are 18 volumes of an epic on a bookshelf arranged serially. In how many ways can five of these book be picked up so that no consecutive volumes are selected?

1.45 Show that for all $n \geq 3$, we have, $2^n > n^2$.

1.46 I have 11 distinct books in three different languages: 5 in English, 4 in Marathi and 2 in Hindi. If I wish to arrange these books on a shelf in such a manner that the books of the same language are put together, in how many ways can I do this?

1.47 Consider the set of all the words of length n generated from the alphabet $\{0, 1, 2\}$. Show that the number of words in each of which the digit 0 appears an even number of times is $\frac{3^n+1}{2}$.

1.48 Show that for any positive integers n and t:

$$t^n = \sum \binom{n}{n_1, n_2, \ldots, n_t}$$

where the sum is taken over all the t-tuples $(n_1, n_2, \ldots, n_t)$ of non-negative integers whose sum is n.

1.49 There are seven types of candies available in boxes of 20 candies each. In all, 240 candies are to selected by choosing 12 boxes. Any given box contains candies of only one type. For each type, at least one box has to be chosen, but no more than 5 boxes of that type are to be chosen. Find the number of ways of doing this.

1.50 What is the least number of dice needed to be rolled to get a sum of 25? What is the largest number of dice that can be used to obtain the same sum? How many ways are there to get a sum of 25 when 10 dice are rolled?

1.51 Let ω be a primitive m-th root of unity where m is a fixed natural number ≥ 2. Let

$$f(x) = \sum_{j=0}^{\infty} f_j x^j$$

be a series with f_j's complex numbers. Let r be an integer such that $0 \leq r \leq m-1$. We wish to find a formula for the subseries

$$g(x) = \sum_{k \equiv r \ (mod\ m)} f_k x^k$$

Using properties of ω, first prove that

$$\sum_{t=0}^{m-1} \omega^{st}$$

equals 0 except when $s \equiv 0 \ (mod\ m)$ in which case it is equal to m. Use this to show that

$$g(x) = \sum_{k \equiv r \ (mod\ m)} f_k x^k = \frac{1}{m} \sum_{j=0}^{m-1} f(x w^j) \omega^{-rj}$$

This result is called multisection of the given series; Refer to Goulden and Jackson [26].

1.52 In how many ways can 200 chairs be divided among four conference rooms such that each room will have 20, 40, 60, 80 or 100 chairs?

1.53 Derive an analogue of Pascal identity for multinomial coefficients and then prove it.

Chapter 2

Listing combinatorial objects

This chapter is devoted to the discussion of some algorithmic questions concerning the combinatorial objects we defined in chapter 1. To be more specific, we wish to mainly concentrate on the question of listing of all the permutations and combinations of a given set. Our discussion, is far from complete and is intended merely to serve as an enticement to the reader to take up reading advanced material on the algorithmic aspects mentioned in the list of references. *We repeat that, all throughout the book,* $[n] = \{1, 2, \ldots, n\}$ *where* n *is a natural number.*

2.1 Permutations

Recall that a permutation of a set is a bijection from the set to itself. A typical permutation on $[n]$ is a function $f : [n] \rightarrow [n]$ satisfying the condition $i \neq j$ implies $f(i) \neq f(j)$. Therefore, if we let $f(i) = x_i$, then we may write f in the following *two-line form*

$$\begin{pmatrix} 1 & 2 & \cdots & \cdots & n-1 & n \\ x_1 & x_2 & \cdots & \cdots & x_{n-1} & x_n \end{pmatrix}$$

For example with $n = 5$, the two line form of the permutation β with

$$\beta(1) = 2,\ \beta(2) = 4,\ \beta(3) = 3,\ \beta(4) = 5,\ \beta(5) = 1$$

is given by:

$$\beta = \begin{pmatrix} 1 & 2 & 3 & 4 & 5 \\ 2 & 4 & 3 & 5 & 1 \end{pmatrix}$$

Note that the second line (as a set) is just the set $[5]$ as it should be since we are dealing with a permutation. Though it is a very detailed description of a permutation, the two line representation is not particularly convenient from the storage point of view. If n is known (which is generally the case), then the first line is redundant. It is therefore more common to use just the one line representation of a permutation:

$$x_1 x_2 \cdots x_{n-1} x_n$$

Here, the understanding is that the above representation is to be read as that of permutation for which $f(i)$ equals x_i. For example, with $n = 5$, the one line representation

$$2\ 4\ 3\ 5\ 1$$

corresponds to the permutation f with $f(1) = 2$, $f(2) = 4$, $f(3) = 3$, $f(4) = 5$ and $f(5) = 1$. Similarly, with $n = 6$ and with the permutation f given by $f(1) = 4$, $f(2) = 5$, $f(3) = 6$, $f(4) = 1$, $f(5) = 2$ and $f(6) = 3$, we have the following one-line representation

$$4\ 5\ 6\ 1\ 2\ 3$$

We now present an algorithm that systematically lists all the $n!$ permutations on $[n]$ This algorithm is described in the Brualdi [13] and the discussion is somewhat loose in the sense that we do not give the precise one-to-one correspondence between the $n!$ permutations and the set of numbers from 0 to $n! - 1$. For a given n, we list all the $n!$ permutations on the set $[n]$ starting from 0 and ending at $n! - 1$, i.e., the first permutation will be numbered 0 and the last permutation will be numbered $n! - 1$. The algorithm is recursive in the sense it lists all the $(n - 1)!$ on $[n - 1]$ before listing the permutations on $[n]$. We begin with $n = 2$ and order the permutation 12 as being the zeroth and the permutation 21 as the first. We then attempt inserting 3 in these permutations first listing 123 as being the zeroth, 132 as the first and 312 as the second. Having done this, we observe that 3 has reached an end and hence switch 1 and 2 to get the permutation 321 and call it the third permutation, then 231 is the fourth and 213 is the fifth and the last permutation. Thus the following table gives the required list.

Rank or Order	Permutation	Rank or Order	Permutation
0	123	3	321
1	132	4	231
2	312	5	213

For $n = 4$, the table on the next page lists the serial order of the permutation and the permutation. Notice that any permutation is obtained from the previous one by switching two adjacent entries. This is computationally a desired property.

Rank or Order	Permutation	Rank or Order	Permutation
0	1234	12	4321
1	1243	13	3421
2	1423	14	3241
3	4123	15	3214
4	4132	16	2314
5	1432	17	2341
6	1342	18	2431
7	1324	19	4231
8	3124	20	4213
9	3142	21	2413
10	3412	22	2143
11	4312	23	2134

A convenient way of identifying a permutation is by knowing as to how far it differs from the identity permutation. Fix an n. Then the identity permutation (on $[n]$) has all the numbers written in their natural order (in the one-line representation). For a permutation such as 132 on $[3]$, the number 2 is not in its natural position (which should be the second position). We express this by saying that (32) is an inversion in the permutation 132. To be precise we have the following.

Definition 2.1.1. Let $\pi = x_1 x_2 \cdots x_{n-1} x_n$ be a permutation of the set $[n]$. *A pair (k, m) with $k > m$ is called an inversion of π if for some i and some j with $i < j$, we have $x_i = k$ and $x_j = m$. (Notice that it is the pair (k, m) which is called an inversion and not the pair (i, j)).*

Example 2.1.2. For $n = 5$ and with $\pi = 24351$, the following pairs are inversions of $\pi : (2, 1), (4, 1), (3, 1), (5, 1), (4, 3)$. Observe that there are no inversions that end in 5, none ending in 4, one ending in 3, none ending in 2 and 4 ending in 1. This sequence of numbers is called an inversion sequence of the permutation π. If $\sigma = 12534$, then the only inversions σ has are $(5, 3)$ and $(5, 4)$.

Definition 2.1.3. Let $\pi = x_1 x_2 \cdots x_n$ be a permutation on $[n]$. For $m = 1, 2, \ldots, n$ let $b_m = b_m(\pi)$ denote the number of inversions that end in m, i.e., those inversions (k, m) where $k > m$. *The sequence $(b_1, b_2, \ldots, b_{n-1}, b_n)$ is called the inversion sequence of π and is denoted by $Inv(\pi)$.*

Example 2.1.4. Let $n = 7$ and let $\tau = 5123764$. Then the inversions are: $(5, 1), (5, 2), (5, 3), (5, 4), (7, 4), (6, 4), (7, 6)$. Hence, $Inv(\tau) = (1, 1, 1, 3, 0, 1, 0)$. For σ and π given in Example 2.1.2, the inversion sequences are:
$Inv(\sigma) = (0, 0, 1, 1, 0)$ and $Inv(\pi) = (4, 0, 1, 0, 0)$.

Let π be a permutation on $[n]$ (for a fixed n). Observe that there can be no inversion that ends in n. The only inversion that can end in $n-1$ is $(n, n-1)$ (if it is there in π). Hence, $b_n = 0$ and $b_{n-1} \leq 1$. In general, for any k between 1 and n, the only possible inversions ending in k are (j, k), where $j > k$, so that $b_k \leq n-k$. Thus b_k can assume at the most $(n-k+1)$ values. It is therefore clear that the total number of inversion sequences is at the most

$$\prod_{k=1}^{k=n}(n-k+1) = \prod_{r=0}^{r=n-1}(r+1) = n!$$

We now show that:

Theorem 2.1.5. *Given a sequence* $(b_1, b_2, \ldots, b_{n-1}, b_n)$ *satisfying* $0 \leq b_k \leq n-k$ *for all* k*, there exists a unique permutation* π *such that*

$$Inv(\pi) = (b_1, b_2, \ldots, b_{n-1}, b_n)$$

Proof The proof is an algorithm and is therefore constructive. Suppose π with the given sequence as the inversion sequence is constructed. Fix a k between 1 and n. How many integers to the left of k (in the one-line representation of π) are greater than k? This number, by definition is b_k. The proof, therefore, proceeds in steps, partially completing the required permutation starting from the last integer n scanned first. We follow an insertion procedure, the integer k being inserted at step $(n+1-k)$.
Step 1 : Write $\pi_1 = n$.
Step 2 : Write $\pi_2 = n-1, n$ if $b_{n-1} = 0$ and $\pi_2 = n, n-1$ if $b_{n-1} = 1$.
At step $n-k$, we would have obtained π_{n-k} which is a partial permutation consisting of the integers from $k+1$ to n.
Step $n+1-k$ (General step) : We have to insert k in the partial permutation π_{n-k} (without disturbing the earlier order given by π_{n-k}). We insert k after the first b_k integers in the partial permutation π_{n-k}.
Since at any further step, the relative order of the integers $k, k+1, \ldots, n$ is not to be disturbed, it follows that in the final permutation $\pi_n = \pi$, the value $b_k(\pi)$ equals b_k. Thus step n completes the construction of π. □

Example 2.1.6. Let $n = 9$ and consider the sequence

$$(b_1, b_2, \ldots, b_8, b_9) = (5, 4, 2, 0, 1, 2, 2, 1, 0)$$

which satisfies the requirement $b_k \leq n-k$ for all k. Following the procedure outlined in the proof of Theorem 2.1.5, we have:

Step number	π_j	reason
1	$\pi_1 = 9$	$b_9 = 0$
2	$\pi_2 = 98$	$b_8 = 1$
3	$\pi_3 = 987$	$b_7 = 2$
4	$\pi_4 = 9867$	$b_6 = 2$
5	$\pi_5 = 95867$	$b_5 = 1$
6	$\pi_6 = 495867$	$b_4 = 0$
7	$\pi_7 = 4935867$	$b_3 = 2$
8	$\pi_8 = 49352867$	$b_2 = 4$
9	$\pi_9 = 493521867$	$b_1 = 5$

The procedure works because any integer k is inserted in such a way that precisely b_k integers to the left of k are greater than k. It is then clear that a permutation with the inversion sequence $(0, 0, \ ldots, 0)$ must be the identity permutation: $1, 2, \ldots, n$ and a permutation with the inversion sequence $(n-1,\ n-2,\ \ldots,\ 1, 0)$ must be the permutation $n,\ n-1,\ \ldots,\ 2, 1$. Here is an example with $n = 8$.

Example 2.1.7. The inversion sequence of $\alpha = 37618452$ is $(3, 6, 0, 3, 3, 1, 0, 0)$ and the permutation β that corresponds to the inversion sequence $(3, 0, 3, 0, 1, 1, 1, 0)$ is sequentially obtained by

$$8, 87, 867, 8567, 48567, 485367, 2485367, 24815367$$

giving $\beta = 24815367$.

2.2 Listing combinations

Definition 2.2.1. A binary relation $<$ on a set X is called an *order relation* if $\forall\, x, y \in X, x \neq y$ we have either $x < y$ or $y < x$ but not both and $x < y,\ y < z$ implies $x < z$. A set X with order relation $<$ is denoted by $(X, <)$ and X is called *an ordered set (*under the order relation $<$).

The standard examples of ordered sets that we encounter in mathematics are $\mathbb{N}$, $\mathbb{Z}$ and $\mathbb{R}$; clearly any finite set can be ordered in many (how many?) ways.

Definition 2.2.2. Let $(X, <)$ be an ordered set and let $x \in X$. An element $z \in X$ said to be an *immediate predecessor* of x if $z < x$ and there is no y such that $z < y < x$. The definition of an *immediate successor* of x (if it exists) is made similarly. An element α is called *smallest element* (if it exists) if for every $\beta \in X$ with $\beta \neq \alpha$, we have $\alpha < \beta$; *largest element* is similarly defined.

We now discuss the question of listing all the k-combinations of $[n]$. A standard method of doing this is what is called the *Lex (lexicographic) or dictionary order.*

For example, Lex listing of all the 3-combinations of $[6]$ is:

$$\{123\}\{124\}\{125\}\{126\}\{134\}\{135\}\{136\}\{145\}\{146\}\{156\}$$

$$\{234\}\{235\}\{236\}\{245\}\{246\}\{256\}\{345\}\{346\}\{356\}\{456\}$$

All through this section, we write a k-subset $\{a_1, a_2, \ldots, a_k\}$ of $[n]$ following the convention that $a_1 < a_2 < \cdots < a_k$. We have the following:

Definition 2.2.3. *(Lex Order)* Let n and k be fixed positive integers with $k \leq n$. Let $A = \{a_1, a_2, \ldots, a_k\}$ and $B = \{b_1, b_2, \ldots, b_k\}$ be two distinct k-subsets of $[n]$. In the Lex order,

$$\{a_1, a_2, \ldots, a_k\} < \{b_1, b_2, \ldots, b_k\}$$

if either $a_1 < b_1$ or $a_1 = b_1,\ a_2 = b_2,\ \ldots, a_{j-1} = b_{j-1}$ and $a_j < b_j$. We thus look for the smallest j such that $a_j \neq b_j$ and for that j, we must have $a_j < b_j$.

Now fix a k-subset $\{a_1, a_2, \ldots, a_k\}$ of $[n]$. How many k-subsets $\{b_1, b_2, \ldots, b_k\}$ succeed the given k-subset $\{a_1, a_2, \ldots, a_k\}$ in the lex order? For this to happen, one possibility is $b_1 > a_1$. i.e., $\{b_1, b_2, \ldots, b_k\}$ is a subset of $\{a_1 + 1,\ a_1 + 2,\ \ldots, n\}$ and hence the number of such k-subsets equals $\binom{n-a_1}{k}$. Otherwise, $b_1 = a_1$ and $b_2 > a_2$. Thus, $\{b_2, \ldots, b_k\}$ is a subset of $\{a_2 + 1, a_2 + 2, \ldots, n\}$ and hence can be chosen in $\binom{n-a_2}{k-1}$ ways. In general, we have, for some (fixed) j between 1 and k: $b_i = a_i$ for all i from 1 to $j - 1$, and $b_j > a_j$. Then $\{b_j, \ldots, b_k\}$ is a subset of $\{a_j + 1, a_j + 2,\ \ldots, n\}$ and hence can be chosen in $\binom{n-a_j}{k-j+1}$ ways. Now for any k-subset $\{a_1, a_2, \ldots, a_k\}$ of $[n]$, the number of k-subsets that succeed it in the lex order is uniquely determined by that k-subset as we just saw. Since the number of k-subsets succeeding a given k-subset can take values from 0 to $\binom{n}{k-1}$, the following theorem is indirectly proved using bijection:

Theorem 2.2.4. *Let n and k be fixed positive integers with $k < n$. Let m be a number between 0 and $\binom{n}{k} - 1$. Then there exist unique integers α_i, $i = 1, 2, \ldots, k$ such that*

$$n - 1 \geq \alpha_1 > \alpha_2 > \cdots > \alpha_k \geq 0$$

and satisfying:

$$m = \binom{\alpha_1}{k} + \binom{\alpha_2}{k-1} + \cdots + \binom{\alpha_k}{1}$$

Proof In the above discussion, let $\{\alpha_1, \alpha_2, \ldots, \alpha_k\}$ denote the sequence obtained from $\{a_1, a_2, \ldots, a_k\}$ by letting $\alpha_i = n - a_i$. Since there is a one-to-one correspondence between the set of k-subsets succeeding a given k-subset and the numbers from 0 to $\binom{n}{k} - 1$, as explained above, the theorem is proved. □

There is yet another interesting way in which the set of all the k-subsets of $[n]$ can be listed. This is called *the Colex (short for colexicographic) order* and is simply the reverse lexicographic order. Here, writing k-subsets with increasing order of elements, we compare the last elements first, and then the second last and so on. For example, colex listing of all the 3-combinations of $[6]$ is:

$$\{123\}\{124\}\{134\}\{234\}\{125\}\{135\}\{235\}\{145\}\{245\}\{345\}$$

$$\{126\}\{136\}\{236\}\{146\}\{246\}\{346\}\{156\}\{256\}\{356\}\{456\}$$

Definition 2.2.5. *(Colex order)* Let n and k be fixed positive integers with $k \leq n$. Let $\{c_1, c_2, \ldots, c_k\}$ and $\{d_1, d_2, \ldots, d_k\}$ be two distinct k-subsets of $[n]$. *In the Colex order,*

$$\{c_1, c_2, \ldots, c_k\} < \{d_1, d_2, \ldots, d_k\}$$

if either $c_k < d_k$ or $c_k = d_k$, $c_{k-1} = d_{k-1}, \ldots, c_{j+1} = d_{j+1}$ and $c_j < d_j$ for some j from $k-1$ to 1.

For convenience, we order the members of the set of all the k-combinations of the n-set $[n]$ from 0 to $\binom{n}{k} - 1$ (instead of 1 to $\binom{n}{k}$). The advantage of this is that if the rank (order) of a k-subset in the colex list is m, then the number of k-subsets preceding it is also m. In the colex order, given a k-subset $\{d_1, d_2, \ldots, d_k\}$, how many k-subsets $\{c_1, c_2, \ldots, c_k\}$ precede it? We note that this can happen if $c_k \leq d_k - 1$. Number of such k-subsets $\{c_1, c_2, \ldots, c_k\}$ is evidently $\binom{d_k - 1}{k}$. If not, then $c_k = d_k$ and $c_{k-1} \leq d_{k-1} - 1$. Hence, such a k-subset can be picked up in $\binom{d_{k-1}-1}{k-1}$ ways. Continuing in this manner, we see that the number of k-subsets preceding $\{d_1, d_2, \ldots, d_k\}$ which is simply the rank of $\{d_1, d_2, \ldots, d_k\}$ equals

$$\sum_{j=1}^{j=k} \binom{d_j - 1}{j}$$

Finally, writing $\beta_j = d_j - 1$, we have a second proof of Theorem 2.2.4 using the colex order:
Let n and k be fixed positive integers with $k < n$. Then for any number m between 0 and $\binom{n}{k} - 1$ there exist unique integers β_i, $i = 1, 2, \ldots, k$ such that

$$0 \leq \beta_1 < \beta_2 < \cdots < \beta_k \leq n - 1$$

and satisfying:

$$m = \binom{\beta_k}{k} + \binom{\beta_{k-1}}{k-1} + \cdots + \binom{\beta_1}{1}$$

Theorem 2.2.4 is an extremely useful combinatorial statement required in counting faces of a simplicial polytope through the famous Kruskal Katona theorem. Though both the proofs of the theorem we indicated above are correct, they are indirect as they are based on bijections. In the concluding part of this chapter, we develop some results that obtain a proof of Theorem 2.2.4 in a more direct manner. This will also help in obtaining the colex and lex orders of k-subsets of $[n]$.

Lemma 2.2.6. *Let n and k be fixed positive integers with $k < n$. Then :*

$$\binom{n}{k} - 1 = \binom{n-1}{k} + \binom{n-2}{k-1} + \cdots + \binom{n-k+1}{2} + \binom{n-k}{1}$$

Proof Though the proof is immediate from Theorem 2.2.4 (what is the last k-subset in the colex list of $[n]$?), we give another combinatorial proof. Let X be an n-set and fix a k-subset $\{x_1, x_2, \ldots, x_k\}$ of X. We make a two-way counting of the set of all the k-subsets of X other than $\{x_1, x_2, \ldots, x_k\}$. If such a subset does not contain x_1, then it can be chosen in $\binom{n-1}{k}$ ways. Suppose it contains x_1 but does not contain x_2. Since x_1 is already chosen, we have to pick up $k-1$ elements out of the remaining $n-2$ elements, a task that can be performed in $\binom{n-2}{k-1}$ ways. Then look at those k-subsets that contain both x_1 and x_2 but not x_3. Proceeding in this manner, we finally count the number of those subsets that contain all of $x_1, x_2, \ldots, x_{k-1}$ but not x_k. We then have to choose only one element out of the remaining $n-k$ elements (other than $(x_1, x_2, \ldots, x_k)$ and obviously that can be done in $\binom{n-k}{1}$ ways. This finishes proof of the lemma. □

We now obtain a third proof of Theorem 2.2.4. Make induction on k, the statement being obviously true for $k = 1$. Let $k \geq 2$ and let m be some integer between 0 and $\binom{n}{k} - 1$. We wish to obtain a sequence $\beta_i,\ i = 1, 2, \ldots, k$ such that

$$0 \leq \beta_1 < \beta_2 < \cdots < \beta_k \leq n-1$$

and satisfying :

$$m = \binom{\beta_k}{k} + \binom{\beta_{k-1}}{k-1} + \cdots + \binom{\beta_1}{1}$$

We first show that β_k is determined uniquely. Let β be the unique integer for which $\binom{\beta}{k} \leq m < \binom{\beta+1}{k} - 1$. Then β_k can not exceed β. On the other hand, if β_k is strictly less than β, then the sum

$$\binom{\beta_k}{k} + \binom{\beta_{k-1}}{k-1} + \cdots + \binom{\beta_1}{1}$$

under the stipulation

$$0 \leq \beta_1 < \beta_2 < \cdots < \beta_k \leq n-1$$

can not exceed

$$\binom{\beta-1}{k} + \binom{\beta-2}{k-1} + \cdots + \binom{\beta-k+1}{2} + \binom{\beta-k}{1}$$

which, using Lemma 2.2.6 equals $\binom{\beta}{k} - 1$. This contradicts the assumption that $\binom{\beta}{k} \leq m$. Hence $\beta_k = \beta$ and β_k is uniquely determined. Having determined $\beta_k = \beta$, we now let m' denote the integer $m - \binom{\beta}{k}$. Clearly, $m' \geq 0$. Can $m' \geq \binom{\beta}{k-1}$? In that case,

$$m \geq \binom{\beta}{k-1} + \binom{\beta}{k} = \binom{\beta+1}{k}$$

which is a contradiction. Hence, $0 \le m' \le \binom{\beta}{k-1} - 1$ and inductively, we find unique integers $\beta_1, \beta_2 \cdots, \beta_{k-1}$ satisfying

$$0 \le \beta_1 < \beta_2 < \cdots < \beta_{k-1} \le \beta - 1$$

such that

$$m' = \binom{\beta_{k-1}}{k-1} + \binom{\beta_{k-2}}{k-2} + \cdots + \binom{\beta_1}{1}$$

Since β equals β_k, we have proved the required result. $\square$

Example 2.2.7. We look at the Lex and Colex listing of all the 3-subsets of $[7]$.

(a) Find the 26-th subset in the colex order.

(b) Find the 18-th subset in the lex order.

(c) Find the rank of the subset $\{1, 4, 7\}$ in the colex order.

(d) Find the rank of the subset $\{1, 4, 7\}$ in the lex order.

Here are the solutions.

(a) Notice successively that $26 = \binom{6}{3} + 6,\ 6 = \binom{4}{2} + 0$ and finally $0 = \binom{0}{1}$ so that

$$26 = \binom{6}{3} + \binom{4}{2} + \binom{0}{1}$$

Hence the the 26-th subset (in colex order) is $\{1, 5, 7\}$.

(b) The number of subsets succeeding the subset in question is 17 since $\binom{7}{3} = 35$. We have, $17 = \binom{5}{3} + 7,\ 7 = \binom{4}{2} + 1$ and finally $1 = \binom{1}{1}$ so that

$$17 = \binom{5}{3} + \binom{4}{2} + \binom{1}{1}$$

We therefore have $a_1 = 7 - 5 = 2,\ a_2 = 7 - 2 = 5,\ a_3 = 7 - 1 = 6$. Thus the required subset is $\{2, 5, 6\}$.

(c) The required number is

$$\binom{6}{3} + \binom{3}{2} + \binom{0}{1} = 23.$$

(d) We find that $\alpha_3 = 0$ because $a_3 = 7$. Similarly, $\alpha_2 = 3$ because $a_2 = 4$ and $\alpha_1 = 6$ because $a_1 = 1$. Hence there are

$$\binom{6}{3} + \binom{3}{2} + \binom{0}{1} = 23$$

succeeding the given subset. Therefore, it follows that the lex rank of the subset $\{1, 4, 7\}$ is 12.

In each one of the lex and the colex orders, how do we find immediate successors and immediate predecessors (Definition 2.2.2) of the given k-subsets of $[n]$? We finally give a procedure for doing this and leave a proof to the reader (Exercise 2.20).

Theorem 2.2.8. *Let n be a positive integer ≥ 3 and let $2 \leq k \leq n-1$. Let $\mathbf{K}$ denote the set of all the k-subsets of $[n]$.*

(a) (Lex order; Immediate Successor) Let $A = \{a_1, a_2, \ldots, a_k\} \in \mathbf{K}$. Define $a_{k+1} := n+1$. Let j be the largest number such that $a_j + 1 < a_{j+1}$. Let $b_j = a_j + 1$ and $b_r = b_j + (r-j)$ for $r > j$. Let $b_i = a_i \; \forall \; i = 1, 2, \ldots, j-1$. Then $B = \{b_1, b_2, \ldots, b_k\}$ is the immediate successor of A (in $\mathbf{K}$). If there is no such j, then $A = \{n-k+1, \ldots, n-1, n\}$ and A is the largest member of $\mathbf{K}$.

(b) (Lex order; Immediate predecessor) Let $B = \{b_1, b_2, \ldots, b_k\} \in \mathbf{K}$. Define $b_0 := 0$. Let j be the largest number such that $b_j - 1 > b_{j-1}$. Let $a_j = b_j - 1$ and $a_r = n - k + r$ for $r > j$. Let $a_i = b_i \; \forall \; i = 1, 2, \ldots, j-1$. Then $A = \{a_1, a_2, \ldots, a_k\}$ is the immediate predecessor of B. If there is no such j, then $B = \{1, 2, \ldots, k\}$ and B is the smallest member of $\mathbf{K}$.

(c) (Colex order; Immediate successor) Let $A = \{a_1, a_2, \ldots, a_k\} \in \mathbf{K}$. Define $a_{k+1} := n+1$. Let j be the smallest number such that $a_j + 1 < a_{j+1}$. Let $b_j = a_j + 1$ and $b_r = a_r$ for $r > j$. Let $b_i = i \; \forall \; i = 1, 2, \ldots, j-1$. Then $B = \{b_1, b_2, \ldots, b_k\}$ is the immediate successor of A. If there is no such j, then $A = \{n-k+1, \ldots, n-1, n\}$ and A is the largest member of $\mathbf{K}$.

(d) (Colex order;Immediate predecessor) Let $B = \{b_1, b_2, \ldots, b_k\} \in \mathbf{K}$. Define $b_0 := 0$. Let j be the smallest number such that $b_j - 1 > b_{j-1}$. Let $a_j = b_j - 1$ and $a_r = b_r$ for $r > j$. Let $a_i = a_j + (i-j) = a_j - (j-i) \; \forall \; i = 1, 2, \ldots, j-1$. Then $A = \{a_1, a_2, \ldots, a_k\}$ is the immediate predecessor of B. If there is no such j, then $B = \{1, 2, \ldots, k\}$ and B is the smallest member of $\mathbf{K}$.

2.3 Exercises for Chapter 2

2.1 (*tedious!*) Obtain a listing of all the 120 permutations of $[5]$ using the the listing of all the 24 permutations on $[4]$ obtained in this chapter. Where does the permutation 13245 appear in this list? Which permutation is at the 73rd position? Can one get a general formula?

2.2 How many permutations are there on the set $[6]$ that have the largest number of inversions? What are these? What is a general result?

2.3 Consider $[6]$. How many permutations have exactly 14 inversions? 13 inversions? 12 inversions?

2.4 For a given permutation σ on $[n]$, we construct a sequence $(t_1, t_2, \ldots, t_n)$ where t_i is the number of inversions of the form (i, j) (that begin with i, instead of ending with i). Show that $t_i \leq i - 1$ holds for all i and given any sequence $(t_1, t_2, \ldots, t_n)$ satisfying this stipulation, there is a unique permutation σ on $[n]$ whose associated sequence is $(t_1, t_2, \ldots, t_n)$. What is the relationship of this sequence with the inversion sequence of σ?

2.5 This exercise presumes the knowledge of the definition of the parity of a permutation and the permutation group S_n; reader may refer to Chapter 7. Let $\sigma \in S_n$ and let $(b_1, b_2, \ldots, b_n)$ be the inversion sequence of σ. Let $k = b_1 + b_2 + \cdots + b_n$. Prove that σ is an odd permutation or an even permutation according as k is odd or even.

2.6 Let $\sigma \in S_n$. Let $b_1, b_2, \ldots, b_n$ be the inversion sequence of the permutation σ and let $c_1, c_2, \ldots, c_n$ be the inversion sequence of the permutation σ^{-1}. Show that :

$$b_1 + + b_2 + \cdots + b_n = c_1 + + c_2 + \cdots + c_n$$

2.7 Determine the inversion sequence of the two permutations 35168274 and 83476215 of $[8]$.

2.8 Write the permutation 83521437 of $[8]$ as a product of minimum number of transpositions of the type $(i, i+1)$. Proving this in general is not very easy. That is, if π is any permutation on $[n]$ and if $i(\pi)$ denotes the number of inversions of n, then the least number of adjacent switches of the type $(j, j+1)$ required to bring the identity permutation to π is $i(\pi)$. Prove this by making induction on n. This result is of fundamental importance in the study of *reflection groups and Coxeter groups*.

2.9 For the set $[9]$, find the inversion sequences of the following permutations:

(a) 361782459

(b) 127985634

(c) 321789456

2.10 For the set $[9]$, find the permutations corresponding to the following inversion sequences:

(a) $(6, 2, 0, 3, 0, 2, 2, 1, 0)$

(b) $(0, 2, 3, 4, 0, 1, 2, 0, 0)$

(c) $(0, 4, 5, 0, 0, 3, 2, 0, 0)$

2.11 What are the permutations on the set $[8]$ whose inversion sequences are $(2, 5, 5, 0, 2, 1, 1, 0)$ and $(6, 6, 1, 4, 2, 1, 0, 0)$?

2.12 What is the rank of $\{2, 3, 4\}$ in the Colex listing of the subsets of $[6]$?

2.13 Determine the 6-combination of $[10]$ that immediately follows $\{2, 3, 4, 6, 9, 10\}$ and also the one that immediately precedes $\{2, 3, 4, 6, 9, 10\}$ in both the Lex and Colex ordering.

2.14 Determine the permutation of $[8]$ whose inversion sequence is $(5, 3, 4, 0, 2, 1, 1, 0)$.

2.15 Determine the 7-combinations that immediately follow $\{1, 2, 4, 6, 8, 14, 15\}$ in the lex and colex ordering of the 7-subsets of $[15]$.

2.16 In the Lexicographic ordering of 3-subsets of the set $[7]$, what is the rank of $\{2, 5, 6\}$? What is its rank in the colex list?

2.17 Is there a 3-subset of $[7]$ with the same lex and colex rank? If so, find it.

2.18 Begin with the 3-subset $\{3, 4, 6\}$ of $[7]$ and find its lex order. Then find the 3-subset with that colex order. Then find the lex order of the resulting new 3-subset and continue in that manner. Where does this end?

2.19 Using the one-to-one correspondence between permutations and inversion sequences (Theorem 2.1.5), prove the following identity.

$$\sum_{\pi \in S_n} q^{i(\pi)} = (1+q)(1+q+q^2)\cdots(1+q+q^2+\cdots+q^{n-1})$$

You can also use induction on n.

2.20 Prove Theorem 2.2.8.

2.21 Consider the lex listing of all the 5-subsets of $[9]$.

(a) What is the immediate predecessor of $\{2, 3, 6, 7, 9\}$?

(b) What is the immediate predecessor of $\{1, 2, 7, 8, 9\}$?

(c) What is the immediate successor of $\{2, 3, 5, 7, 9\}$?

(d) What is the immediate successor of $\{1, 4, 5, 6, 9\}$?

2.22 Consider the colex listing of all the 5-subsets of $[9]$.

(a) What is the immediate predecessor of $\{1, 2, 3, 8, 9\}$?

(b) What is the immediate predecessor of $\{1, 6, 7, 8, 9\}$?

(c) What is the immediate successor of $\{3, 4, 5, 6, 8\}$?

(d) What is the immediate successor of $\{2, 3, 4, 5, 9\}$?

2.23 Given two ordered sets $(X, <_X)$ and $(Y, <_Y)$, a bijective function $f : X \to Y$ is said to be an *order isomorphism* if for all $x_1, x_2 \in X$, we have $x_1 <_X x_2$ iff $f(x_1) <_Y f(x_2)$ and then we call $(X, <_X)$ and $(Y, <_Y)$ to be *order isomorphic.*

(a) Show that an order isomorphism preserves all of the following: immediate predecessor, immediate successor, smallest and largest elements.

(b) Let X be a finite set. Show that under any two orders $<$ and $<'$, we have an order isomorphism between $(X, <)$ and $(X, <')$.

(c) Show that under the natural orders on them, no two of the $\mathbb{N}, \mathbb{Z}$ or $\mathbb{Q}$ are order isomorphic.

(d) Let $X = \mathbb{N} \times \mathbb{N}$ and define the following three orders $<_i$ for $i = 1,, 2, 3$ on X. Let $(x_1, x_2), (y_1, y_2) \in X$.

(i) $(x_1, x_2) <_1 (y_1, y_2)$ if $x_1 < y_1$ or $x_1 = y_1$ and $x_2 < y_2$.

(ii) $(x_1, x_2) <_2 (y_1, y_2)$ if $x_1 - x_2 < y_1 - y_2$ or $x_1 - x_2 = y_1 - y_2$ and $x_1 < y_1$.

(iii) $(x_1, x_2) <_3 (y_1, y_2)$ if $x_1 + x_2 < y_1 + y_2$ or $x_1 + x_2 = y_1 + y_2$ and $x_1 < y_1$.

Show that $(X, <_i)$ are pairwise not order isomorphic for $i = 1, 2, 3$.

2.24 (Continuation of Exercise 2.23) Let $k \geq 2$ be a fixed integer and let $X = \{A : A \subset \mathbb{N} \textit{ and } |A| = k\}$ denote the set of all the k-subsets of $\mathbb{N}$. We then have two different orders the Lex and the Colex order on X. Show that under both the orders, every member has an immediate successor but in the Lex order there are infinitely many members that have no immediate predecessor while in the Colex order each member except one has an immediate predecessor. Conclude that the Lex and Colex orders give ordered sets that are not order isomorphic.

Chapter 3

Permutations

3.1 Combinatorial representations of a permutation

Permutations form the single most important and basic object that combinatorics deals with. This chapter, therefore, is intended to make a somewhat detailed study of permutations. For a positive integer n, let $\pi = a_1a_2\cdots a_n$ represent a permutation on $[n] = \{1, 2, \ldots, n\}$. This simply means π takes i to a_i for every i. For a better combinatorial insight of a permutation the two line or one line representation is not very useful. We therefore need a different representation called the *cycle decomposition*. Suppose π takes m to j, j to k and so on. then we may chase this path as follows. If m is called c_1, then π takes c_1 to c_2, c_2 to c_3 (say), and then we get a sequence $(c_1, c_2, \ldots, c_r)$ which must come back to where we started (i.e. for some r, π must take c_r to c_1). Such a sequence is called a cycle and is written in the form $(c_1c_2\cdots c_r)$. This means $\pi(c_j) = c_{j+1}$ where the subscripts are read modulo r. Notice that the same cycle could also have been written in the form $(c_j, c_{j+1}\cdots c_rc_1\cdots c_{j-1})$. In fact, if we define the equivalence relation on $[n]$ by $x \sim y$ if there is a sequence $x = d_1, d_2, \ldots, d_n = y$ with $\pi(d_i) = d_{i+1}$ then the equivalence classes are just the cycles. Notice however, that cycles are not just sets. They are cyclically ordered in a particular manner. Also, a cycle of length 1 is just an element which is fixed under π. It is now clear that the given permutation π itself can be read in the form $C_1C_2\cdots C_p$ where p is the number of cycles in π and the action of π on any element is to be found by reading the next element in the cycle containing it.

Example 3.1.1. Let $n = 9$ and let $\pi = 325894671$ in the one line notation. Chasing the path of 1, we find that the cycle containing 1 is (1359). The other cycles are (2) and (4876). Hence in the cycle decomposition notation, $\pi = (1359)(2)(4876)$. We could also have written this as $\pi = (2)(4876)(1359)$.

Definition 3.1.2. For a permutation π, an expression that gives it as a product of disjoint cycles is called its *cycle decomposition*. Singleton cycles of π are called its *fixed points* and cycles of length (size) 2 are called its *transpositions*.

A cycle decomposition is not unique but up to permuting the cycles, decomposition is unique. It is also a convention not to write singleton cycles (fixed points). Thus the permutation π in the above example also equals $(4876)(1359)$ or $(1359)(4876)$. The word "cycle (or full cycle)" is sometimes used to mean a permutation with a single cycle (whose length then must be n). This is also sometimes called a circular permutation (Exercise 1.4). In a more general sense, the word cycle of length k (on a set of order $n \geq k$) means a permutation on a set of order n whose cycle decomposition consists of a single cycle of length k and the remaining points are fixed. Disjoint cycles commute and therefore a cycle decomposition of π is also an expression that gives π as a product of disjoint cycles. *A cycle of length k is called a k-cycle.* Thus in Example 3.1.1, π has two 4-cycles and one 1-cycle (or one fixed point). If we add all the lengths of the cycles, then we must get n, the cardinality of the set on which the permutation acts.

Example 3.1.3. Let $n = 9$ and consider the following three permutations.

(a) Let $\tau = 372548196$ (in the one line notation). Since 1 goes to 3, 3 goes to 2, 2 to 7 and 7 back to 1, we see that (1327) is a cycle. Similarly chasing other elements of $[9]$ gives us $\tau = (1327)(45)(689)$ in its cycle decomposition form.

(b) Let $\alpha = 123789456$. Arguing exactly in the same manner, we get $\alpha = (47)(58)(69)$ and thus α has three fixed points and three involutions.

(c) Let $\sigma = (376)(25)(18)$. Here the permutation is given in the cycle decomposition form with the points 4 and 9 fixed. In one line form, we have $\sigma = 857423619$.

Since the cycle decomposition is unique up to permutations of the cycles and cyclical shifts of the elements in a cycle, we have the following.

Definition 3.1.4. Let π be a permutation on $[n]$. Suppose π has b_i cycles of length i where $i = 1, 2, \ldots, n$ (evidently, $b_1 + 2b_2 + \cdots = n$). Then π is said to have *cycle type or type* $(b_1, b_2, \ldots)$. If we know that $b_i = 0$ for all $i > k$, then we also write the type of π as $(b_1, b_2, \ldots, b_k)$ (after deleting all the 0's to the right of b_k).

Thus, the permutation π given in Example 3.1.1 has type $(1, 0, 0, 2)$. The three permutations τ, α and σ of Example 3.1.3 have types $(0, 1, 1, 1)$, $(3, 3)$ and $(2, 2, 1)$ respectively.

Theorem 3.1.5. *Let $(b_1, b_2, \ldots)$ be a sequence of non-negative integers such that*

$$b_1 + 2b_2 + 3b_3 + \cdots = n$$

Then the number of permutations on $[n]$ of type $(b_1, b_2, \ldots)$ is given by:

$$\frac{n!}{\prod_i i^{b_i} b_i!}$$

Proof Write $c_k = kb_k$. Then any permutation of the desired type is constructed by first partitioning the given set of order n into distinct sets such that the k-th set has c_k elements. Evidently this can be done in A ways where A is given by the multinomial coefficient

$$\binom{n}{c_1, c_2, \ldots, c_k, \ldots} = \frac{n!}{\prod_k (c_k)!} = \frac{n!}{\prod_k (kb_k)!}$$

Having done this, the elements in the k-th set need to be divided into b_k disjoint sets, each of size k. However, there is no specific order between two distinct k-cycles, which is why we need to divide by $(b_k)!$ to get rid of the order among the k-cycles. Hence, for each k, this can be separately achieved in B_k ways, where B_k is given by the multinomial coefficient

$$\frac{\binom{kb_k}{k,k,\ldots}}{(b_k)!} = \frac{(kb_k)!}{(k!)^{b_k}(b_k)!}$$

Finally, for each subset of order k thus obtained, we can make a cycle of it in $(k-1)!$ ways. Hence the number of ways, in which the chosen $c_k = kb_k$ elements can be made into b_k cycles of length k each is given by C_k where C_k equals

$$[(k-1)!]^{b_k} B_k = \frac{(c_k)!}{(b_k)!k^{b_k}}$$

Multiplying A and all the C_k's obtains the desired expression. □

Definition 3.1.6. $c(n,k)$ denotes the number of permutations on n letters that have exactly k distinct cycles.

Observe that $c(n,1)$ is just the number of cycles of length n (or full cycles) which we have already discussed. and exercise 1.4 gives $c(n,1) = (n-1)!$ The computation of $c(n,k)$ is facilitated by the use of the following recursion, analogous to Pascal identity (Theorem 1.2.6).

Lemma 3.1.7. *Let $n \geq 2$ and let $k \leq n$. Then*

$$c(n,k) = c(n-1,k-1) + (n-1)c(n-1,k)$$

Proof Any permutation on n letters with exactly k cycles either has n in a singleton cycle (that is, n is a fixed point) or has n in a cycle of length $r \geq 2$. In the former case, we may remove n to get a permutation on $n-1$ letters that has $k-1$ cycles (and this process is reversible). In the latter case also, we can remove n from the cycle decomposition of the permutation to get a permutation on $n-1$ letters with exactly k cycles. (for example, in the permutation π given in Example 3.1.1, the new permutation on 8 letters is $(2)(4876)(135)$) However, to insert n in a cycle decomposition of some permutation on $n-1$ letters (with k cycles), we have $n-1$ possibilities (this is easily checked; inserting n in a cycle of length r can be done in r ways). This completes the proof. □

Theorem 3.1.8. *For a positive integer n, we have*

$$\sum_k c(n,k)x^k = x(x+1)(x+2)\cdots(x+n-1) = [x]^n$$

Proof Denote the left hand side, which is a polynomial of degree n in x by $f_n(x)$. Then $f_n(x)$ equals the r.h.s. function for $n = 1$. We now make induction on n. Using Lemma 3.1.7, $f_n(x) = xf_{n-1}(x) + (n-1)f_{n-1}(x)$. Since $f_n(x) = (x+n-1)f_{n-1}(x)$, the result is proved by induction. □

Is it possible to make the cycle decomposition of a permutation unique by introducing an order on the cycles as well as on the elements inside a cycle? There are two ways of doing it. Call the smallest and the largest elements of a cycle its *guard* and *leader* respectively. Consider π of Example 3.1.1. Here $n = 9$ and $\pi = (2)(4876)(1359)$. The three cycles have guards $2, 4$ and 1 respectively while the leaders are $2, 8$ and 9. (Observe here that 1 will always be the guard of some cycle and n will always be the leader of some cycle if we are dealing with a permutation on $[n]$.)

Definition 3.1.9. *(Guard representation)* The smallest number in each cycle is the *guard* of that cycle. Write each cycle with the guard of the cycle at the end of the cycle. Then write the cycles in increasing order of magnitude of the guards.

Definition 3.1.10. *(Leader representation)* The largest number in each cycle is the *leader* of that cycle. Write each cycle with the leader of the cycle as the first element of the cycle. Then write the cycles in increasing order of the leaders.

Notice that in the guard representation,the first cycle must end in 1 while in the leader representation, the last cycle must begin with n. In Example 3.1.1, the guard representation of π is $(3591)(2)(8764)$ while the leader representation of the same permutation is $(2)(8764)(9135)$ As another example, with $n = 9$, let $\sigma = (413)(695)(278)$. then in the guard representation, this should read $\sigma = (341)(782)(695)$, while in the leader representation, the same permutation would read $\sigma = (413)(827)(956)$.

Definition 3.1.11. Let $\pi = a_1a_2\cdots a_n$ be a permutation on $[n]$ (in its one line representation). Let $i \in [n]$.

(a) π is said to have *a left to right minimum at* a_i if $\forall\, j > i$ we have $a_i < a_j$.

(b) π is said to have *a left to right maximum at* a_i if $\forall\, j > i$ we have $a_i > a_j$.

(c) π is said to have *a right to left minimum at* a_i if $\forall\, s < i$ we have $a_i < a_s$.

(d) π is said to have *a right to left maximum at* a_i *if* $\forall\, s < i$ *we have* $a_i > a_s$.

Note that by definition π has a left to right minimum at 1 (wherever it occurs in π) as also at a_n. Similarly, π has a left to right maximum at n (wherever it occurs in π) as also at a_n. Similar considerations also apply to right to left minima and maxima. Thus π has a right to left minimum (also maximum) at a_1.

Example 3.1.12. Let $n = 9$ and let $\pi = 371892564$. Then π has left to right minima at $1, 2, 4$ and has left to right maxima at $9, 6, 4$; π has right to left minima at $1, 3$ and has right to left maxima at $9, 8, 7, 3$.

We are now ready to describe two clever combinatorial bijections. Let S_n denote the set of all the $n!$ permutations on $[n]$ (this is called the *symmetric group* since there is a group structure on it and we shall see it more closely in Chapter 14). By convention, suppose we write the members of S_n in a one line notation $a_1a_2\cdots a_n$. This simply means that 1 goes to a_1, 2 goes to a_2 etc. Now assume that $\overline{S_n}$ denotes a copy of S_n where permutations are written in cycle decomposition with the (unique) guard representation. Then the bijection is given by the following: Let, $\pi = a_1a_2\cdots a_n$ be in S_n. Let $1 = i_1 < i_2 < \cdots < i_k$ denote the *positions denoting the left to right minima of* π (note that $i_1, i_2, \cdots, i_k$ are uniquely determined and we allow the possibility of $k = 1$ in which case $a_n = 1$ as we have already seen it). Now let $\overline{\pi}$ be obtained by writing $(a_1a_2\cdots a_{i_1})$ as the first cycle (that ends in its smallest element $a_{i_1} = 1$). Then we have $(a_{i_1}+1\cdots a_{i_2})$ as the second cycle (which ends in its smallest element a_{i_2}) and so on. We thus get a permutation with exactly k cycles and with the guards $a_{i_1}, a_{i_2}, \ldots, a_{i_k}$ respectively. Setting up the bijective map in the other direction is much easier. Simply write a permutation with its cycle decomposition written in the guard representation and then just drop the parentheses to be able to read the permutation in its one-line notation.

Example 3.1.13. Let $n = 9$ and let $\pi = 752138946$. Then $\overline{\pi} = (7521)(3)(894)(6)$ in its cycle decomposition with the guard representation. Let $\sigma = 187529364$. Then $\overline{\sigma} = (1)(8752)(93)(64)$. Notice also that γ (in one-line notation) has 1 at the end if and only if $\overline{\gamma}$ has exactly one cycle (which has 1 as its last element). If $\beta = 984312576$, then $\overline{\beta} = (98431)(2)(5)(76)$. In the guard representation, we must write every cycle (including the fixed points). *It is thus clear that, obtaining $\overline{\tau}$ from τ involves putting right parenthesis immediately after each left to right minimum and then adjusting the left parenthesis accordingly.*

Proposition 3.1.14. *The correspondence $\sigma \rightarrow \overline{\sigma}$ given in the above procedure is a well-defined bijection.*

Corollary 3.1.15. *With the notation introduced above, let π be a permutation on $[n]$ with $\pi = a_1a_2\cdots a_n$. Let $k \in [n]$. Then the following statements are equivalent.*

(a) $a_n \leq k$.

(b) Each cycle of $\overline{\pi}$ contains at least one element from the set $[k]$.

Proof Let (a) hold. Then $a_n = j \leq k$. Since the correspondence $\pi \rightarrow \overline{\pi}$ looks for the left to right minima, we see that each guard (of every cycle) of $\overline{\pi}$ is a number $i \leq j$ (and the last cycle has the guard j). Hence (b) holds. Conversely let (b) hold. Then each cycle in $\overline{\pi}$ has a guard $i \leq k$ and, in particular, the last cycle of $\overline{\pi}$ must have a guard $j \leq k$. Then $a_n = j \leq k$ proving (a). □

It is also possible to set up a bijection between S_n and ${S_n}^*$, the set of permutations written with cycle decompositions in the leader representation. To do this, let $\pi = a_1a_2\cdots a_n$ be in S_n. We put a left parenthesis at each a_i (before every such a_i) which is a right to left maximum. This constructs the permutation π^* in ${S_n}^*$ after we put right parentheses accordingly. For example, if $\beta = 147652983$, then we have right to left maxima at $1, 4, 7$ and at 9 and hence we put left parentheses there: (1(4(7652(983. Adjusting the right parentheses obtains $\beta^* = (1)(4)(7652)(983)$. Similarly, with $\sigma = 187529364$, we have $\sigma^* = (1)(8752)(9364)$ and with $\pi = 752138946$ we have $\pi^* = (75213)(8)(946)$ We thus have,

Proposition 3.1.16. *Let $k \leq n$ where n is a natural number. There is a one to one correspondence between any two of the following.*

(a) The set of permutations on $[n]$ with k left to right maxima.

(b) The set of permutations on $[n]$ with k left to right minima.

(c) The set of permutations on $[n]$ with k right to left minima.

(d) The set of permutations on $[n]$ with k right to left maxima.

(e) The set of permutations on $[n]$ with exactly k cycles. .

This number is equal to $c(n,k)$.

Proof Given a permutation π on $[n]$, we define two other permutations π' and π'' as follows. Let $\pi = a_1a_2\cdots a_n$. Then $\pi' = b_1b_2\cdots b_n$ and $\pi'' = c_1c_2\cdots c_n$ where $b_i = a_{n+1-i}\ \forall\ i$ and $c_i = (n+1) - a_i\ \forall\ i$. For example, with $n = 9$ and with $\pi = 371892564$, we have $\pi' = 465298173$ and $\pi'' = 739218546$. Just observe that the bijection $\pi \longleftrightarrow \pi'$ gives a bijection between (a) and (c) and between (b) and (d) while the bijection $\pi \longleftrightarrow \pi''$ gives a bijection between (a) and (b) (refer to Definition 3.1.11). □

3.2 Descents and the Eulerian polynomial

The material in this and the next section is based on Stanley [49]. We again stick to the one-line notation and make the following definition.

Definition 3.2.1. Let $\pi = a_1a_2\cdots a_n$ be a permutation on $[n]$. An integer i is called *a descent* of π if $a_i > a_{i+1}$. The set consisting of all the descents of π is called its *descent set* and is denoted by $D(\pi)$. We denote the number of descents of π by $d(\pi)$.

By convention, we do not allow $n \in D(\pi)$. Thus the descent set of $\pi = 346719852$ is $\{4,6,7,8\}$. An immediate question is to find the number of permutations with a given descent set. Instead, we answer the following weaker question.

Theorem 3.2.2. *Let $S = \{s_1, s_2, \ldots, s_k\}$ be a subset of $[n-1]$ with $s_1 < s_2 < \cdots < s_k$. Then the number of permutations π whose descent set $D(\pi)$ is contained in S is given by*

$$\alpha(S) = \binom{n}{s_1, s_2 - s_1, \ldots, s_k - s_{k-1}, n - s_k}$$

Proof Only those permutation π whose descent set is contained in S qualify for being counted. This amounts to saying that at any i not in S, we must have $a_i < a_{i+1}$. Thus there is no descent at the first s_1 elements except possibly at s_1. We may therefore, arbitrarily choose a subset of order s_1 and write its elements in the form $a_1 < a_2 < \cdots < a_{s_1}$ to get the first s_1 elements (in the one-line notation). Having done that, we are left with $n - s_1$ elements from which we choose the next set of $s_2 - s_1$ elements in $\binom{n-s_1}{n-s_2}$ ways and write the next segment:

$a_{s_1+1} < a_{s_1+2} < \cdots < a_{s_2}$. Continuing in this manner, we obtain the desired result (after using the multiplication principle!):

$$\binom{n}{s_1} \times \binom{n-s_1}{s_2-s_1} \times \cdots \times \binom{n-s_{k-1}}{s_k-s_{k-1}}$$

which equals

$$\binom{n}{s_1, s_2-s_1, \ldots, s_k-s_{k-1}, n-s_k}$$

as desired. □

It will certainly be nice to know the number $\beta(S)$ of permutations for which the descent set is exactly S (and not merely contained in S). This is postponed to Chapter 4.

Proposition 3.2.3. *Let $\pi = a_1 a_2 \cdots a_n$ have k descents. Then $\sigma = a_n a_{n-1} \cdots a_2 a_1$ has $n-k-1$ descents.*

Proof This follows immediately because for any i with $1 \le i \le n-1$, if i is in $D(\pi)$, then $n-i$ is not in $D(\sigma)$ and vice versa. Hence the number of places where σ has no descent is k. □

Definition 3.2.4. Let $\pi = a_1 a_2 \cdots a_n$ be a permutation on $[n]$. Then the permutation $a_n a_{n-1} \cdots a_2 a_1$ is called the *reverse* of π.

Proposition 3.2.3 allows us to see the symmetry of the following definition.

Definition 3.2.5. Recall that S_n refers to the set of all the permutations of $[n]$. The *Eulerian polynomial* $A_n(x)$ is a polynomial of degree n that is given by the formula

$$A_n(x) = \sum_{\pi \in S_n} x^{1 + d(\pi)}$$

The coefficient of x^k in $A_n(x)$ is denoted by $A(n,k)$ and is called *an Eulerian number.*

Evidently $A(n,k)$ counts the number of permutations in S_n with $k-1$ descents. Proposition 3.2.3 shows that $A(n,k) = A(n, n+1-k)$. We list the first four Eulerian polynomials below.

$$A_1(x) = x$$

$$A_2(x) = x + x^2$$

$$A_3(x) = x + 4x^2 + x^3$$

$$A_4(x) = x + 11x^2 + 11x^3 + x^4$$

Some properties of Eulerian numbers will be discussed in the exercises (Exercises 3.11 through 3.15). In particular, Eulerian numbers satisfy a recurrence relation similar to the Pascal identity for combinations (Exercise 3.14).

3.3 Tree representations for permutations

This section discusses two tree representations for permutations. *The treatment here, as in Section 2, follows Stanley [49].* Readers are expected to have some knowledge of elementary graph theory. In particular, a tree is a graph that has no cycles (circuits). A labeled tree is a tree with each vertex receiving a different label. A rooted tree is one which has a distinguished vertex called the root (the entire tree is read from this vertex). An increasing tree is a labeled tree in which the labels are increasing as one moves along any path beginning from the root. In a rooted tree, the successors (or children) of any non-root vertex v are the $d(v) - 1$ vertices u adjacent to v such that u is not on the unique path from the root to v. This, of course means that the root r has $d(r)$ successors. Here $d(v)$ denotes the degree of v. An oriented binary tree is a tree in which each vertex has at the most two successors, left and right and *we make distinction between the left and the right child.*

We are now ready to describe a procedure for obtaining an oriented binary increasing tree from a permutation on $[n]$. Let $\pi = a_1a_2 \cdots a_n$ be a permutation on $[n]$. The procedure is recursive. At any stage, we scan a sequence of positive integers say $b_1b_2 \cdots b_r$ (this is a subsequence of $a_1a_2 \cdots a_n$ that has consecutive entries and hence equals $a_m a_{m+1} \cdots a_{m+t}$ for some m and t) and locate the smallest integer say b_i in this sequence. Remove b_i from the sequence, making it an ancestor of all the remaining numbers in the sequence and divide the sequence into two parts: the left subsequence $b_1b_2 \cdots b_{i-1}$and the right subsequence $b_{i+1} \cdots b_r$ (one or both of these could be empty and in the latter case, b_i is an end vertex or a leaf of the tree). We then repeat this process on the two subsequences, the left and right subsequence and locate their minimum elements and so on. Thus to start with, we locate the smallest integer in the sequence $a_1a_2 \cdots a_n$ which is 1. This integer say $a_i = 1$ is the root of the binary tree. Removal of this integer divides the sequence into two subsequences: the left subsequence and the right subsequence. The procedure is now repeated for each of these subsequences. That is, if a_j is the smallest number in the left subsequence and if a_k is the smallest number in the right subsequence, then the root a_i has two children: a_j is th left child and a_k is the right child. If any of the subsequences is empty then the root has only one child. Treating a_j as the root of a tree with vertices given by the integers in the left subsequence (left progeny of the root) and a_k as the root of a tree with vertices given by the integers in the right subsequence (right progeny of the root) and repeating the procedure sufficiently many times (until all the integers are exhausted), obtains the desired increasing binary tree.

We illustrate this procedure through the following examples. First let $\pi = 356218497$. Root the tree at 1. This makes the two subsequences 3562 and 8497. Hence 2 is the left child and 4 is the right child of the root. Observe then that 2 has no right child as there is no integer to the right of 2 in the subsequence 3562 and 3 is the left child of 2. Finally 3 has only the right child 5 and 5 has only the right child 6. Turning to the other subtree, 4 has 8 as the left child and 7 as the right child. Finally 9 is the left child of 7. The entire tree looks like the first diagram in Figure 3.1. Similarly, the binary tree corresponding to the permutation 364197258 is shown in the second diagram of Figure 3.1.

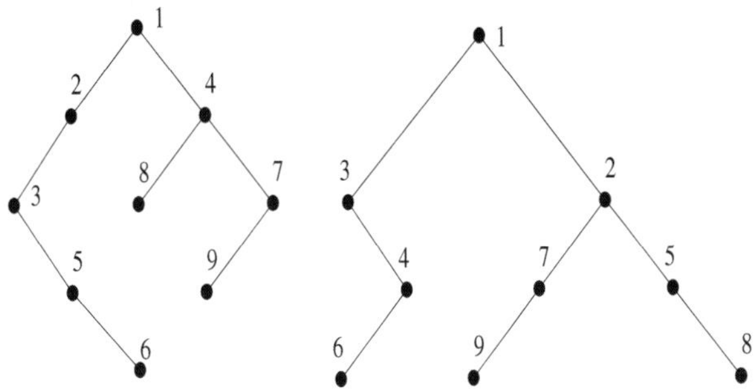

Figure 3.1: Binary, increasing, oriented trees

How do we obtain a permutation corresponding to the following binary, oriented and increasing tree?

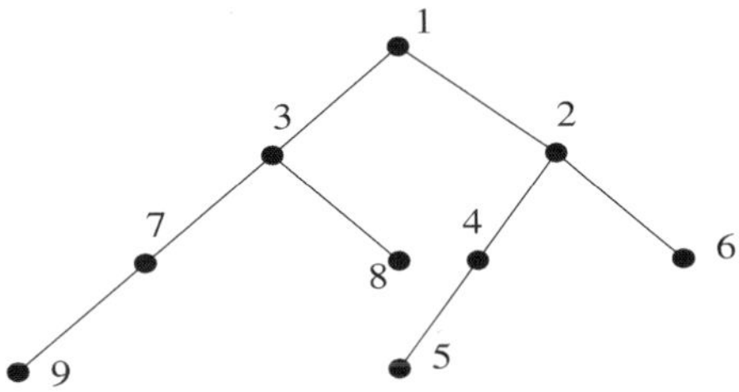

Figure 3.2: Increasing, oriented, binary tree

We proceed through the following sequence of partial permutations as follows. We have

$$312 \rightarrow 73812 \rightarrow 973812 \rightarrow 97381426 \rightarrow 973815426$$

These illustrations establish a bijective correspondence between the set of all the permutations on $[n]$ and the set of all the binary, oriented and increasing trees on the set $[n]$. Several useful conclusions can be drawn from the above binary tree representation. First observe that a vertex a_i has a left child only if a_{i-1} is not scanned (as we move along a path from the root) before a_i. This simply means that a_i has a left child if and only if $a_{i-1} > a_i$. Thus a_i has a left child for $i = 2, 3, \ldots, n$ if and only if $i-1$ is in the descent of the given permutation $\pi = a_1 a_2 \cdots a_n$. Hence the number of permutations whose binary tree has k vertices that have left children is thus same as $A(n, k+1)$. For a permutation π as given above, we say that we have a *rise, fall, valley and peak* at a_i as per the following table.

rise	$a_{i-1} < a_i < a_{i+1}$
fall	$a_{i-1} > a_i > a_{i+1}$
valley	$a_{i-1} > a_i < a_{i+1}$
peak	$a_{i-1} < a_i > a_{i+1}$

Suppose we have a rise at a_i. Then, in scanning the numbers, a_{i-1} will be scanned before a_i and both a_i and a_{i+1} will be on the right subtree (relative to a_{i-1}). Subsequently, when a_i is scanned (this will happen before a_{i+1}) we observe that a_i has only right subtree (which will house a_{i+1}). It then follows that a_i has only a right child and no left child. Other situations are similar. If there is a fall at a_i, then a_i has only the left child and no right child. Finally, in case of a valley, a_i has two children while in the case of a peak, a_i has no children, i.e., a_i is an end vertex of the tree.

Notice that if we let σ denote the permutation $b_1 b_2 \cdots b_n$ with $b_j = n + 1 - a_j$, then the rises of σ correspond to the falls of π and conversely. Similarly, the peaks of σ correspond to the valleys of π and conversely. Finally a binary tree as above is called *complete* if each vertex has either two children or is an end point. A complete binary tree must have an odd number of vertices (make an induction on the number of vertices). For such a tree, the corresponding permutation has only peaks or valleys. If a_1 was not an end vertex, then it would have only the right child which is not allowed. Since peaks and valleys must alternate in such a permutation, the permutation must be of the form $a_1 > a_2 < \cdots > a_{2m} < a_{2m+1}$, where $n = 2m + 1$. Such permutations are called *alternating permutations*. These observations are summed up in the following theorem.

Theorem 3.3.1. *There is a one-to-one correspondence between the set of all the permutations on $[n]$ and the set of all the oriented binary trees on $[n]$. Further, we have:*

(a) The number of such trees in which exactly k vertices have left children is the Eulerian number $A(n, k+1)$.

(b) The number of such trees with r vertices that have two children is same as the number of such trees with r vertices that have no children.

(c) The number of alternating permutations on $2m + 1$ elements is equal to the number of increasing complete binary trees on $2m + 1$ vertices.

We now describe a different tree representation for permutations. This representation sets up a bijection between the set of all rooted increasing trees on $n + 1$ vertices consisting of the vertices in $[n]$ and the single vertex 0 (which must be the root) and the set of all the permutations on $[n]$. *Note that the tree here is rooted at* 0, *is increasing but is not oriented.* It works as follows. Let $\pi = a_1 a_2 \cdots a_n$ be a permutation on $[n]$. We describe the predecessor (father or mother) of any vertex a_i in the rooted tree. This is the rightmost a_j such that both $j < i$ and $a_j < a_i$ hold. If there is no such a_j then make the vertex a_i adjacent to the vertex 0 (in particular, a_1 as well as the integer 1, wherever it appears have 0 as the predecessor). The children of the root 0 are thus all the a_i that are right to left minima of the permutation. We can, of course describe successors (children) of any vertex a_i as follows. These are all the a_k such that $k > i$ and $a_i < a_k$ with the additional requirement that every integer between a_i and a_k is strictly larger than a_k. This procedure will now be illustrated with the following examples.

Begin with the same example given earlier: $\pi = 356218497$ and observe that for each of the integers, 3, 2 and 1 there is no integer to the left which is smaller. Hence these vertices are precisely the children of the root 0. Then observe that 3 must be the predecessor of 5 and 5 must be the predecessor of 6. Finally, 8 and 4 both have 1 as the predecessor while 4 would have both 9 and 7 as successors. The increasing, rooted tree is drawn as the first tree in Figure 3.3. Likewise, the second tree in Figure 3.3 corresponds to the permutation 364197258.

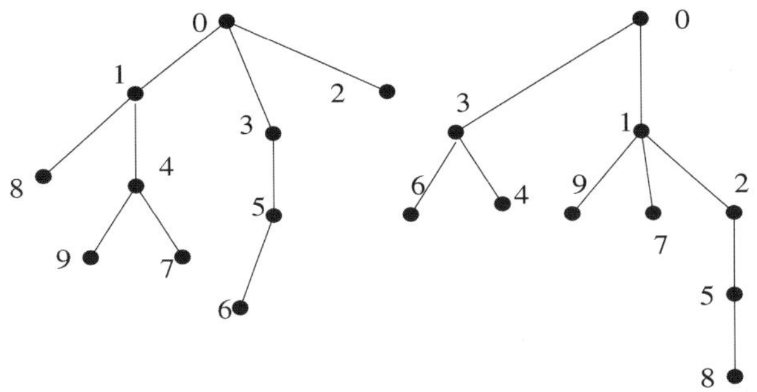

Figure 3.3: Increasing trees rooted at zero

Similarly, what is the permutation that corresponds to the rooted increasing tree drawn in Figure 3.4?

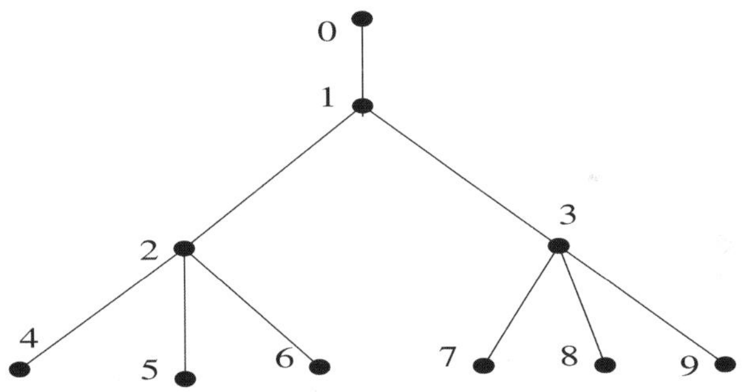

Figure 3.4: Increasing tree rooted at zero

We proceed through the following sequence of partial permutations as follows. We have

$$1 \to 132 \to 139872 \to 139872654$$

These examples establish the fact that we have a bijection between the set of all the $n!$ permutations and the set of all the increasing rooted trees on the set $[n] \cup \{0\}$. In the above procedure, observe that an integer a_i is made adjacent to the root if and only if it is a *right to left minimum*, that is, it has the property that it is smaller than all the integers to its left. Hence using Proposition 3.1.16, the number of rooted trees that in which the root has k children equals $c(n, k)$, the number of permutations with exactly k cycles. Similarly, when is some a_i an end vertex of the tree? This will happen only when $i = n$ or when $a_{i+1} < a_i$. Hence a tree with k end-vertices corresponds to a permutation with $k - 1$ descents. The following theorem is thus proved.

Theorem 3.3.2. *There are exactly* $n!$ *rooted increasing trees on* $n+1$ *vertices. Further, the number of such trees in which the root has degree* k *equals the number* $c(n,k)$ *while the number of those trees that have exactly* k *end vertices equals the Eulerian number* $A(n,k)$.

3.4 Exercises for Chapter 3

3.1 The following permutations on $[9]$ are given in their one line notation. Write them down in their cycle decomposition form (with the least number of cycles). Then write the same permutations in the guard representations and the leader representations respectively.

(a) 372149568

(b) 123789564

(c) 278596431

(d) 918274653

3.2 The following permutations on $[9]$ are given in their cycle decomposition forms. Write them down in the one line notations.

(a) $(372)(148)(69)$

(b) $(169234)(78)$

(c) $(147)(258)$

(d) $(248)(15)(69)(37)$

3.3 For each one of the following four permutations σ_i on $[9]$ that are written in their one line notations, find the corresponding permutations $\overline{\sigma_i}$ in the guard representations.

(a) $\sigma_1 = 372418965$

(b) $\sigma_2 = 419635827$

(c) $\sigma_3 = 182497635$

(d) $\sigma_4 = 827594613$

3.4 For each one of the following four permutations α_i on $[9]$ that are written in their one line notations, find the corresponding permutations $\alpha_i{}^*$ in the leader representations.

(a) $\alpha_1 = 725498163$

(b) $\alpha_2 = 259137864$

(c) $\alpha_3 = 128649537$

(d) $\alpha_4 = 293645718$

3.5 Let $n = 9$ and let $\overline{\pi}$ denote the image of permutation π (in its one line notation) written in the guard representation. Write $\overline{\pi}$ where π equals

(a) 724198356

(b) 298716345

(c) 437198256

(d) 146732985

(e) 739824156

In each case, write down $\overline{\pi}$ in one line form and find the number of descents of $\overline{\pi}$.

3.6 Let $n = 9$ and let π^* denote the image of permutation π (in its one line notation) written in the leader representation. Write π^* where π equals

(a) 423968157

(b) 941683257

(c) 827651394

(d) 613782594

(e) 432167985

In each case, write down π^* in one line form and find the number of descents of π^*.

3.7 All the examples as well as exercises in this chapter have permutations on the set $[9]$. Can you think of a reason?

3.8 Let n be a fixed positive integer ≥ 2 and let $1 \leq k \leq n$ be a fixed integer. Let $m = n - k$.

(a) Show that the sum $\sum a_1 a_2 \cdots a_m$ where each summand is a product of m distinct integers in the set $[n-1]$ and the sum is taken over all the m-subsets of $[n-1]$ is equal to $c(n, k)$ (hint: use Lemma 3.1.6).

(b) Show that the sum in (a) actually counts the number of pairs (S, f) where S is a subset of $[n-1]$ with $|S| = n - k$ and $f : S \to [n-1]$ is a function such that $f(i) \leq i$ for all i.

3.9 (Continuation of Exercise 3.8) Let n and k be as in Exercise 3.5.

(a) Let Ω denote the set of all the pairs (S, f) as in Exercise 3.8(b). Let Γ denote the set of all the permutations on $[n]$ with exactly k disjoint cycles. Define $\phi : \Omega \to \Gamma$ as follows. Given (S, f) in Ω with $1 \leq a_1 < a_2 < \cdots < a_m \leq n - 1$ and let $T = \{j \in [n] : n - j \notin S\}$. Then T has k elements and the elements of $[n] - T$ can be written in the form: $b_1 > b_2 > \cdots > b_m$. We write the member π of Γ corresponding to the above pair in its leader representation as follows. In each cycle of π the leader (largest element) is an element of T, one in each cycle and arranged in the leader representation. The b_i's are then distributed in these cycles so that for every i, the number of elements of π preceding b_i and larger than b_i is exactly equal to $f(a_i)$. This is illustrated through the following example. Let $n = 9$ and $k = 3$. Let $S = \{2, 4, 6, 7, 8\}$ and let the function f be given by

j	2	4	6	7	8
$f(j)$	1	2	2	4	3

Then, $[9] - T$ is the set $\{7 > 5 > 3 > 2 > 1\}$ and hence $T = \{4, 6, 8, 9\}$. We then have π equal to

$$(4)(631)(8257)(9)$$

Show that ϕ gives a bijection.

(b) Using Exercise 3.8(b), and part (a), give an alternative proof of Lemma 3.1.7.

3.10 What is the number of permutations on $[n]$ that contain 1 in a k-cycle? Show the somewhat surprising result that this does not depend on k (but depends only on n). You may use the guard representation or prove the result by direct counting. How many permutations contain any j in a k-cycle? What is the number of permutations that contain 1 and 2 in the same cycle?

3.11 Observe that the only permutation on $[n]$ that has no descents is the identity permutation. Use this to show that the number of permutations with exactly one descent $= A(n, 2) = 2^n - (n + 1)$.

3.12 (a) How many permutations on $[n]$ have exactly $n - 1$ descents? How many have exactly $n - 2$ descents?

(b) Given a permutation $\pi = a_1 a_2 \cdots a_n$ on $[n]$, define the permutations $\sigma = b_1 b_2 \cdots b_n$ by $b_i = n + 1 - a_i$. Show that π ha a descent at r iff σ does not have a descent at r.

(c) Let $1 \leq r \leq n - 1$. Show that the total number of permutations on $[n]$ that have descent at r is $\frac{n!}{2}$.

(d) Show that the $A_n(1) = n!$ and

$$\left\{ \frac{x{A_n}'(x) - A_n(x)}{x^2} \right\} \Big| (x = 1) = \frac{n!(n-1)}{2}$$

where $A_n(x)$ is the Eulerian polynomial.

3.13 Use Exercise 3.11 to find out the Eulerian polynomials $A_5(x)$ and $A_6(x)$.

3.14 Prove that for all k such that $2 \leq k \leq n - 1$, we have

$$A(n, k) = kA(n - 1, k) + (n + 1 - k)A(n - 1, k - 1)$$

3.15 Use Exercises 3.11 and 3.13 to find $A_7(x)$.

3.16 Let $n = 9$. For each one of the following permutations, write down the corresponding increasing oriented binary trees.

(a) 361728945

(b) 672984135

(c) 456123789

3.17 For each of the the permutations in Exercises 3.4 and 3.5, write the corresponding rooted, increasing tree on 10 vertices $\{0, 1, \ldots, 9\}$.

3.18 Show that the number of increasing, oriented binary trees in which each vertex has either no or two children is the same as the number of alternating permutations.

3.19 For the following two increasing oriented binary trees on $[9]$, find the corresponding permutation.

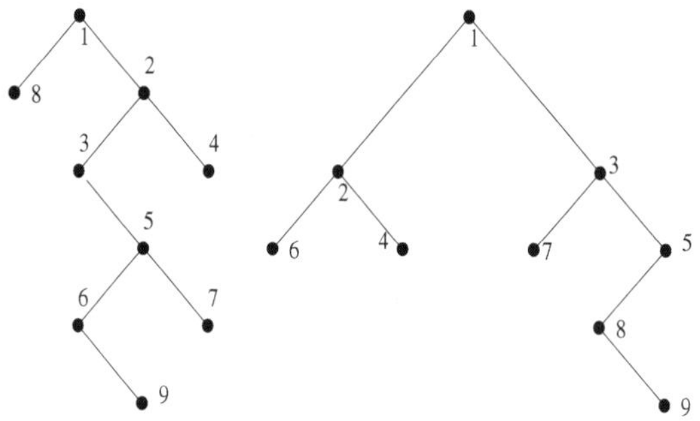

Figure 3.5: Increasing, oriented binary trees

3.20 For the increasing rooted trees on the set$\{0, 1, \ldots, 9\}$in Figure 3.6, find the corresponding permutations.

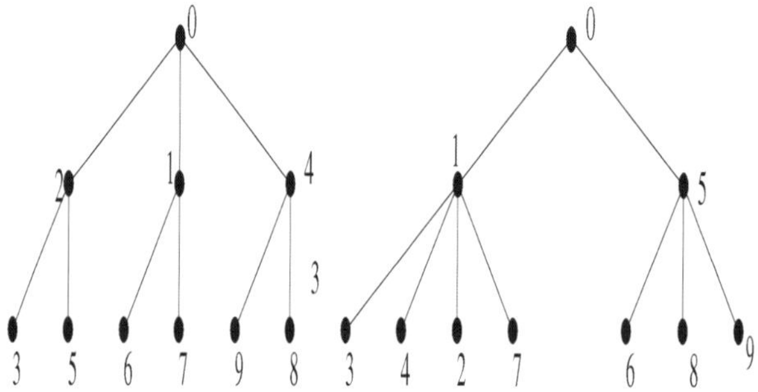

Figure 3.6: Increasing trees rooted at 0

3.21 For the three oriented, rooted and increasing binary trees on $[9]$ that are drawn in Figure 3.7, find the corresponding permutations and then draw the corresponding rooted increasing trees on $\{0\} \cup [9]$.

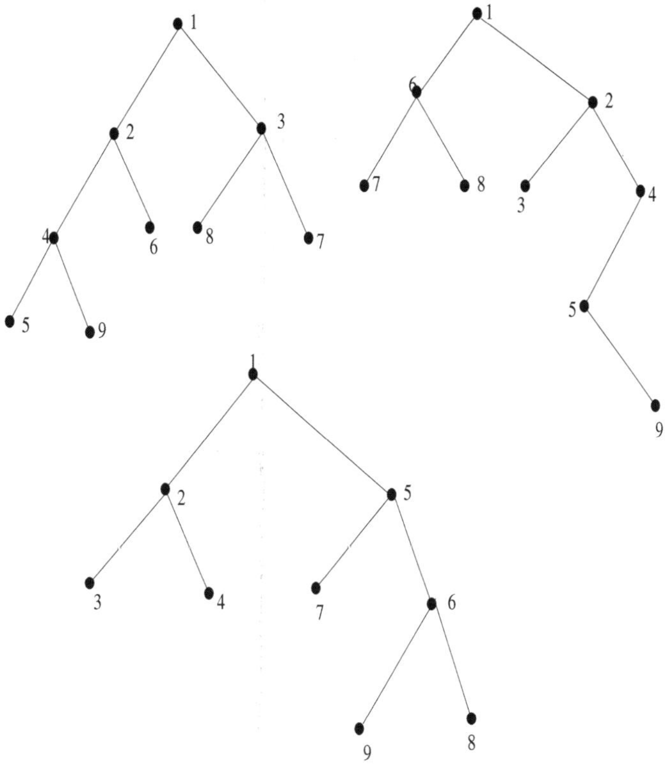

Figure 3.7: Increasing,binary, rooted, oriented trees

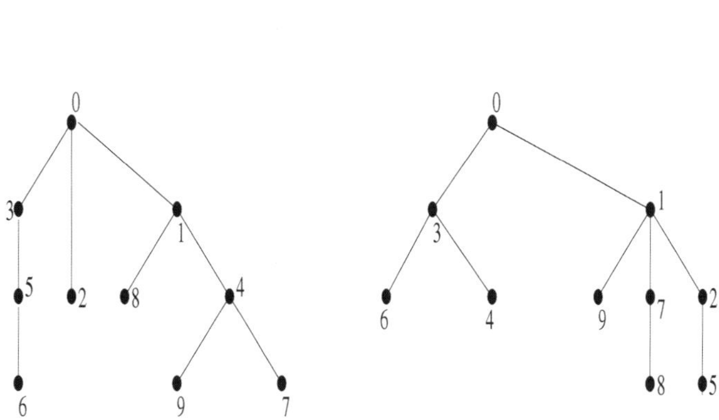

Figure 3.8: Increasing oriented trees rooted at 0

3.22 For the two rooted and increasing trees on $\{0\} \cup [9]$ (given in Figure 3.8), find the corresponding permutations and then draw the corresponding rooted binary oriented increasing trees on $[9]$.

3.23 (Stanley [49]) We are given n square envelopes of n distinct sizes (so that given any two, the smaller can be put inside the larger). Let a_n denote the number of arrangements of the envelopes satisfying the condition that no bigger envelope is put inside a smaller envelope. Let the envelopes be labeled from 1 to n with 1 representing the largest envelope and n representing the smallest. That is, envelope j can be put inside envelope i if and only if $i > j$. With $n = 2$, the possible arrangements are 2 inside 1 or 1 and 2 kept separately. Thus $a_2 = 2$. For $n = 3$, we have six possible arrangements. These are: (1) 3 in 2 (2) 3 in 1 (3) 2 in 1 (4) 3 in 2 and 2 in 1 (5) both 3 and 2 are inside 1 but 3 is not inside 2 (6) 1, 2 and 3 are put separately (none containing any other). Thus $a_3 = 6$. Prove, in general that $a_n = n!$ You may use induction (this is boring and uninteresting). A clever argument: We can set up a bijection [49] between the set of all the arrangements of n envelopes and the set of all the rooted increasing trees on $\{0, 1, \ldots, n\}$.

Chapter 4

The inclusion-exclusion principle

4.1 The principle and some applications

The inclusion-exclusion principle, is among the most basic techniques of combinatorics. Suppose we have a set X with subsets A and B. Then the number of elements that are in A or B (or both) i.e., the cardinality of $A \cup B$ is given by $|A| + |B| - |A \cap B|$: The elements that are in both A and B were counted twice. To get rid of the over-counting, we must subtract. If $\overline{A}$ and $\overline{B}$ denote the complements of A and B respectively, then how many elements does the set $\overline{A} \cup \overline{B}$ have? This number is $|X| - |A| - |B| + |A \cap B|$. The explanation is as before. From the set of all the elements we get rid of those that are in A or B. In doing so, we subtracted the elements of $A \cap B$ twice. This has to be corrected by adding such elements once. *Essentially, this way of over-counting (inclusion), correcting it using under-counting (exclusion) and again correcting (overcorrection) and so on is referred to as the inclusion-exclusion technique.* As another example, consider the question of finding how many positive integers up to 100 are not divisible by 2, 3 or 5. We see that there are 50 integers that are multiple of 2, 33 that are multiples of 3 and 20 that are multiples of 5. This certainly amounts to over-counting as there are integers that are divisible by two of the given three numbers 2, 3 and 5. In fact, the number of integers divisible by both 2 and 3 is 16, the number of integers divisible by both 2 and 5, that is divisible by 10 is 10. The number of integers divisible by both 3 and 5 is the number of integers below 100 and divisible by 15 and that number is 6. Finally the number of integers divisible by all of 2, 3 and 5 is just 3. Hence the number of integers not divisible by any one of 2, 3 or 5 is

$$100 - (50 + 33 + 20) + (16 + 10 + 6) - 3 = 26$$

The technique used above is called a *sieve method.* This technique was known to the Greeks and is in fact, known as the Sieve of Eratosthenes. Inclusion-exclusion principle, in its simplest form, is the following.

Theorem 4.1.1. *(The inclusion-exclusion principle) Let X be a finite set and let P_i ; $i = 1, 2, \ldots, n$ be a set of n properties satisfied by (some of) the elements* *of X. Let A_i denote the set of those elements of X that satisfy the property P_i. Then the size of the set $\overline{A_1} \cap \overline{A_2} \cap \cdots \cap \overline{A_n}$ of all those elements that do not satisfy any one of these properties is given by*

$$\begin{aligned}\overline{A_1} \cap \overline{A_2} \cap \cdots \cap \overline{A_n} = |X| - \sum_{i=1}^{n} |A_i| + \sum_{1 \le i < j \le n} |A_i \cap A_j| - \cdots \\ + \{(-1)^k \sum_{1 \le i_1 < i_2 < \cdots < i_k \le n} |A_{i_1} \cap A_{i_2} \cdots \cap A_{i_k}|\} \\ + \cdots + (-1)^n |A_1 \cap A_2 \cdots \cap A_n|\end{aligned} \tag{4.1}$$

Proof The proof will show that every object in the set X is counted the same number of times on both the sides. Suppose $x \in X$ and assume that x is an element of the set on the left hand side of equation (4.1). Then x has none of the properties P_i. We need to show that in this case, x is counted only once on the right hand side. This is obvious since x is not in any of the A_i and $x \in X$. Thus x is counted only once in the first summand and is not counted in any other summand since $x \notin A_i$ for all i. Now let x have k properties say $P_{i_1}, P_{i_2}, \ldots, P_{i_k}$ (and no others). Then x is counted once in X. In the next sum, x is occurs in k summands (since x is in k of the subsets A_j). In the next sum, x occurs $\binom{k}{2}$ times and so on. Thus, on the right hand side, x is counted precisely

$$\binom{k}{0} - \binom{k}{1} + \binom{k}{2} + \cdots + (-1)^k \binom{k}{k}$$

times. Using the binomial theorem (Theorem 1.2.5), this sum is $(1-1)^k$ which is 0 and hence x is not counted on the right hand side. This completes the proof. □

There are more sophisticated versions of the inclusion-exclusion principle, that achieve something more. We refrain from giving these versions and merely refer the reader to the list of references. In particular, Stanley [49] has a vector space version of the inclusion-exclusion principle. A convenient way of writing equation (4.1) in Theorem 4.1.1 is the following. Let H_k denote the quantity

$$H_k = \sum_{1 \le i_1 < i_2 < \cdots < i_k \le n} |A_{i_1} \cap A_{i_2} \cdots \cap A_{i_k}|$$

Then equation (4.1) can be rewritten as

$$|\overline{A_1} \cap \overline{A_2} \cap \cdots \cap \overline{A_n}| = H_0 - H_1 + H_2 - \cdots + (-1)^k H_k + \cdots + (-1)^n H_n \tag{4.2}$$

This is convenient since many times it is much easier to calculate H_k directly (by making use of some symmetry in the problem under consideration). We begin the list of applications with the canonical question of derangements. Suppose n people leave their umbrellas outside a hall (where a music concert is going on) and while coming out each one picks up an umbrella which is not his. In how many ways can that happen? Essentially, this amounts to finding the total number of permutations in S_n that have no fixed points. Such permutations are called *derangements*.

Definition 4.1.2. *A derangement* is a permutation π such that $\pi(i) \neq i$ for every i. We denote the number of derangements on an n-set by D_n.

The two-line notation is to be preferred for the understanding of the problem. Clearly, the values of D_1, D_2 and D_3 are 0, 1 and 2 respectively. The two derangements on [3] are

$$\begin{pmatrix} 1 & 2 & 3 \\ 2 & 3 & 1 \end{pmatrix}, \begin{pmatrix} 1 & 2 & 3 \\ 3 & 1 & 2 \end{pmatrix}$$

while it is easily checked that D_4 equals 9. *By convention, we let D_0 equal* 1.

Proposition 4.1.3. *The number of derangements D_n on n letters is given by*

$$D_n = n! \left\{ 1 - 1 + \frac{1}{2!} - \cdots + (-1)^n \frac{1}{n!} \right\} \tag{4.3}$$

Proof With the notations as in Theorem 4.1.1 and equation (4.2), first note that theset X consists of all the $n!$ permutations. Let the property P_j be defined on the set of all the permutations by the requirement $\pi(j) = j$. Since any permutation satisfying P_j can move any of the remaining letters (among themselves), it follows that the cardinality of the set A_j is $(n-1)!$ so that $H_1 = n \times (n-1)!$. It is now clear that the same pattern repeats. For $1 \leq i_1 < i_2 < \cdots < i_k \leq n$ the set $A_{i_1} \cap A_{i_2} \cap \cdots \cap A_{i_k}$ consists of those permutations that fix the points $i_1, i_2, \cdots, i_k$ (and move the remaining $n-k$ points among themselves). Clearly this set contains $(n-k)!$ permutations. Since the numbers $1 \leq i_1 < i_2 < \cdots < i_k \leq n$ can be chosen in $\binom{n}{k}$ ways, it follows that $H_k = \binom{n}{k}(n-k)!$. Using equation (4.2), we have

$$D_n = n! + \sum_{k=1}^{n} (-1)^k \binom{n}{k} \times (n-k)!$$

Using the value of $\binom{n}{k}$, this gives

$$D_n = n! \sum_{k=0}^{n} \frac{(-1)^k}{k!} \tag{4.4}$$

as desired. □

The proof of Proposition 4.1.3 is not the only available derivation. We indicate one more here and a third one using generating functions will be given later in Chapter 12. Let $n \geq 2$ and observe that the number of derangements that take 1 to 2 is the same as the number of derangements that take 1 to j for any $j \geq 2$. This is obvious since we may multiply such a permutation on both the sides by a permutation that switches 2 and j. The resulting permutation is still fixed point free. Let then the number of derangements that take 1 to 2 be denoted by R. Then $D_n = (n-1) \times R$ since there are $n-1$ such j's. To evaluate R, we make two cases. Let π be such a derangement where π is given by

$$\pi = \begin{pmatrix} 1 & 2 & \cdots & \cdots & n \\ 2 & x_2 & \cdots & \cdots & x_n \end{pmatrix}$$

Since π takes 1 to 2, it *must take* 2 *to a number other than* 2 *say to* $x_2 = m$.
Case 1 $x_2 = m = 1$. Let π' denote the permutation on the set $T = \{3, 4, \ldots, n\}$ obtained from π by deleting the first two entries in π:

$$\pi' = \begin{pmatrix} 3 & 4 & \cdots & \cdots & n \\ x_3 & x_4 & \cdots & \cdots & x_n \end{pmatrix}$$

Then π' is a derangement on T and conversely any such derangement on T can be uniquely extended to give a derangement of the set $[n]$. We actually obtain π' from π by just reading the portion obtained by deleting the first two entries in the two-line notation. Conversely, given a derangement π' on the set T, we stick two additional columns

$$\begin{matrix} 1 & 2 \\ 2 & 1 \end{matrix}$$

on its left to get a derangement π of the required kind. This is clearly a bijection between the set of derangements that take 1 to 2, 2 to 1 and the set of all the derangements on T.
Case 2 $m \neq 1$. Then for some $k \neq 1, 2$, we must have $x_k = \pi(k) = 1$. In the two-line representation of π we replace 1 on the second line by 2 to get a permutation π^* on the set $U = \{2, 3, \ldots, n\}$.

$$\pi^* = \begin{pmatrix} 2 & \cdots & k & \cdots & n \\ x_2 & \cdots & 2 & \cdots & x_n \end{pmatrix}$$

Conversely, given any such permutation π^* which is a derangement on U, π^* does not take 2 to 1 and replacing 2 by 1 on the second line obtains the required permutation π (by augmenting on the left with $\pi(1) = 2$). Thus there is a bijection between the set of all the derangements that take 1 to 2, 2 to a number other than 1 and the set of all the derangements on the set U.
The number of derangements in Case 1 is D_{n-2} while it is D_{n-1} in Case 2. Hence $R = D_{n-1} + D_{n-2}$. We thus have

$$D_n = (n-1)(D_{n-1} + D_{n-2}) \tag{4.5}$$

Writing $E_n = D_n - nD_{n-1}$, equation (4.5) gives

$$\begin{aligned} E_n &= D_n - nD_{n-1} \\ &= -D_{n-1} + (n-1)D_{n-2} \\ &= -(D_{n-1} - (n-1)D_{n-2}) \\ &= -E_{n-1} \end{aligned}$$

which holds for all $n \geq 2$. Thus, by iterating the above recurrence relation, we obtain $E_n = (-1)^{n-2}E_2$. But $E_2 = 1$ so that $E_n = (-1)^n$ is established. We thus have

$$D_n = nD_{n-1} + (-1)^n \tag{4.6}$$

We then use induction on n and obtain an explicit formula (equation (4.3) or (4.4)) for derangements that was obtained earlier.

We now discuss a forbidden position problem which has the following description. Let $\pi = a_1a_2 \cdots a_n$ be a permutation on $[n]$ (in its one-line notation). We are looking for the set of all the permutations π that have the property that for any i, i and $i+1$ do not occur as consecutive integers (*in the one-line notation of* π) for any i from 1 to $n-1$. This means that $a_j + 1 \neq a_{j+1}$ for any $j = 1, 2, \ldots, n-1$. Let Q_n denote the number of such permutations on the set $[n]$. Clearly, we have $Q_1 = Q_2 = 1$ while $132, 213$ and 321 are the only permissible permutations on 3 letters so that $Q_3 = 3$. We seek a general formula for Q_n. To that end, represent a permutation in one-line notation and let P_i be the property that the permutation contains i and $i+1$ as consecutive integers. We then have $n-1$ such properties and *we desire to find the number of those permutations that have none of the these properties.* Call a pair $[i, i+1]$ a snag. The number of permutations that have such a snag, i.e., the number of permutations that contain i and $i+1$ one after the other can be found by considering the snag as a single element. Thus we have only $n-1$ elements (including the pair given by the snag, treated as a single element). For example, with $i = 1$, we are actually finding all the permutations on the set $\{[12], 3, 4, \ldots, n\}$. Hence the number of permutations that have property P_i is $(n-1)!$ Thus, in the notation of of Theorem 4.1.1, $H_1 = \binom{n-1}{1} \times (n-1)!$. How many permutations have properties P_i and P_j where we let w.l.o.g. $i < j$? There are thus two snags $[i, i+1]$ and $[j, j+1]$. If $i+1 < j$, then as above, we consider these snags as single elements. Along with $n-4$ elements not in the snags, this gives an effective total of $n-2$ elements, on which we get $(n-2)!$ permutations. Otherwise, $i+1 = j$ and we obtain a *trail* $[i, i+1, i+2]$ whose symbols must occur one after the other. In this case again, we are effectively looking at all the permutations on $(n-3)+1 = n-2$ letters. Thus, in both the cases, we obtain $(n-2)!$ permutations. For example, let $n = 9$. If $i = 3$ and $j = 6$, we are dealing with all the permutations on the 7-set $\{1, 2, [34], 5, [67], 8, 9\}$. If $i = 3$ and $j = 4$, then we are dealing with all the permutations on the 7-set $\{1, 2, [345], 6, 7, 8, 9\}$. Thus, in both the cases, we have a permutation on $n-2 = 7$ letters. Hence $H_2 = \binom{n-1}{2} \times (n-2)!$

The general situation is now clear. Suppose snags are to occur at k places. Convert these into trails that are, say, m in number and are given by

$$[b_1, b_1+1, \ldots, b_1+r_1]; [b_2, b_2+1, \ldots, b_2+r_2]; \cdots ; [b_m, b_m+1, \ldots, b_m+r_m].$$

Here, each r_j is a positive integer. Also, $b_j + r_j < b_{j+1}$ for all $j = 1, 2, \ldots, m-1$. In the j-th trail we have exactly r_j snags. Since the total number of snags is k, we have

$$r_1 + r_2 + \cdots + r_m = k$$

Each trail is clearly to be treated as a single point. But the j-th trail contains $r_j + 1$ numbers. Hence the number of elements not in any trail is

$$(n - m) - (r_1 + r_2 + \cdots + r_m) = (n - m) - k$$

Since each trail is to be treated as a single point, the effective number of points on which we seek a permutation is $[(n - m) - k] + m = n - k$. Thus in the situation of k snags, we get $(n - k)!$ permutations (*irrespective of where the snags occur*). Therefore, $H_k = \binom{n-1}{k} \times (n - k)!$ We have proved the following.

Theorem 4.1.4. *The number Q_n of permutations $\pi = a_1 a_2 \cdots a_n$ on $[n]$ such that $a_j + 1 \neq a_{j+1} \; \forall \; j = 1, 2, \ldots, n-1$ is given by*

$$Q_n = \sum_{k=0}^{n-1} \binom{n-1}{k} (n - k)!$$

4.2 Use of Rook polynomials

Placement of non-attacking rooks on an $n \times n$ chessboard makes a reasonable setting for many combinatorial problems that involve counting permutations with certain forbidden positions. A typical example of this is the problem of derangements as we shall soon see. By a *board*, we mean a set B such that $B \subset [n] \times [n]$. By $r_k(B) = r_k$, we denote the number of ways of placing k non-attacking rooks on the board B. For those who do not know the game of chess, this amounts to picking up k cells (squares) of B such that no two cells are on the same row or the same column. By definition, $r_0(B) = 1$ for all the boards. Let

$$R(x; B) = r_0 + r_1 x + r_2 x^2 + \cdots .$$

Since B is contained in the $n \times n$ board it follows that $r_j(B) = 0$ for all $j > n$. Hence $R(x; B)$ is a polynomial. This polynomial is called the *Rook polynomial of the board B*.

We compute the Rook polynomial of the board B, shown in Figure 4.1 as follows. We have $r_0(B) = 1$ and since there are 6 cells in B, $r_1(B) = 6$. To compute $r_2(B)$, the number of ways of placing 2 non-attacking rooks, we argue as follows. If we place a rook on a, then we may place the second rook on any one of d, e or f in three distinct ways. This also holds if we place a rook on b. This gives 6 possibilities in all. Else, we may place a rook on c so that the second rook will have to be put on one of e or f. If we do not place a rook on the bottom row at all, then the only way to place two rooks is by putting them on d and f. We thus have, $r_2(B) = 6 + 2 + 1 = 9$. Finally, if we need to place three non attacking rooks on B, then we have to place one on each of the three rows. This means we must place a rook on f and hence must place another rook on d. This leaves us with the possibility of placing the third rook on the bottom row at a or b. Hence, we have $r_3(B) = 2$. Therefore, the rook polynomial of the board B is given by

$$R(x; B) = 1 + 6x + 9x^2 + 2x^3$$

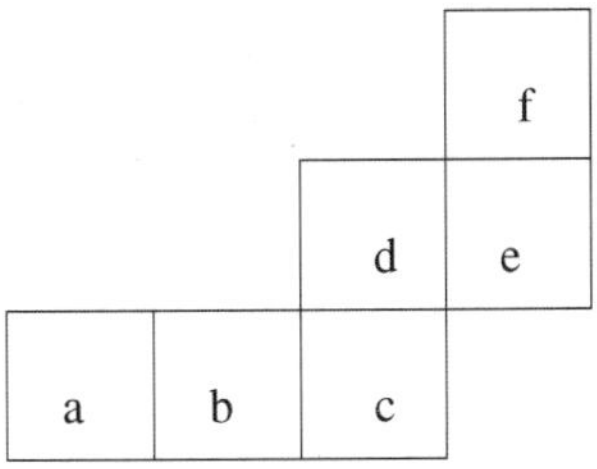

Figure 4.1: Chessboard B

Fix any one cell say C in a non-empty board B and let B' denote the board obtained from B by deleting the cell C. Let B'' be the board obtained from B by deleting all the cells in the row and column of C (including C). In placing k non-taking rooks on B we either place a rook on C or do not place it on C. In the former case, we still have to put $k - 1$ non-attacking rooks on the board obtained from B by deleting the row and column of C, which is just B'', while in the latter case we may as well delete C and try putting k non-taking rooks on the resulting board B'. Thus $r_k(B) = r_k(B') + r_{k-1}(B'')$. Multiplying by x^k and summing over all k obtains the following Lemma, called the expansion formula.

Lemma 4.2.1. *(Expansion formula) With the above notations and terminology, we have*

$$R(x; B) = R(x; B') + xR(x; B'')$$

The expansion formula is very useful in computing the rook polynomial $R(x; B)$ recursively. We illustrate this with the following example.

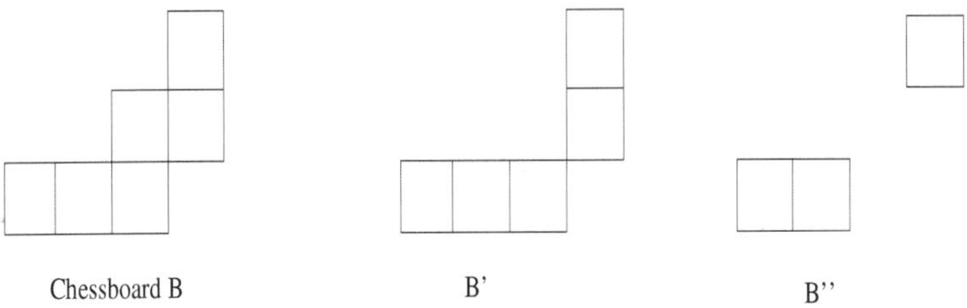

Figure 4.2: Use of expansion formula

In Figure 4.2, the board B has C as the designated cell. The board B' is obtained by removing the cell C and the board B'' is obtained by removing two other cells that are in the row and column of C. The Rook polynomial of the board B' is

$$R(x; B') = (1 + 3x)(1 + 2x) = 1 + 5x + 6x^2$$

and an equally easy calculation gives the Rook polynomial of the board B'' to be

$$R(x; B'') = (1 + 2x)(1 + x) = 1 + 3x + 2x^2$$

Therefore the expansion formula gives the Rook polynomial of the board B:

$$\begin{aligned} R(x; B) &= (1 + 5x + 6x^2) + x(1 + 3x + 2x^2) \\ &= 1 + 6x + 9x^2 + 2x^3 \end{aligned}$$

Let E denote an $[n] \times [n]$ chessboard. We briefly describe as to how a permutation on $[n]$ is represented on the chessboard E. Any permutation π on $[n]$ can be viewed as a collection of n cells of E with no two on the same row or column. Such an arrangement is also called a *transversal*. Thus a transversal is just a set $T = \{(i, \sigma(i)) : i = 1, 2, \ldots, n\}$ where σ is some permutation. If we make a binary (i.e. a $(0, 1)$) matrix of order n by putting 1's in the positions where a cell is chosen and 0 elsewhere, then the resulting matrix is called a *permutation matrix* . In other words, a permutation matrix is a matrix with entries 1 or 0 such that in any row or column precisely one entry is 1. Given a permutation π we can form a permutation matrix $A = [a_{ij}]$ by letting $a_{ij} = 1$ if $\pi(i) = j$ and $a_{ij} = 0$, otherwise. We can then make a transversal on E by picking up exactly those cells (i, j) where $a_{ij} = 1$. *In short, permutations, permutation matrices, transversals and arrangements of n non-attacking rooks on an $[n] \times [n]$ chessboard are equivalent; they are just different manifestations of the same thing.*

Let now $D \subset [n] \times [n]$ be viewed as being a part of an $[n] \times [n]$ chessboard say E. This situation is illustrated in Figure 4.3. Note that in the figure on the right, the shaded squares or cells are the forbidden positions, that form a chessboard say B.
Many questions on permutations can be handled through this set-up. Given a permutation π on $[n]$, the corresponding arrangement of n non-attacking rooks may have some cells that are in B and some that are not in B. In fact, we treat B as a set of forbidden positions. For a permutation π, let $G(\pi)$ denote the set $\{(i, \pi(i)) : i \in [n]\}$ of all the cells that are in the transversal of π.

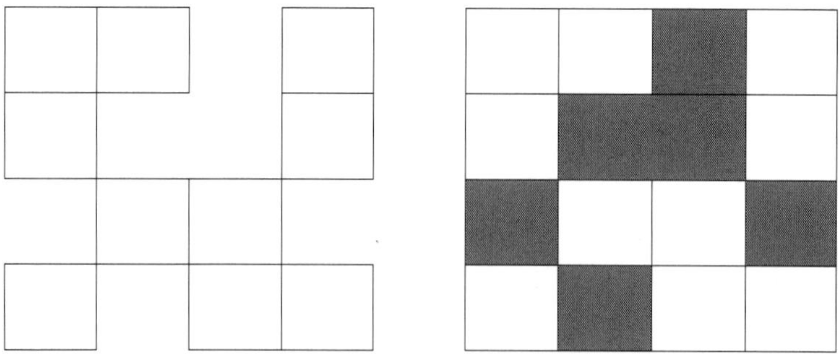

Figure 4.3: The board and its embedding

The question handled in Theorem 4.2.2 is the following. We wish to place n non-attacking rooks on an $[n] \times [n]$. A part of this board consists of good cells that we denote by D and the other part consist of forbidden cells that constitute the board B.

Let N_j denote the number of ways of placing n non-attacking rooks on the (full) board in such a manner that exactly j are in the board B. If we denote the Rook polynomial of B by $R(x) = \sum_{k=0}^{n} r_k x^k$, then is it possible to find the number N_j of permutations π for which the set $G(\pi)$ contains exactly j cells of B? In particular, we would like to ask the question as to how many permutations have no cells in the forbidden position.

Theorem 4.2.2. *Let N_j denote the number of permutations for which the transversal has exactly j cells in the forbidden positions (that is in B). Then,*

$$\sum_{j=0}^{n} N_j x^j = \sum_{k=0}^{n} r_k (n-k)!(x-1)^k$$

Proof Fixing k, we find the cardinality of the set

$$\{(\pi, C) : |C| = k \ and \ C \subset G(\pi) \cap B\}$$

in two ways. For any j, N_j is the number of permutations that have exactly j cells in B, Hence for any such permutation, C can be chosen $\binom{j}{k}$ ways. Therefore the set in question has cardinality $N_j\binom{j}{k}$. On the other hand, we may first choose C in r_k ways and (ignoring the rows and columns of the cells in C) extend it (which amounts to looking at the remaining $(n-k) \times (n-k)$ board) to π in $(n-k)!$ ways. Multiplying both sides by y^k and summing over k obtains

$$\sum_{k=0}^{n} \sum_{j=0}^{n} N_j \binom{j}{k} y^k = \sum_{k=0}^{n} r_k (n-k)! y^k$$

On the left hand side, change the order of summation (first sum over k) and use the binomial theorem (Theorem 1.2.5) to get

$$\sum_{j=0}^{n} N_j (1+y)^j = \sum_{k=0}^{n} r_k (n-k)! y^k$$

Finally putting $x = y+1$ obtains the required result. □

Substitution of $x = 0$ in Theorem 4.2.2, yields the following important corollary.

Corollary 4.2.3. *The number of ways of placing n non-attacking rooks so that none is in a forbidden position is equal to*

$$\sum_{k=0}^{n} (-1)^k r_k(B)(n-k)!$$

The derangement problem can again be solved as a prototypical application of the Corollary 4.2.3. Here, the chessboard B, of all the forbidden cells consisting of all the diagonal cells $\{(i,i) : i = 1, 2, \ldots, n\}$ forms the forbidden board since a derangement π has $\pi(i) \neq i$ for every i.

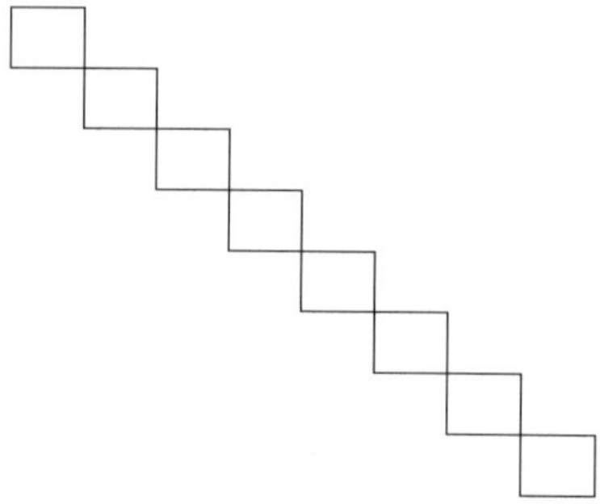

Figure 4.4: The diagonal board of forbidden positions

The Rook polynomial of B is easily found: we have $r_k = \binom{n}{k}$. Hence the number of derangements D_n using Corollary 4.2.3 is

$$D_n = n! \sum_{k=0}^{n} \frac{(-1)^k}{k!}$$

which agrees with the earlier finding (Equation (4.4)).

We end this section by describing a very clever application of the inclusion-exclusion principle. This famous problem, called the *Menage problem* is as follows: n couples, that is, n husbands and n wives are to be seated for dinner around a (round) table consisting of $2n$ chairs so that men and women occupy chairs alternately and no man sits adjacent, i.e., either to the right or left of his wife.One can, of course, make the ladies sit first alternately (on chairs numbered $1, 3, 5, \ldots, 2n-1$ say) and also assume that their numbers are $1, 2, \ldots, n$ in the clockwise order. This can be done in $2(n!)$ ways. Assuming that the ladies have been seated in that way, the problem then is to find seats for the men satisfying the given condition. suppose that the (vacant) seat on the right hand side of the lady number i is denoted by i, the problem is equivalent to making n men sit on these vacant chairs so that man j does not occupy the seats j or $j+1$. Here, one must read numbers modulo n. Thus man n should not occupy either of the chairs n or 1. Call this number $M(n)$. Then $M(1) = M(2) = 0$ and $M(3) = 1$. It is also clear from the above discussion that $M(n)$ is simply the number of permutations π on $[n]$ such that $\pi(j) \neq j, j+1$ where the integers are read modulo n. Equivalently, we are trying to find the number N_0 of placing n rooks on an $[n] \times [n]$ chessboard such that no rook is on the forbidden board B where B consists of the following cells $(1,1), (1,2), (2,2), (2,3), \ldots, (n-1,n), (n,n), (n,1)$. This board is drawn in Figure 4.5 and it corresponds to $n = 8$.

We then use the inclusion-exclusion principle and Lemma 4.2.4, to prove the final result.

Lemma 4.2.4. *Suppose r points on a circle are labeled from 1 to r in a clockwise manner. Let $f(r,k)$ be the number of ways of picking up k of these points so that no two points that have been picked up are consecutive (on the circle). Then*

$$f(r,k) = \frac{r}{r-k}\binom{r-k}{k}$$

Proof Let $g(r,k)$ denote the number of ways of picking up k points out of r points on a line so that no two points that have been picked up are consecutive. Using Exercise 1.14, we see that $g(r,k) = \binom{r-k+1}{k}$. It now suffices to obtain a connection between $g(r,k)$ and $f(r,k)$. If the point labeled 1 on the circle is picked up, then we can not pick up any of the two points adjacent to it. Removing these three points obtains $r-3$ points on a line of which we need to pick up $k-1$. This can be done in $g(r-3,k-1) = \binom{r-k-1}{k-1}$ ways. Otherwise, the point labeled 1 on the circle is not picked up. Breaking the circle at that point, we still have to pick up k points out of $r-1$ points on a line, with no two adjacent. This can be done in $g(r-1,k) = \binom{r-k}{k}$ ways. So,

$$\begin{aligned} f(r,k) &= \binom{r-k-1}{k-1} + \binom{r-k}{k} \\ &= \left[\frac{k}{r-k} + 1\right]\binom{r-k}{k} \\ &= \frac{r}{r-k}\binom{r-k}{k} \end{aligned}$$

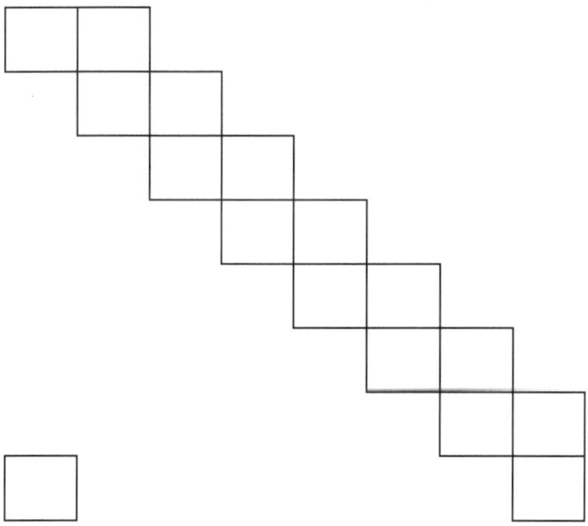

Figure 4.5: Menage Board: forbidden positions in Menage problem

The inclusion-exclusion principle enters in the final part of the evaluation of $M(n)$ as follows. Using Corollary 4.2.3, it suffices to find r_k in order to determine the number of ways of putting n rooks on the $[n] \times [n]$ chessboard so that no rook is on the forbidden board. The number r_k is the number of ways of putting k rooks on the peculiar board B of forbidden positions. Connecting the cells in an obvious manner so that the cell (i,i) is connected to $(i,i+1)$ and the cell $(i,i+1)$ is connected to the cell $(i+1,i+1)$. The cell (n,n) is connected to $(n,1)$) and finally the cell $(n,1)$ to $(1,1)$. This produces an arrangement of $2n$ points on a circle. If k non-taking rooks are to be placed on the board B, then the cells occupied by these rooks can not be in the same row or column. This is equivalent to the requirement that we pick up k point out of $2n$ points on the circle so that no two points that have been picked up are adjacent. Hence r_k equal $f(2n, k)$. Substituting the value of $f(2n, k)$ from the Lemma 4.2.4 and using Corollary 4.2.3, we obtain the menage number.

Theorem 4.2.5. *The menage number $M(n)$ is given by*

$$M(n) = \sum_{k=0}^{n}(-1)^k \frac{2n}{2n-k}\binom{2n-k}{k}(n-k)!$$

It is possible to avoid the setting of Rook polynomials and obtain the menage numbers, though use of Lemma 4.2.4 (in some form) seems necessary. The reader is asked to do this in exercise 4.39. Consider a chessboard with n^2 cells but with the number of cells increasing from top to bottom. This is in analogy with the Ferrers diagram for a partition of a positive integer, which we study in a later chapter (Chapter 13). Using a clever combinatorial argument, Stanley [49] shows that the number of Ferrers boards that have the same Rook polynomial as the standard $[n] \times [n]$ board is 3^{n-1}. He also shows that among these, there is a unique board with increasing row lengths, the lengths being $1, 3, \ldots, 2n-1$ (Exercise 4.40).

4.3 Some arithmetic and the Möbius function

Historically, inclusion exclusion principle arose as a sieve formula in arithmetic. In the beginning of this chapter, we gave an illustration of this. The sieve of Eratosthenes can check if a given number n is prime or not. To do this we look at all the numbers less than or equal to $\sqrt{n}$ and try to divide n by all such numbers.

Definition 4.3.1. Let n be a positive integer. Then $\phi(n)$ *called the Euler's totient function* is the number of positive integers less than n and coprime to n.

For example, since $1, 3, 7$ and 9 are coprime to 10 we have $\phi(10) = 4$ and if p is a prime, then $\phi(p) = p - 1$. We base ourselves on the prime power factorization of an integer and obtain the following.

Proposition 4.3.2. *Let n be a positive integer with a prime power factorization $n = {p_1}^{a_1}{p_2}^{a_2}\cdots{p_r}^{a_r}$, where $p_1, p_2, \ldots, p_r$ are distinct primes dividing n. Then*

$$\phi(n) = n\prod_{i=1}^{r}\left\{1 - \frac{1}{p_i}\right\}$$

Proof Define P_i to be the property on the set of all positive integers unto n given by: k satisfies P_i if p_i divides k. Then $\phi(n)$ is just the cardinality of the set of integers not having any of these properties, for i from 1 to r. If $p_{i_1}, p_{i_2}, \ldots, p_{i_j}$ are distinct primes (among the above r primes dividing n), then the number of integers divisible by all of them is simply

$$\frac{n}{p_{i_1}p_{i_2}\cdots p_{i_j}}$$

So, in the notation of Theorem 4.1.1 and equation (4.2), we have

$$H_j = n \sum_{1 \leq i_1 < \cdots < i_j \leq r} \frac{1}{p_{i_1}p_{i_2}\cdots p_{i_j}}.$$

Hence, we obtain

$$\phi(n) = n\sum_{j=0}^{r}(-1)^j \sum_{1 \leq i_1 < \cdots < i_j \leq r} \frac{1}{p_{i_1}p_{i_2}\cdots p_{i_j}}$$

which is the desired expression (rearrange the terms). □

We also note the following result where the underlying group theoretic assertion is that a group of order n is cyclic if and only if for every divisor d of n there is a unique (or at most one) subgroup of order d. The reader is asked to prove that result as an exercise.

Proposition 4.3.3. *Let n be a positive integer. Then*

$$\sum_{d|n}\phi(d) = n$$

Proof Let m be any positive integer $m \leq n$. Let d be the g.c.d. of m and n. Then $n = n_1 d$ and $m = m_1 d$ where m_1 and n_1 are coprime. The number of such integers m for which the g.c.d. of m and n is d is then given by the number of integers coprime to n_1, which is just $\phi(n_1)$. Since this holds for all d and since $n_1 = \frac{n}{d}$, we obtain $n = \sum_{d|n}\phi(\frac{n}{d})$ which is the same as what we have to prove. □

It is time now to introduce the Möbius function. This function is *defined over the set of all the natural numbers and takes values* 0 *or* ± 1.

Definition 4.3.4. *The Möbius function* is defined as follows. $\mu(1) = 1$. If $n \neq 1$, then let $n = p_1{}^{a_1}p_2{}^{a_2}\cdots p_r{}^{a_r}$ be the prime factorization of n.

$$\mu(n) = \begin{cases} (-1)^r & \text{if } a_1 = a_2 = \cdots = a_r = 1 \\ 0 & \text{if } a_i > 1, \text{ for some } i \end{cases}$$

Lemma 4.3.5. *Let n be a positive integer. Then*

$$\sum_{d|n}\mu(d) = \begin{cases} 1 & \text{if } n = 1, \\ 0 & \text{otherwise} \end{cases}$$

Proof For $n = 1$, the result is obvious. Let $n \geq 2$ and suppose the prime factorization of n is $n = p_1{}^{a_1} p_2{}^{a_2} \cdots p_r{}^{a_r}$. Since $\mu(k)$ is zero if a square of a prime divides k, we need to only consider the sum $\sum_{d|m} \mu(d)$ where $m = p_1 p_2 \cdots p_r$. If d is a product of j distinct primes among $p_1, p_2 \cdots, p_r$ then $\mu(d) = (-1)^j$ and the number of such divisors d that are products of j primes is clearly $\binom{r}{j}$ and hence the required sum is

$$\sum_{j=0}^{r} (-1)^j \binom{r}{j} = (1-1)^r = 0$$

using the binomial theorem (Theorem 1.2.25). □

Theorem 4.3.6. *(Möbius inversion formula) Let f and g be functions whose domain is the set of all natural numbers and whose range is the field of complex numbers. Assume that for all natural numbers n,*

$$g(n) = \sum_{d|n} f(d)$$

Then f can be expressed in terms of g by the following formula.

$$f(n) = \sum_{d|n} \mu(d) g\left(\frac{n}{d}\right)$$

Proof We have,

$$\begin{aligned} R.H.S. &= \sum_{d|n} \mu\left(\frac{n}{d}\right) g(d) \\ &= \sum_{d|n} \mu\left(\frac{n}{d}\right) \sum_{e|d} f(e) \\ &= \sum_{e|n} \left(\sum_{e|d|n} \left[\mu\left(\frac{n}{d}\right)\right] f(e) \right) \end{aligned}$$

Now the terms in the parentheses can be simplified as follows. Writing $e' = \frac{n}{e}$ and $d' = \frac{n}{d}$, we obtain

$$\sum_{e|d|n} \mu\left(\frac{n}{d}\right) = \sum_{d'|e'} \mu(d')$$

This sum is 0 by the Lemma 4.3.4 except when e' equals 1. Hence, to get a non-zero summand, we must have $e = n$ so that the right hand side is just $f(n)$. □

The inverse of Möbius function is the classical Riemann zeta function. This is explored in Exercise 4.26. We now present two classical applications of Möbius inversion formula. The first one gives the existence of irreducible polynomials of arbitrary degree

n over any finite field $GF(q)$ of order q where q is a prime power. This depends on the factorization of $x^{q^n} - x$ as a product of irreducible polynomials of degree d. Here d must divide n. Letting $M(d)$ denote the number of irreducible polynomials of degree d, equating the degrees of the terms of highest degree obtains the following basic equation.

$$q^n = \sum_{d|n} dM(d)$$

The Möbius inversion formula comes into play here. We let $g(n) = q^n$ and $f(n) = nM(n)$. Then the formula gives

Theorem 4.3.7. *The number of monic irreducible polynomials over a finite field $GF(q)$ is given by*

$$M(n) = \frac{1}{n}\sum_{d|n} \mu(d)q^{n/d} \tag{4.7}$$

The right hand side is never zero (Exercise 4.25). This proves that for every n, there are irreducible polynomials of degree n over $GF(q)$ and consequently fields of orders q^n exist for all n.

Our second application concerns counting of necklaces with beads of m distinct colors such that the necklace has n beads (that are regularly placed on the circle). If we ignore symmetries, then the number of necklaces that can be made is just m^n since this is the totality of functions that can be defined from an n-set to an m-set. We label the n points on the circle by numbers $1, 2, \ldots, n$ (in counterclockwise manner) and color each position j in one of the available m colors. Given a necklace in which the beads are colored, a *block* is a *minimal set* of consecutive positions $i, i+1, \ldots, i+j-1$ such that for any r with $0 \leq r \leq j-1$ the color of the bead $i+r+mj$ is the same as that of $i+r$ where m is any integer and the numbers are read modulo n. Note that a block is not unique but the length of a block is. For example, if R, B, Y refer to red, blue and yellow colours, then Figure 4.6 shows a colouring where $(1, 2, 3), (4, 5, 6), (7, 8, 9)$ and $10, 11, 12)$ are blocks RBY.

Note that a block in a given colouring is not unique. In Figure 4.6, there is also a block BYR given to positions $(2, 3, 4)$ and cyclically moving in counterclockwise direction, the other three blocks at $(5, 6, 7), (8, 9, 10)$ and $(11, 12, 1)$. Also block is a segment with minimal length. For example the same 12 bead necklace with all beads coloured red has a segment RR as well as RRR repeating. But the block length in this colouring is 1. Finally, note that a block exists for every colouring: the colouring for the 12 beads numbered from 1 through 12 given by $RRBRRRBRRBBB$ (in that order) has a block of size 12. *To sum up, blocks exist but are not unique but their length is.* It is easily seen that the length of a block divides n and that associated with every necklace (with beads colored) is a unique integer j such that j is the the length of (any) block. *In this case, we say that the necklace has period j.* By shifting the beads cyclically, we see that a necklace with period d has d distinct necklaces in its equivalence class. Finally, a necklace of length n with period n is called *indecomposable.* Fix n, the length of the necklace as also the number of colours m. let $M(d)$ denote the number of necklaces with period d. This also means that if we pick up any block of a necklace with period d and make it into a necklace with d beads then that necklace (with d beads) is indecomposable.

Our discussion so far has made it amply clear that the number of necklaces, which is same as the number of equivalence classes is equal to the number of equivalence classes of necklaces with period d, where d divides n. The latter number is $M(d)$, as we already saw. It now follows that the required number is: $N_n = \sum_{d|n} M(d)$. It is therefore sufficient to find what $M(d)$ is. To that end, we observe that there are d necklaces in the equivalence class of each necklace with period d. Hence counting the total number of necklaces obtains

$$m^n = \sum_{k|n} kM(k)$$

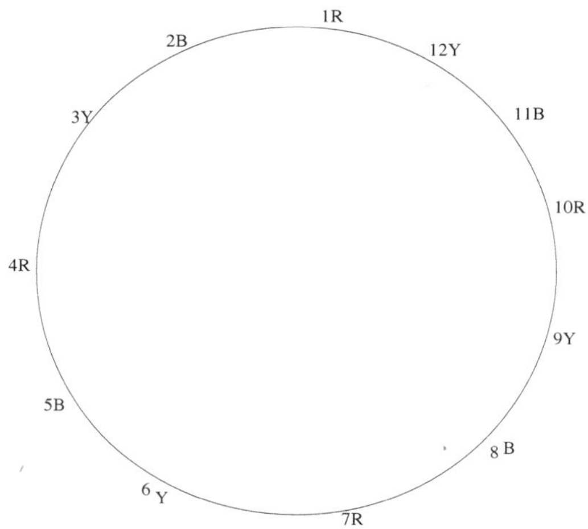

Figure 4.6: 12 beads and period 3

We can now use Möbius inversion formula (Theorem 4.3.6) letting $g(n) = m^n$ and $f(k) = kM(k)$ to get

$$kM(k) = \sum_{d|k} \mu(d)m^{k/d}$$

Therefore, we have

$$\begin{aligned}
N_n &= \sum_{k|n} M(k) \\
&= \sum_{k|n} \frac{1}{k} \sum_{d|k} \mu\left(\frac{k}{d}\right) m^d \\
&= \sum_{d|n} \left\{ \sum_{d|k|n} \frac{1}{k} \mu\left(\frac{k}{d}\right) \right\} m^d \\
&= \sum_{d|n} \left\{ \sum_{k'|\frac{n}{d}} \frac{1}{k'd} \mu(k') \right\} m^d
\end{aligned}$$

Using Exercise 4.11, we get

$$\begin{aligned}
N_n &= \sum_{d|n} \left\{ \sum_{k'|\frac{n}{d}} \frac{\mu(k')}{k'} \right\} \frac{m^d}{d} \\
&= \sum_{d|n} \frac{\phi(n/d)}{n/d} \frac{m^d}{d} \\
&= \frac{1}{n} \sum_{r|n} \phi(r) m^{n/r}
\end{aligned}$$

We have thus proved the following.

Theorem 4.3.8. *Consider two circular necklaces to be identical if one can be obtained from the other by a rotation. The number of circular necklaces with n beads colored in m colors is*

$$N_n = \frac{1}{n} \sum_{r|n} \phi(r) m^{n/r}$$

We close this chapter after informing the reader that the Möbius function and the inversion formula afford the most natural and abstract setting for the inclusion-exclusion principle. Rota extended the notion of Möbius function to partially ordered sets. This sophisticated machinery achieves some results that are not obvious otherwise. The applications include chromatic polynomials of graphs and many others. For all these, we refer the reader to the book by van Lint and Wilson [57].

4.4 Exercises for Chapter 4

4.1 In an M.Sc. class consisting of 25 students, most of the students know either of the three languages: Marathi, Hindi and Tamil. 14 students know Hindi, 12 students know Marathi, 6 students know both Hindi and Marathi, 5 students know Hindi and Tamil and 2 students know all the three languages. There are six students who know Tamil and any student who knows Tamil knows some other language, Marathi or Hindi. How many students do not know any of the three languages, Marathi, Hindi or Tamil?

4.2 We wish to prove Theorem 4.1.1 by induction. To that end, assume that the assertion holds for n and hence equation (4.1) is true. Let P_{n+1} be $(n+1)$-th property and A_{n+1} the set of elements of X that satisfy P_{n+1}. Prove the following.

(a) Using the induction hypothesis obtain the following expression for $|\overline{A_1} \cap \overline{A_2} \cap \cdots \cap \overline{A_n} \cap A_{n+1}|$:

$$|A_{n+1}| - \sum_{1 \le i \le n} |A_i \cap A_{n+1}| + \cdots +$$

$$\sum_r (-1)^r \sum_{1 \le i_1 < \cdots < i_r \le n} |(A_{i_1} \cap \cdots \cap A_{i_r}) \cap A_{n+1}|$$

(b) Subtract the expression in (a) from equation (4.1) to prove the assertion for $n+1$.

4.3 The function f is defined from the set of natural numbers to the set of real numbers by the following equation:

$$\sum_{d|n} f(d) = \log n$$

Find f.

4.4 Three workmen a, b, c have to be assigned a job each out of the available jobs $1, 2, 3, 4$. Further, a is not qualified for jobs 2 and 3, b is not qualified for the jobs 2 and 4 and c is not qualified for the job 1. Using Rook polynomials, find the number of ways in which each of them is assigned a distinct job for which he is qualified.

4.5 Among the numbers $1, 2, \ldots, 500$ how many are not divisible by 2, 3 and 11?

4.6 Determine the number of permutations of $[8]$ in which no even integer is in its natural position.

4.7 Show that the numbers Q_n defined in this chapter can be rewritten in the form

$$Q_n = (n-1)!\left(n - \frac{n-1}{1!} + \frac{n-2}{2!} - \frac{n-3}{3!} + \cdots + \frac{(-1)^n}{(n-1)!}\right)$$

4.8 Show that

$$(-1)^k \frac{n-k}{k!} = (-1)^k \frac{n}{k!} + (-1)^{k-1} \frac{1}{(k-1)!}$$

Use this identity and Exercise 4.7 to prove that

$$Q_n = D_n + D_{n-1}.$$

4.9 Let the set $[n]$ equal

$$\{\alpha_1, \alpha_2, \ldots, \alpha_n\} = \{\beta_1, \beta_2, \ldots, \beta_n\}$$

Find the number of permutations π with the property that for every i, π does not map α_i to β_i.

4.10 If m and n are coprime then prove that

$$\phi(mn) = \phi(m)\phi(n)$$

4.11 Prove that

$$\frac{\phi(n)}{n} = \sum_{d|n} \frac{\mu(d)}{d}$$

4.12 In how many ways can the integers $1, 2, \ldots, 9$ be permuted so that no odd integer will be in its original position?

4.13 For a natural number n , define $\lambda(n) = log\ p,$ if $n = p^m$ for some prime p and some positive integer m and $\lambda(n) = 0$, otherwise. Prove that

$$\lambda(n) = \sum_{d|n} \mu(d) \log\left(\frac{n}{d}\right)$$

4.14 At a party, 7 gentlemen check their hats. In how many ways can their hats be returned to them so that exactly two of the gentlemen receive their own hats?

4.15 In a recent survey of 400 students on Vidyanagari Campus, it was found that half the students spoke English language, half spoke Marathi and half spoke Hindi and every student spoke at least one of the three languages. There were 104 students who spoke both Marathi and English, 116 students spoke Marathi and Hindi and 93 students spoke English and Hindi. Find the number of students who can speak all the three languages.

4.16 In a certain factory with 800 employees, 300 are men, 552 are union members and 424 are married. Further, 188 men are union members, 166 men are married and there are 208 married union members. Find the number of single (unmarried), non-union women employees.

4.17 Find the number of derangements of $\{x_1, x_2, \ldots, x_{12}\}$ in which the first six elements $x_1, x_2, \ldots, x_6$ are mapped onto $x_1, x_2, \ldots, x_6$ in some order. Also find the number of derangements in which the first six elements $x_1, x_2, \ldots, x_6$ are mapped onto the last six elements $x_7, x_8, \ldots, x_{12}$ in some order.

4.18 Find the Rook polynomial of the ordinary 8 by 8 chess board. *Do not simplify the expression.*

4.19 (a) Write down the permutation matrix corresponding to the following permutation.

$$\sigma = \begin{pmatrix} 1 & 2 & 3 & 4 \\ 2 & 3 & 4 & 1 \end{pmatrix}$$

(b) Write down the permutation corresponding to the following permutation matrix.

$$\begin{pmatrix} 0 & 1 & 0 & 0 \\ 1 & 0 & 0 & 0 \\ 0 & 0 & 0 & 1 \\ 0 & 0 & 1 & 0 \end{pmatrix}$$

4.20 Solve the following equations in integers.

(a) $x + y + z + w = 20$, subject to the conditions $1 \leq x \leq 6$, $0 \leq y \leq 7$, $4 \leq z \leq 8$, $2 \leq w \leq 6$.

(b) $z + x + y + \;\; + t = 18$, subject to the conditions $1 \leq x \leq 5$, $1 \leq y \leq 6$, $2 \leq z \leq 7$, $4 \leq t \leq 10$.

4.21 Determine the number of 12-combinations of the multi-set $\{4.a, 3.b, 4.c, 5.d\}$.

4.22 Determine the number of 10-combinations of the multi-set

$$S = \{\infty.a,\ 3.b,\ 5.c,\ 7.d\}$$

4.23 Determine the number of all the permutations of the multi-set $\{3.a,\ 4.b,\ 2.c\}$ in which all the letters of any one kind do not appear consecutively.

4.24 State and prove the converse of Möbius inversion formula.

4.25 Prove that if q is a prime power then the expression for the number of irreducible polynomials over $GF(q)$ given in equation (4.7) is never 0.

4.26 A Dirichlet generating function $A(s)$ for a sequence a_n of numbers is a formal series

$$A(s) = \sum_{n \geq 1} \frac{a_n}{n^s}$$

If $B(s)$ is another Dirichlet generating function corresponding to the sequence b_n of numbers then define $A(s) + B(s) = C(s)$ and $A(s)B(s) = D(s)$ where $C(s)$ corresponds to the sequence $c_n = a_n + b_n$ and $D(s)$ corresponds to the sequence d_n, where d_n equals the sum of all the terms of the type $a_k b_r$ where k is a divisor of n and $kr = n$.

(a) Show that the set of all the Dirichlet generating functions forms a commutative ring with unity.

(b) Show that if $a_1 \neq 0$, then $A(s)$ has a multiplicative inverse.

(c) The Dirichlet generating function $\zeta(s) = \sum \frac{1}{n^s}$ is called the Riemann zeta function. Show that the multiplicative inverse of this function is given by $\sum \frac{\mu(n)}{n^s}$.

4.27 Find the following.

$$\sum_{d|n} \mu(d)\phi(d).$$

$$\sum_{d|n} \frac{\mu(d)}{\phi(d)}$$

4.28 Suppose n people leave their hats as well as bags outside a hall and while stepping out of the hall, each person picks up one hat and one bag. What is the number of ways in which every person neither gets back his own hat nor his own bag?

4.29 How many ways are there to roll 10 distinct dice so that all the 6 faces appear?

4.30 How many positive integers less than 720 are coprime to 720?

4.31 Refer to Figure 4.6 and in each one of the following cases, when the 12 beads of a circular necklaces are consecutively coloured in one of R, B (red, blue), find the period of the colouring.

(a) $RRBBRRBBRRBB$

(b) $RBRBRBRBRBRB$

(c) $RBRRBRRBRRBR$

(d) $RRRRBBRBBBBB$

4.32 Find the total number of circular necklaces with 10 beads in three colours red, blue and yellow.

4.33 What is the number of permutations $a_1 a_2 \cdots a_n$ (in the one-line notation) on $[n]$ for which j is not immediately followed by $j+1$ as also n is not immediately followed by 1? (This is slightly more restrictive than the forbidden position problem we considered in this chapter.)

4.34 Show that

$$\binom{n-m}{n-k} = \binom{m}{0}\binom{n}{k} - \binom{m}{1}\binom{n-1}{k} + \binom{m}{2}\binom{n-2}{k} - \cdots + (-1)^m \binom{m}{m}\binom{n-m}{k}$$

4.35 Consider the polynomial $N(x) = \sum_{j=0}^{n} N_j x^j$ introduced in this chapter. How would you interpret the number $\frac{N(1) + N(-1)}{2}$?

4.36 Use the inclusion-exclusion principle to prove the following identities.

$$\sum_{k=m}^{n} (-1)^{k-m} \binom{n}{k} = \binom{n-1}{m-1}$$

$$n = \sum_{k=1}^{n} (-1)^{k-1} k \binom{n}{k} 2^{n-k}$$

4.37 Find the Rook polynomial of the following boards. (Note: Forbidden positions are shaded).

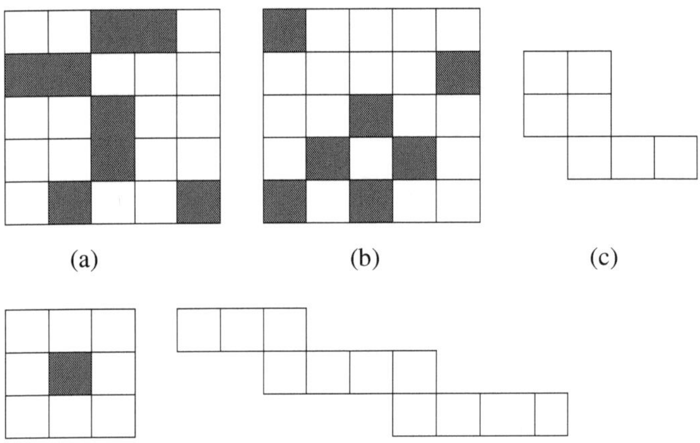

(a) (b) (c)

(d) (e)

Figure 4.7: Forbidden positions shaded

4.38 Let $R_{n,m}(x)$ denote the Rook polynomial of an $n \times m$ board. Prove the following.

$$R_{n,m}(x) = R_{n-1,m}(x) + mxR_{n-1,m-1}(x)$$

$$\frac{d}{dx} R_{n,m}(x) = nmR_{n-1,m-1}(x)$$

4.39 Denote by A_{2n-1} to be the chessboard with $2n - 1$ squares as in staircase. This is obtained by deleting the square at $(n, 1)$ from the Menage board. Similarly let B_{2n} denote a staircase chessboard with an even number of squares. These are shown on the left and right respectively of Figure 4.8.

(a) Prove the following.

$$A_{2n-1}(x) = xA_{2n-3}(x) + B_{2n-2}(x)$$

$$B_{2n}(x) = xB_{2n-2}(x) + A_{2n-1}(x)$$

(b) Prove the following boundary conditions.

$$A_1(x) = 1 + x; \;\; A_3(x) = 1 + 3x + x^2$$

$$B_0(x) = 1; \;\; B_2(x) = 1 + 2x$$

(c) Use induction to show that

$$A_{2n-1}(x) = \sum_{k=0}^{n} \binom{2n-k}{k} x^k$$

$$B_{2n}(x) = \sum_{k=0}^{n} \binom{2n-k+1}{k} x^k$$

(d) Show that the Rook polynomial of the Menage board is given by

$$M_n(x) = xA_{2n-3}(x) + A_{2n-1}(x)$$

(e) Use the above result to solve the Menage problem (Theorem 4.2.5).

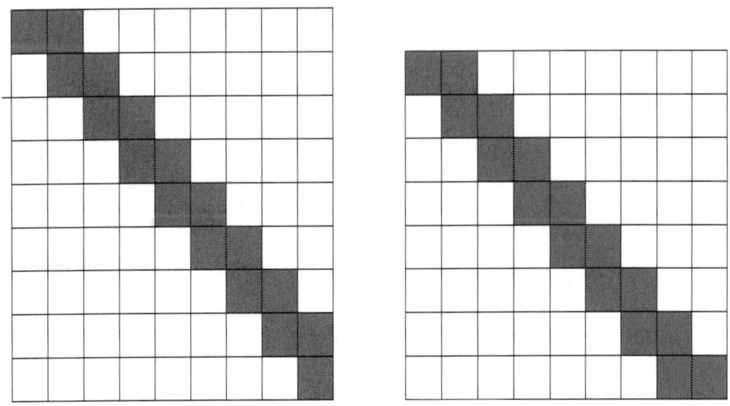

Staircase Boaard with 17 Squares   Staircase Board with 16 Squares

Figure 4.8: Staircase Chessboards

4.40 Show that the Rook polynomial of the board with rows of lengths $1, 3, 5, \ldots, 2n-1$ (aligned at the left column) is the same as the usual $n \times n$ chessboard (Refer to Stanley [49]).

Chapter 5

Basic probability

5.1 Introduction

At this stage of the development of the book, it becomes imperative to introduce some elementary ideas from the theory of probability to the reader. There are two independent reasons for introducing probability at this stage. One obvious reason is that probability is a good ground for applying all the combinatorics that has been learned so far and will also be learned in later chapters. In fact, serious study of combinatorial problems is many times undertaken in order to achieve something in a different field, probability being the most important among them. This is the second reason for studying the basics of probability. As outlined in the preface, serious attempts to develop combinatorial techniques were made by probabilists such as de Moivre and Laplace. Intuitive notions about probability have existed in the minds of human beings since the antiquity. For example, in a lottery is it better to purchase two tickets for the same lottery or one ticket in each of two different lotteries? It does appear intuitively to be case that it does not matter. However, all such questions require rigorous mathematical answers, which is what we attempt to do. Some other questions can also be asked. Is it more likely or less likely to obtain two heads in four tosses of a coin than obtaining one head in two tosses of a coin? In a bridge hand, what is more likely distribution of suits? Probability theory tries to make a concrete mathematical model of such situations so that it becomes available for being applied. We develop the basic probability model through the following definitions.

Definition 5.1.1. The basic building material here consists of a finite set which is now given the name *sample space* and is denoted by Ω. The subsets of Ω are called *events.* Two events are called *(mutually) exclusive* if they are disjoint. *A probability (or probability measure)* on the sample space Ω is a real valued function P defined on the events such that the following axioms hold.

(1) For all the events A, $P(A) \geq 0$.

(2) $P(\Omega) = 1$.

(3) $P(A \cup B) = P(A) + P(B)$, if A and B are exclusive events.

We now derive some obvious consequences of the definition.

Lemma 5.1.2. *Let P be a probability measure on Ω. Then:*

(a) For any event A and its complement A', we have $P(A) + P(A') = 1$.

(b) $P(\emptyset) = 0$.

(c) If $A \subset B$, then $P(A) \leq P(B)$.

(d) For any two events A and B, we have $P(A \cup B) \leq P(A) + P(B)$.

Proof (a) follows from Definition 5.1.1(3). Then (b) follows by taking $A = \emptyset$ and using Definition 5.1.1(2). For (c), note that $B = A \cup \{B - A\}$ where we have union of disjoint sets. Hence, using Definition 5.1.1(3) and 5.1.1(1), we have the required result. Finally, (d) follows from (c). □

The terminology of probability is something we should get used to, for the sake of understanding the essence of the theory. For example, instead of saying that an element is (or is not) in the set, we often change the language to, "an event has occurred (or has not occurred)".

Lemma 5.1.3. *Let $\Gamma = \{A_1, A_2, \ldots, A_m\}$ be a set of mutually exclusive events (that is, $A_i \cap A_j = \emptyset$ for all $i \neq j$). Then*

$$P(\cup_{i=1}^m A_i) = \sum_{i=1}^m P(A_i)$$

The proof follows easily by making induction on m and using (3) in the definition of probability measure. The definition of probability tells us that we know the probability measure completely if we prescribe $P(A)$ for all the singleton sets A. Thus, if $A = \{a_1, a_2, \ldots, a_n\}$ is an event then the probability of A is obtained by simply adding $P(\{a_i\})$ for i from 1 to n. It is for this reason that singleton sets are called *atomic events.* By a slight abuse of notations *we denote $\{a\}$ by just a*, dropping the parentheses.

Definition 5.1.4. Let $\Omega = \{a_1, a_2, \ldots, a_r\}$. The probability measure P for which $P(a_i) = \frac{1}{r}$ is called *the equally likely probability measure* on Ω. This is also called *the uniform distribution on Ω.*

Lemma 5.1.5. *Let P be the equally likely probability measure on Ω and let A be any event. Then $P(A)$ equals $\dfrac{|A|}{|\Omega|}$.*

There are three main sources of our examples. Since the reader may not be familiar with some of them (possibly the cards) we briefly explain the set-ups.

(1) *Tossing of a coin:* When a coin is tossed once, there are two possible outcomes, Heads and Tails and hence the sample space $\Omega = \{H, T\}$. Here H refers to heads and T refers to the tails. When a coin is tossed two times, we have the sample space $\Omega = \{HH, HT, TH, TT\}$. In general, when the coin is repeatedly tossed n times, the sample space consists of all the 2^n sequences of length

n with each entry equal to H or T. Clearly then the equally likely probability measure prescribes probability 2^{-n} to each atomic event of the sample space.

(2) *Throw of a die:* A regular die (this means the die is a cube and on the six faces of the die are written six different numbers from 1 to 6 so that the numbers on the opposite faces add to 7) is thrown (and the number on the top is read). The sample space consists of 6 possible outcomes $1, 2, 3, 4, 5, 6$. Two successive throws of a die (or two distinct dice thrown once) result in the sample space consisting of 36 outcomes in all. Clearly n throws of a die result in a sample space whose elements are sequences of the type $i_1, i_2, \ldots, i_n$ where i_j is an integer from 1 to 6. An equally likely probability measure on the sample space would have the assignment of 6^{-n} on each atomic event.

(3) *Pack of cards:* A pack of cards consists of 52 cards with 13 cards of each of the suits spades, hearts, diamonds and clubs $\spadesuit, \heartsuit, \diamondsuit, \clubsuit$ (that is the hierarchy in the game of bridge). Concentrating on the game of bridge, we have the picture cards in each suit consisting the ace, the king, the queen and the jack. A bridge hand consists of thirteen cards. When the cards are dealt, each of north, south, east and west receives 13 cards. Here, north and south are partners and so are east and west. Various sample spaces can be built on this model. Assuming equally likely probability measure, it is just enough to find the size of the event under consideration relative to the size of the sample space, as outlined in Lemma 5.1.5. For example, the sample space consisting of all the hands (of 13 cards) has size $\binom{52}{13}$. The number of ways in which cards can be dealt to all the four players is the multinomial coefficient $\binom{52}{13,13,13,13}$.

Example 5.1.6. We now give a number of examples that will enable our reader with a discrete mind-set to get used to the thought of a probability set-up. In all the examples below, assume that we are dealing with the equally likely probability measure. In the situation of a coin tossing experiment or the die throwing experiment, we refer to this by saying that the coin is unbiased or is fair.

(a) The probability of getting a particular hand (consisting of 13 cards) is

$$\frac{1}{\binom{52}{13}}$$

(b) The probability of getting at least one head in two tosses of a fair coin is $\frac{3}{4}$ since there are 3 out of 4 ways in which we get one or two heads: HH, HT, TH.

(c) Similarly, the probability of the event A of getting at least two heads in four tosses of a coin is $\frac{11}{16}$: If B denotes the complementary event, then B can occur in 5 ways since there is only one way of not getting H, and $\binom{4}{1}$ ways of getting exactly one H. Use Lemma 5.1.2(a) to obtain $P(A)$. Observe that $P(A)$ is not the same as the one in the previous example (but is slightly lower).

(d) Suppose n (distinct) balls are to be distributed among n distinct boxes. Then the probability that each box receives exactly one ball is $\frac{n!}{n^n}$. Even for moderately small values of n, this number is small (it decreases with n). Thus, for $n = 7$, this number is less than 0.2; see Feller [24]. This observation has surprising consequences. Suppose we know that in a week 7 accidents have occurred. Then the probability that on each day, there is one accident is small (less than 0.2). So, it is more likely that there are at least two or more accidents on a day than one accident per day. *In short, if we know that an accident has occurred on a particular day, then it is more likely that one more will occur the same day!*

(e) In a hand of thirteen cards, the probability of getting exactly one ace is

$$\frac{\binom{4}{1}\binom{48}{12}}{\binom{52}{13}}$$

(f) If there are r particles to be distributed to n cells (energy levels) so that no cell is occupied by more than one particle (Pauli exclusion principle), then the probability of obtaining a particular configuration is

$$\frac{1}{\binom{n}{r}}$$

We have dealt with this example earlier. It is the model for the Fermi-Dirac Statistic (Chapter 1, 1.3.8(c)).

(g) If there are r identical particles to be distributed among n distinct cells , then the probability of obtaining a particular configuration is

$$\frac{1}{\binom{r+n-1}{r}}$$

This is the Bose -Einstein Statistic (Chapter 1, 1.3.8(b)).

(h) We throw a die two times. The sample space consists of $6 \times 6 = 36$ outcomes. The probability of getting a sum of 7 is $\frac{1}{6}$ since there are 6 ways of getting the sum 7.

(i) Out of n objects in the collection, r are known to be defective. If we pick up m objects then the probability that k will be defective is

$$\frac{\binom{r}{k}\binom{n-r}{m-k}}{\binom{n}{m}}$$

This is called the *hypergeometric distribution.*

(j) *Birthday problem* We ask n different people their birthdays. Assuming that nobody is born on February 29 of a leap year, the probability that two people have the same birthday is high even for small values of n. The argument here is very similar to that we made in the earlier example on accidents in a week. Assuming that p_n is the required probability, let $q_n = 1 - p_n$, the probability that no two people have the same birthday. With $m = 365$, we have

$$q_n = \frac{[m]_n}{m^n} = \left(1 - \frac{1}{m}\right)\left(1 - \frac{2}{m}\right)\cdots\left(1 - \frac{n-1}{m}\right)$$

It is clear that $q_{n+1} = (1 - \frac{n}{m})q_n$ and for $n = 23$, we obtain $q_{23} \leq \frac{1}{2}$. It is worth noting that q_n's form a decreasing sequence and with $n = 30$, we have $q_n \leq 0.3$ and with $n = 50$ we get $q_{50} \leq 0.03$ so that the probability that two people have identical birthdays if we have a sample of 50 or more people is higher than 0.97!

(k) *Probability of a derangement* Using the formula for derangement obtained in Chapter 4 and giving equally likely probability measure to the sample space of all the $n!$ outcomes (consisting of all the permutations on an n-set), we obtain (using Proposition 4.1.3) p_n, the probability of the occurrence of a derangement to be

$$p_n = \frac{D_n}{n!} = 1 - 1 + \frac{1}{2!} - \cdots + \frac{(-1)^n}{n!}$$

The expression on the right hand side is the truncated Taylor expansion of $\frac{1}{e}$ and since the series is alternating the convergence is fast. Thus, for $n \geq 10$, we see that p_n is very close to $\frac{1}{e}$, which is approximately 0.37.

5.2 The ballot problem

In this problem, two candidates A and B contest an election where A wins by securing m votes while B secures n (which is less than m) votes. While counting is going on, what is the probability that at any stage of counting, A leads over B? The total number of ways in which votes can be counted is just the total number of sequences of A's and B's (write A if the vote is for A and B if the vote is for B) of length $m+n$ of which exactly m are A's and remaining n are B's. This number is $\binom{m+n}{m}$. We represent all the ways of counting by graphs in the real plane that begin at $(0,0)$ and end at $(m+n, m-n)$. The first co-ordinate here refers to the total number of votes counted and the y co-ordinate refers to the difference between the votes secured by A and those secured by B (at that stage of counting). A typical graph then is a path that follows is a zigzag pattern making an angle of $\pm 45^o$ with the x-axis. We move $m+n$ units in the x-direction because that is the total number of votes counted. At each $x = r$ we ask the question if the the next, i.e. $r+1$-th vote cast is in favor of A or B. In the former case, the portion of the graph from $x = r$ to $x = r+1$ is a line segment that makes an angle of 45° with the horizontal and in the latter case, it is a line segment that makes an angle of -45° with the x-axis (turning downward). In this manner, each of the $\binom{m+n}{m}$ paths that go from $(0,0)$ to $(m+n, m-n)$ correspond to a unique way of counting votes (with precisely m votes to A and n votes to B). To count the number of patterns of vote counting in which A always leads over B (at every stage of counting), it suffices to count the total number of paths as above that are *invalid*. Necessarily then, an invalid path must touch the x-axis at some stage of counting. That is, *valid paths are exactly those that stay above the x-axis at all the stages of counting.* We actually count the total number of invalid paths by using a trick called the *reflection principle.* This principle, discovered by Kelvin is to be found in many texts on combinatorics, though the first appearance in a text seems to be in the book by Feller [24].

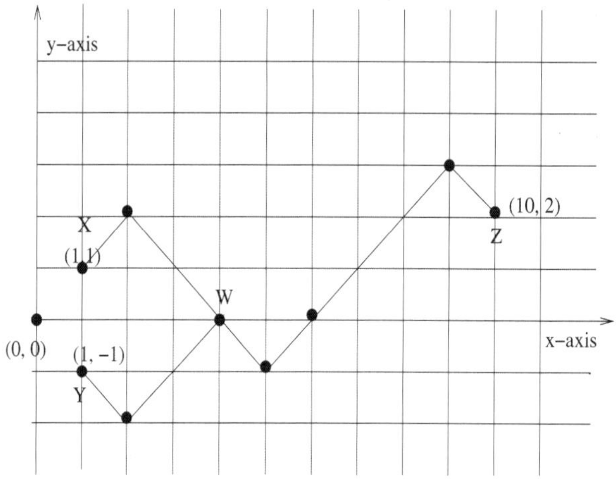

Figure 5.1: An Invalid Path and its Reflected Path

The paths we wish to count must have the the point $(1, 1)$ on them since the first vote counted must be in favor of A. The total number of such paths (satisfying the above description) is evidently $\binom{m+n-1}{m-1}$ (this is clear: only $m + n - 1$ votes remain to be counted, of which $m - 1$ are in favor of A). What is the number of invalid paths that begin at $(1, 1)$ and end at $(m + n, m - n)$? For any such path, there is a first place where it touches (meets) the x-axis. Such a place corresponds to the first time when A and B have equal votes and must then correspond to a point $(2r, 0)$ for some $r > 0$. *We call it a snag.* Reflect this portion of the graph into the x-axis, and retain the remaining portion as it was. This creates a path from $(1, -1)$ to $(m + n, m - n)$. Conversely, given any such path that goes from $(1, -1)$ to $(m + n, m - n)$, we can get an invalid path by reflecting such a path into x-axis. This is also easy to see. The two paths that are mirror images of each other (in the x-axis) actually meet each other at the first place of the form $(2r, 0)$ where we also have the first point of their meeting the x-axis. Hence, to count the number of invalid paths from $(1, 1)$ to $(m + n, m - n)$, it suffices to just count the total number of paths from $(1, -1)$ to $(m + n, m - n)$. As an illustration, consider the following invalid path which begins at X, has a snag at W and ends at Z, where X is $(1, 1)$, W is $(6, 0)$ and Z is $(12, 2)$. The invalid path and the reflected path are both shown in Figure 5.1. The actual counting of votes proceeds as follows: $AAABBBABAABA$.

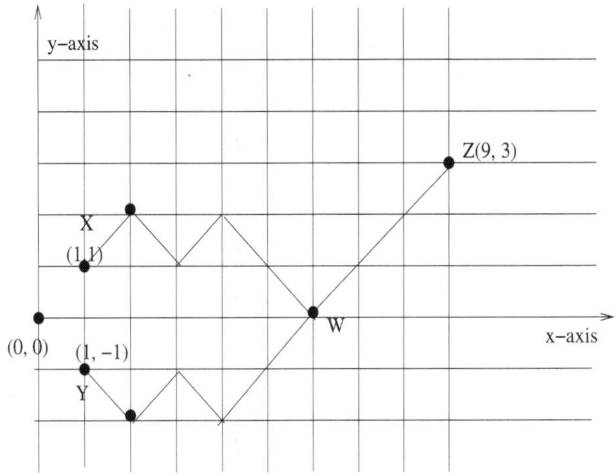

Figure 5.2: Another Invalid Path and its Reflected Path

Let the point $(1, 1)$, where the invalid path begins be called X. Let the point $(1, -1)$, which is a mirror reflection (in the X-axis) of the point X be called Y. Let the point $(m+n, m-n)$ which is the final destination of paths be called Z. The total number of paths that go from X to Z with exactly $m-1$ votes for A from the total of $m+n-1$ votes (since one vote has already been counted) is α where α equals $\binom{m+n-1}{m-1}$. We then count the number of invalid paths β. For any invalid path, we have a first meeting point with the x-axis say W where, as explained above, W has coordinates $(2r, 0)$ for some positive number r. Again as we already explained we reflect the portion of the invalid path from X to W to get a portion of the new path from Y to W; the remaining part of the path from W to Z is retained as in the original invalid path. Observe that upto W, A got $r-1$ votes while B got r votes. Hence, in the reflected portion A gets r votes. Notice also that in the portion from W to Z, A got $m-r$ votes which goes to show that in the new path that begins from Y, A got m votes in all. The bijection between invalid paths going from X to Z and the paths going from Y to Z that we just set up, shows that the number of invalid paths is just $\beta = \binom{m+n-1}{m}$.

Hence the number of ways in which A gets more votes than B at any stage of counting is obtained by subtracting β from α. This number is therefore equal to

$$\binom{m+n-1}{n} - \binom{m+n-1}{m}$$

Dividing this number by $\binom{m+n}{m}$, the total number of ways in which the votes can be counted, we get (after simplification), the proof of Theorem 5.2.1.

Theorem 5.2.1. *Let $m > n$. Let A and B get m and n votes respectively. Then the probability that at every stage of counting A has more votes than B is equal to $\frac{m-n}{m+n}$.*

Figure 5.2 is another example corresponding to the invalid vote counting pattern $AABBBAAAAB$ where A got 6 and B got 4 votes.

5.3 Conditional probability and Bayes' theorem

In many situations of interest, we have a prior knowledge of the occurrence (or non-occurrence) of some event. This creates problems in the sense that we do not know what sample space we are looking at. For example, in tossing a coin two times, let the event A consist of obtaining H on the first toss, the event B obtaining H on the second toss and the event C consists of obtaining H on both the first and second toss. Does the occurrence of A have any effect on the occurrence of B? The answer is clearly a flat "No". However, if we know that A has occurred, the chances of the occurrence of C get enhanced from a mere $\frac{1}{4}$ to $\frac{1}{2}$. Essentially, we are not dealing with the original sample space but we are dealing with a revised sample space where the new sample space equals a particular event. As another example, consider an experiment where we are trying to open a lock using a set of n keys of which only one will open the lock. If the keys are successively tried, then the probability that any one key will open the lock is just $\frac{1}{n}$. However, if it is known beforehand, that the first key did not open the lock, the probability that the second key will open the lock is $\frac{1}{n-1}$ (and not $\frac{1}{n}$ as many might believe): If we consider A to be the event that the first key did not open the lock, then $P(A)$ is just $\frac{n-1}{n}$. Now the probability of the event B that the second key would open the lock (given that A has occurred) is simply $\frac{P(A \cap B)}{P(A)}$. Since the event B is contained in A, the numerator is just $P(B)$, which is $\frac{1}{n}$. This gives the required theoretical justification. We are ready to state the following definition.

Definition 5.3.1. Let A and B two events of the sample space Ω. Suppose $P(A) \neq 0$. Then *the conditional probability of B given A, denoted by $P_A(B)$* is defined to be $\frac{P(A \cap B)}{P(A)}$.

Example 5.3.2. We now give some examples that will make the concept of conditional probability clear.

(a) Let Ω be the sample space consisting of all the 13-card hands (in a pack of 52 cards). Let A be the event that the hand has in it the queen of ♠. Let B be the event that the hand contains all the four queens. Let C be the event that the hand contains all the four kings. We find that

$$P(B) = \frac{13 \times 12 \times 11 \times 10}{52 \times 51 \times 50 \times 49}\ ; P_A(B) = \frac{12 \times 11 \times 10}{51 \times 50 \times 49}$$

Notice that $P_A(B)$ is actually 4 times larger than $P(B)$. Similarly, $P_A(C)$ is smaller than $P(C)$. The values are

$$P(C) = \frac{13 \times 12 \times 11 \times 10}{52 \times 51 \times 50 \times 49}\ ; P_A(C) = \frac{12 \times 11 \times 10 \times 9}{51 \times 50 \times 49 \times 48}$$

(b) Three dice are rolled. Let A be the event that no two dice show the same face. A can occur in $6 \times 5 \times 4$ ways while the totality of outcomes is 6^3. Hence

$P(A) = \frac{5}{9}$. Now let B be the event that some face shows an ace (i.e., shows 6). Using inclusion-exclusion or otherwise, it is seen that B can occur in 91 ways. We then compute $P(A \cap B) = \frac{5}{18}$. Hence, $P_A(B) = \frac{1}{2}$ which is higher than $P(B)$. An intuitive understanding: it is more likely an ace would occur if one has a prior knowledge that the three faces show up distinct numbers.

(c) We toss a coin seven times. Let B be the event that we obtain at least 4 heads. It is clear that B has probability $\frac{1}{2}$. If A is the event that the first two tosses resulted in heads, then $P(A) = \frac{1}{4}$ and $P(A \cap B) = \frac{13}{64}$ so that $P_A(B)$ equals $\frac{13}{16}$. This is also intuitively clear: it is more likely to get at least four heads (in all) if the first two tosses have already resulted in heads. On the other hand, suppose that C is the event that the first two tosses resulted in tails. Then $P(C)$ also equals $\frac{1}{4}$ while $P(C \cap B) = \frac{3}{64}$ and hence $P_C(B)$ equals $\frac{3}{16}$. It is clear that the occurrence of two tails on the first two tosses considerably hampers the possibility of the occurrence of B. Finally, how does the occurrences A or C have an effect on the other? We leave this to the reader.

(d) (Feller) I have a friend whom I never visited in the last two decades. I know that he has two children but I know nothing about their ages or genders. I visit the friend and find that a child of his who opened the door is a girl. What is the probability that the other child is also a girl? Assuming the equally likely probability measure on the sample space $\Omega = \{BB, BG, GB, GG\}$ (where B refers to a boy and G to a girl), we see that the event A that the friend has a girl child can occur in three ways while the event that both children are girls can occur in one among these three ways. Hence the required conditional probability is $\frac{1}{3}$ and not $\frac{1}{2}$ as many people might believe. On the other hand, the same question has a different answer if we know that the child who opened the door is the elder child. In that case, the probability that the other child is also a girl is just $\frac{1}{2}$.

Definition 5.3.3. Two events A and B are said to be *independent* if $P(A \cap B) = P(A)P(B)$. Using the definition of conditional probability, we see that this is equivalent to $P_A(B) = P(B)$ provided $P(A)$ is not zero. Even in a case where conditional probability is not defined, the independence of two events is defined. We also say that two events are *dependent* if they are *not independent.*

Observe that if A and B are independent then the occurrence of one of them has no effect on the other. A word of caution for a reader at an elementary level will not be out of order here. Many times, we confuse between exclusiveness and independence of events. These are very different notions. In fact, exclusive (i.e., disjoint) events are actually dependent!

Example 5.3.4. We begin with the following three examples.

(a) If one card is drawn at random then the sample space consists of 52 outcomes. Let the event A be drawing the ace (of some suit) and let the event B be drawing a card in the spade suit. We have $P(A) = \frac{1}{13}$, $P(B) = \frac{1}{4}$ and $P(A \cap B) = \frac{1}{52}$ so that these events are independent as should be expected.

(b) Let Ω be the sample space consisting of all the hands (with 13 cards). Let A be the event that the hand has in it the ace of spades and let B be the event that the hand contains all the cards of the same suit. Then

$$P(A) = \frac{\binom{51}{12}}{\binom{52}{13}} = \frac{1}{4}; P(B) = \frac{4}{\binom{52}{13}}$$

so that

$$P(A)P(B) = \frac{1}{\binom{52}{13}} = P(A \cap B)$$

Hence these two events are independent.

(c) Let the experiment consists of tossing a coin 7 times. Let A be the event that we obtain heads on the first toss and let B be the event that the total number of heads is even. Then the probability of A and B is $\frac{1}{2}$ each while $P(A \cap B) = \frac{1}{4}$. Hence these two events are independent.

It is possible to generalize the notion of independence of two events to three or more events in the following manner.

Definition 5.3.5. Let $A_1, A_2, \ldots, A_n$ be events. We say that these events are *pairwise independent* if for every $i \neq j$ with $i, j = 1, 2, \ldots, n$, we have $P(A_i \cap A_j) = P(A_i)P(A_j)$. These events are called *mutually independent* if for every choice of $i_1, i_2, \ldots, i_k$ with $1 \leq i_1 < i_2 < \cdots < i_k \leq n$, we have

$$P(A_{i_1} \cap A_{i_2} \cap \cdots \cap A_{i_k}) = P(A_{i_1})P(A_{i_2}) \cdots P(A_{i_k})$$

It is clear that mutual independence implies pairwise independence. The converse is not true as the following somewhat artificial example shows.

Example 5.3.6. Let the sample space Ω consist of the six permutations on $[3]$ (treated as triples in the one-line notation) along with three (special) additional triples $111, 222$ and 333. On this sample space of 9 elements, let us have the equally likely probability measure. Let A_i be the event that at the i-th place in the triple, we have i. Then A_i can occur in three ways and hence $P(A_i) = \frac{1}{3}$ for each i. If $i \neq j$, then $P(A_i \cap A_j) = \frac{1}{9}$ since $A_1 \cap A_2$ consists of the single outcome 123. Hence these events are pairwise independent. However, the event $A_1 \cap A_2 \cap A_3$ also consists of the single outcome 123 and hence has probability $\frac{1}{9}$ and not $\frac{1}{27}$ which show that these three events are not mutually independent. However, the following n events are *mutually independent*: On the sample space of n tosses of a coin, let the event A_i be obtaining heads on the i-th toss of the coin.

If we are given $P_A(B)$ as well as $P(A)$, it should be possible to determine $P_B(A)$. The probability $P(A)$ which is revised to get $P_B(A)$ is called *a priori probability*. The latter revised probability $P_B(A)$ is called *a posterior probability*. In order to motivate the discussion in the right direction, we make the following definition.

Definition 5.3.7. Let $\mathbf{A} = \{A_1, A_2, \ldots, A_n\}$ be a set of events. $\mathbf{A}$ is called *a partition of the sample space* Ω if these n events are mutually exclusive and exhaustive (that is, $A_1 \cup A_2 \cup \cdots \cup A_n = \Omega$ and $A_i \cap A_j = \emptyset$ for all $i \neq j$).

Suppose India and Sri Lanka play three one day international cricket matches at three locations, Delhi, Mumbai and Bangalore. An avid follower of the game has his own judgment (guess) on who would win the match if it is played at any of the three cities, with probability of India winning the match clearly determined in his mind. Assume further, for the sake of simplicity that there is no tie and hence one of the two teams must win and other must lose. In addition, the cricket fan has some probabilities assigned to the matches being played at the three cities (the most natural assignment being $\frac{1}{3}$ to each city). Being a non-resident Indian however, he has not been able keep track of the matches. Suddenly, he hears that India has won. What is the probability that the match was played at Bangalore? The initial probability that our friend assigned to the match being played at Bangalore is the a priori probability. The revised probability is the posterior probability. Computation of posterior probabilities is a task that is performed using the following.

Theorem 5.3.8. *(Bayes' Theorem) Let $\{A_1, A_2, \ldots, A_n\}$ be a partition of the sample space Ω. Let B be any event. If the a priori probabilities $P(A_k)$ are known, then the a posteriori probabilities $P_B(A_k)$ can be computed using the following formula*

$$P_B(A_k) = \frac{P_{A_k}(B)P(A_k)}{\sum_{j=1}^{n} P_{A_j}(B)P(A_j)}$$

Proof Denote the R.H.S. by $\frac{N}{D}$ where N refers to the numerator and D refers to the denominator. Using the formula for conditional probabilities, we see that $N = P(A_k \cap B)$ and $D = \sum_{j=1}^{n} P(A_j \cap B)$. Our assumption implies that the set of events $\{A_j \cap B : j = 1, 2, \ldots, n\}$ forms a partition of B and hence $D = P(B)$. Therefore, R.H.S. is equal to

$$\frac{P(A_k \cap B)}{P(B)} = P_B(A_k)$$

which is the L.H.S. □

5.4 Examples based on conditional probability and Bayes' theorem

We begin by proving the following theorem useful in many computations. Note also that this theorem has been used in the previous section in the proof of Bayes' theorem (Theorem 5.3.8).

Theorem 5.4.1. *(Theorem on total probabilities) Let $\{C_1, C_2, \ldots, C_m\}$ be a partition of the sample space Ω such that $P(C_i) \neq 0$ for all $i = 1, 2, \ldots, m$. Let B be any event. Then*

$$P(B) = \sum_{i=1}^{m} P_{C_i}(B)P(C_i)$$

Proof R.H.S. equals

$$\sum_{i=1}^{m} \frac{P(B \cap C_i)}{P(C_i)} \times P(C_i)$$

which equals $\sum_{i=1}^{m} P(B \cap C_i) = P(B)$ using Lemma 5.1.3. □

We now give a number of examples that use Bayes' theorem and conditional probabilities.

Example 5.4.2. In the example preceding Bayes' theorem, we let A_1, A_2 and A_3 respectively denote the events play in Mumbai, Delhi and Bangalore. Let B be the event that India wins. Suppose we assign the a priori probabilities:

$$P_{A_1}(B) = \frac{1}{3};\ P_{A_2}(B) = \frac{1}{2};\ P_{A_3}(B) = \frac{2}{3}$$

Also assume that the probability that the match will be played at a particular city is the same for all the three cities. Hence we assign the value $\frac{1}{3}$ to each one of $P(A_1)$, $P(A_2)$ and $P(A_3)$. Using Bayes' theorem, we compute a posteriori probabilities:

$$\begin{aligned} P_B(A_3) &= \frac{P_{A_3}(B)P(A_3)}{P_{A_1}(B)P(A_1) + P_{A_2}(B)P(A_2) + P_{A_3}(B)P(A_3)} \\ &= \frac{2/3 \times 1/3}{(1/3 + 1/2 + 1/3) \times 1/3} = \frac{2/3}{3/2} = \frac{4}{9} \end{aligned}$$

Notice that the posterior probability is higher than the a priori probability which is just $\frac{1}{3}$. Similarly the probability $P_B(A_1)$ that the match was played at Mumbai is $\frac{2}{9}$ which is lower than $\frac{1}{3}$, the a priori probability and $P_B(A_2)$, the probability that the match was played at Delhi is $\frac{1/2}{3/2}$ which equals $\frac{1}{3}$, the same as the a priori probability.

Example 5.4.3. Urn models are the most appropriate prototypical applications of conditional probabilities and Bayes' theorem. Consider the situation of three distinct urns U_1, U_2 and U_3. Assume that U_1 contains 3 red balls and 2 black balls, U_2 contains 4 red and 3 black balls while U_3 contains 3 red and 3 black balls. The experiment consists of choosing one of the three urns and picking up a ball from that urn whose color is noted. Clearly, the sample space Ω consists of 6 members and Ω is, in fact, partitioned by the events $A_i = \{(U_i, R), (U_i, B)\}$, where R refers to choosing a red ball and B to choosing a black ball. In the absence of any other information, we consider the probability of choosing any urn to be $\frac{1}{3}$, though it is not essential to do so. Hence $P(A_i) = \frac{1}{3}$ for all i. Let R (respectively B) stand for the event of choosing a red (respectively a black) ball. Then it is clear, (looking at the balls the three urns contain), that, with equally likely probability measure, we have:

$$P_{A_1}(R) = \frac{3}{5},\ P_{A_1}(B) = \frac{2}{5},\ P_{A_2}(R) = \frac{4}{7},$$

$$P_{A_2}(B) = \frac{3}{7},\ P_{A_3}(R) = \frac{1}{2} \textit{ and } P_{A_3}(B) = \frac{1}{2}$$

Hence using Bayes' theorem (Theorem 5.3.8), the posteriori probabilities are:

$$P_R(A_1) = \frac{\frac{1}{3}\times\frac{3}{5}}{\frac{1}{3}\times\frac{3}{5}+\frac{1}{3}\times\frac{4}{7}+\frac{1}{3}\times\frac{1}{2}} = \frac{\frac{3}{5}}{\frac{3}{5}+\frac{4}{7}+\frac{1}{3}} = \frac{42}{117}$$

Similarly,

$$P_R(A_2) = \frac{\frac{4}{7}}{\frac{3}{5}+\frac{4}{7}+\frac{1}{2}} = \frac{40}{117},$$

and

$$P_R(A_3) = \frac{\frac{1}{2}}{\frac{3}{5}+\frac{4}{7}+\frac{1}{2}} = \frac{35}{117}$$

Notice that the posteriori probabilities of A_1 and A_2 show a slight increase from the earlier value $\frac{1}{3}$ while that of A_3 shows a slight decrease. In a very similar manner, we get $P_B(A_1) = \frac{28}{93}$, $P_B(A_2) = \frac{30}{93}$ and $P_B(A_3) = \frac{35}{93}$.

Example 5.4.4. We now discuss a general question of *sampling without replacement* of the following kind. An urn contains b black and r red balls. Balls are successively drawn from the urn and are thrown out. Let B_1 be the event in which the first draw results in a black ball. Clearly $P(B_1) = \frac{b}{b+r}$. If B_2 denotes black at the second draw, then $P_{B_1}(B_2) = \frac{b}{b+r} \times \frac{b-1}{b+r-1}$ and $P_{R_1}(B_2) = \frac{r}{b+r} \times \frac{b}{b+r-1}$ Hence using the theorem on total probabilities (Theorem 5.4.1), wc gct

$$\begin{aligned} P(B_2) &= P_{R_1}(B_2)P(R_1) + P_{B_1}(B_2)P(B_1) \\ &= \frac{b}{b+r-1}\times\frac{r}{b+r} + \frac{b-1}{b+r-1}\times\frac{b}{b+r} \\ &= \frac{b}{b+r} \end{aligned}$$

Let R_j (respectively B_j) refer to red (respectively black) at the j-th draw. It is, in fact, true that the probability of a black draw *at any stage* is $\frac{b}{b+r}$ as the following theorem shows.

Theorem 5.4.5. *(Poisson's theorem) For all n with $1 \le n \le b+r$, we have*

$$P(B_n) = \frac{b}{b+r}$$

Proof There are two ways of proving this. First is a brute force method where we find out all the ways in which B_{n+1} can occur. Let Y_j be the event in which the first n draws resulted in exactly j black balls (and $n-j$ red balls). Clearly, $P_{Y_j}(B_{n+1})$ equals $\frac{b-j}{b+r-n}$ since j black balls and n balls in all have already been chosen. It is also easy to see that the event Y_j takes place in

$$\frac{\binom{b}{j}\binom{r}{n-j}}{\binom{b+r}{n}}$$

ways. So, using

$$\binom{b}{j} = \frac{b}{b-j}\binom{b-1}{j}$$

and

$$\binom{b+r}{n} = \frac{b+r}{b+r-n} \times \binom{b+r-1}{n-1}$$

we obtain (using the theorem on total probabilities, Theorem 5.4.1),

$$\begin{aligned} P(B_{n+1}) &= \sum_j P_{Y_j}(B_{n+1})P(Y_j) \\ &= \sum_j \frac{b-j}{b+r-n} \times \frac{\binom{b}{j}\binom{r}{n-j}}{\binom{b+r}{n}} \\ &= \frac{b}{b+r} \times \frac{\sum_j \binom{b-1}{j}\binom{r}{n-j}}{\binom{b+r-1}{n}} \end{aligned}$$

A standard counting argument from Chapter 1 will prove that the sum in the numerator of the second term above equals the denominator. Hence the result.

The second proof is much shorter but it requires a degree of ingenuity. Imagine as if the black balls are labeled from 1 to b and the red balls are labeled from $b+1$ to $b+r$. We use the falling factorial notation. We have a black ball on the $(n+1)$-th draw and this can be (independently) done in b ways. This leaves us with $b+r-1$ balls of which n are to be selected (and also ranked) which is clearly doable in $[b+r-1]_n$ ways. Without the constraint on the $(n+1)$-th ball, we can perform the task in $[b+r]_{n+1}$ ways. So, we have

$$P(B_{n+1}) = \frac{b[b+r-1]_n}{[b+r]_{n+1}} = \frac{b}{b+r}$$

Example 5.4.6. The following is a simple model of two urns that sheds some light on our empirical understanding of the notion of probability. We leave some details to the reader. Two urns U_1 and U_2 are given such that U_1 contains 2 black and 3 red balls while U_2 contains 3 black and 2 red balls. The experiment consists of randomly picking up one urn and repeatedly drawing balls from the same urn, noting the color and *replacing the ball.* With *equally likely probability measures*, we have $P(U_1) = P(U_2) = \frac{1}{2}$. The question we are interested in is: if a large number of successive draws resulted in a black ball, is it more or very much more likely that the urn chosen was U_2 and not U_1? We have the following argument. Let B_n denote the event that the n-th draw resulted in a black ball. As an example, $B_1R_2B_3$ refers to the event of black on draw 1, red on draw 2 and black on draw 3. Thus, $B_1B_2\cdots B_n$ is the event of drawing black on all the n draws (it is the intersection of the events B_j where B_j is black on the j-th draw). Since the occurrence of B_2 given that B_1 has occurred, depends on the choice of one of U_1 or U_2, we get (using the Theorem 5.4.1 on total probabilities)

$$
\begin{aligned}
P_{B_1}(B_2) &= \frac{P(B_1 \cap B_2)}{P(B_1)} \\
&= \frac{P(B_1 B_2)}{P(B_1)} \\
&= \frac{\frac{1}{2} \times [\frac{2}{5} \times \frac{2}{5} + \frac{3}{5} \times \frac{3}{5}]}{\frac{1}{2}} \\
&= \frac{13}{25}
\end{aligned}
$$

This should be compared with

$$
P(B_n) = P(U_1)P_{U_1}(B_n) + P(U_2)P_{U_2}(B_n) = \frac{1}{2}
$$

We also have

$$
\begin{aligned}
P(B_1 B_2 \cdots B_n) &= P(U_1)P_{U_1}(B_1 B_2 \cdots B_n) + P(U_2)P_{U_2}(B_1 B_2 \cdots B_n) \\
&= \frac{1}{2}\left\{\left(\frac{2}{5}\right)^n + \left(\frac{3}{5}\right)^n\right\}
\end{aligned}
$$

Therefore,

$$
\begin{aligned}
P_{B_1 B_2 \cdots B_n}(B_{n+1}) &= \frac{P(B_1 B_2 \cdots B_n B_{n+1})}{P(B_1 B_2 \cdots B_n)} \\
&= \frac{\left(\frac{2}{5}\right)^{n+1} + \left(\frac{3}{5}\right)^{n+1}}{\left(\frac{2}{5}\right)^n + \left(\frac{3}{5}\right)^n} \\
&= \frac{3}{5} \times \frac{(\frac{2}{3})^{n+1} + 1}{\left(\frac{2}{3}\right)^n + 1}
\end{aligned}
$$

and observe that as $n \to \infty$, we have

$$
P_{B_1 B_2 \cdots B_n}(B_{n+1}) \to \frac{3}{5} = P_{U_2}(B_{n+1})
$$

Finally, we have

$$
\begin{aligned}
P_{B_1 B_2 \cdots B_n}(U_1) &= \frac{P_{U_1}(B_1 B_2 \cdots B_n)P(U_1)}{P_{U_1}(B_1 B_2 \cdots B_n)P(U_1) + P_{U_2}(B_1 B_2 \cdots B_n)P(U_2)} \\
&= \frac{(\frac{2}{5})^n}{(\frac{2}{5})^n + (\frac{3}{5})^n} \\
&= \frac{1}{1 + (\frac{3}{2})^n}
\end{aligned}
$$

Thus $P_{B_1 B_2 \cdots B_n}(U_1)$ tends to 0 as n tends to ∞ while a similar calculation shows that a posteriori probability of U_2 given $B_1 B_2 \cdots B_n$ tends to 1. We record the findings:

$$\lim_{n\to\infty} P_{B_1B_2\cdots B_n}(B_{n+1}) = \frac{3}{5} = P_{U_2}(B_j),$$

and

$$\lim_{n\to\infty} P_{B_1B_2\cdots B_n}(U_2) = 1$$

In other words, if a large number of successive draws resulted in balls of black colour, then the probability of getting a black ball on the next draw is $\frac{3}{5}$ (since it is more likely that the draws are from the second urn). This is also supported by the second limit which tells us that if a large number of draws were black, then one is almost certain that the second urn was used. These calculations of probabilities and limits is highly corroborative with our intuitive notions of probability.

Example 5.4.7. A sufficiently general urn scheme is described as follows. An urn contains b black and r red balls. A ball is drawn at random, its color is noted and it is put back. We also put c balls of the same color and d balls of the opposite color in the urn. The experiment is repeated n times. Some special cases of this model are highly interesting. With $c = 0$ and $d > 0$, we obtain the model called *the safety campaign*. Treat the red balls as accidents and the black balls are safety measures. Whenever an accident occurs safety measures are stepped up and when no accident occurs the safety campaign slackens and more accidents occur. *The Polya urn scheme* is characterized by $c > 0$ and $d = 0$. This is a model for *contagious disease*. The occurrence of a case of disease enhances the chances of finding more affected people. If in the n successive draws, we have k black balls first followed by m red balls (where $k + m = n$), then the probability of this event A is

$$\frac{b(b+c)(b+2c)\cdots(b+(k-1)c)\times r(r+c)\cdots(r+(m-1)c)}{(b+r)(b+r+c)(b+r+2c)\cdots(b+r+(n-1)c)}$$

The most elegant point which makes the Polya urn scheme simple is the following. Let B be the event in which n balls are drawn of which k are black and m are red but not necessarily in that order (but in a specific order). Then we get the same expression as above but the terms may get permuted. In other words, we have $P(B) = P(A)$. In the above expression, we may divide throughout by c^n and then divide the numerator and denominator by suitable factorials to get

$$P(A) = \frac{\binom{-b/c}{k}\binom{-r/c}{m}}{\binom{-(b+r)/c}{n}} \times \frac{1}{\binom{n}{k}}$$

If we now let X denote the event of obtaining exactly k black balls, then

$$P(X) = \frac{\binom{-b/c}{k}\binom{-r/c}{m}}{\binom{-(b+r)/c}{n}}$$

Example 5.4.8. It will be in the fitness of the things to wind up this Chapter with a discussion on the *Laplace law of succession*. We have a collection of $N + 1$ urns labeled $U_0, U_1, \ldots, U_N$ such that the all the urns contain N balls of two colors red and white. Assume further that the urn U_k contains k red and $N - k$ white balls. The experiment consists of picking up any one urn and repeatedly drawing balls from the same urn *with replacement*. In what follows, we assume that N is sufficiently large so that approximations can be duly carried out. Let A_n denote the event that n successive draws resulted in all the balls of the red color. If the k-th urn is chosen then we may repeatedly make n draws of a red ball (with replacement) from U_k with probability $(\frac{k}{N})^n$ (since U_k has k red balls). But k can take any value from 1 to N and the urn can be initially chosen in $N + 1$ equally likely ways. Thus, we have

$$\begin{aligned} P(A_n) &= \frac{1}{N+1}\sum_{k=0}^{N}(\frac{k}{N})^n \\ &= \frac{1^n + 2^n + \cdots + N^n}{N^n(N+1)} \end{aligned}$$

Let I_n stand for the integral

$$\begin{aligned} \frac{1}{N^n(N+1)}\int_0^N x^n dx &= \frac{N}{N+1} \times \frac{1}{n+1} \\ &\approx \frac{1}{n+1} \end{aligned}$$

Observe that $P(A_n)$ is just the Riemann sum approximation for I_n. Hence

$$P(A_n) \approx I_n \approx \frac{1}{n+1}$$

Now let A_{n+1} be the event in which we get $n+1$ red balls in succession. Since A_{n+1} is contained in A_n, we have

$$\begin{aligned} P_{A_n}(A_{n+1}) &= \frac{P(A_{n+1})}{P(A_n)} \\ &\approx \frac{\frac{1}{n+2}}{\frac{1}{n+1}} \\ &= \frac{n+1}{n+2} \end{aligned}$$

This leads to the Laplace law of succession: the probability that $(n+1)$-th ball is red given that n draws have resulted in red is very large and tends to 1 as n tends to ∞. What we have derived here is the discrete (combinatorial) version of the Laplace law. For the continuous version of this law and a nice discussion on it refer to Chung [17] or Feller [24]. Using Laplace law of succession, *Laplace was willing to give odds of* $1,826,214$ *(which is the number of days in* $5,000$ *years), to* 1 *in favor of the assertion that the sun will rise tomorrow, given that it has risen for the last* $5,000$ *years!* This is seen by putting n equal to the requisite number of days in the above formula. For a discussion on how and why this has to be taken with a pinch of salt the reader is referred to the discussion in Feller [24].

5.5 Exercises for Chapter 5

5.1 Five cards are drawn, one by one, from a pack of 52 cards. What is the probability that an ace will appear for the first time on the fifth turn?

5.2 A is the event that consists of all the hands of 13 cards with exactly k aces (from a pack of 52 cards) and B is the event that consists of all the ways of distributing 13 cards to each of west, east, south and north so that west receives k aces. Show that A and B have the same probability (The sample spaces of the two events are not the same).

5.3 An experiment consists of a coin and two dice A an B of which A has four red faces and two blue faces while B has two red faces and four blue faces. The coin is first tossed (equally likely probability measure). If it falls heads, then die A is used repeatedly (die B is not used) and thrown many times. On the other hand, if the coin falls tails then die B is used repeatedly (die A is not used) and thrown many times.

(a) Show that the probability of red at any throw is $\frac{1}{2}$.

(b) If the first two throws resulted in red what is the probability of getting red on the third throw?

(c) If red turns up at the first n throws of a die, what is the probability that die A was used?

5.4 In rolling of an unbiased die 12 times, what is the probability of getting each face (number) exactly two times?

5.5 A thirteen card hand is dealt from a standard deck of cards. Let A denote the event that it has the ace of spades, B the event that it has at least one ace and C the event that it has at least two aces. Find the following:

$$P_A(B), P_A(C), P_B(A), P_B(C), P_C(A)$$

5.6 A red and a green die are thrown simultaneously. If we have an equally likely probability measure on the sample space of 36 outcomes then find the conditional probabilities of the following events:

(a) A sum of faces of 7 given that the number on the red die is odd.

(b) The number on the red die is odd given that the sum of faces is 7.

(c) A sum of faces is 7 given that the difference between the number on the red die and the number on the green die is 1.

5.7 In a hardware store containing 100 bulbs, 6 bulbs are known to be defective. If 9 bulbs are drawn at random (without replacement), what is the probability that at the most one of them will be defective?

5.8 A total of $m + n$ people, of which m are men and n are women, form a line from left to right with all the permutations equally likely.

(a) What is the probability that the oldest man is between two women?

(b) What is the probability that the oldest man has a woman to his left?

5.9 Let p be equally likely probability measure on the sample space of descriptions of all the 7-card hands from a standard pack of 52 cards. Which of the following pairs of events is independent?

(a) Having the ace of spades and having exactly 4 black cards.

(b) Having the ace of spades and having 4 cards of the same denomination.

(c) Having exactly two spades, two hearts and having more black cards than red.

(d) Having at most two cards in each suit and having the jack of hearts.

(e) Having no more than two hearts and having the ace of spades.

(f) Having the queen of spades and having no red card.

(g) Having at lest three kings and having at least three queens.

5.10 In a community of $1,000$ people, two percent are exceptionally intelligent. A journalist meets 20 people in this community. What is the probability that he meets exactly 2 exceptionally intelligent people?

5.11 We have $2m$ coins of which m are fair and the remaining m biased with probability of heads equal to p. A coin is selected at random and it shows heads. What is the probability that it was a fair coin? Does the answer depend on m?

5.12 A student has to sit for an examination that consists of 3 questions selected out of a known set of 100 questions. To pass the test, he has to answer all the questions correctly. The student knows correct answers to 90 questions. What the the probability that the student will pass the examination?

5.13 A survey is carried out to find the percentage of people who sing in the bathroom. Because some people may be too embarrassed to admit openly that they are bathroom singers, each person is asked to roll a die in secret and answer NO if the number shown is 1 and YES if it is 6, no matter what the true answer is, but to tell the truth (YES or NO) if $2, 3, 4$ or 5 comes out. Because, the number shown on the die is not revealed, it is impossible to tell from the answer given if the person is a bathroom singer or not. Suppose the probability of getting the answer YES in this survey is $\frac{2}{3}$. What is the probability of being a bathroom singer?

5.14 When is an event A independent of itself?

5.15 In a set of 10 people, one is a couple consisting of Mr. and Mrs. A. If we randomly make these 10 people on a round table, what is the probability that the couple occupies adjacent seats?

5.16 Fifty two cards from a standard deck are distributed equally among four players A,B,C and D. What is the probability that A and B get all the spade cards among them?

5.17 We have two dice, a red die and a white die. The red die is loaded so that the probability of 6 is $\frac{1}{3}$ and all the other outcomes from 1 to 5 are equally likely among themselves. The white die is fair. We can pick up one die and leave the other die to the opponent. The dice are rolled and the player with higher score wins. In case of a tie, the player with the white die wins. Which die should one choose to have a better chance of winning?

5.18 In a hardware store containing 100 bulbs, 5 bulbs are known to be defective. If 8 bulbs are drawn at random (without replacement), what is the probability that at the most one of them will be defective?

5.19 Two dice are thrown. What is the probability that both the faces are 4 given that the sum is 7 or 8.

5.20 Two dice are simultaneously thrown six times. Let A be the event "sum seven on first throw" and let B be the event "sum seven on each of the last 4 throws". Find the probability of $A \cup B$.

5.21 Three unbiased coins are flipped. What is the probability that the third coin shows heads given that the first two coins agree?

5.22 Show that it is more likely to get at least one six in 4 throws of single die than to get at least a pair of sixes in 24 throws of two dice. *This is called de Mere's paradox (cf. Feller [24]).*

5.23 Five cards are drawn at random (without replacement) from a standard pack of 52 cards one at a time. Find the probability that the fifth card is a spade, given that the first four cards were spades.

5.24 A student writes a test in which the answer to each question is true or false (true/false test). The student marks the correct answer when he knows it and flips an unbiased coin when he does not know the correct answer. It is known that the probability that he knows the correct answer is $\frac{3}{5}$. What is the probability that he knew the answer if his answer was found to be correct (by the examiner)?

5.25 Suppose 5 men out of 100 and 25 women out of $5,000$ are color-blind. A person chosen at random was found to color-blind. What is the probability that the person is a male?

5.26 In a bolt factory, machines A,B and C manufacture respectively 25,35 and 40 percent of the total production of bolts. Of their outputs, respectively 5,4 and 2 percent are defective. A bolt drawn at random was found to be defective. What is the probability that it was manufactured by B?

5.27 A man throws a die three times in succession. Assuming equally likely probability measure, what is the conditional probability of the sum of three tosses to be 16 given that at least two tosses resulted in the same number. What is the conditional probability that the sum of three tosses is at least ten given that the first throw resulted in a two?

5.28 There are six cards of which three bear the letter D, two the letter O and one bears the letter X. The cards are thoroughly shuffled and the top four cards are turned face up (without changing the order in which they lay on top of the shuffled pack). What is the probability that they will spell the word $DODO$?

5.29 In a certain town, there are $10,000$ bicycles each of which is assigned a license number from 1 to $10,000$ (no two bicycles have the same license number). What is the probability that the first bicycle one encounters on a street has a number that has no 8 among its digits?

5.30 If the probability of hitting a target is $\frac{1}{5}$, and ten shots are fired independently, what is the probability of the target being hit at least twice?

5.31 Given that a throw of ten dice produced at least one six, what is the probability of at least two sixes?

5.32 Suppose a coin is tossed n times with the probability of heads equal to p on any toss and probability of tails equal to q on any toss. What is the probability that at least k tosses will result in heads?

5.33 In the ballot problem, assume that the candidate A scored 8 and the candidate B scored 6 votes. For each of the following counting patterns, find out if the pattern is valid or not. For an invalid pattern, plot the corresponding invalid path and then show its reflected paths.

(a) $ABABAABAABAABB$

(b) $AAABBBBAAAABBA$

5.34 (continuation of Exercise 5.33) For the following two reflected paths, obtain the corresponding invalid paths (given under the bijection set up by the reflection principle) and then construct the corresponding invalid sequences.

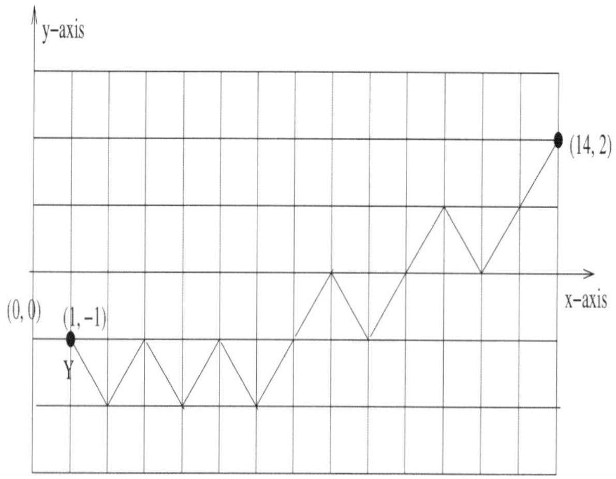

Figure 5.3: Figure for Exercise 5.34

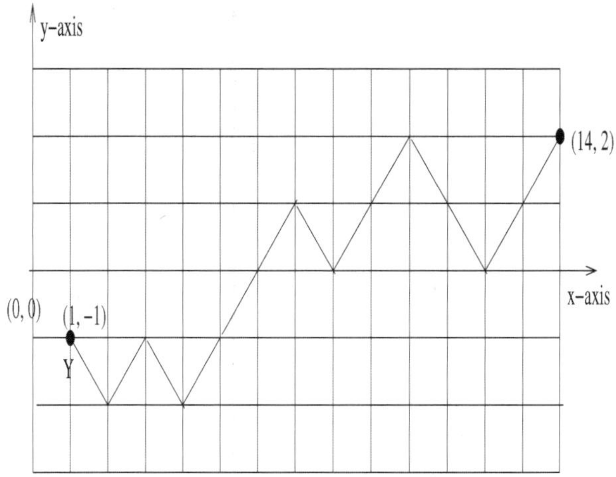

Figure 5.4: Figure for Exercise 5.34

5.35 In the ballot problem, candidate A scored m and B scored n votes where $m > n$. In the sample space of all the counting patterns, find the conditional probability that A was leading over B at all the stages of counting given that the first counted vote went in favour of A. Also find the conditional probability of the same event given that the first two votes counted were in favour of A.

5.36 Let $\Omega = [n]$ denote the sample space of the first n natural numbers with the equally likely probability measure. Let A and B be two events that contain i and j elements of Ω respectively. Show that if A and B are independent events then

$$j = k \times \frac{n}{g}$$

where g equals the g.c.d. of i and n and k is some number in $[g]$.

5.37 *We discuss the famous mailbox problem through this exercise.* We have n mailboxes each with a different key. Label the boxes by numbers from 1 to n and a box can be opened only by its own key (and by no other). Either inadvertently or by intention, the n keys have been randomly put in the boxes, one in each box. In order to open all the boxes, we adopt the following policy. We break open the lock of some box, find a key in that box, use that key to open possibly another box and then use that key to open a third box and so on. We wish to know the probability that we can open all the boxes. Suppose the key found in the box B_i is the key for B_j, then we open that box and find possibly a different key in that box and so on. We can thus assume that we have a unique cycle decomposition in the given arrangement of keys in the boxes. For example, if the box B_i contains its own key, then we have a singleton cycle (B_i).

(a) Show that if the lock of any one box is randomly opened by breaking it, then the probability that we can open all the boxes using the above procedure is $\frac{1}{n}$.

(b) Suppose now that instead of just one box randomly opened by breaking the lock, we break open the locks of k boxes and suppose w.l.o.g. that these boxes are numbered from 1 to k. Show that using the k keys found in these boxes, we can open all the boxes only if we have at the most k cycles in the permutation and each cycle contains at least one key from the set $[k]$.

(c) Using the correspondence under the guard representation given by Corollary 3.1.15, show that the accomplishment of opening of all the boxes is equivalent to having a permutation π on $[n]$ such that $\pi(n) = j$ where $j \in [k]$.

(d) Conclude that if the locks of k boxes are broken, then the probability of opening all the boxes using the k keys found in these boxes is $\frac{k}{n}$.

Chapter 6

Random variables

6.1 Random variables, mean and variance

The model of a probability space we introduced in chapter 5 lacks something serious from the practical point of view. A person who is applying the results in that chapter is not really interested in such abstract notions that are more based on set theory and less on the (harsh) reality of life, which, to him is reflected in terms of the real numbers. This passage from abstract probability space to the world of real numbers is made possible through a device called the *random variable*. For example, consider a simple experiment of tossing a biased coin with where the probability of obtaining heads is $\frac{1}{3}$ and the probability of obtaining tails is $\frac{2}{3}$. If a child tosses the coin and obtains heads, she is given 12 chocolates while if she obtains tails, she is given 6 chocolates. In a way that will be made more precise later, her average gain is $(12 \times \frac{1}{3}) + (6 \times \frac{2}{3}) = 8$ chocolates. What we have to point out is that the child is more interested in "how many chocolates" than the question of probabilities of heads and tails. Note that the emphasis is now shifting from the sample space Ω to something else. Consider one more example. This time, we toss an unbiased coin 20 times, where A pays B Rs. 4 if heads appear and B pays A Rs. 2 if tails appear on the toss. If we are only interested in how much money B lost or gained after the experiment, the range of values should be the set of all the integers in the interval $[-40, 80]$. A sample point (atomic outcome) is a sequence of length 20 consisting of heads and tails whose detailed description is of no interest to A or B. We are ready to make the following definition.

Definition 6.1.1. Let Ω be a sample space. Any real valued function $X : \Omega \to \mathbb{R}$ is called *a random variable and is abbreviated r.v.*

Example 6.1.2. Let p and q be fixed positive real numbers such that $p + q = 1$. Fix a positive integer n. On the sample space of all the possibilities of tossing a coin n times with probability of heads p and probability of tails q on any toss, we define a random variable X as follows. Let ω be any sequence of length n in the sample space Ω. Then $X(\omega) = k$ if ω has in it exactly k heads (and $n - k$ tails). This random variable is rather special and is called the *binomial random variable* or sometimes also repeated

Bernoulli trials. The probability that this random variable takes the value k is given by

$$b(k;n,p) \quad = Pr(X=k) = \binom{n}{k} p^k q^{n-k}$$

The range of this random variable is the set of all the integers between 0 and n.

We reserve capital letters X, Y, Z etc. for random variables. The notion of a random variable (r.v.) is sufficiently general. This allows us to exploit the algebra of random variables in the following manner. Given two r.v.s X and Y, we can add them. The resulting r.v. $X+Y$ is defined by $(X+Y)(\omega) = X(\omega) + Y(\omega)$ for all $\omega \in \Omega$. We can multiply X and Y to get an new r.v. XY which takes value $X(\omega)Y(\omega)$ for all $\omega \in \Omega$. We also have constant r.v.s that are just the constant real-valued functions on Ω and in particular, we also have a zero r.v. and a unit r.v. that take constant values 0 and 1 respectively. It is also clear that for any real number r and any r.v. X, the function rX defined by $(rX)(\omega) = r \times (X(\omega))$ is an r.v. In short random variables can be added, subtracted, multiplied, and even divided if the operation does not involve division by zero. In mathematical terms, this is expressed by saying that the r.v.s form a *vector space*. Further, they also form a ring so that we actually have an *algebra* of random variables. In particular, if X is an r.v. and f is any polynomial function, then $f(X)$ is also an r.v. Advantage of this understanding is the following. For an event A in the sample space, define an *indicator r.v.* $X = X_A$ by $X(\omega) = 1$ if $\omega \in A$ and $X(\omega) = 0$, otherwise. In particular, if the sample space Ω consists of n points $a_1, a_2, \ldots, a_n$ then we may define the indicator r.v. X_i by $X_i(a_j) = \delta_{ij}$ where δ_{ij} is the Kronecker delta, i.e., $\delta_{ij} = 1$ if $i = j$ and equals 0 otherwise.

Theorem 6.1.3. *The set* $\{X_i : i = 1, 2, \ldots, n\}$ *is a basis for the vector space of all the random variables on the given sample space* $\Omega = \{a_1, a_2, \ldots, a_n\}$.

Proof If $\sum_{i=1}^{n} r_i X_i$ is the zero r.v., then its evaluation on a_j leads to $r_j = 0$. and shows that we have a linearly independent set. Let X be any r.v. and let $X(a_i) = s_i$ for $i = 1, 2, \ldots, n$. Then X equals the r.v. $\sum_{i=1}^{n} s_i X_i$. □

For a random variable X on a sample space Ω, we let $\{x_1, x_2, \ldots, x_k\}$ to be the set of all the distinct real numbers that constitute the range of the r.v. X. Let $A_i = \{\omega \in \Omega : X(\omega) = x_i\}$. Then the set $\{A_1, A_2, \ldots, A_k\}$ forms a partition of the sample space Ω. This is called the *partition induced by X and will be denoted by* Θ_X. Here we also write $Pr(X = x_i)$ to mean the probability of the set A_i defined by $A_i = X^{-1}(x_i)$. Notice that $\sum_{i=1}^{k} P(X = x_i) = 1$ since A_i's form a partition of the sample space Ω. For real numbers α and β, by $P(\alpha \le X \le \beta)$ we mean the probability of the event A which consists of precisely those points of the sample space that are mapped to the interval $[\alpha, \beta]$ of the real line (by X). This is best expressed in terms of the idea of probability distribution function (p.d.f) which we now define.

Definition 6.1.4. Let X be a random variable. The function $f_X : \mathbb{R} \to [0, 1]$ defined by $f_X(x) = Pr(X = x)$ is called *the probability distribution function of the random variable* X.

In some sense, the probability distribution function (p.d.f.) of a random variable (r.v) X catches everything that we need to know about X. We go back to the experiment of tossing a coin n times with the probability of getting heads p on any toss. The corresponding distribution is called the *binomial distribution* and is denoted by $b(k; n, p)$. Here we assume that n and p are fixed. We give another example which occurs equally often. We have N balls of which k are red and $N - k$ are white. The experiment consists of n draws of balls without replacement. The random variable X we have in mind takes value r where r is the number of red balls drawn. Assuming equally likely probability measure on the sample space (the reader is encouraged to describe the sample space concretely), the distribution is given by

Definition 6.1.5. *The hypergeometric distribution* is given by

$$H(r; N, k, n) = f_X(r) = \frac{\binom{k}{r} \times \binom{N-k}{n-r}}{\binom{N}{n}}$$

Definition 6.1.6. Let $\mathbf{A} = \{A_1, A_2, \ldots, A_k\}$ and $\mathbf{B} = \{B_1, B_2, \ldots, B_m\}$ be two partitions of a sample space Ω. Then $\mathbf{A}$ and $\mathbf{B}$ are called *independent partitions* if for every $i = 1, 2, \ldots, k$ and for very $j = 1, 2, \ldots, m$ we have: A_i and B_j are independent events. Recalling the definition in Chapter 5, this means
$P(A_i \cap B_j) = P(A_i)P(B_j)$.

Example 6.1.7. Let the sample space Ω consist of the set of the 52 cards in a pack. Let the partition $\mathbf{A}$ consist of the 13 events A_i where A_i refers to the cards of denomination i. We also conveniently assume that A_1 refers to aces, A_{11} to jacks A_{12} to queens and A_{13} to kings. Clearly $P(A_i) = \frac{1}{13}$ for all i. Now let B_j refer to the four events of the card from the suit spades, hearts, diamonds and clubs respectively. Since any suit has 13 cards, $\mathbf{B}$ gives a partition of Ω where $P(B_j) = \frac{13}{52} = \frac{1}{4}$. Here $A_i \cap B_j$ refers to a card of a specific suit and specific denomination and is therefore a specific card. Thus, for all i and j, we must have $P(A_i \cap B_j) = P(A_i)P(B_j)$ so that $\mathbf{A}$ and $\mathbf{B}$ are independent partitions.

We will require the notion of independent partitions when we study two random variables. We now define a notion which in some sense is a weighted average.

Definition 6.1.8. Let X be a random variable on a sample space Ω. *The expectation or mean of* X is denoted by $E(X)$ *or* μ_X and is defined by

$$E(X) = \sum_{\omega \in \Omega} X(\omega)P(\omega)$$

Note that if the range of X is $\{x_1, x_2, \ldots, x_k\}$, then $E(X)$ can also be expressed as

$$
\begin{aligned}
E(X) &= \sum_{i=1}^{k} x_i f_X(x_i) \\
&= \sum_{i=1}^{k} x_i Pr(X = x_i) \\
&= \sum_{i=1}^{k} x_i P(A_i)
\end{aligned}
$$

In the above expression, A_i denotes $X^{-1}(x_i)$ so that $\mathbf{A} = \{A_1, A_2, \ldots, A_k\}$ is the partition induced by the r.v. X. Let us go back to the earlier example of chocolates. A toss of heads (with probability $\frac{1}{3}$) gives the child 12 chocolates while a toss of tails (with probability $\frac{2}{3}$) gives the child 6 chocolates. So, the expectation of the random variable is $(12 \times \frac{1}{3}) + (6 \times \frac{2}{3}) = 8$. Hence the child should be expected to get about 8 chocolates per toss (averaged over a large number of flips of the coin). Next consider the binomial r.v. whose distribution is $b(k; n, p) = \binom{n}{k} p^k q^{n-k}$. The expectation of this random variable X is

$$
\begin{aligned}
E(X) &= \sum_{k=0}^{n} k b(k; n, p) \\
&= \sum_{k=0}^{n} k \times \binom{n}{k} p^k q^{n-k} \\
&= np \times \sum_{k=1}^{n} \binom{n-1}{k-1} p^{k-1} q^{n-k} \\
&= np \times (p+q)^{n-1} \\
&= np
\end{aligned}
$$

This just says that if the probability of heads is p, then in n flips of a coin, we should expect to get np heads. How does the mean of the r.v. $X + Y$ relate to the means of the r.v.s X and Y? We have

$$
\begin{aligned}
E(X+Y) &= \sum_{\omega \in \Omega} (X+Y)(\omega) P(\omega) \\
&= \sum_{\omega \in \Omega} (X(\omega) + Y(\omega)) P(\omega) \\
&= \sum_{\omega \in \Omega} X(\omega) P(\omega) + \sum_{\omega \in \Omega} Y(\omega) P(\omega) \\
&= E(X) + E(Y)
\end{aligned}
$$

We summarize this in the following Lemma.

Lemma 6.1.9. *The following statements are true.*

(a) If X and Y are r.v.s on the sample space Ω, then

$$E(X + Y) = E(X) + E(Y)$$

(b) Linearity of the expectation: If $\{X_1, X_2, \ldots, X_n\}$ is any set of random variables on the sample space Ω, then

$$E(X_1 + X_2 \cdots + X_n) = E(X_1) + E(X_2) + \cdots + E(X_n)$$

(c) If r is any real number, then $E(rX) = rE(X)$.

As an application, again consider the binomial random variable X with binomial probability distribution function $b(k; n, p)$ described earlier. This is an important example from the point view of all the discussion in this Chapter and the reader is advised to keep it in mind all the time. X can be expressed as a sum $X_1 + X_2 + \cdots + X_n$ where X_i refers to the r.v. that takes value 1 if we have success (heads) on the i-th flip of the coin. Note that X_i is indifferent to the outcome of the other flips of the coin. Since X_i takes only two values 1 and 0 with probabilities p and q respectively, the expectation of X_i is $(1 \times p) + (0 \times q) = p$. It then follows from Lemma 6.1.9 that the expectation of the binomial r.v. X is $\underbrace{p + \cdots + p}_{n \text{ times}}$ which is just np. This reproves the earlier result. We can now use the machinery developed to prove a nice result on the expected number of cycles in a randomly chosen permutation π in the symmetric group on $[n]$. Let the sample space be the set of all the permutations on $[n]$ and for fixed k and for fixed i in $[n]$, denote by $X_{k,i}$ the random variable which takes the only values 1 and 0 with $X_{k,i}(\pi) = 1$ if and only if π contains i in a cycle of length k. Also let

$$X_k = \frac{1}{k} \sum_{i=1}^{n} X_{k,i}$$

so that the r.v. X_k counts the number of k-cycles in a given permutation. As we already saw (using the guard representation in Chapter 3), we have

$$E(X_{k,i}) = E(X_{k,1}) = \frac{(n-1)!}{n!} = \frac{1}{n}$$

Hence Lemma 6.1.9 shows that $E(X_k) = \frac{1}{k}$. Now let X be the r.v. which is obtained by summing all the r.v.s X_k (from 1 to n). Then another use of the Lemma 6.1.9 gives

$$E(X) = 1 + \frac{1}{2} + \cdots + \frac{1}{n}$$

We have thus proved:

Theorem 6.1.10. *A randomly chosen permutation on n letters is expected to have $1 + \frac{1}{2} + \cdots + \frac{1}{n}$ number of cycles.*

We have reproved a result of Chapter 3 (Exercise 3.9) in this new setting. For those with some idea of the sum of the harmonic series, we have the following interesting consequence : Since log n approximates the above sum (upto the Euler constant), the expected number of cycles in a randomly chosen permutation on n letters is very well approximated by the function log n .
The expectation has been shown to be linear. Is it also multiplicative? This is not true in general and we now attempt to make a set-up that will give a sufficient condition for $E(XY)$ to be equal to $E(X)E(Y)$. To that end, assume that X and Y are r.v.s on the sample space Ω. Let the ranges of X and Y be respectively $\{x_1, x_2, \ldots, x_k\}$ and $\{y_1, y_2, \ldots, y_m\}$. Let $A_i = X^{-1}(x_i)$ and $B_j = Y^{-1}(y_j)$ so that $\Theta_X = \{A_1, A_2, \ldots, A_k\}$ and $\Theta_Y = \{B_1, B_2, \ldots, B_m\}$ are the partitions of Ω induced by X and Y respectively.

Definition 6.1.11. The random variables X and Y are called *independent* if the partitions Θ_X and Θ_Y induced by X and Y respectively are independent partitions.

As an example, consider the experiment of tossing a fair coin n times. Let X be the random variable such that X takes value 1 if the sequence of tosses involves heads on the first toss and takes value 0 otherwise. Let Y be the random variable on the same sample space which takes value 1 if the number of heads is even and takes value 0 if the number of heads is odd. We then have $\mathbf{A} = \{A_1, A_2\}$ $\mathbf{B} = \{B_1, B_2\}$ where

$$\begin{aligned} A_1 &= \{\omega : \omega \; has \; H \; on \; first \; toss\} \\ A_2 &= \{\omega : \omega \; has \; T \; on \; first \; toss\} \\ B_1 &= \{\omega : \omega \; has \; even \; number \; of \; H\} \\ B_2 &= \{\omega : \omega \; has \; odd \; number \; of \; H\} \end{aligned}$$

Then we have

$$P(A_1 \cap B_1) = \frac{\sum_j \binom{n-1}{2j+1}}{2^n} = \frac{2^{n-2}}{2^n} = \frac{1}{4}$$

and

$$P(A_1 \cap B_2) = \frac{\sum_j \binom{n-1}{2j}}{2^n} = \frac{2^{n-2}}{2^n} = \frac{1}{4}$$

Similarly, $P(A_2 \cap B_1) = P(A_2 \cap B_2) = \frac{1}{4}$. Also $P(A_i) = P(B_j) = \frac{1}{2}$ for $i, j = 1, 2$ showing that X and Y are independent r.v.s.
In the set-up preceding the definition of independent of the r.v.s X and Y, we see that the independence of the r.v.s X and Y just means that for all $i = 1, 2, \ldots, k$ and $j = 1, 2, \ldots, m$, A_i and B_j are independent events, that is, $P(A_i \cap B_j)$ equals $P(A_i)P(B_j)$ where $\mathbf{A} = \{A_1, A_2, \ldots, A_k\}$ and $\mathbf{B} = \{B_1, B_2, \ldots, B_m\}$ are partitions of Ω induced by X and Y respectively. The ideas we wish to put across are better understood in terms of the joint probability distribution function of two r.v.s.

Definition 6.1.12. Let X and Y be random variables. We define the *joint probability distribution function (joint p.d.f)* of the r.v.s X and Y to be the function $f_{X,Y} : \mathbb{R}^2 \to [0, 1]$ given by

$$\begin{aligned} f_{X,Y}(x,y) &= Pr(X = x, Y = y) \\ &= P(X^{-1}(x) \cap Y^{-1}(y)) \end{aligned}$$

Theorem 6.1.13. *Let X and Y be independent random variables. Then the joint probability distribution function $f_{X,Y}$ satisfies $f_{X,Y}(x,y) = f_X(x) \times f_Y(y)$ for any real numbers x and y, where f_X and f_Y are the p.d.f.'s of X and Y respectively. Moreover,*

$$E(XY) = E(X)E(Y)$$

Proof Let $\mathbf{A} = \{A_1, A_2, \ldots, A_k\}$ and $\mathbf{B} = \{B_1, B_2, \ldots, B_m\}$ be the partitions of Ω induced by X and Y respectively. Write $C_{i,j} = A_i \cap B_j$ so that the value taken by the r.v. XY on the set $C_{i,j}$ is $x_i y_j$. Hence, we have

$$\begin{aligned} E(XY) &= \sum_{\omega \in \Omega} (XY)(\omega) P(\omega) \\ &= \sum_{\omega \in \Omega} X(\omega) Y(\omega) P(\omega) \\ &= \sum_{i=1}^{k} \sum_{j=1}^{m} x_i y_j P(C_{i,j}) \\ &= \sum_{i=1}^{k} \sum_{j=1}^{m} x_i y_j P(A_i \cap B_j) \end{aligned}$$

But the independence of X and Y implies that $P(A_i \cap B_j)$ equals $P(A_i)P(B_j)$. Hence,

$$\begin{aligned} E(XY) &= \sum_{i=1}^{k} \sum_{j=1}^{m} x_i y_j P(A_i) P(B_j) \\ &= \sum_{i=1}^{k} x_i P(A_i) \sum_{j=1}^{m} y_j P(B_j) \\ &= E(X)E(Y) \end{aligned}$$

□

Let X be an r.v. and let μ_X denote the mean $E(X)$ of X. We would like to find out how the values taken by X (or observations) are spread around the mean. A quantity such as $E(X - \mu_X)$ is useless since by the linearity of expectation, that function is identically zero. A quantity such as $E(|X - \mu_X|)$ would be of some use but cumbersome to compute. This prompts the following definition.

Definition 6.1.14. Let X be a random variable with mean μ_X. The expectation of the random variable $(X - \mu_X)^2$ is called the *variance* of X and is denoted by $V(X)$. We have, $V(X) = E((X - \mu_X)^2)$. The square root of $V(X)$ is called the *standard deviation* of X and is usually denoted by $\sigma = \sigma_X$. Thus $V(X) = \sigma^2$.

Lemma 6.1.15. *We have the following formula for* $V(X)$.

$$V(X) = E(X^2) - (\mu_X)^2$$

Proof Since μ_X is a constant, its expectation is the same number. So, from definition 6.1.15 the L.H.S. equals

$$E(X^2 - \mu_X \times 2X + (\mu_X)^2)$$

which is easily seen (by the linearity of E) to be

$$E(X^2) - 2\mu_X E(X) + (\mu_X)^2$$

But $E(X) = \mu_X$ and the required result follows. □

Unfortunately, the variance is not a linear function. That is, in general, we do not have $V(X+Y) = V(X) + V(Y)$.

Definition 6.1.16. Let X and Y be r.v.s with means μ_X and μ_Y respectively. Then the *covariance of* X *and* Y is denoted by $Cov(X,Y)$ and is defined by

$$Cov(X,Y) = E((X - \mu_X)(Y - \mu_Y))$$

Theorem 6.1.17. *Let* X *and* Y *be random variables. Then*

(a) $V(X+Y) = V(X) + V(Y) + 2Cov(X,Y)$.

(b) If X *and* Y *are independent r.v.s then* $Cov(X,Y) = 0$ *and hence* $V(X+Y) = V(X) + V(Y)$.

Proof We have $V(X+Y) = E(((X+Y) - (\mu_X + \mu_Y))^2)$ since $\mu_X + \mu_Y$ is $E(X+Y)$. Rearranging the terms, we get

$$V(X+Y) = E((X - \mu_X)^2 + (Y - \mu_Y)^2 + 2(X - \mu_X)(Y - \mu_Y))$$

so that (a) is obtained by using the linearity of the expectation. For (b), the linearity of the expectation again shows that $Cov(X,Y)$ equals $E(XY) - \mu_X E(Y) - \mu_Y E(X) + \mu_X\mu_Y$ which is just $E(XY) - \mu_X\mu_Y$. Now if X and Y are independent, then $E(XY) = E(X)E(Y) = \mu_X\mu_Y$ so that (b) is proved. □

Theorem 6.1.18. *Let* $X_1, X_2, \ldots, X_n$ *be independent r.v.s on a sample space* Ω. *Then*

$$V(X_1 + X_2 + \cdots + X_n) = V(X_1) + V(X_2) + \cdots + V(X_n)$$

We leave the proof to the reader. Induction is one way of proving it though it can also be proved directly. As an application, let X be the binomial random variable with parameters n and p. Then, $X = X_1 + X_2 + \cdots + X_n$ where X_k is the r.v. for which $X_k(\omega) = 1$ if ω has heads on the k-th toss and $X_k(\omega) = 0$, if ω has tails on the k-th toss. It is easily seen that X_i and X_j are independent if $i \neq j$. For a fixed i we have $E({X_i}^2) = E(X_i) = p$ so that $V(X_i) = p - p^2 = pq$. Hence it follows that

Corollary 6.1.19. *The binomial r.v. with parameters n and p has mean np and variance npq where $q = 1 - p$.*

Observe that the the variance is the largest when $p = \frac{1}{2}$. In some sense, this amounts to saying that in flipping a fair coin n times, the observations are more spread out (than flipping of an unfair coin). As another very interesting example, consider the following which is due to Graham, Knuth and Patashnik [27]. In a lottery, 100 tickets are sold each worth one rupee. A single prize of rupees 100 is given to the lucky number. An interested person decides to purchase two tickets. Which is a better policy? Should he purchase both the tickets in the same lottery (and thereby increase his chance of winning) or should he purchase one ticket each in two different lotteries (thereby give himself some chance of winning more money)? If X denote the random variable for the first situation and Y for the second situation, it is easy to see that Y is actually a sum of two independent r.v.s Y_1 and Y_2, each of which has expectation $100 \times \frac{1}{100} + 0 \times \frac{1}{100} = 1$ and hence $E(Y) = 1 + 1 = 2$. The sample space in the first situation consists of (unordered) pairs of tickets, 50×99 in number, of which only 99 pairs contain the lucky ticket. Hence, even in this case, the expectation is $E(X) = 100 \times \frac{99}{50 \times 99} = 2$. Thus the expectations are the same in both the situations. But $E(X^2) = 100^2 \times \frac{99}{50 \times 99} = 200$ so that $V(X) = 200 - 2^2 = 196$. A similar calculation for Y_i shows that $E({Y_i}^2) = 100$ so that $V(Y_i) = 99$ and hence $V(Y) = 198$ showing that Y has a larger variance. Thus the gains are more spread around the mean for the second situation than the first. Which of the two situations is better can not really be answered by mathematics alone. An adventurous person will probably adopt the second policy, trying to get a deviation from the mean while a conservative approach would be to go for the first situation not allowing the outcomes to deviate too heavily from the mean.

6.2 Chebyshev inequality

If a fair coin is tossed independently 100 times, and the number of heads recorded is α, then how far would $\frac{\alpha}{100}$ deviate from $\frac{1}{2}$? Can the probability of this number deviating from the mean by more than .01 be bounded (in terms of the variance of the r.v. which can be computed)? Given a sample space and a r.v. X on it, how far would the values taken by X deviate from the mean $E(X)$? These questions are answered in the following basic inequality.

Theorem 6.2.1. *(Chebyshev inequality) Let P be a probability measure on the sample space Ω and let X be a random variable with mean μ and variance V. Let a be any*

positive real real number. Then

$$Pr(|\, X - \mu \,| \geq a) \leq \frac{V}{a^2}$$

Proof Let A be the event representing the L.H.S. That is, A is the set of all the outcomes ω for which the random variable on the L.H.S. takes value greater than or equal to a. Thus A consists of all those ω for which $|X(\omega) - \mu| \geq a$. Let A' denote the complement of A. We have to prove that $a^2P(A) \leq V$. Calculation of V gives

$$\begin{aligned} V &= \sum_{\omega\in\Omega}(X(\omega) - \mu)^2P(\omega) \\ &= \sum_{\omega\in A}(X(\omega) - \mu)^2P(\omega) + \sum_{\omega\in A'}(X(\omega) - \mu)^2P(\omega) \\ &\geq \sum_{\omega\in A}(X(\omega) - \mu)^2P(\omega) \end{aligned}$$

since the second sum is non-negative. By assumption, for every ω in A, we have $(X(\omega) - \mu)^2 \geq a^2$ and hence

$$\begin{aligned} V &\geq \sum_{\omega\in A} a^2P(\omega) \\ &= a^2\sum_{\omega\in A} P(\omega) \\ &= a^2P(A) \end{aligned}$$

proving the desired assertion. □

Example 6.2.2. In an examination with maximum marks 100 given to a class of 100 students, the class average was found to be 65 and the variance was 5. Then, using Chebyshev inequality (Theorem 6.2.1), we have

$$Pr(|\, X - 65 \,| \geq 5) \leq \frac{5}{5^2} = \frac{5}{25} = 0.2$$

Therefore, the probability that a randomly chosen answer script scores between 61 and 69 marks is at least 80%. As a second example, let a fair coin be tossed $10,000$ times and the number of heads be noted. Then the binomial r.v. X has mean $5,000$ and variance $2,500$. Hence the probability that the number of heads will differ from $5,000$ by ≥ 200 is at the most $\frac{2,500}{200^2}$ which is 6.25%.

Chebyshev inequality is of great theoretical importance. It is also applied in branches of combinatorics such as graph theory to obtain some very surprising "existence results". The theoretical importance of Chebyshev inequality could be seen on the basis of a couple of results that we now discuss. Chebyshev inequality paves way to these results. Let Y_n refer to the r.v. with the binomial p.d.f. $b(k; n, p)$. That is, suppose at each toss of a coin we have the probability of heads (success) given by p and the probability of tails (failures) given by q, where p and q are fixed positive real numbers with $p + q = 1$. Consider the simplest case (for convenience) say $p = q = \frac{1}{2}$ and suppose we flip an unbiased coin a large number of times say 100. How far are we sure that we obtain 50 heads? If the flips are $1,000$ do we get 500 heads? Are we more sure? Can we, in some sense, prove this? The answer to these questions, is provided in the following theorem of *Bernoulli*, for which reason the above experiment is also called *repeated Bernoulli trials.*

Theorem 6.2.3. *(Bernoulli's theorem) Let Y_n denote the binomial random variable with parameters n and p. That is, we have n trials with probability of success (heads) p at each trial where p is a fixed positive real number (between 0 and 1). Let $\epsilon > 0$ be a real number. Then*

$$\lim_{n\to\infty} Pr\left(\left|\frac{Y_n}{n} - p\right| > \epsilon\right) = 0$$

Further, the convergence is uniform in the following sense: given $\epsilon > 0$ and $\alpha > 0$, there is an n_0 (independent of p) such that for every $n \geq n_0$, we have

$$Pr\left(\left|\frac{Y_n}{n} - p\right| > \epsilon\right) < \alpha$$

Proof Let B_n be the event described by $\left|\frac{Y_n}{n} - p\right| > \epsilon$. We already saw that $E(Y_n) = np$ and $V(Y_n) = npq$. If we now let Z_n denote the r.v. $\frac{Y_n}{n}$, then $E(Z_n) = p$ and $V(Z_n) = \frac{npq}{n^2} = \frac{pq}{n}$. Hence applying Chebyshev inequality (Theorem 6.2.1) to the r.v. Z_n, we get

$$\begin{aligned} Pr(|\, Z_n - p \,| \geq \epsilon) &\leq \frac{V(Z_n)}{\epsilon^2} \\ &= \frac{pq}{n\epsilon^2} \\ &\leq \frac{1}{4n\epsilon^2} \end{aligned}$$

The assertion is now obvious since the R.H.S. goes to 0 with n independent of p. □

We now apply Bernoulli's theorem, (Theorem 6.2.3) to prove *Weierstrass approximation Theorem.* To that end recall the following definition from (real) analysis.

Definition 6.2.4. Let (f_n) be a sequence of real valued functions defined on the closed interval $[0, 1]$. Then (f_n) *converges to a function f uniformly on* $[0, 1]$ *(written $(f_n) \to f$)* if for any given $\epsilon > 0$ there is an n_0 such that for all $x \in [0, 1]$ and $\forall\, n \geq n_0$, we have $|f_n(x) - f(x)| < \epsilon$.

Definition 6.2.5. Let f be a (given) real valued function. Let n be a natural number. *The Bernstein polynomial $P_n(x)$ of degree n* is defined by

$$P_n(x) = \sum_{k=0}^{n} f\left(\frac{k}{n}\right)\binom{n}{k} x^k (1-x)^{n-k}$$

Theorem 6.2.6. *(Weierstrass approximation theorem) Let $f : [0,1] \to \mathbb{R}$ be a continuous function. Then $\exists$ a sequence $(P_n(x))$ (where $P_n(x)$ is a polynomial of degree n) such that $(P_n(x)) \to f(x)$ uniformly on $[0,1]$.*

Proof Since f is continuous on the closed interval $[0,1]$, it is also *uniformly continuous*. Let $\epsilon > 0$ be given. Then $\exists\, \delta > 0$ such that

$$|x_1 - x_2| < \delta \Rightarrow |f(x_1) - f(x_2)| < \epsilon$$

Also, since f is continuous on the closed interval $[0,1]$, it is bounded and has a Supremum say M. Let $Z_n = \frac{Y_n}{n}$ where the r.v. Y_n is defined by $X_1 + \cdots + X_n$ (in the set up of Bernoulli's theorem 6.2.3). Thus $f(Z_n)$ is also a r.v. and we can compute expectation of that r.v. as follows.

$$\begin{aligned} E(f(Z_n)) &= \sum_{k=0}^{n} f\left(\frac{k}{n}\right) Pr(Y_n = k) \\ &= \sum_{k=0}^{n} \binom{n}{k} f\left(\frac{k}{n}\right) p^k (1-p)^{n-k} \end{aligned}$$

where the second equality follows from the fact that Y_n has the binomial distribution $b(k; n, p)$. Using Bernoulli's theorem (Theorem 6.2.3), $\exists\, n_0$ such that for all $n \geq n_0$, we have

$$Pr\left(\left|\frac{Y_n}{n} - p\right| \geq p\right) < \epsilon$$

Let $n \geq n_0$. Then

$$\begin{aligned} \left|E\left(f\left(\frac{Y_n}{n} - f(p)\right)\right)\right| &= \left|\sum_{k=0}^{n} \left(f\left(\frac{k}{n}\right) - f(p)\right) Pr(Y_n = k)\right| \\ &\leq \sum_{|\frac{k}{n} - p| < \delta} \left|f\left(\frac{k}{n}\right) - f(p)\right| Pr(Y_n = k) \\ &\quad + \sum_{|\frac{k}{n} - p| \geq \delta} \left|f\left(\frac{k}{n}\right) - f(p)\right| Pr(Y_n = k) \\ &= A_1 + A_2 \end{aligned}$$

say, where A_1 refers to the first sum and A_2 to the second sum. Then using uniform continuity of f, we see that $A_1 < \epsilon$. Consider A_2. We have

$$\begin{aligned} A_2 &\leq 2M[\sum_{|\frac{k}{n} - p| \geq \delta} Pr(Y_n = k)] \\ &= 2M\left[\sum_{|\frac{k}{n} - p| \geq \delta} Pr\left(Z_n = \frac{k}{n}\right)\right] \\ &= 2M \times Pr(|Z_n - p| \geq \delta) \\ &\leq 2M\epsilon \end{aligned}$$

where the last inequality follows using Bernoulli's theorem (Theorem 6.2.3). So,

$$E\left(f\left(\frac{Y_n}{n}\right) - f(p)\right) < \epsilon(1 + 2M)$$

holds for every $n \geq n_0$. Replacing p by x we get

$$E\left(f\left(\frac{Y_n}{n}\right)\right) = \sum_{k=0}^{n} \binom{n}{k} f\left(\frac{k}{n}\right) x^k (1-x)^{n-k} = P_n(x)$$

Hence $\forall\, n \geq n_0$, we have

$$|P_n(x) - f(x)| < \epsilon(1 + 2M)$$

showing that $(P_n(x)) \to f(x)$ uniformly on $[0, 1]$. □

A subtle point to be noted about the proof of Bernoulli's theorem (Theorem 6.2.3) is that B_n is an event in the sample space Ω_n of all the sequences (of heads and tails) of length n. Thus different B_n's are events in different sample spaces. Also, though all the sample spaces are finite, they are becoming very large with increasing n. We will indicate a generalization of Bernoulli's theorem. The setting is of the following kind. We repeat an experiment a large number of times. Each trial is a random variable say, the i-th trial is the r.v. X_i. The set-up is so made that X_i can be considered as an r.v. on the entire sample space (of all the trials) though X_i concentrates only on the i-th trial and is indifferent to the outcome of the other trials. We thus assume that the result of X_i has nothing to do with the result of X_j for different i and j. However, we assume that all the X_i are otherwise identical in every sense. Thus all the X_i's have the same mean μ and variance σ^2. Such a set of random variables is called *independently and identically distributed random variables* (i.i.d. r.v.s) in the statistical parlance. Then the assertion analogous to Bernoulli's theorem should hold for the r.v. $X_1 + X_2 + \cdots + X_n$ (in a limiting situation). This assertion is called *the weak law of large numbers* and its discussion is beyond our scope.

6.3 Some more discrete r.v.s

Our consideration in Bernoulli's theorem (Theorem 6.2.3) and the weak law of large numbers following it) has been a sequence of sample spaces that are finite but become very large. There are also other considerations that compel us to look at sample spaces with infinitely many points. In the situation of the binomial distribution, we are interested in a new random variable X whose task is just to find out when the first success occurred (and stop flipping the coin at that stage). That is X takes value k if the first success occurs at the k-th toss of the coin. Then the sample space we look at can not be finite since X may take any natural number k as its value. A second reason for looking at infinite sample spaces is computational. In a way, it is easier to compute exponentials than factorials or binomial coefficients. For example, in the binomial distribution $b(k; n, p) = \binom{n}{k} p^k q^{n-k}$, even for moderately large values of n, the computation can be very tedious. It would be better to replace the above computation by a quantity which is easier to compute as well as represents the situation sufficiently accurately in terms of approximation. There are three different r.v.s on *countably infinite* spaces that we consider here.

Example 6.3.1. The *geometric r.v.* is a r.v. X whose range is the set of all natural numbers with distribution given by:

$$Pr(X = k) = G(k; p) = q^{k-1}p$$

In the (independent) Bernoulli trials, we ask the question as to when the first success appears. We have $Pr(X = 1) = p$, $Pr(X = 2) = qp$ and $Pr(X = 3) = q^2p$ and so on. Let us write $G(k; p)$ for the p.d.f. That is $G(k; p) = Pr(X = k)$. To find $G(k; p)$, observe that we must have failures at the first $k - 1$ trials, the probability of which is q^{k-1} and a success at the k-th trial, the probability of which is p. Hence, we have $G(k; p) = q^{k-1}p$. This distribution is called the *Geometric distribution* because of its connection with the obvious geometric series. Note that $G(k; p)$ is a "waiting time distribution" in the sense that it gives us the probability that we have to wait upto k trials before the occurrence of a success. Since $\sum_{k=1}^{\infty} pq^{k-1} = 1$, we confirm that things are in order.

Proposition 6.3.2. *The mean and variance of the geometric distribution $G(k; p)$ are given by $\mu = \frac{1}{p}$ and $\sigma^2 = \frac{q}{p^2}$.*

Proof We prove the expression for the mean leaving the other computation to the reader. Writing $\frac{d}{dq}$ to denote the derivative, we have

$$\begin{aligned}
\mu &= \sum_{k \geq 1} kq^{k-1}p \\
&= p\frac{d}{dq}(\sum_{k \geq 0} q^k) \\
&= p\frac{d}{dq}(1-q)^{-1} \\
&= \frac{p}{(1-q)^2} \\
&= \frac{1}{p}
\end{aligned}$$

□

We can obtain a generalization of the geometric distribution by asking the probability of the r-th success where $r \geq 1$ is fixed. Let us denote this probability distribution by $G(k; r, p)$. That is, $G(k; r, p)$ is the probability that r-th success occurs on the k-th trial. Then $(r - 1)$ successes must have occurred in the first $k - 1$ trials and the k-th trial is also a success. So, we have the following.

Definition 6.3.3. The *negative binomial r.v.* X has natural numbers ($\geq r$) as its range and its distribution is given by

$$G(k; r, p) = \binom{k-1}{r-1} q^{k-r} p^r$$

Writing $m = k - r$, the R.H.S. is equal to $\binom{m+r-1}{r-1} q^m p^r$. Since

$$\binom{m+r-1}{r-1} = \binom{m+r-1}{m} = (-1)^m \binom{-r}{m}$$

we have

$$G(m+r; r, p) = p^r (-1)^m \binom{-r}{m} q^m$$

From this, it follows that

$$\begin{aligned} \sum_{m=0}^{\infty} G(m+r; r, p) &= p^r (1-q)^{-r} \\ &= p^r p^{-r} = 1 \end{aligned}$$

so that the probability setting is in order. The proof of the following makes it clear as to why the distribution is called the negative binomial distribution.

Proposition 6.3.4. *Let X be a r.v. with the negative binomial distribution. Then the mean of X is $\frac{r}{p}$.*

Proof We have

$$\begin{aligned} \sum_{m=0}^{\infty} (m+r) G(m+r; r, p) &= p^r \sum_{m=0}^{\infty} (m+r) \binom{m+r-1}{r-1} q^m \\ &= rp' \sum_{m=0}^{\infty} \binom{m+r}{r} q^m \\ &= rp^r \sum_{m=0}^{\infty} \binom{-(r+1)}{m} (-q)^m \\ &= rp^r (1-q)^{-(r+1)} \\ &= \frac{r}{p} \end{aligned}$$

□

Since the geometric distribution is a special case of the negative binomial distribution with $r = 1$, we obtain the mean of the geometric distribution as a special case. A slightly tedious computation along the same lines can be done to find the variance. We leave its derivation as an exercise. Alternatively, we may view the negative binomial r.v. X as a sum of r.v.s X_i's each with a geometric distribution. How does this work?

After the first success (this is given by X_1) occurring say at n_1, we may suppose that the second success occurs at some trial numbered $n_1 + n_2$ where n_2 is some positive integer. Then the third success occurs at $n_1 + n_2 + n_3$ where n_3 is some positive integer and so on. Once we see things that way, $X = X_1 + X_2 + \cdots + X_r$ where each X_i has a geometric distribution $G(k;p)$ and X_i's are independent. Hence the earlier calculation of the mean of X_i tells us that the mean of the negative binomial r.v. is $\frac{r}{p}$ and the variance is $\frac{rq}{p^2}$ since the variance of the geometric r.v. is $\frac{q}{p^2}$.

The experiment with repeated Bernoulli trials has two parameters n and p. Moreover, even for moderately large values of n and k, the computation of $b(k;n,p)$ is difficult due to the fact that factorial expressions are difficult to compute. It is, therefore, desirable that we should have a distribution which works as good as a large number of repeated Bernoulli trials and also reasonably reflects the reality! Both these provisions are made in an important distribution called the *Poisson distribution*. For this distribution, we have a (fixed) positive real number λ as a parameter. Here, the random variable X takes non-negative integers as values.

Definition 6.3.5. *(Poisson distribution)* Let X be a r.v. X is said to have the *Poisson distribution* if X takes non-negative integers as values and

$$Pr(X = k) = P(k;\lambda) = \frac{\lambda^k}{k!}e^{-\lambda}$$

Since

$$\begin{aligned}\sum_{k=0}^{\infty} P(k;\lambda) &= \sum_{k=0}^{\infty} \frac{\lambda^k}{k!}e^{-\lambda} \\ &= \left(\sum_{k=0}^{\infty} \frac{\lambda^k}{k!}\right)e^{-\lambda} \\ &= 1\end{aligned}$$

the setting is in order.

We now show that in a limiting sense the binomial distribution can be approximated by the Poisson distribution.

Theorem 6.3.6. *(Poisson limit theorem) Assume that, in the binomial distribution, p is small while n is large (and tends to infinity), though the product $np = \lambda$ is held fixed. Then*

$$\lim_{n\to\infty} b(k;n,p) = P(k;\lambda)$$

Proof We have

$$\begin{aligned}b(k;n,p) &= \binom{n}{k}p^k(1-p)^{n-k} \\ &= \frac{n(n-1)\cdots(n-k+1)}{k!}\frac{\lambda^k}{n^k}\frac{(1-p)^n}{(1-p)^k} \\ &= \frac{\lambda^k}{k!}\frac{n(n-1)\cdots(n-k+1)}{n^k}(1-\frac{\lambda}{n})^{-k}\left(1-\frac{\lambda}{n}\right)^n\end{aligned}$$

The R.H.S. has the form $\frac{\lambda^k}{k!}ABC$, where

$$A = \left(1 - \frac{1}{n}\right)\left(1 - \frac{2}{n}\right)\cdots\left(1 - \frac{k-1}{n}\right),$$

$$B = \left(1 - \frac{\lambda}{n}\right)^{-k},$$

$$C = \left(1 - \frac{\lambda}{n}\right)^{n}$$

By assumption, k is fixed and so is λ while $n \to \infty$. So, $A \approx 1$, $B \approx 1$ and $C \to e^{-\lambda}$. Hence the limiting value of the R.H.S. is $\frac{\lambda^k}{k!}e^{-\lambda}$ which is the probability of obtaining k successes. Thus the limiting value of the R.H.S. is $P(k; \lambda)$. □

Similar arguments are used to prove the "Central limit theorems" of which Poisson limit theorem 6.3.6 a one special case. Essentially, a central limit theorem tells us that under nice assumptions limiting probability distribution functions take nice forms. The Poisson distribution is important both from the theoretical and practical point of view. As an example, suppose that, at an average, there are $10,000$ vehicles passing over a given road in a day and at an average, 5 accidents are reported per day. Suppose we observe $4,000$ vehicles passing on the road (in a day). Then the probability set up has $p = \frac{5}{10,000}$ while the number of observations n is $4,000$. If we now need to find the probability that k accidents occur, it is going to be a difficult task, if we use the binomial distribution. Instead, we use the Poisson distribution with $\lambda = 4,000 \times p = 2$. Hence, the probability that k accidents occur on any given day is $P(k; 2) = \frac{2^k}{k!}e^{-2}$. An r.v. which is a sum of (several) Poisson r.v.s.is called a "Poisson process" and understanding it is very important to the study of "Stochastic Processes". For all this, we refer to the book by Chung [17]. We wind up the discussion with the computation of the mean and variance of the Poisson distribution.

Theorem 6.3.7. *Let X be the Poisson r.v. with parameter λ. Then X has both the mean and variance equal to λ.*

Proof We have

$$\begin{aligned} E(X) &= \sum_{k=0}^{\infty} k\frac{\lambda^k}{k!}e^{-\lambda} \\ &= \lambda e^{-\lambda}\sum_{k=1}^{\infty}\frac{\lambda^{k-1}}{(k-1)!} \\ &= \lambda e^{-\lambda}e^{\lambda} \\ &= \lambda \end{aligned}$$

Similarly,

$$\begin{aligned} E(X^2) &= \sum_{k=0}^{\infty} k^2 \frac{\lambda^k}{k!} e^{-\lambda} \\ &= e^{-\lambda} \left\{ \sum_{k=1}^{\infty} \frac{k(k-1)}{k!} + \sum_{k=1}^{\infty} k \frac{\lambda^k}{k!} \right\} \\ &= e^{-\lambda}(\lambda^2 e^{\lambda} + \lambda e^{\lambda}) \\ &= \lambda^2 + \lambda \end{aligned}$$

Therefore,

$$\begin{aligned} V(X) &= \lambda^2 + \lambda - (E(X))^2 \\ &= \lambda^2 + \lambda - \lambda^2 \\ &= \lambda \end{aligned}$$

□

6.4 Random walk on a line and gambler's ruin

Definition 6.4.1. By *a random walk* we mean the following situation. Fix positive real numbers p and q such that $p+q=1$. A person A occupies position at any given time t on the elements of the set $\mathbb{Z}$. If, at a given point of time t the person A is at position j, then he tosses a coin with probability of heads equal to p and probability of tails equal to q. At time $t+1$, he moves to $j+1$ if the coin shows heads and moves to $j-1$ if the coin shows tails. Thus, at time $t+1$, he moves to $j+1$ with probability p and to $j-1$ with probability q.

Example 6.4.2. Suppose A is at 0 at time 0 and the coin successively shows $HHTTTHHTHTT$. Then the man will be at the positions

$$1, 2, 1, 0, -1, 0, 1, 0, 1, 0, -1$$

in that order. If A is at 4 and if the coin successively shows $HTTTT$, then he will be at the positions $5, 4, 3, 2, 1$ in that order.

Theorem 6.4.3. *Consider the following restricted form of random walk. Let n be a fixed positive integer. A person A begins at some integer point in the closed interval $[0, n]$ and performs a random walk and moves to his right (by one unit) with probability p and moves to his left (by one unit) with probability q. The random walk stops when he hits either the left or right wall of the interval, that is, the random walk stops when he reaches 0 or n. Then, independent of the position j (in the interval) where A starts the random walk, the probability that the A will reach either of the two ends is one.*

Proof (Chung [17]) Let u_j denote the probability that starting at j the person reaches the end 0 (before reaching the right end n). From the j-th position, the movement of A can be either to his right with probability p or to his left with probability q. In the former case, the probability of his hitting the left end is u_{j+1} while in the latter case, the probability of his hitting the left end is u_{j-1}. Therefore, using the theorem on total probabilities (Theorem 5.4.1), we have

$$u_j = pu_{j+1} + qu_{j-1} \ \forall\, j = 1, 2, \ldots, n-1$$

subject to $u_0 = 1$ and $u_n = 0$. Since $1 = p+q$, we get $p(u_j - u_{j+1}) = q(u_{j-1} - u_j)$. Write $u_j - u_{j+1} = \alpha_j$ for j from 0 to $n-1$ and let $r = \frac{q}{p}$. Then we get $\alpha_j = r\alpha_{j-1}$ and iteration gives $\alpha_j = r^j \alpha_0$. Now write $\beta_j = \sum_{i=0}^{j} \alpha_i$. Then, for all j from 1 to n, we have $\beta_j = u_0 - u_j = 1 - u_j$ and, in particular, we have $\beta_n = 1$. On the other hand, we also have

$$\beta_j = \left(\sum_{i=0}^{j} r^i \right) \alpha_0 = \frac{1 - r^j}{1 - r} \alpha_0$$

and $1 = \beta_n = \frac{1-r^n}{1-r}\alpha_0$ giving $\alpha_0 = \frac{1-r}{1-r^n}$. Therefore, $\beta_j = \frac{1-r^j}{1-r^n}$ which gives

$$u_j = 1 - \frac{1 - r^j}{1 - r^n} = \frac{r^j - r^n}{1 - r^n}$$

Now let v_j denote the probability that starting at position j the man reaches the right end n. Let s denote $r^{-1} = \frac{p}{q}$. Then, by symmetry, the formula for v_j can be obtained from that of u_j as follows. We replace r by s and j by $n-j$. Thus we have

$$v_j = \frac{s^{n-j} - s^n}{1 - s^n} = \frac{s^{-j} - 1}{r^n - 1} = \frac{1 - r^j}{1 - r^n}$$

The proof is completed by observing that $u_j + v_j = 1$. □

Remark 6.4.4. The area of probability theory that we have touched upon is discrete probability and have not dealt with continuous distributions. The most important example of a continuous distribution is the *normal distribution* (with its bell shaped curve). In defining a probability measure on an uncountable set such as $\mathbb{R}$ one can not be as simplistic as we have been in the finite set up. Here, the probabilities have to be assigned to only some (measurable) subsets of the real line. These are called Borel sets. In particular, the probability of the occurrence of a singleton (or any finite set) has to be assigned the value 0. It is expected that an interested reader will take up looking at this serious aspect of probability theory. We refer the interested reader to [9].

6.5 Exercises for Chapter 6

6.1 On an average, the percentage of left-handers in a random sample is one. Use Poisson approximation to the binomial distribution to find the probability of having at least four left- handers among 200 people.

6.2 Two people toss an unbiased coin n times and independently. What is the probability they get the same number of heads?

6.3 Two people independently toss a fair coin 11 times each. A wins if he has more heads than those obtained by B. What is probability that A wins the game?

6.4 For two independent throws of a balanced die, find the expectation of the random variable X whose value is the first score minus the second score.

6.5 For two independent throws of a balanced die, find the expectations of the following random variables:

(a) The score on the first throw.

(b) The total score.

(c) The total number of fives and sixes.

6.6 Let N be a set of order n and assume equally likely probability measure on the sample space Ω consisting of all the subsets of N. Suppose X is a random variable defined on the sample space $\Omega \times \Omega$ and given by

$$\begin{aligned} X(A, B) &= 1, \; if \;\; A \subset B, \;\; and \\ &= 0, \; otherwise \end{aligned}$$

Show that the $E(X) = (\frac{3}{4})^n$.

6.7 Use the Poisson distribution to calculate the probability that a box of 100 fuses has at the most 2 defective fuses if 3 percent of the fuses are known to be defective.

6.8 Suppose n fair dice are rolled. Let X denote the random variable which counts the sum of the numbers shown on the n faces. Find the mean and variance of X.

6.9 By using Theorem 3.1.8 give an alternative proof of Theorem 6.1.10.

6.10 Prove the *Markov inequality* : Let X be a r.v. whose range is contained in the set of all non-negative real numbers. Let $a > 0$ be a real number. Then

$$Pr(X \geq a) \leq \frac{E(X)}{a}$$

6.11 Let X be a r.v. whose range is the set of all non-negative integers.

(a) Suppose X has both the mean and variance equal to 1. Show that for $k = 3, 4, \ldots, n$ we have

$$Pr(X \geq k) \leq \frac{1}{(k-1)^2}$$

(b) Let X denote the r.v. on the set of all the $n!$ permutations of $[n]$ defined by $X(\alpha)$ equal to the number of fixed points of α. Let X_i denote the r.v. on the same set that takes value 1 when α fixes i and X_i equals 0 otherwise. Find the mean and variance of X_i and hence find the mean and variance of X and thus conclude that a randomly chosen permutation has 1 fixed point.

(c) Use this to show that for a randomly chosen permutation on n letters with $n \geq 11$, the probability that the permutation has at least 11 fixed points is no more than one percent!

6.12 An urn contains 8 balls of which three are red and five are black. Four balls are drawn from the urn without replacement. Let X denote the r.v. that counts the number of red balls and Y the r.v. that counts the number of black balls. Find the probability distributions of X and Y. Then find the joint the probability distribution of these r.v.'s.

6.13 Let X and Y be two r.v.'s. Show that X and Y are independent if and only if X and rY are independent where r is a non-zero real number.

6.14 Suppose the sample space consists of all the 13-card hands from a pack of 52 cards. Assume equally likely probability measure. Consider the following four r.v.s and find out if any pair among them is dependent or independent.

(a) X is the number of heart cards drawn.

(b) Y is the number of red cards drawn.

(c) Z is the number of kings drawn.

(d) W is the number of black cards drawn.

6.15 On any sample space Ω, consider the r.v. I which takes value 1 at all the points of Ω. Let X be any r.v. Show that X and I are independent.

6.16 For the binomial r.v. with distribution $b(k; n, p)$, find the the value of k for which $b(k; n, p)$ is the largest. Use this to show that the binomial coefficient $\binom{n}{k}$ is the largest for $k = [\frac{n}{2}]$.

6.17 Suppose we have $n + m$ Bernoulli trials with a common probability of success p. Let X be the r.v. that counts the number of successes in the first n trials and Y the r.v. that counts the number of successes in the last m trials. Show that X and Y are independent. Find the probability distribution of $X + Y$ and use this to show that

$$\sum_{k=0}^{n} b(k; n, p) b(r - k; m, p) = b(r; n + m, p)$$

Put $p = \frac{1}{2}$ to obtain a combinatorial identity.

6.18 (continuation) Let X be a random variable which has value 1 if at the most k trials (out of n) result in success and X takes value 0 otherwise. Show that

$$B(k;n,p) = f_X(1) = \sum_{r=0}^{k} b(r;n,p)$$

Further, show that

$$B(k;n+1,p) = b(k;n,p) - pb(k;n,p)' \text{ and}$$

$$B(k+1;n+1,p) = b(k;n,p) + qb(k+1;n,p)$$

6.19 (continuation) Use the fundamental theorem of calculus to show that

$$B(k;n,p) = (n-k)\binom{n}{k}\int_0^q t^{n-k-1}(1-t)^k dt.$$

6.20 Consider a random sample of 13 cards out of a pack of 52 cards. Find the expectations and variances of each of the following random variables. (Hint: Write the given r.v. as a sum of more elementary r.v.s whose expectations and variances can be found easily.)

(a) The number of aces drawn.

(b) The number of black cards drawn

(c) The number of hearts drawn.

(d) The number of heart kings drawn.

(e) The number of black cards minus the number of red cards. drawn.

6.21 A game by two persons is called a fair game if the expectation of returns is zero for both the players. Suppose in tossing an unbiased die, player A gives player B Rs. 12 if the score on the die is less than or equal to 2. How much should player B give player A if the score is greater than 2 if we want to have a fair game?

6.22 Show that for real numbers r and s, and for r.v. s X and Y, we have

$$Cov(rX, sY) = rsCov(X,Y) \text{ and}$$

$$Cov(X+r, Y+s) = Cov(X,Y)$$

6.23 Show that $V(X - c)$ is minimized if $c = \mu = E(X)$. A random variable is said to be standardized if the mean is zero and the standard deviation is 1. Show that if σ is the standard deviation of the r.v. X, then the r.v. $Y = \frac{X-\mu}{\sigma}$ is standardized.

6.24 If a machine produces defective items with a probability of .01, what is the expected number of defective items in a random sample of size $1,000$? What is the variance of the number of defective items?

6.25 Consider the r.v. X with the hypergeometric distribution $H(r; n, k, N)$. Show that the mean and variance of this r.v. are given by

$$E(X) = \frac{nk}{N},$$
$$V(X) = \frac{nk(N-k)}{N^2}\frac{N-n}{N-1}$$

6.26 It is observed that a student entering a college chooses commerce with probability $\frac{2}{5}$ and other majors (arts or science) with probability $\frac{3}{5}$. What is the approximate probability that among 500 students seeking admission to the college, between 175 and 225 will opt for commerce?

6.27 A fair coin is tossed 400 times. Determine a number x such that the probability that the number of heads is between $200 - x$ and $200 + x$ is approximately 0.85.

6.28 If the probability that a person lives beyond the age of 90 is 0.02, what is the approximate probability (use Poisson approximation) that among a sample of 300 mathematics teachers at the most two live past the age of 90?

6.29 A fair coin is tossed n times. Let X be the r.v. that counts the difference between the number of heads and the number of tails. Find the mean and expectation of X.

6.30 n bins are initially empty and balls are thrown at random in the bins with each bin equally likely. What is the expected number of throws for any one bin to contain a ball? What is the expected number of throws for any bin to contain two balls?

6.31 A sequence (x_n) of real numbers is said to converge to some x in Euler sense (*called Euler convergent to* x) if $\exists\ s \in (0, 1)$ such that

$$\lim_{n\to\infty} \sum_{j=0}^{n} \binom{n}{j} s^j (1-s)^{n-j} x_j = x$$

(a) Show that the sequence $(0, 1, 0, 1, 0, \ldots)$ is Euler convergent to $\frac{1}{2}$ (compare this with Bernoulli trials).

(b) Show that the sequence $(1, -1, 1, -1, \ldots)$ is Euler convergent to zero.

These examples show that Euler convergence does not imply convergence.

6.32 (continuation) We wish to prove that if $(x_n) \to x$, then (x_n) is also Euler convergent to x. Proceed through the following steps.

(a) Show that this statement is equivalent to: if (y_n) converges to 0, then it is also Euler convergent to 0.

(b) (y_n) is convergent to 0, and hence given $\epsilon > 0$, there is a M such that $|y_k| < \epsilon\ \forall\ k > M$.

(c) Let λ denote the maximum of $|y_k|$ (over all k). Let $n > M$. Then

$$\left|\sum_{j=0}^{n}\binom{n}{j}s^j(1-s)^{n-j}y_j\right| < \lambda\sum_{j=0}^{M}\binom{n}{j}s^j(1-s)^{n-j}+\epsilon$$

(d) We wish to make the first sum on the R.H.S. very small. For a given ϵ, M and $s > \epsilon$, we choose n so large that $n > \frac{M}{s-\epsilon}$. Then $ns - j \geq n\epsilon \ \forall \ j = 0, 1, \ldots, M$.

(e) Now consider the binomial r.v. X with probability of success s where X counts the number of heads. Then show that

$$\sum_{j=0}^{M}\binom{n}{j}s^j(1-s)^{n-j} = Pr(X \leq M) \leq Pr(|X - ns| \geq n\epsilon)$$

(f) Now use Chebyshev inequality (Theorem 6.2.1) to conclude that

$$\sum_{j=0}^{M}\binom{n}{j}s^j(1-s)^{n-j} \leq \frac{ns(1-s)}{n^2\epsilon^2} \leq \frac{1}{4n\epsilon^2}$$

and hence by choosing n very large, we can indeed bound this expression.

(g) Conclude that convergence implies Euler convergence.

6.33 *St. Petersburg paradox* Let X denote a geometric r.v. defined as follows. A tosses a fair coin repeatedly he stops tossing the coin when the first success appears. If the the first success is on the k-th toss then A receives 2^k rupees. Is it possible to define the expectation of this r.v.?

6.34 *Gambler's ruin* Two player's Peter and Paul play the following simple gambling game. To start with Peter has a rupees with him and Paul has b rupees with him. A coin (with probability of heads equal to p and that of tails equal to q) is tossed. If it shows heads then Paul gives Peter one rupee and if it shows tails, then Peter gives Paul one rupee. The game stops when one of the player becomes bankrupt (is left with no money). Show that the probability that the game will stop (irrespective of what a and b are) is one.

Chapter 7

Parity

7.1 Introduction

A man living alone has only one front door to his apartment opening to the walkway that leads to the road outside. He is very particular about opening (and immediately closing) the front door only when he enters or leaves his apartment but his job permits him to visit his apartment any number of times throughout the day. Merely using this information, we can conclude that independent of how many times he made trips to the outside world throughout the day, if the man starts from home in the morning and returns at night (including the possibility of his not having ventured out at all), the number of times the door was opened must be an even number. Arguments that use ideas of this kind are called *parity based arguments.* We begin our tour into the applications of parity based arguments by first studying the following examples.

Example 7.1.1. Let $x_1, x_2, \ldots, x_n$ be n numbers, each equal to ± 1. Consider the sum

$$x_1x_2x_3x_4 + x_2x_3x_4x_5 + \cdots + x_{n-1}x_nx_1x_2 + x_nx_1x_2x_3$$

where we consider four consecutive terms at a time and move in a cyclic manner. Suppose the above sum is zero. We wish to prove that n is a multiple of 4.

For a solution, denote the above kind of sum by S and suppose we change the sign of some x_i. This will change the signs of exactly four summands that involve x_i. Hence some a summands change from positive to negative and b summands change from negative to positive so that S changes to $S + 2(b - a)$. Since $b + a$ equals 4 and since $b - a$ has the same parity (mod 2) as $b + a$, it follows that the new sum S' is also a multiple of 4 (S was a multiple of 4). We can now change the sign of each x_i that equals -1 one by one. This way, we transform the given sum through a sequence S', S'' etc. so that each such sum is a multiple of 4. The final sum say T is also then, a multiple of 4. But the sum T has each summand positive so that T must equal n. Hence n is a multiple of 4.

Example 7.1.2. Consider an $n \times n$ chessboard. A *ruling* is a horizontal or vertical line drawn parallel to the edges of the board that is used to divide the board into n^2 cells (or squares). We thus have $n-1$ horizontal and $n-1$ vertical rulings (the edges are not considered rulings). A domino is a 2×1 piece that occupies either two horizontal squares (called a horizontal domino) or two vertical squares (called a vertical domino) as shown in the following figure.

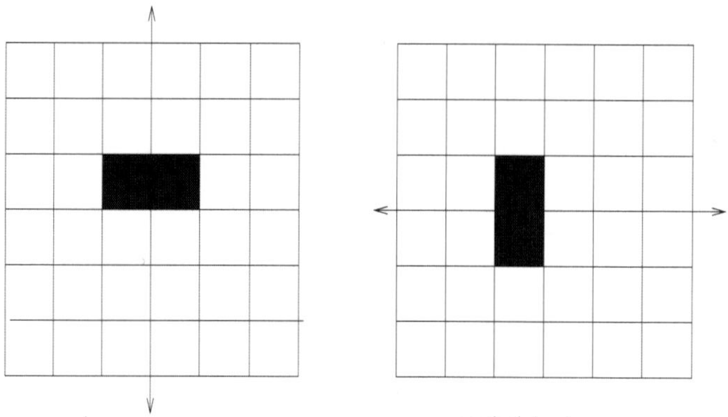

Figure 7.1: Domino and the ruling it cuts

A covering of a chessboard by dominoes is an arrangement of dominoes on the chessboard so that no two dominoes are overlapping and all the squares are covered. Evidently, n must be an even number for us to have a covering of the $n \times n$ chessboard. Even for a moderately large even numbers n, the total number of coverings is very large. For $n = 8$ this number is $2^4 \times (901)^2 = 12,966,816$. We wish to consider $n = 6$ and say that a domino crosses a ruling if the straight line dividing the domino into two 1×1 pieces is aligned with the ruling. Clearly, a horizontal domino crosses a vertical ruling and a vertical domino crosses a horizontal ruling as shown in Figure 7.1. We wish to prove the following: *In any covering of the* 6×6 *chessboard by dominoes, there will be some ruling that is not crossed by any domino.*

A solution is as follows. If a domino is horizontal then the vertical ruling it crosses divides the chessboard into two smaller chessboards. If there is no other domino that cuts the same ruling (that is a domino placed parallel to the given domino but in a different row), then the remaining dominoes must cover the two chessboards separately. But these two chessboards have odd number of cells each (because one cell is lost to the given domino) and hence cannot be covered by dominoes. We thus see that if a ruling is crossed by a domino, then there is one more domino that crosses the same ruling. Hence a ruling is either not crossed at all or is crossed by at least two dominoes. Since the number of rulings is 10, if each ruling is to be crossed by some domino, we would require as many as $2 \times 10 = 20$ dominoes but we actually have only $36/2 = 18$ dominoes proving that there is a ruling not crossed by any domino.

7.2 Parity in graph theory

We now recall some terminology (Example 1.1.1) from graph theory that we introduced in Chapter 1 and make some more definitions.

Definition 7.2.1. Let G be a graph. A *walk* W from a vertex x to a vertex y is a sequence $v_0e_1v_1e_2\cdots v_{i-1}e_iv_i\cdots e_mv_m$ where v_j's are vertices and e_j's are edges and the edge e_i is the edge $(v_{i-1}v_i)$ and $v_0 = x$ and $v_m = y$. We call such a walk W *an* (x, y)*-walk. The length of* W is the number of edges m in W. A walk W in which edges are not repeated (no edge occurs more than once but the same vertex may occur number of times) is called a *tour.* A *path* is a tour in which *all* the vertices are distinct. A tour $W : x_0e_1x_1\cdots x_{m-1}e_mx_m$ (of length m) in which the first and the last vertex are equal, i.e. $x_m = x_0$ is called a *circuit* (or a closed tour). A circuit in which all the vertices (except the end vertices) are distinct is called a *cycle of length* m if there are m edges. We also call this an m*-cycle.* Thus a 1*-cycle* is a *loop* and a 2*-cycle* is a pair of multiple edges xe_1ye_2x as shown in Figure 7.2. Notice that an m-cycle in a simple graph must have $m \geq 3$. A graph is called a *connected graph* if there is a walk joining every two distinct vertices x and y.

If W is an (x, y)-walk as above and if for some $i < j$, we have $v_i = v_j$, then we may remove the portion between the two occurrences of $v = v_i = v_j$ to get a shorter (x, y)-walk W'. Continuing in this manner, we can remove all the repetitions of vertices and obtain an (x, y)-path. Since every path is a walk, we may define a graph G to be connected if we have (x, y)-path between every two distinct vertices x and y. For distinct vertices x and y that have a path joining them, the distance $d(x, y)$ is the length of a shortest path that joins x to y. The two examples in Figure 7.3 illustrate simple graphs.
In the first graph in Figure 7.3, $v_1v_2v_4v_5$ and $v_1v_2v_3v_7v_{12}v_5$ are both paths that connect the vertices v_1 and v_5. "Being connected by a path" is both symmetric and transitive relation. Assuming reflexivity, this forms an equivalence relation, the equivalence classes under which are called the *components* or *connected components.* For example, the subgraphs G_1 on the vertex set $\{w_1, w_2, w_3, w_4\}$ and the subgraph G_2 on the vertex set $\{w_5, w_6, w_7, w_8, w_9\}$ form two connected components of the second graph in Figure 7.3. Thus a graph is called *connected* if it has only one component. Equivalently, this just means that between any pair of distinct vertices, we have a path (connecting them). As an example, the first graph in Figure 7.3 is connected while the second is not. In the first graph, $(v_2v_3v_7v_{12}v_5v_4v_2)$ as well as $(v_7v_8v_9v_{12}v_7)$ are cycles. Similarly, in the second graph, $(w_5w_6w_7w_5)$ is a cycle.

Definition 7.2.2. A connected simple graph without cycles is called a *tree.* A *leaf* in a simple graph is a vertex of degree one.

Lemma 7.2.3. *Let* $T = (V, E)$ *be a tree such such that* T *has* $|V| = n \geq 2$ *vertices and* $|E| = m$ *edges. Then the following assertions hold.*

(a) *Given any two distinct vertices, there is a unique path joining them in* T.

(b) T *has at least two leaves.*

(c) $m = n - 1$.

Proof For any two vertices x and y there is a path joining x to y since T is connected and such a path is unique (for, if not, we get a closed walk and hence a cycle that contains x and y, a contradiction). This proves (a). In the given tree T let x and y be two vertices for which $d(x, y)$ is the largest. Let the path joining x and y be $x_0x_1 \cdots x_d$ where $x = x_0$ and $y = x_d$. Suppose that, besides x_{d-1}, y is also adjacent to some other vertex z. Since the given path is the *only* (x, y)-path it follows that such a z cannot equal x_i for $i < d - 1$. If z is not on the (x, y)-path then the uniqueness of the path joining two vertices tells us that we have an (x, z)-path $x_0x_1 \cdots x_{d-1}yz$ and hence $d(x, z) = d + 1$, a contradiction. So the degree of y is 1. The same argument also shows that degree of x is 1. Finally consider (c). This is easily seen to hold for $n = 2$. Let $n \geq 3$. By the second assertion, we have a leaf say x in T and suppose x is adjacent to v. Then deleting the edge (xv) leaves us with a tree with $n' = n - 1$ and $m' = m - 1$. By induction $m' = n' - 1$. □

Some special classes of simple graphs deserve more attention. We assume that n denotes the number of vertices. Let k be an integer between 0 and $n - 1$.

Definition 7.2.4. A *k-regular* graph G is a graph all of whose vertices have degree k. Let G have n vertices. Notice that if n is odd, then k must be even. A k-regular graph with $k = n - 1$ is called the *complete graph K_n* . In this case, every pair of vertices is adjacent. A subset S of vertices is called an *independent set* if no two vertices of S are adjacent. A *bipartite graph* $G = (V, E)$ has a partition on the vertex set V in the form $V = X \cup Y$ such that X and Y are disjoint and are both independent sets. Thus we have no loops and also have no edges from any vertex of X to a vertex of X nor from any vertex of Y to a vertex of Y. A bipartite graph is called a *complete bipartite graph $K_{m,n}$* if $|X| = m$, $|Y| = n$ and E consists of all the pairs (xy) with $x \in X$ and $y \in Y$.

A bipartite graph models many real life situations. For example, if X represents the set of men and Y the set of women, then the edges can be easily interpreted to mean acquaintances between men and women. Similarly, X can be taken to represent the set of applicants and Y the set of jobs so that the edges amount to suitability of the applicants for specific jobs. Figure 7.4 illustrates complete graphs on 3, 4 and 5 vertices. The first diagram in Figure 7.5 illustrates a tree on 12 vertices and the second graph is $K_{3,3}$ which is also called the *utility graph.*

Lemma 7.2.5. *If G has a closed walk of odd length $l(W)$, then it also has an odd cycle.*

Proof This is true when $l(W) = 1$ since a closed walk of length one is just a loop and is therefore a 1-cycle. We now make induction on $l(W)$ assuming the result to be true if we have a closed walk W' with $l(W') < l(W)$ and $l(W')$ an odd number. Let $l(W)$ be an odd number ≥ 3. Let W be given by $W : x_0e_1x_1e_2 \cdots e_{2m+1}x_{2m+1}$ where $x_{2m+1} = x_0$. If all the vertices except the first and the last are distinct, then we indeed have a $(2m + 1)$-cycle and we are done. Otherwise, for some $i < j$ and $(i, j) \neq (0, 2m + 1)$, we must have $x_j = x_i$. We can then modify the walk W

to get two walks W' and W'' as follows. $W' : x_0e_1x_1 \cdots x_ie_{j+1}x_{j+1} \cdots x_{2m+1}$ and $W'' : x_ie_{i+1}x_{i+1} \cdots e_jx_j$. Both W' and W'' are closed walks and we have $l(W') + l(W'') = 2m + 1$ so that one of $l(W')$ and $l(W'')$ is both odd and strictly less than $l(W)$ and hence we are done by induction. □

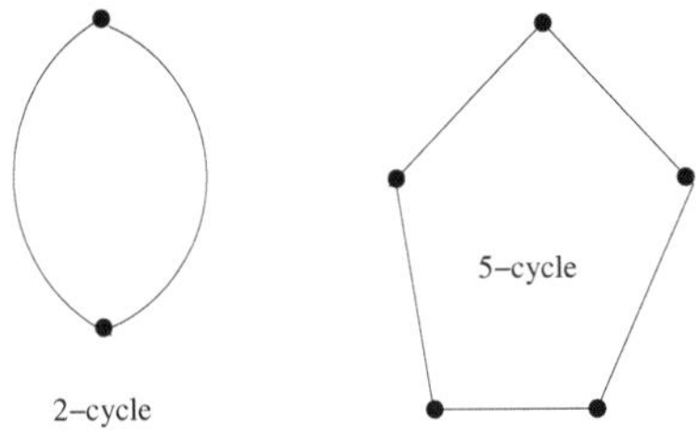

Figure 7.2: cycles

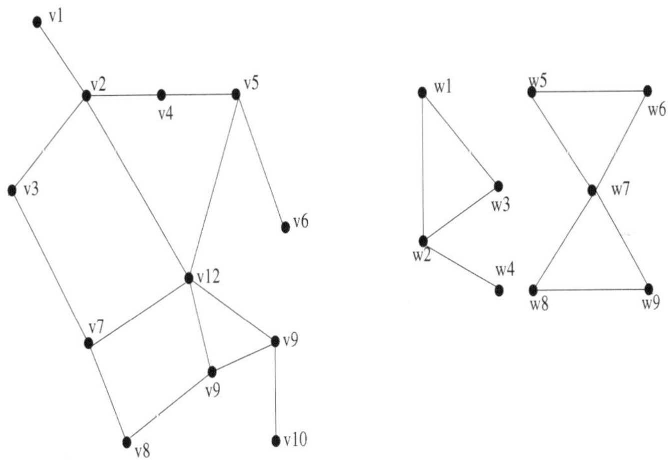

Figure 7.3: a connected and disconnected graph

Theorem 7.2.6. *A graph G is bipartite if and only if there are no odd cycles (cycles of odd length) in G.*

Proof Let G have bipartition (X, Y). Then a cycle starting at a vertex in X must alternate $X, Y, X, Y, \ldots$. Since the first and the last vertex are in X, any cycle in G must have an even length (observe that this also shows that we can not have a loop since a loop will join a vertex of X or Y with itself which is not allowed by the definition of a bipartite graph). Now let G be a graph that has no odd cycles. *Then, using Lemma 7.2.5, G has no closed walk of odd length.* We can assume, w.l.o.g. that G is connected: for a disconnected G, the components G_i will have bipartitions (X_i, Y_i) which can be combined to obtain $X = \cup X_i$ and $Y = \cup Y_i$ so that we get a bipartition of G. For two distinct vertices x and y, let $d = d(x, y)$ where $d(x, y)$ is the distance between x and y. Let Q be an (x, y)-path and let P be a shortest length (x, y)-path. Then combining Q with P (in the reverse direction) produces a closed walk W (that begins and ends in x). By assumption $l(W)$ is even and hence $l(Q)+l(P)$ is even. This shows that $l(Q)$ and $l(P) = d$ have the same parity (modulo 2). Thus $l(Q)$ is even (respectively odd) iff d is even (respectively odd). Fix a vertex x_0. We now partition the vertex set V of G into two parts X and Y as follows: $X = \{z : d(x_0, z)\ is\ even\}$ and $Y = \{z : d(x_0, z)\ is\ odd\}$. Then x_0 is in X and using connectedness all the vertices are in $X \cup Y$. If there is an edge between two distinct vertices z and w in X then combining an (x_0, z)- path, the edge (zw) and the (w, x_0)-path, we get a closed walk W that begins and ends in x_0. The two paths have both even lengths by assumption and hence $l(W)$ is odd, a contradiction. Thus there are no edges in X and similarly there are no edges in Y. This shows that both X and Y are independent sets completing the proof. □

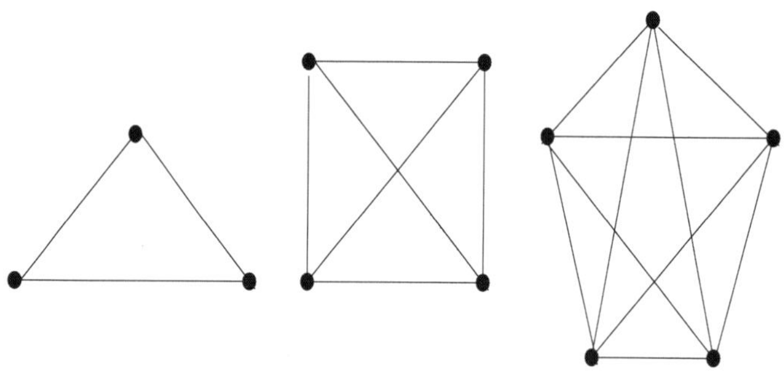

Figure 7.4: Complete graphs on 3, 4 and 5 vertices

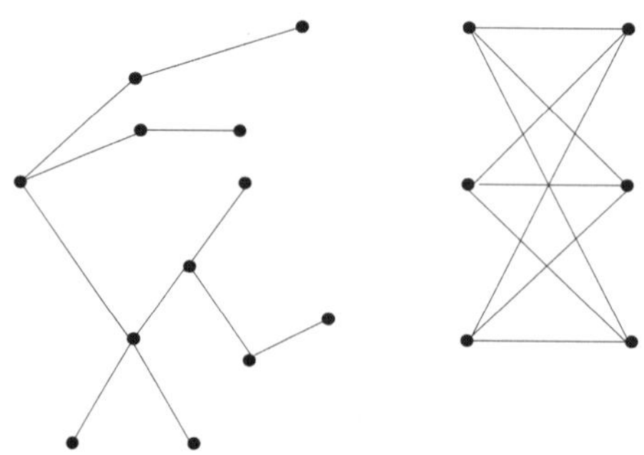

Figure 7.5: A tree and the utility graph

7.3 Eulerian circuits in graphs

Recall (Definition 7.2.1) that an (x, y)-tour of length m in a graph G is an alternating sequence $x_0 e_1 x_1 e_2 \cdots x_{m-1} e_m x_m$ where $x = x_0, y = x_m$ and all the edges $e_1, e_2, \ldots, e_m$ are distinct.

Definition 7.3.1. An (x, y)-tour is called an *Eulerian tour* if every edge occurs somewhere (at a unique place) on the tour. If $x \neq y$, then we call this an *open Eulerian tour* while if $x = y$ then it is called a closed Eulerian tour or *an Eulerian circuit.* Finally a graph G is called *an Eulerian graph* if it has an Eulerian circuit.

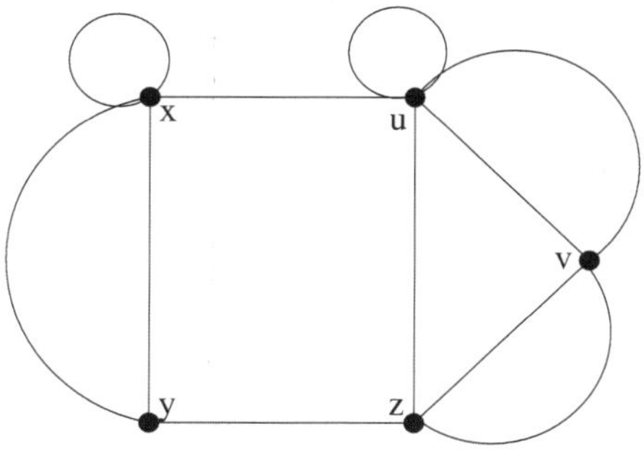

Figure 7.6: A connected graph with odd and even vertices

For example, the graphs K_3 and K_5 are both Eulerian. Not all connected graphs have Eulerian circuits. For example, K_4 is not Eulerian. An Eulerian graph may be disconnected but in that case, it will have components that consist of a single vertex. *To avoid this situation, we assume that our graph G is connected.* An obvious requirement for a simple connected graph to be Eulerian is that the degree of each vertex be even (since the number of times we enter a vertex on an Eulerian tour equals the number of times we exit it). In order to extend this to *all graphs*, we make a simple assumption that loops contribute 2 to the degree of a vertex. For example in Figure 7.6, the degrees of vertices x, y, z, u, v are $5, 3, 4, 6, 4$ respectively.
Historically, the origin of graph theory dates back to the famous Koenigsberg bridge problem solved by Euler in 1736. We have a river in which there are two islands say *I1* and *I2* and the two banks of the river. These are joined by 7 bridges. The question posed to Euler was to find if there is a way in which we can start at any one of these destinations, go over all the seven bridges exactly once and come back to the initial destination. In our terminology, we are trying to find an Eulerian circuit in the graph of figure 7.7. As was rightly observed by Euler, this is not possible, since each vertex, in fact has an odd degree (the degrees are 5 and 3 for the islands *I1* and *I2* respectively while for the banks $B1$ and $B2$ they are 3 each).

Theorem 7.3.2. *Let G be a connected graph. Then G is Eulerian iff each vertex has an even degree.*

Proof Let G be a connected graph in which every vertex has an even degree. Let T denote a longest possible tour (largest possible length) and let T be equal to

$$x_0 e_1 x_2 e_2 \cdots e_m x_m$$

Let x be the starting point and y the end point of T.

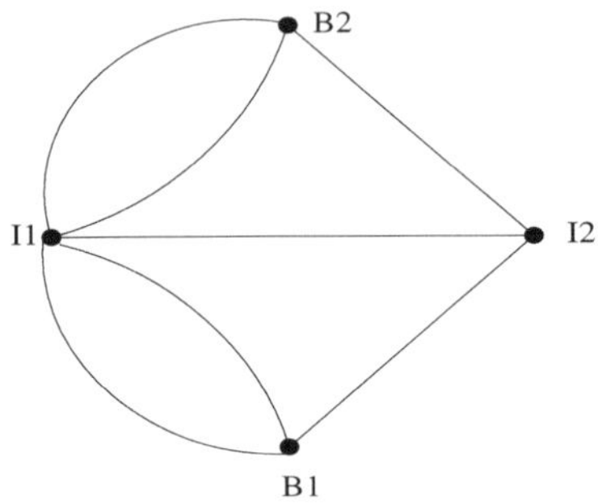

Figure 7.7: A graphical representation of the Koenigsberg problem

We claim that x and y are the same. If not, then the the number of times the tour enters y is one more than the number of times it leaves y and hence there is some edge say e' at y which is not on our tour. Augmenting T by e' then produces a longer tour, a contradiction to the fact that T is longest. Hence T is a circuit. We next claim that if z is some vertex occurring on the circuit, then all the edges incident at z occur on the circuit T. Suppose not. We may get a new circuit T' from T, that begins and ends at z (this is clear; simply begin the circuit at the first appearance of z, trace the edges along z getting to the initial vertex x and then trace the initial portion from x to z). Let e be an edge that does not occur on T' and is incident at z. Then augmenting T' by e produces a tour longer than T' (and hence T), which is a contradiction. Therefore, every vertex that occurs on the circuit T has all the edges incident with it occurring on T. Finally suppose some vertex w does not occur on T. Since G is connected, there is a tour $z_0 f_1 z_1 f_2 \cdots f_r z_r$ such that the tour begins at some vertex $z = z_0$ on T and ends at $w = z_r$. Since all the edges incident at z are covered on T, f_1 is covered on T and hence the other end point z_1 of f_1 occurs on T. We now sequentially capture end points of all the f_i's continuing in the same manner. We see that (after r steps), w occurs on T. This shows that T is an Eulerian tour, completing the proof. □

7.4 Eulerian circuits in digraphs and de Bruijn circuits

For many practical applications, it is desirable to deal with directed graphs or digraphs. Here we have edges that are directed and hence adjacencies are not symmetric. A directed edge e is an ordered pair $(\overrightarrow{xy})$ and we draw this edge by an arrow with x as its tail and y as its head. We also allow the possibility that $y = x$ and we then have a directed loop at x. The out-degree $d^+(x)$ is the number of edges whose tail is x and the in-degree $d^-(x)$ is the number of edges whose head is x (clearly with a loop at x contributing 1 each to the out-degree and in-degree). A (directed) walk in a digraph D is an alternating sequence of vertices and edges $W : v_0e_1v_1e_2 \cdots v_{i-1}e_iv_i \cdots e_mv_m$ where v_j's are vertices and e_j's are edges and the edge e_i is the edge $\overrightarrow{(v_{i-1}v_i)}$ and a directed path from v_0 to v_m is a walk in which vertices are all distinct. Most of the treatment we gave to (undirected) graphs extends easily to digraphs. Thus we have an (x, y)-directed path iff we have (x, y)-directed walk.

Let D be a digraph and G the underlying undirected graph obtained from D by removing the directions on the edges (but retaining all the edges). Then D called *weakly connected* if the underlying graph G is connected and D is called *connected* if for every two vertices x and y in D, there is a directed path from x to y. In Figure 7.8, we have a digraph, D on 5 vertices v_1, v_2, v_3, v_4, v_5. The degree pairs $(d^-(v), d^+(v))$ are $(3, 1), (0, 2), (0, 3), (1, 1), (3, 0)$ respectively. Notice that the sum of all the in-degrees equals the sum of all the out-degrees. D is weakly connected (since its underlying graph is a 5-cycle with an extra loop added) but is not connected since there is no directed path from v_4 to v_2.

We define *an Eulerian circuit* in a digraph D to be to be a directed walk $W : x_0e_1x_1e_2 \cdots x_{i-1}e_ix_i \cdots e_mx_m$ where $x_m = x_0$ and each edge of D is covered exactly once on W. We have the following analogue of Theorem 7.3.2 for directed graphs.

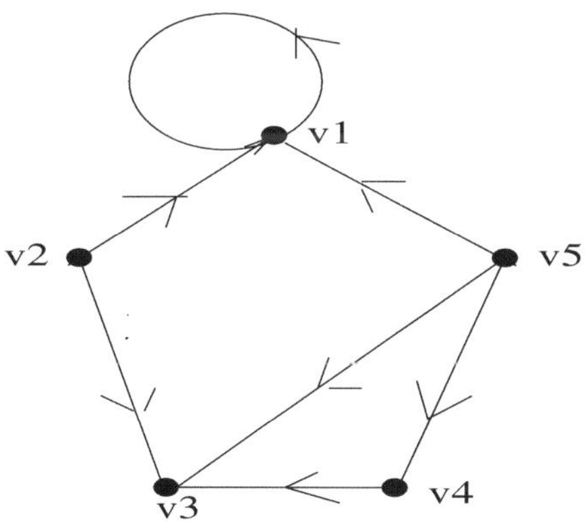

Figure 7.8: A digraph

Theorem 7.4.1. *Let D be a connected directed graph. Then G has an Eulerian circuit iff $d^+(x) = d^-(x)$ for all the vertices x in D.*

Definition 7.4.2. Let n be a natural number. *A de Bruijn sequence M_n* is a cyclic binary sequence of length 2^n satisfying the following property:

Let $M_n = (x_0x_1x_2 \cdots x_{2^n-1})$. Then for any i from 0 to $2^n - 1$, the sequence $m_i = x_ix_{i+1} \cdots x_{i+n-1}$ consisting of n consecutive entries (where the subscripts are read modulo 2^n) has the property that all the m_i's are different and hence *every binary sequence of length* n occurs as m_i for some i.

For example, $M_2 = (0011)$ is a de Bruijn sequence of length $2^2 = 4$ since $m_0 = 00, m_1 = 01, m_2 = 11, m_3 = 10$ are all distinct and hence every possible binary sequence of length 2 occurs as a unique m_i. Similarly $M_3 = (00010111)$ is a de Bruijn sequence of length 2^3. The following observation is most crucial for the construction of a de Bruijn sequence: If $m_i = x_i x_{i+1} \cdots x_{i+n-1}$, then $m_{i+1} = x_{i+1} x_{i+2} \cdots x_{i+n-1} x_{i+n}$. Hence m_{i+1} is obtained from m_i by removing its leftmost digit (i.e. x_i) and replacing it by a (new) rightmost digit (i.e. x_{i+n}). Let D_n denote a digraph whose vertex set V_n consists of all the 2^{n-1} binary sequences of length $n-1$. Given two vertices $a = (a_1 a_2 \cdots a_{n-1})$ and $b = (b_1 b_2 \cdots b_{n-1})$, we make a directed edge $e = \overrightarrow{(ab)}$ if $b_1 = a_2, b_2 = a_3, \ldots, b_{n-2} = a_{n-1}$ and write the value b_{n-1} (which is equal to 0 or 1) on this edge (on top of the arrow). For example, with $n = 4$, we have $(010) \xrightarrow{1} (101)$ and there is a loop at (000) with label 0 on it. Given a vertex $a = (a_1 a_2 \cdots a_{n-1})$ of D_n there are two out-edges from a and these clearly correspond to $b_{n-1} = 0$ and $b_{n-1} = 1$. Hence each vertex has out-degree 2 and similarly, each vertex also has in-degree 2. For $n = 3$, we have the following picture of D_3.

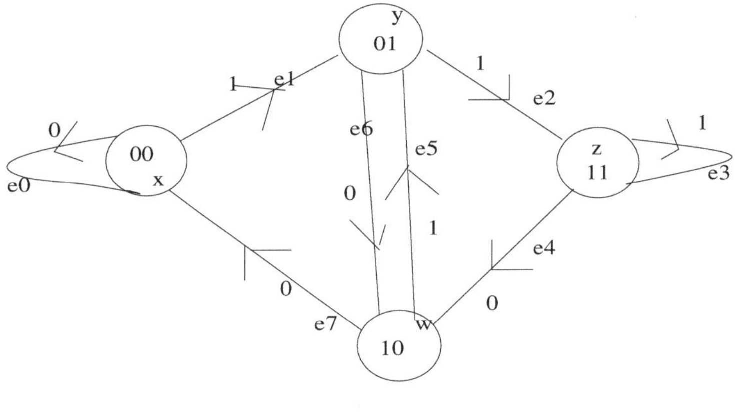

Figure 7.9: The graph D_3

Since the out-degree is 2, the number of edges of D_n is $2 \times 2^{n-1} = 2^n$. Given any two vertices $x = (x_1 x_2 \cdots x_{n-1})$ and $y = (y_1 y_2 \cdots y_{n-1})$, we can reach from x to y using a directed path with at the most $n-1$ edges (keep pushing the left co-ordinates of y to the right of x).

Theorem 7.4.3. *D_n has an Eulerian circuit. Hence, de Bruijn circuits exist for all possible orders 2^n.*

Proof Given two vertices x and y, there is at the most one edge with tail x and head y (this is true even when $y = x$). Therefore, a given Eulerian circuit in D_n can be described as a sequence of vertices (2^n in number with each vertex appearing twice): $\cdots vw \cdots$ where $e = (\overrightarrow{vw})$ is an edge and there is also an edge from the last vertex to the first. For example, with $n = 3$, let $x = (00), y = (01), z = (11), w = (10)$. Then we have an Eulerian circuit $xxyzzwyw$. Alternatively, we could also describe an

Eulerian circuit as a sequence of 2^n distinct edges of D_n as they occur on the Eulerian circuit (in succession) : $e_0e_1e_2\cdots e_{2^n-1}$ where the head of e_i equals the tail of e_{i+1} (and the subscripts are read modulo 2^n). Thus the same Eulerian circuit in D_3 can be written as

$$e_0=(\overset{0}{\overrightarrow{xx}}),e_1=(\overset{1}{\overrightarrow{xy}}),e_2=(\overset{1}{\overrightarrow{yz}}),e_3=(\overset{1}{\overrightarrow{zz}}),e_4=(\overset{0}{\overrightarrow{zw}})$$

$$e_5=(\overset{1}{\overrightarrow{wy}}),e_6=(\overset{0}{\overrightarrow{yw}}),e_7=(\overset{0}{\overrightarrow{wx}})$$

giving rise to the de Bruijn circuit (01110100) when the edge labels are written in that order. Since each edge corresponds to 0 or 1 (the label on the edge), it follows that we get a binary sequence M_n of length 2^n. We claim that this binary sequence is a *de Bruijn sequence.* To see this, define a flag to be an ordered pair (x, e) where x is a vertex and e an edge with x the tail of e. The manner in which we have defined adjacencies in D_n tells us that a subsequence $x_ix_{i+1}\cdots x_{i+n-1}$ in M_n consists of n consecutive edges in D_n and hence actually a flag (x, e) where $x = (x_ix_{i+1}\cdots x_{i+n-2})$ is a vertex and $e = x_{i+n-1}$ is the edge going out of x. Two such subsequences of length n occurring at different locations can never be equal for if they did, then we have the same flag, which amounts to saying that we repeat the vertex as also the edge on the circuit, a contradiction to the definition of an Eulerian circuit. □

7.5 Hypercubes and Gray codes

The vertices in the graph D_n in a de Bruijn sequence come from the object called a hypercube. An n-dimensional cube (or a hypercube) Q_n is the analogue of a cube in higher dimensions. For our purpose we construct this object recursively as a graph with 2^n vertices. Thus the graph Q_1 consists of two vertices (0) and (1) and a single edge joining them. To construct Q_2 we take two copies of Q_1 and join the corresponding vertices as shown in Figure 7.10. Adjoining 0 to the left of vertices in the first copy and 1 to the vertices in the second copy obtains the graph Q_2. This graph is just the familiar 4-cycle C_4 and is shown in Figure 7.10. To obtain Q_3, we take two copies $Q_2^{(0)}$ and $Q_2^{(1)}$, attach 0 to the left of all the vertices in $Q_2^{(0)}$ and 1 to the left of all the vertices in $Q_2^{(1)}$ of Q_2, and make the corresponding vertices adjacent as shown in Figure 7.12.
This graph is just the graph of the usual cube in 3-dimensions. Q_n is constructed by taking two copies $Q_{n-1}^{(0)}$ and $Q_{n-1}^{(1)}$. To each vertex of Q_{n-1}, we attach i on the left to get $Q_{n-1}^{(i)}$ where $i = 0, 1$ (thus creating a binary sequence of length n). A vertex $\underline{x}$ of $Q_{n-1}^{(0)}$ and a vertex $\underline{y}$ of $Q_{n-1}^{(1)}$ are adjacent iff deleting their leftmost digits (which are 0 and 1 respectively), the $(n-1)$-tuples we get are identical and hence correspond to the same vertex in Q_{n-1}. We observe that Q_n has as vertex set all the vertices that are binary n-tuples $(v_{n-1}v_{n-2}\cdots v_1v_0)$ with two vertices $(x_{n-1}x_{n-2}\cdots x_1x_0)$ and $(y_{n-1}y_{n-2}\cdots y_1y_0)$ adjacent iff for some i we have $y_i \neq x_i$ and $y_j = x_j\ \forall\ j \neq i$. This follows by induction: If $x_{n-1} \neq y_{n-1}$ then these two vertices are in different

copies say $Q_{n-1}^{(0)}$ and $Q_{n-1}^{(1)}$ respectively. Otherwise, $x_{n-1} = y_{n-1}$ and we are in the same copy say $Q_{n-1}^{(0)}$ and the result is true because it holds true in $Q_{n-1}^{(0)}$.

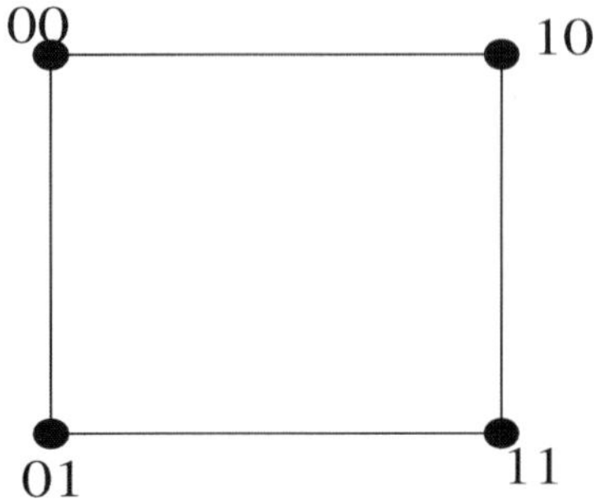

Figure 7.10: The graph Q_2

Given Q_n, we wish to start at the vertex $(00\cdots0)$ visit every vertex moving over adjacent vertices and return back to the starting vertex $(00\cdots0)$ so that we have a full cycle of length 2^n. This is certainly true when $n = 2$ since we can move in the order $(00), (01), (11), (10)$ and return back to the original vertex (00). In order to construct such a cycle for Q_n, we assume that we have already constructed one for Q_{n-1} and use it as follows. We first move in the copy $Q_{n-1}^{(0)}$ starting from the vertex $(00\cdots00)$ and instead of returning back to the same vertex, break the cycle at the last vertex $(0100\cdots0)$. We then enter the other copy $Q_{n-1}^{(1)}$ at the vertex $(110\cdots0)$ (which is the last vertex in the original full cycle in Q_{n-1}) and traverse the full cycle of Q_{n-1} backwards ending the journey in $Q_{n-1}^{(1)}$ at the vertex $(10\cdots0)$ and finally end the cycle at $(00\cdots0)$ as desired. The transition from $n = 2$ to $n = 3$ is shown Figures 7.12 and 7.13.

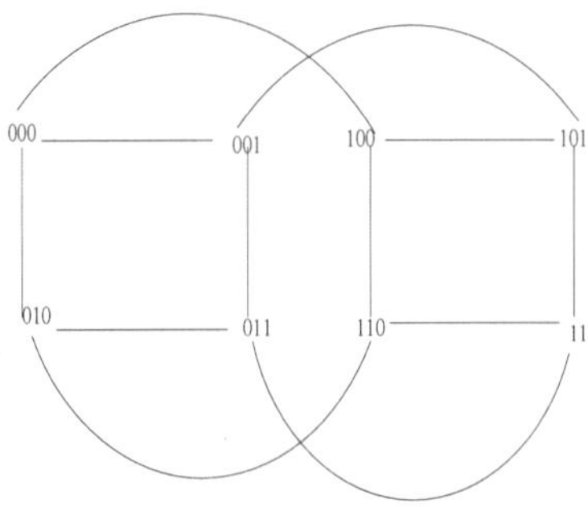

Figure 7.11: The graph Q_3

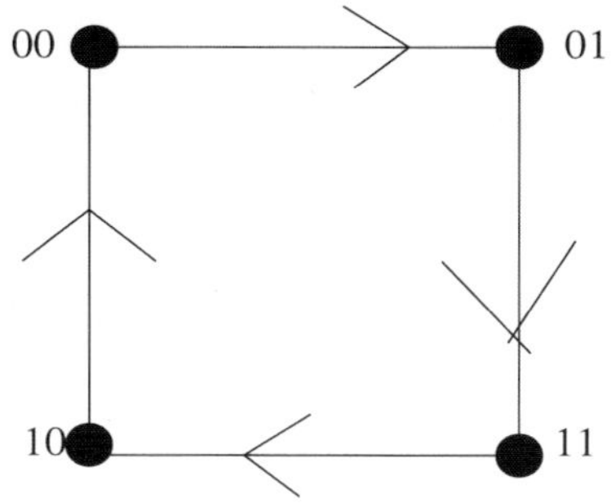

Figure 7.12: The Gray Code of Q_2

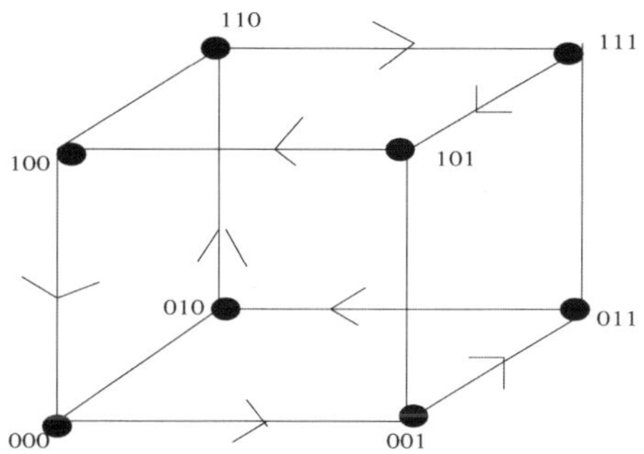

Figure 7.13: The Gray Code of Q_3

Let us construct the full cycle in Q_4 using the one we just constructed for Q_3:

$$(0000) \to (0001) \to (0011) \to (0010) \to (0110) \to (0111) \to (0101)$$
$$\to (0100) \to (1100) \to (1101) \to (1111) \to (1110)$$
$$\to (1010) \to (1011) \to (1001) \to (1000) \to (0000)$$

Definition 7.5.1. The full cycle in Q_n recursively constructed as above is called the *Gray Code of* Q_n.

Every binary length n sequence $(a_{n-1}a_{n-2}\cdots a_1a_0)$ is uniquely associated with the non-negative integer $a = \sum_{i=0}^{n-1} a_i 2^i$, we have a natural bijection from the set of all binary length n sequences and the set $\{0, 1, \ldots, 2^n - 1\}$. The position of the vertex $(00\cdots 0)$ is clearly $(00\cdots 0)$. Under this identification, we could ask the two questions: where does a certain number a occur in the Gray code (i.e. at which position) and which number occupies a given position. These questions are clearly inverses of each other. So, let $f_n : \{0, 1, 2, \ldots, 2^n - 1\} \to \{0, 1, 2, \ldots, 2^n - 1\}$ be defined by $f_n(a) = b$ if the integer $a = (a_{n-1}a_{n-2}\cdots a_0)$ is at the position $b = (b_{n-1}b_{n-2}\cdots b_0)$ on the Gray code. For $n = 3$, the function f_3 has the following table. Here $y = f_3(x)$.

binary x	numerical x	binary y	numerical
000	0	000	0
001	1	001	1
011	3	010	2
010	2	011	3
110	6	100	4
111	7	101	5
101	5	110	6
100	4	111	7

Theorem 7.5.2. *Let the function f_n give the location $y = (y_{n-1}y_{n-2}\cdots y_0)$ of the integer $x = (x_{n-1}x_{n-2}\cdots x_0)$ on the Gray code. Then*

$$\begin{aligned} y_0 &= x_0 + x_1 + x_2 + \cdots + x_{n-1} \\ y_1 &= x_1 + x_2 + \cdots + x_{n-1} \\ y_2 &= x_2 + \cdots + x_{n-1} \\ \cdots &= \cdots\cdots\cdots\cdots\cdots\cdots \\ \cdots &= \cdots\cdots\cdots\cdots\cdots\cdots \\ y_{n-1} &= x_{n-1} \end{aligned}$$

where all the computations are performed modulo 2.

Proof This is easily checked for $n = 1, 2$. We see that the Gray code of Q_{n+1} is constructed from that of Q_n as follows. For the first 2^n vertices in $Q_n^{(0)}$ we just attach 0 to the left and have the same ordering as in Q_n and for the next 2^n vertices we attach 1 to the left but follow the cycle in $Q_n^{(1)}$ backwards. This amounts to interchanging 0's and 1's in the remaining co-ordinates. Specifically, let

$$f_{n+1}(x_n x_{n-1}\cdots x_0) = (y_n y_{n-1}\cdots y_0)$$

Then y_n equals 0 iff $x_n = 0$. Let

$$f_n(x_{n-1}x_{n-2}\cdots x_0) = (z_{n-1}z_{n-2}\cdots z_0)$$

If $x_n = 0$, then $(y_n y_{n-1} y_{n-2}\cdots y_0) = (0z_{n-1}z_{n-2}\cdots z_0)$ and if $x_n = 1$, then $(y_n y_{n-1}y_{n-2}\cdots y_0) = (1y_{n-1}y_{n-2}\cdots y_0)$ is obtained by interchanging 0's and 1's in $(z_{n-1}z_{n-2}\cdots z_0)$ (this is necessary since the cycle in $Q_n^{(1)}$ is traced backwards). Then, by induction, we have $z_j = x_j + x_{j+1} + \cdots + x_{n-1}$ and in both the cases, we have $y_j = z_j + x_n$ which gives $y_j = x_j + x_{j+1} + \cdots + x_{n-1} + x_n$ proving the assertion.

Since the function f_n is bijective, it must have an inverse g_n. If $g_n(y_{n-1}y_{n-2}\cdots y_0) = (x_{n-1}x_{n-2}\cdots x_0)$, then $x_j = y_j + y_{j+1}$, $j = 0, 1, \ldots, n-1$ (interpret this with

$y_n = 0$). This can either be see by solving the system of linear equations or by looking at the inverse of the coefficient matrix of the upper triangular matrix on the R.H.S. □

Example 7.5.3. In the Gray code of Q_5, where does the integer 17 occur? We have $17 = (10001) = (x_4x_3x_2x_1x_0)$. Hence if $f_4(x_4x_3x_2x_1x_0) = (y_4y_3y_2y_1y_0)$, then the formula gives $(y_4y_3y_2y_1y_0) = (11110) = 30$. So the number 17 occurs at position 30. As a second example, which integer is at the position 105 in the Gray code of Q_7? Here $105 = 2^6 + 2^5 + 2^3 + 2^0 = (1101001)$ and $g_7(1101001) = (1011101) = 2^6 + 2^4 + 2^3 + 2^2 + 2^0 = 91$.

We now describe a puzzle called the "instant insanity puzzle". We are given four (identical) unit cubes C_1, C_2, C_3, C_4. All the six faces of each cube are coloured in one of the four colours red, blue green and yellow (in an arbitrary manner). We have to arrange the cubes one on top of the other so that they form a $1 \times 1 \times 4$ tower in such a manner that on each of the four sides (front, back, left and right) of the tower, each colour appears on some cube exactly once. The freedom we have here is that of rotating the cubes. When the cubes are arranged so as to form such a tower, two faces of each cube (the top and the bottom) are not used but we do not beforehand know as to which pair of opposite faces is to be kept unused. A solution to the instant insanity puzzle (suggested by Tutte, the founder figure of graph theory, who wrote under the pseudonym Carte Blanche) is the following [56]. For each cube C_i make an undirected graph G_i whose vertex set $V = \{R, B, G, Y\}$ one vertex for each colour. We draw an edge between two vertices if there is a pair of opposite faces with the corresponding two colours appearing on the opposite faces. We also label the edges of G_i by the number i just for book keeping purpose of knowing that that edge came from G_i and hence that face pair is from the cube C_i. For example, if the six faces of the cube C_1 are coloured as shown on the left of Figure 7.14. The corresponding graph G_1 (with three edges) is drawn on the right. *This requires a little interpretation. The middle R means that the bottom is coloured red while the back is red and the front is yellow. The left wall is blue while the right wall is green and finally the top is also green.*
We always have three edges in each G_i and assuming that a loop contributes 2 to the degree of a vertex, the sum of all the degrees in G_i is 6. Now suppose that the remaining cubes C_2, C_3 and C_4 are coloured as shown in Figure 7.15. The corresponding graphs G_2, G_3 and G_4 are shown in Figure 7.16.
Notice that the final solution to the puzzle if there is one, must mark one pair for the front-back and one for the right-left. Since each colour must appear on each side in the final solution, we are looking for two graphs H and H' which come from these four given graphs G_1, G_2, G_3, G_4 such that in each one of H and H' the degree of any vertex is 2 (since it must show on both of the opposing sides of the tower). We now superpose these four graphs to get a graph G^* on four vertices and 12 edges and try to find two edge disjoint subgraphs H, H' of G^* so that both H and H' are 2-regular with four edges each bearing a distinct label. For our example, the graph G^* is drawn on the left of the following figure and on the right, we see the two graphs H, H' satisfying the required conditions.

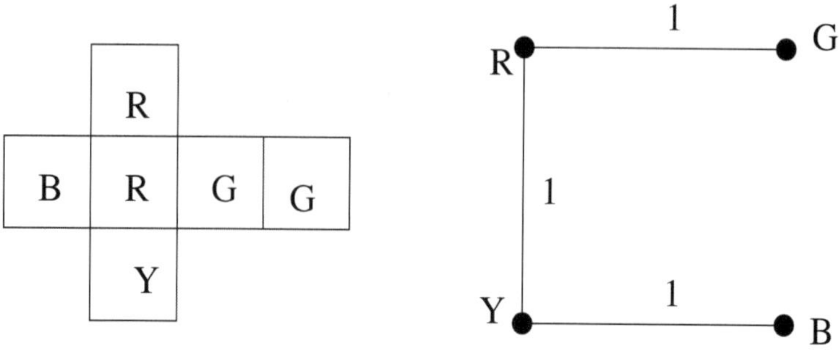

Figure 7.14: The cube C_1 and its graph G_1

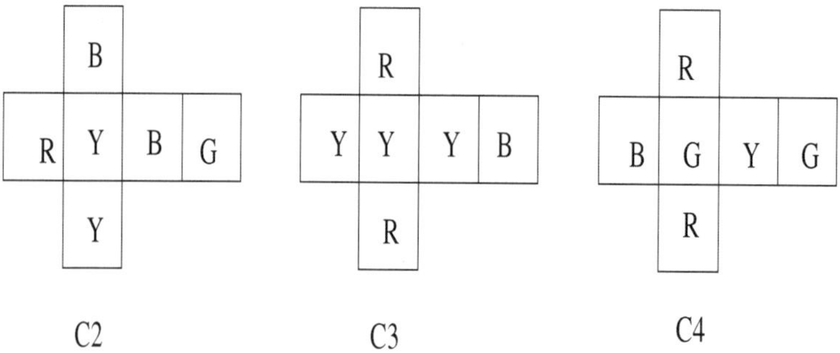

Figure 7.15: Remaining cubes $C2, C3, C4$

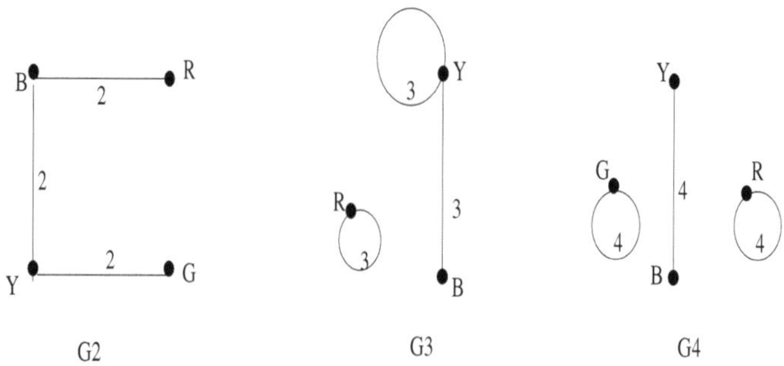

Figure 7.16: The graphs G_2, G_3, G_4

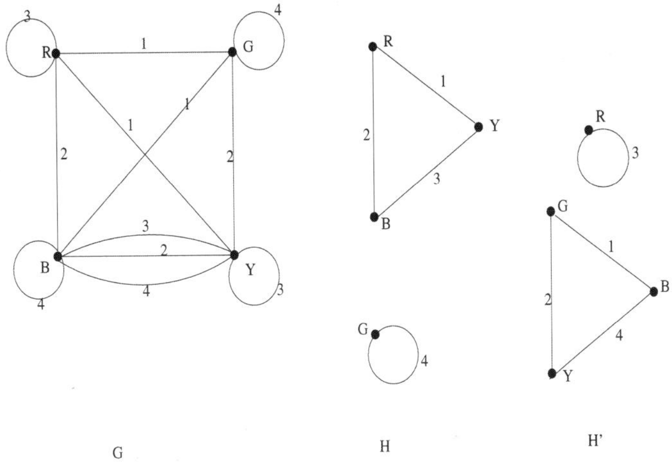

Figure 7.17: The full graph G and the graphs H and H'

The graphs H and H' give us the following array for a pair of opposite colours on the cubes.

$$\begin{array}{cccc} Y & R & B & G \\ R & B & G & Y \\ B & Y & R & R \\ G & G & Y & B \end{array}$$

The arrangement of cubes is shown in Figure 7.18 (on the next page) with cubes numbered 1 to 4 from the top to the bottom. The back wall colour is not shown. These colours on the back wall are G, Y, R, B in that order. The reader can now write down a condition that is both necessary and sufficient for the existence of a solution to the instant insanity puzzle! These are explored in the exercises.

7.6 Winning a Nim game

We now describe a nice application of parity (mod 2 parity, to be specific) in the form of a solution to the classical Nim game. In order to understand the question properly, it is essential to look at some basic framework of mathematical game theory and a general idea of a two person game. Two players A and B take turns and make 'moves' on a set of certain positions. A move consists of going from a given position to some other position. Not all positions are reachable from a given position (there are rules determining possible moves at a given stage and these are operative all through the progress of the game). We also have some positions (at least one) that are designated as 'winning positions'. As soon as any player makes a move to a winning position, the game ends and the player who moved to a winning position is declared the winner

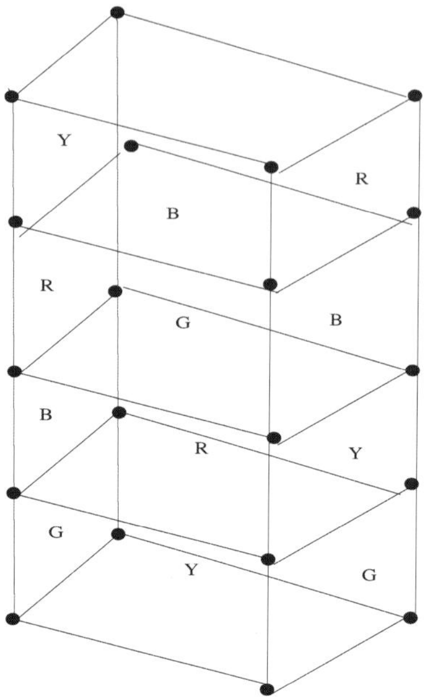

Figure 7.18: An arrangement of the four cubes

and the other player is the loser. The objective of both the players is to reach some winning position (and equivalently prevent the opponent from doing so). Consider the following example. On a table that has initially no chips on it, two players A and B take turns and put a certain number of chips on the table. The number of chips a player can put on the table is between 1 and 10 and hence at every stage, the number of chips on the table increases. The object of the game is to get a total of 100 chips on the table; the game ends there and the player who moved last is the winner. Mathematical problem here is to find a 'good move', that is a sequence of moves that will ensure a win regardless of how intelligent the opponent is. To that end, we draw a digraph of the game with each position represented by a vertex and we have a directed edge from a vertex x to y if we can move from x to y. Thus for the game we just described, the set of vertices is all the natural numbers upto 100 with an edge $\overrightarrow{(xy)}$ if $x < y$ and $y - x \leq 10$.

A *progressively finite game* is a game that has finitely many potions with a number of valid moves at each position (which are known beforehand) with the following additional stipulations: No position can ever get repeated during the course of the game and the game (regardless of the choice of one or both the players) ends after a finite number of moves. In terms of the game digraph D, this translates into not having a directed cycle in D. A standard analysis of a progressively finite game involves the idea of *backtracking*. Suppose player A wishes to win the game just described. Then he must get to 100 and therefore his opponent B must have been on positions from 90 to 99 (but not 100) before that. For this to happen, it is sufficient for A to reach 89. After this no matter how clever B is, he is forced to move to a number from 90 to 99 (but not 100) and A can then safely reach 100 on his next move. Hence if A is at position where the table has 89 chips on it (and asks B to make a move), then A is the winner. So 89 is a good number. In order for A to be able to get to 89, it is sufficient if he is at position 78 (then regardless of how many chips B adds A can always make the total on the table 89 but B cannot do this since $78 + 11 = 89$). Backtracking in this manner, we see that the set of all good vertices is $K = \{100, 89, 78, 67, 56, 45, 34, 23, 12, 1\}$. Hence if player A begins the game by putting one chip on the table, then regardless of how B plays (how many chips B adds on the table), he can always add a certain number of chips from 1 to 10 and get to one of the numbers in K. The set K consisting of all the 'good positions' is called *kernel* of the game.

Definition 7.6.1. Let D be the game graph of a progressively finite game. Let K be a set of vertices of D. The K is called *Kernel of D* if it satisfies the following three properties: (i) all the winning vertices are in K (ii) there is no edge from any vertex in K to any (other) vertex in K (iii) from every vertex not in K, there is an edge to some vertex in K.

If we can find kernel of a progressively finite game, then we have *a winning strategy*: A always tries to reach a vertex in the kernel K forcing B to make a move that takes him outside K from which A can reach a vertex in K and so on. Since the winning vertices are in K, this strategy will ensure a win for A. In the game we described if A foolishly began by putting 3 chips on the table (instead of 1) then B can follow the same strategy: B then adds 9 chips to the total to make it to 12 chips on the table and moves on the kernel vertices and is assured a win. The moral of the story is that with correct and intelligent play on the part of both the players, this progressively finite game is a win for A. Thus finding a kernel is the key to solving a progressively finite game. *A theorem of Sprague and Grundy proved in the first half of the twentieth century asserts that every progressively finite game has a unique kernel* [56].

We now describe the classical Nim game [56]. For some $m \geq 2$ we have m piles of chips with the j-th pile containing $a^{(j)} \geq 0$ chips. Two players A and B take turn and make following moves. A player whose turn it is, chooses some nonempty pile say the j-th pile, removes a certain (positive) number of chips from that pile and leaves the remaining piles unchanged. From the sequence $(a^{(1)}, (a^{(2)}, \ldots, a^{(m)})$, we thus get new sequence $(b^{(1)}, (b^{(2)}, \ldots, b^{(m)})$ where $b^{(j)} < a^{(j)}$ and for all $i \neq j$, we have $b^{(i)} = a^{(i)}$. Now we have a new sequence of m piles with the number of chips given by the new sequence $(b^{(1)}, b^{(2)}, \ldots, b^{(m)})$. The next player again chooses some nonempty pile, removes any non-zero number of chips from that pile and the game goes on. The game ends when there are no chips left (i.e. the sequence has become $(0, 0, \ldots, 0)$) and the last player to move is the winner. Here is the simplest situation when $m = 2$. If we have two piles both containing the same number a of chips, then the second player can win by imitating the moves of the first player in the other pile. Thus if A takes off d chips from the first pile, then B also takes off d chips from the second pile making the situation that both the piles have the same number of chips. It follows that the second player is assured a win if the game begins with the two piles having the same number of chips while the first player can exactly follow the same

strategy and win the game if we have two piles with different number of chips. Hence the kernel K for this game consists of the set $\{(a,a) : a \in \mathbb{Z}^+\}$, that is, the set of positions where both the piles have the same number of chips.
This simple minded analysis that works in the situation of two piles does not extend easily to the situation of larger number of piles. To accomplish the latter task, we define a few more terms. *Let G denote the set of all finite binary sequences $a = (a_m, a_{m-1}, \ldots, a_1, a_0)$. Here each a_i is 0 or 1 and by the k-th term of such a sequence, we mean the number a_k.* Also, by convention,

$$(0, 0, \ldots, 0, a_m, a_{m-1}, \ldots, a_1, a_0) = (a_m, a_{m-1}, \ldots, a_1, a_0)$$

and hence we may add any number of zeros to the left. Thanks to this convention, we can define a sum $\oplus$ on the set G by

$$(a_m, a_{m-1}, \ldots, a_1, a_0) \oplus (b_m, b_{m-1}, \ldots, b_1, b_0) = (c_m, c_{m-1}, \ldots, c_1, c_0)$$

where $c_i = a_i + b_i \ (mod\ 2)$ *and* $c_i = 0, 1$ and hence if $a \oplus b = c$ with $a, b \in G$, then $c \in G$. We also have

$$a \oplus b = c \Rightarrow b \oplus c = a \Rightarrow a \oplus c = b$$

as well as $a \oplus a = 0$ where 0 refers to the binary sequence with all the terms zero. This defines an additive group structure on G (which is even commutative and each element is an idempotent). We now use the natural one-to-one correspondence between $\mathbb{Z}^+$ and G to obtain an isomorphic group structure on $\mathbb{Z}^+$. Given a non-negative integer α, we have $\alpha = \sum_{i=0}^{m} a_i 2^i$ the unique binary expansion of α. Let $a = (a_m, a_{m-1}, \ldots, a_1, a_0)$. Then we associate α with $a \in G$. The uniqueness of the binary expansion of a non-negative integer ensures that this is a well-defined bijection and we are therefore, in a position to translate the digital sum operation to $\mathbb{Z}^+$ in a natural manner. Let a and b be the binary sequences associated with non-negative integers α and β respectively. Let $c = a \oplus b$. Then $\alpha \oplus \beta = \gamma$ where the binary sequence associated with γ is c. Thus

$$59 \oplus 23 = (111011) \oplus (10111) = (111011) \oplus (010111) = (101100) = 44$$

Note also that $\alpha \oplus \alpha = 0$.Now suppose we have n piles with the j-th pile containing $a^{(j)}$ chips where $a^{(j)}$ equals $(a_m^{(j)}, a_{m-1}^{(j)}, \ldots, a_1^{(j)}, a_0^{(j)})$ for $j = 1, 2, \ldots, n$. Write $\oplus \sum$ to denote the digital sum (of more than two numbers) and let $\oplus \sum_{j=1}^{n} a^{(j)} = e \neq 0$. *In this case, we wish to show that chips from one of the piles can be removed so that the new digital sum is* 0. Since $e = \sum_{i=0}^{m} e_i 2^i$, there is a largest r such that $e_r = 1$ and $e_i = 0 \ \forall\ i > r$. We can then write

$$\begin{aligned} a^{(j)} &= (a_m^{(j)}, a_{m-1}^{(j)}, \ldots, a_r^{(j)}, \ldots, a_0^{(j)}) \\ &= (a_m^{(j)}, a_{m-1}^{(j)}, \ldots, a_{r+1}^{(j)}, 0, \ldots, 0, 0) \oplus (0, 0, \ldots, 0, a_r^{(j)}, \ldots, a_0^{(j)}) \\ &= b^{(j)} \oplus c^{(j)} \end{aligned}$$

say, where $b^{(j)} = (a_m^{(j)}, a_{m-1}^{(j)}, \ldots, a_{r+1}^{(j)}, 0, \ldots, 0, 0)$ and $c^{(j)} = (0, 0, \ldots, a_r^{(j)}, \ldots, a_0^{(j)}) = (a_r^{(j)}, \ldots, a_0^{(j)})$. Thus $\forall\ i > r,\ \sum_{j=1}^{n} a_i^{(j)} \equiv 0\ (mod\ 2)$ and $\forall\ i \leq r,\ \sum_{j=1}^{n} a_i^{(j)} \equiv e_i\ (mod\ 2)$. Therefore, $\oplus \sum_{j=1}^{n} b^{(j)} = 0$ and $\oplus \sum_{j=1}^{n} c^{(j)} = e = (e_r, e_{r-1}, \ldots, e_1, e_0)$ (obtained after deleting all the zeros to the left of e_r). Here $e_r = 1$ and hence for some j, we must have $c_r^{(j)} = 1$. Let, w.l.o.g. $c_r^{(1)} = 1$. For each $t = 2, 3, \ldots, n$ let $d^{(t)} = (c_{r-1}^{(t)}, \ldots, c_0^{(t)})$ be obtained by removing the leftmost co-ordinate from each $c^{(t)}$. Let $d = \oplus \sum_{t=2}^{n} d^{(t)}$. Then d has non-zero coefficients only upto $r-1$ and hence $d \leq 2^{r-1} + 2^{r-2} + \cdots + 1 = 2^r - 1$ and $c_r^{(1)} = 1$ implies that $c^{(1)} \geq 2^r$. Hence it is possible to remove $c^{(1)} - d > 0$ chips from the first pile and be left with d chips in the first pile. The first pile now looks $b^{(1)} \oplus d$. But we already have $\oplus \sum_{j=1}^{n} b^{(j)} = 0$ and $\oplus \sum_{t=2}^{n} c^{(t)} = d$. Therefore $(\oplus \sum_{t=2}^{n} b^{(t)}) \oplus d = d \oplus d = 0$ and we indeed get the digital sum equal to 0.
In the preceding arguments we have proved a major part of the following theorem.

Theorem 7.6.2. *Let a position in the Nim game with n piles be described by a sequence of $(x^{(1)}, x^{(2)}, \ldots, x^{(n)})$ where the i-th pile contains $x^{(i)}$ chips. Then the set*

$$K = \{y = (y^{(1)}, (y^{(2)}, \ldots, y^{(n)}) : \oplus \sum_{j=1}^{n} y^{(j)} = 0\}$$

is the kernel of the Nim game.

Proof We have already proved that if we have a vertex v outside K, then there is a directed edge from v to some vertex in K. It just remains to show that there are no edges inside K. That is, we wish to show that if $y = (y^{(1)}, (y^{(2)}, \ldots, y^{(n)})$ is in K and if we have a directed edge from y to $z = (z^{(1)}, z^{(2)}, \ldots, z^{(n)})$, then $z \notin K$. The definition of the digraph implies that there is a unique j such that $z^{(i)} = y^{(i)}\ \forall\ i \neq j$ and $z^{(j)} < y^{(j)}$. We now compute $\sum_{i=1}^{n} z^{(i)}$ and prove that this cannot be zero.

$$\begin{aligned}
\oplus \sum_{i=1}^{n} z^{(i)} &= [\oplus \sum_{i \neq j} z^{(i)}] \oplus z^{(j)} \\
&= [\oplus \sum_{i \neq j} y^{(i)}] \oplus z^{(j)} \\
&= [\oplus \sum_{i \neq j} y^{(i)}] \oplus [y^{(j)} \oplus y^{(j)}] \oplus z^{(j)} \\
&= [\oplus \sum_{i=1}^{n} y^{(i)}] \oplus (y^{(j)} \oplus z^{(j)}).
\end{aligned}$$

On the right hand side the first sum is 0 since $y \in K$ and hence the left hand side equals $y^{(j)} \oplus z^{(j)}$ which cannot equal 0 since $z^{(j)} < y^{(j)}$. □

Example 7.6.3. Let $n = 3$ and the given sequence of piles be $(x^{(1)}, x^{(2)}, x^{(3)}) = (7, 17, 3)$. We have $7 \oplus 17 \oplus 3 = (00111) \oplus (10001) \oplus (00011) = (10101) \neq 0$ and hence the initial position is not in K. Our procedure dictates that we remove chips from the second pile (in our notation, we have $r = 4$). Then $7 \oplus 3 = (111) \oplus (011) = (100)$ and hence we should keep 4 chips in the second pile (and remove 13). If player A makes this move, then the new position is $(7, 4, 3)$ which is in K. Any move on the part of B from this position must take him to a position out of K. Suppose B removes 2 chips from the second pile to create the sequence $(7, 2, 3)$. Then A must remove chips from the first pile and since $2 \oplus 3 = 1$ he should keep only 1 chip in the first pile (and remove 6 chips) creating a new sequence $(1, 2, 3)$ which is in K.

Example 7.6.4. Consider the following example with 6 piles whose initial sequence of chips is

$$(a^{(1)}, a^{(2)}, a^{((3)}, a^{(4)}, a^{(5)}, a^{(6)}) = (17, 29, 34, 23, 44, 9)$$

Then $17 \oplus 29 \oplus 34 \oplus 23 \oplus 44 \oplus 9$ equals

$$(10001) \oplus (11101) \oplus (100010) \oplus (10111) \oplus (101100) \oplus (1001) = (011100) \neq 0$$

Hence the initial position is not in the kernel and to reach a position in the kernel we should locate a pile that has 1 in the fourth place. All of $a^{(1)}, a^{(2)}$ and $a^{(4)}$ have this property and we choose the first pile. The digital sum $a^{(2)} \oplus a^{((3)} \oplus a^{(4)} \oplus a^{(5)} \oplus a^{(6)}$ equals $(1101) = 13$ and hence if we leave 13 in the first pile, then we are in K. This requires that we take off $17 - 13 = 4$ chips from the first pile. Instead of this, we could have chosen the second pile and then the digital sum $a^{(1)} \oplus a^{(3)} \oplus a^{(4)} \oplus a^{(5)} \oplus a^{(6)}$ equals 1 so that we should remove 28 chips from the second pile and create a new sequence $(17, 1, 34, 23, 44, 9)$ which is in the kernel.

7.7 Parity of a permutation

We wish to give two equivalent definitions of parity of a permutation. To that end, our beginning point is the basic observation made in Chapter 3 which tells us that the cycle decomposition of a permutation π on $[n]$ is unique upto shuffling (reordering) of the cycles and cyclically shifting of the numbers in a given cycle. This allows us to represent a permutation graphically as a directed graph in which each vertex has out-degree and in-degree both equal to 1 and hence the graph is a disjoint union of directed cycles. A fixed point x of such a permutation is a loop (xx) and a transposition is a directed 2-cycle (xy) in which we have two edges one going from x to y and the other from y to x. For example, if σ is a permutation on $[9]$ written in the cycle decomposition form as $\sigma = (4638)(25)(19)$, then its diagrammatic representation is the graph drawn in Figure 7.19.
For the following discussion fix a natural number $n \geq 2$ and consider permutations on $[n]$. Call a cycle of even (respectively odd) length to be an *even cycle (respectively an odd cycle)*. Write $e(\pi)$ to mean the number of even cycles of π. The uniqueness of the cycle decomposition of π implies that $e(\pi)$, the number of even cycles in π is a well-defined number. Thus $e(id) = 0$ (here id refers to the identity permutation) and for a transposition θ, $e(\theta) = 1$. The following lemma is most crucial to the further discussion.

Lemma 7.7.1. *Let π be any permutation and let $\theta = (ab)$ be any transposition. Then we have*

$$e(\pi\theta) = e(\pi) \pm 1$$

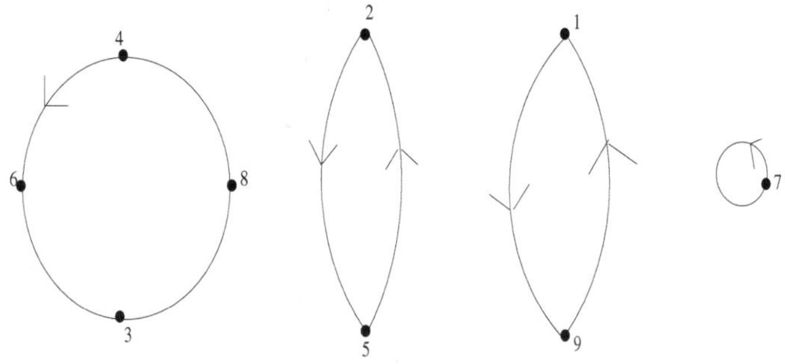

Figure 7.19: Graph of σ

Proof Write π in its cycle decomposition and make the following two cases depending on whether both a and b are in the same cycle of π.

Case 1 Both a and b occur in the same cycle A of π. Since in the multiplication of π with θ, other cycles remain unaffected, we may assume, w.l.o.g. that the cycle containing both a and b is $(a_1 a_2 \cdots a_m)$ where $a_1 = a$ and $a_r = b$ for some r such that $2 \leq r \leq m$. Then

$$(a_1 a_2 \cdots a_{r-1} a_r a_{r+1} \cdots a_m)(a_1 a_r) = (a_1 a_2 \cdots a_{r-1})(a_r a_{r+1} \cdots a_m)$$

so that a single cycle breaks into two cycles $B = (a_1 a_2 \cdots a_{r-1})$ and $B' = (a_r a_{r+1} \cdots a_m)$. Figure 7.20 illustrates the situation when $m = 8$ and $r = 4$. Denoting the lengths $|B|$ and $|B'|$ by k and k' respectively, we see that $m = k + k'$. Clearly, if m is even then either both k, k' are odd or both are even. Since the single even cycle A is replaced by two odd or two even cycles, $e(\pi\theta)$ is either equal to $e(\pi) - 1$ (when both k, k' are odd) or is equal to $e(\pi) + 1$ (when both k, k' are even). We thus get $e(\pi\theta) = e(\pi) \pm 1$. Now suppose m is odd. In this case, precisely one of k and k' is even and the other is odd. Then a single odd cycle (which contributes nothing to $e(\pi)$) is replaced by an even cycle and an odd cycle (which contributes one to $e(\pi\theta)$). Hence we get $e(\pi\theta) = e(\pi) + 1$.

Case 2 Both a and b occur in the different cycles A and B of π. We may assume that these these two cycles of π are $A = (a_1 a_2 \cdots a_r)$ and $B = (b_1 b_2 \cdots b_m)$ respectively where a equals a_1 and b equals b_1. Then

$$(a_1 a_2 \cdots a_r)(b_1 b_2 \cdots b_m)(a_1 b_1) = (a_1 a_2 \cdots a_r b_1 b_2 \cdots b_m)$$

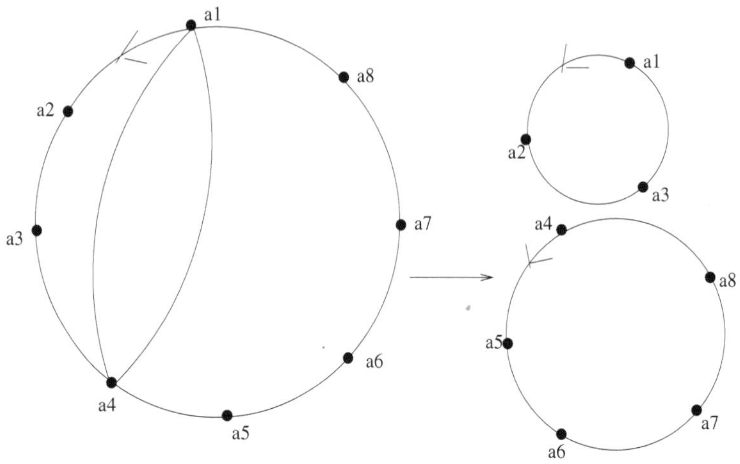

Figure 7.20: One cycle breaks into two

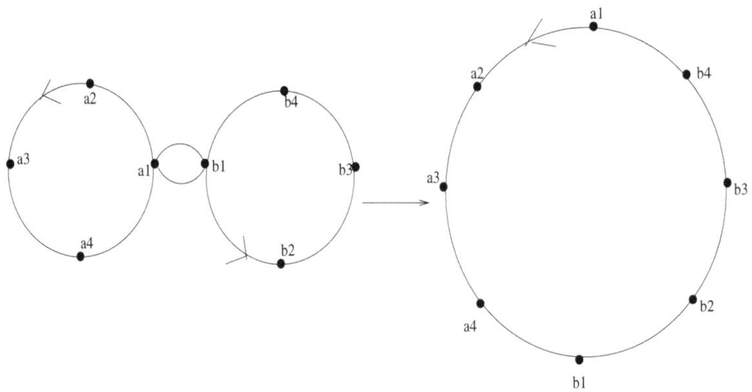

Figure 7.21: Two cycles merge into one

so that two cycles merge into a single cycle C of length $r + m$. Figure 7.21 illustrates the situation when $r = m = 4$. If both r and m are even, then the contribution of A and B to $e(\pi)$ is 2 while that of C (which is also an even cycle) is 1 and we get $e(\pi\theta) = e(\pi) - 1$. If both r and m are odd, then we similarly get $e(\pi\theta) = e(\pi) + 1$. Finally, let one of r and m be even and the other odd. Then the contribution of A and B to $e(\pi)$ is 1 while that of C (which is an odd cycle) is 0 and we get $e(\pi\theta) = e(\pi) - 1$. $\Box$

Lemma 7.7.2. *Every permutation π can be written as a product of transpositions.*

Proof Since π has a (unique) cycle decomposition, it suffices to prove the assertion for a single cycle say $A = (1, 2, \ldots, m)$. It is easily seen by actual multiplication that

$$A = (1, m)(m-1, m)(m-2, m-1) \ldots (3, 2)$$

In fact, we have shown that an m-cycle can be written as a product of $m-1$ transpositions. $\square$

Theorem 7.7.3. *$e(\pi)$ is an even number iff π can be written as a product of an even number of transpositions and no permutation can be written as a product of both an even number of transpositions and an odd number of transpositions.*

Proof The second statement follows from the first since $e(\pi)$ is either even or odd but not both. Let π be written as a product of an even number of transpositions, say

$$\pi = T_1 T_2 \cdots T_{2m}$$

Denoting the product $T_1 T_2 \cdots T_j$ by U_j see that $U_1 = T_1$ and hence $e(U_1) = 1$, $U_{j+1} = U_j T_{j+1}$ and hence by Lemma 7.7.1 we get $e(U_{j+1}) = e(U_j) \pm 1$. Inductively, $e(U_j)$ consists of j summands each of which is ± 1. Finally, $\pi = U_{2m}$ and hence $e(\pi)$ consists of $2m$ summands each of which is ± 1, say k of these are $+1$ each and r of these are -1 each so that $e(\pi) = k - r$. Since $k - r$ and $k + r$ have the same parity modulo 2 and since $k + r = 2m$, it follows that $e(\pi)$ is also even. The same argument works in case π is written as a product of an odd number of permutations to show that in that case, $e(\pi)$ must be an odd number. $\square$

Theorem 7.7.3 shows that Definition 7.7.4 makes sense:

Definition 7.7.4. A permutation π is called an *even permutation* if π can be written as a product of an even number of transpositions. π is called an *odd permutation* if π can be written as a product of an odd number of transpositions.

7.8 Quadratic reciprocity

In this last section on parity, we consider the quadratic reciprocity law, one of the finest discoveries of Gauss. We begin by collecting some facts about prime numbers. Let p be a prime number. Then the set of all residue classes modulo p is a field $\mathbb{F}_p = \mathbb{F}$ and by abuse of language we will assume that the elements of $\mathbb{F}$ come either from the set $\{0, 1, 2, \ldots, p-1\}$ or when p is odd, from the set $\{0, \pm 1, \pm 2, \ldots \pm p'\}$ where p' equals $\frac{p-1}{2}$. Notice that in both the cases, we have a full representative system of residue classes and all the calculations are performed modulo p. The multiplicative group $\mathbb{F}^* = \mathbb{F} - \{0\}$ is *a cyclic group of order* $p - 1$. In view of this, the mapping

$$\mathbb{F}^* \longrightarrow \mathbb{F}^{*2} = \{b^2 : b \in \mathbb{F}^*\}$$

given by $x \longrightarrow x^2$ is a surjective homomorphism. Thus, for an odd prime p, half the elements of $\mathbb{F}^*$ are quadratic residues (squares) and half the elements are quadratic non-residues (non-squares) and every non-zero element y of $\mathbb{F}$ either has no square-root or has exactly two square roots $\pm x$ where $y = x^2$.

Example 7.8.1. For $p = 13$, the quadratic residues are $1, 4, 9, 3, 12, 10$. Here, $3 \equiv 16(mod\ 13)$ and $-1 \equiv 12 \equiv 25(mod\ 13)$ is the justification for including

3 and 12 in the list of quadratic residues. For $p = 19$ the quadratic residues are $1, 4, 9, 16, 6, 17, 11, 7, 5$. Observe that for $p = 13$, $-1 = 12$ is a quadratic residue while for $p = 19$, $-1 = 18$ is not a square. We will show that this is a consequence of the quadratic reciprocity law.

Definition 7.8.2. Let p be an odd prime and let a be an integer coprime to p (thus the residue class of a is in $\mathbb{F}^*$). We define *the Legendre symbol* $\left(\frac{a}{p}\right)$ *by:* $\left(\frac{a}{p}\right) = 1$ if a is a quadratic residue *modulo* p and $\left(\frac{a}{p}\right) = -1$ if a is a quadratic non-residue *modulo* p.

Note that the map $a \to \left(\frac{a}{p}\right)$ is a homomorphism of $\mathbb{F}^*$ onto the multiplicative group $\{1, -1\}$ (whose kernel is the set of all the quadratic residues) and hence we have:

Lemma 7.8.3. *Let a, b be coprime to p where p is an odd prime. Then the following assertions hold.*

(a) $\left(\frac{1}{p}\right) = 1$.

(b) $\left(\frac{ab}{p}\right) = \left(\frac{a}{p}\right)\left(\frac{b}{p}\right)$.

(c) *(Wilson's theorem)* $(p-1)! \equiv -1 (mod\ p)$.

(d) $\left(\frac{a}{p}\right) = a^{\frac{p-1}{2}} (mod\ p)$.

Proof Only the last two assertions need proofs and consider (c) which is true for $p = 3$. Let $p \geq 5$. We collect all possible unordered pairs (k, k') where $kk' \equiv 1(mod\ p)$. Here, 1 and -1 are the only square roots of 1 and the remaining $p-3$ elements in $\mathbb{F}^*$ are paired into pairs of distinct elements k and k'. Since $-1 \equiv (p-1)(mod\ p)$, multiplying these $\frac{p-3}{2}$ equations, we get:

$$2 \times 3 \times \cdots \times (p-2) \equiv 1(mod\ p)$$

Multiplying both the sides of this equation by $p - 1 \equiv \ -1\ (mod\ p)$, we get (c). For (d), observe that if $a = b^2$, then $a^{\frac{p-1}{2}} = b^{p-1} \equiv 1\ (mod\ p)$ because the group $\mathbb{F}^*$ has order $p-1$. Now let a be a quadratic non-residue and again solve $kk' = a$ to obtain pairs (k, k') in $\mathbb{F}^*$. In this case, $k' = k$ is impossible since a is not a quadratic residue. So we get exactly $\frac{p-1}{2}$ such pairs and equally many equations. Multiplying all these equations (we see that each $k \in \mathbb{F}^*$ occurs exactly once), we get $1 \times 2 \times \cdots \times (p-1) \equiv a^{\frac{p-1}{2}} (mod\ p)$ and by (c), the L.H.S. is congruent to -1 modulo p and (d) is proved since $\left(\frac{a}{p}\right) = -1$ in this case. □

For the following discussion, we fix an odd prime p and we also fix a natural number D with $1 \leq D \leq p-1$. Write $p' = \frac{p-1}{2}$. Let $x \in \{1, 2, \ldots, p'\}$. Using the standard Euclidean algorithm (division algorithm), we have the equation:

$$Dx = g_x p + R_x \tag{7.1}$$

where g_x is a non-negative integer, which is the quotient of the division of Dx by p and R_x is the remainder so that $1 \leq R_x \leq p-1$. We now divide the set of all x's in $[p']$

into two disjoint sets P and N where P denotes those x's for which $R_x < \frac{p}{2}$ and N denotes those x's for which $R_x > \frac{p}{2}$. These are exhaustive and exclusive possibilities since p is an odd number. Now define ρ_x and h_x as follows. If $x \in P$, then $h_x = g_x$ and $\rho_x = R_x$. If $x \in N$, then $h_x = g_x + 1$ and $\rho_x = R_x - p$. Note that $1 \leq |\rho_x| \leq p'$ and (7.1) can be rewritten as

$$Dx = h_x p + \rho_x \tag{7.2}$$

Lemma 7.8.4. *(Gauss Lemma) Let p be an odd prime and let $1 \leq D \leq p-1$. Among the values of $x \in [p']$, let exactly n values of x be such that $R_x \geq p' + 1 = \frac{p+1}{2}$ (equivalently n values of x in (7.2) are such that $\rho_x < 0$). Then*

$$\left(\frac{D}{p}\right) = (-1)^n$$

Proof Let $1 \leq x \leq y \leq p'$. We claim that if $|\rho_x| = |\rho_y|$ (in (7.2)), then $x = y$. First assume that both ρ_x and ρ_y are positive. Then by subtraction in (7.2), we get $D(x-y)$ is a multiple of p and since D is coprime to p, we see that $x \equiv y(mod\ p)$ which forces $y = x$. The same argument works if both ρ_x and ρ_y are negative. Finally, let ρ_x be positive and ρ_y be negative. Then by addition, we get p divides $D(x+y)$ (since $\rho_y = -\rho_x$) which is impossible since $x + y \leq 2p' = p - 1$ and D is coprime to p. We have thus proved the claim. Since $|\rho_x|$ takes values only upto p' and since $1 \leq x \leq p'$ the argument shows that all the $|\rho_x|$ are different and hence take every value from 1 to p'. Now let m equations in (7.2) have $\rho_x > 0$ so that $m + n = p'$. Read these equations (7.2) modulo p and multiply them all to get

$$D^{p'}(1 \times 2 \times \cdots \times p') \equiv (-1)^n(1 \times 2 \times \cdots \times p')$$

where the congruence is modulo p. But $(p')!$ is coprime to p and hence may be canceled on both the sides to get $D^{\frac{p-1}{2}} = D^{p'} \equiv (-1)^n(mod\ p)$. Using Lemma 7.8.3(d), we see that $(\frac{D}{p}) \equiv (-1)^n$. Here both the sides are ± 1 and $p \geq 3$. Therefore, we get the equality:$(\frac{D}{p}) = (-1)^n$. □

Corollary 7.8.5. *Let p be an odd prime. Then the following assertions hold.*

(a) $\left(\frac{-1}{p}\right) = (-1)^{\frac{p-1}{2}}$. *Hence if $p \equiv 1(mod\ 4)$, then -1 is a quadratic residue and if $p \equiv 3(mod\ 4)$, then -1 is a quadratic non-residue.*

(b) $\left(\frac{2}{p}\right) = (-1)^{\frac{p^2-1}{8}}$. *Thus $\left(\frac{2}{p}\right)$ equals 1 when $p \equiv \pm 1(mod\ 8)$ and -1 when $p \equiv \pm 3(mod\ 8)$.*

Proof Use of Gauss Lemma (Lemma 7.8.4) with $D = -1$ shows that all the ρ_x in (7.2) are negative (since $1 \leq x \leq p'$) and hence $n = p' = \frac{p-1}{2}$. This proves (a). For (b), take $D = 2$ in the Gauss Lemma. Here $m + n = p'$ and m is the number of terms that are less than or equal to p' in the set $\{2, 4, \ldots, 2p'\}$. Clearly this number is equal to $\left\lfloor \frac{p'}{2} \right\rfloor = \left\lfloor \frac{p-1}{4} \right\rfloor$. So, $n = \frac{p-1}{2} - \left\lfloor \frac{p-1}{4} \right\rfloor$. The remaining part of the proof consists of making cases: $p \equiv 1, 3, 5, 7 (mod\ 8)$. Let $p \equiv 1\ (mod\ 8)$. Then $p = 8t + 1$ so that $n = \frac{p-1}{2} - \lfloor \frac{p-1}{4} \rfloor = 4t - 2t = 2t$ and $\frac{p^2-1}{8} = t(8t+2)$ are both even. The other cases $p \equiv 3, 5, 7 (mod\ 8)$ are similar. □

Theorem 7.8.6. *(Quadratic Reciprocity Law) Let p and q be distinct odd primes. Then*

$$\left(\frac{q}{p}\right)\left(\frac{p}{q}\right) = (-1)^{\frac{(p-1)(q-1)}{4}}$$

Proof In (7.1), take $D = q$ and read the equation modulo 2 to get

$$x \equiv R_x + g_x (mod\ 2)$$

where $R_x = \rho_x$ if $x \in P$ and $R_x = \rho_x + p \equiv \rho_x + 1 (mod\ 2)$ if $x \in N$. After summing both sides from 1 to p', we get

$$\sum_{x=1}^{p'} x \equiv n + \sum_{x=1}^{p'} \rho_x + \sum_{x=1}^{p'} g_x\ (mod\ 2)$$

Since $-\rho_x \equiv |\rho_x|\ (mod\ 2)$ we can change all the negative ρ_x to positive and get

$$\sum_{x=1}^{p'} x \equiv n + \sum_{x=1}^{p'} |\rho_x| + \sum_{x=1}^{p'} g_x\ (mod\ 2)$$

Since $|\rho_x|$'s are all different, this reduces to

$$\sum_{x=1}^{p'} x \equiv n + \sum_{x=1}^{p'} x + \sum_{x=1}^{p'} g_x\ (mod\ 2)$$

and therefore, $n \equiv \sum_{x=1}^{p'} g_x (mod\ 2)$. Since g_x is the quotient on division of qx by p, we see that $g_x = \left\lfloor \frac{qx}{p} \right\rfloor$ and hence $n \equiv \sum_{x=1}^{p'} \left\lfloor \frac{qx}{p} \right\rfloor$ (refer to Chapter 1 for the explanation of the notations). Using Gauss Lemma (Lemma 7.8.4) we get $\left(\frac{q}{p}\right) = (-1)^L$ and exactly using the same arguments $\left(\frac{p}{q}\right) = (-1)^U$ where $L = \sum_{x=1}^{p'} \left\lfloor \frac{qx}{p} \right\rfloor$ and $U = \sum_{y=1}^{q'} \left\lfloor \frac{py}{q} \right\rfloor$ with $p' = \frac{p-1}{2}$ and $q' = \frac{q-1}{2}$. It remains to find out what L and U are. To that end, draw a rectangle in the first quadrant of $\mathbb{R}^2$ with corners at $(0, 0), (\frac{p}{2}, 0), (\frac{p}{2}, \frac{q}{2}), (0, \frac{p}{2})$ as shown in Figure 7.22 so that R has length $\frac{p}{2}$ along the x-axis and $\frac{q}{2}$ along the y-axis.

Call a point (x_0, y_0) a *lattice point* if both x_0 and y_0 are positive integers. Since p, q are both odd, no point on the boundary of R is a lattice point. Also, for a lattice point (x_0, y_0), we must have $1 \leq x_0 \leq \frac{p-1}{2}$ and $1 \leq y_0 \leq \frac{q-1}{2}$ and hence the number of lattice points is $\frac{(p-1)(q-1)}{4}$. Since $py = qx$ has no integer solution (x_0, y_0) in R, no lattice point lies on the diagonal of R. We can thus count the number of points below and above the diagonal of R. For a fixed positive integer $x_0 \leq p' = \frac{p-1}{2}$ this number is just the number of lattice points on the (vertical) line $x = x_0$ that are below the diagonal. Since the diagonal has the equation $qx = py$, we see that the number of lattice points on this line is $\left\lfloor \frac{qx_0}{p} \right\rfloor$. Summing over all $x_0 \leq \frac{p-1}{2}$, we see that the number of lattice points below the diagonal is equal to $\sum_{x=1}^{p'} = \left\lfloor \frac{qx}{p} \right\rfloor$ which is equal to L. Similarly, the number of lattice points above the diagonal is equal to U. This completes the proof of the theorem. □

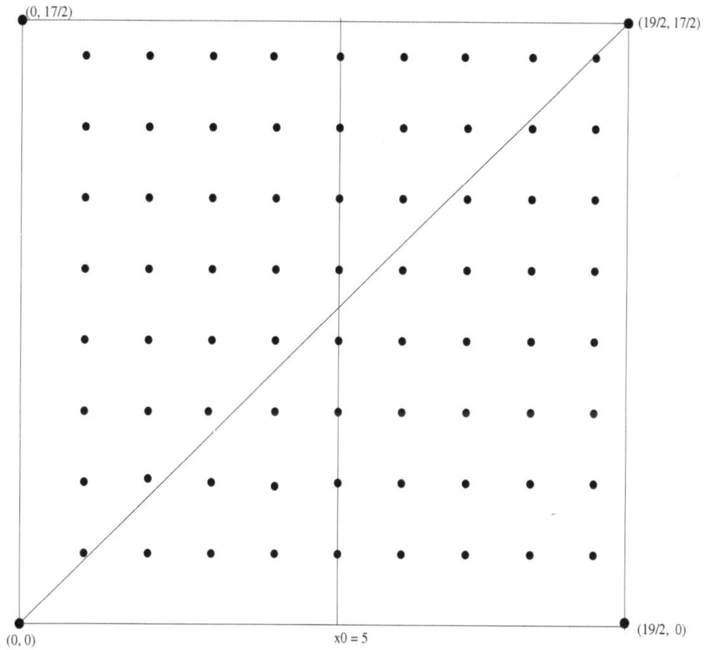

Figure 7.22: The 13×11 rectangle and lattice points

Example 7.8.7. Note that if p and q are odd primes then $(-1)^{\frac{(p-1)(q-1)}{4}} = 1$ except when p and q are both congruent to $3(mod\ 4)$. Consider the following examples.

(a) We wish to find $\left(\frac{101}{103}\right)$. Here $\left(\frac{103}{101}\right) = \left(\frac{2}{101}\right) = 1$ using Corollary 7.8.5. Using the quadratic reciprocity law (Theorem 7.8.6), $\left(\frac{101}{103}\right) \times \left(\frac{103}{101}\right) = 1$. Hence $\left(\frac{101}{103}\right) = -1$. Alternatively, $\left(\frac{101}{103}\right) = \left(\frac{-2}{103}\right) = \left(\frac{-1}{103}\right) \times \left(\frac{2}{103}\right)$. Here the first term is -1 since $103 \equiv 3(mod\ 4)$ and the second term is 1 since $103 \equiv 7(mod\ 8)$.

(b) We wish to find $\left(\frac{682}{257}\right)$. Here $682 = 2 \times 11 \times 31$ and hence $\left(\frac{682}{257}\right) = \left(\frac{2}{257}\right) \times \left(\frac{11}{257}\right) \times \left(\frac{31}{257}\right)$ which is equal to ABC say. We have $A = \left(\frac{2}{257}\right) = 1$ using Corollary 7.8.5. $B = \left(\frac{11}{257}\right)$ and using the quadratic reciprocity law (Theorem 7.8.6), $\left(\frac{11}{257}\right)\left(\frac{257}{11}\right) = -1$. But $\left(\frac{257}{11}\right) = \left(\frac{4}{11}\right) = 1$ and hence $B = -1$. Finally, $C = \left(\frac{31}{257}\right)$ and using the quadratic reciprocity law (7.8.6), $\left(\frac{31}{257}\right)\left(\frac{257}{31}\right) = 1$. Here $\left(\frac{257}{31}\right) = \left(\frac{9}{31}\right) = 1$ and hence $C = \left(\frac{31}{257}\right) = 1$. Therefore $ABC = -1$.

7.9 Exercises for Chapter 7

7.1 Consider the following alternative proof of the assertion in Example 7.1.1. Take the product of all the summands in the given equation to obtain 1. Hence argue that we must have an even number of summands that are -1 and then complete the solution.

7.2 Show that on an 8×8 chessboard, it is possible to construct a domino covering in which every ruling is crossed by some domino. Compare this with the result of Example 7.1.2.

7.3 Prune an 8×8 chessboard at the diametrically opposite ends (the cells $a1$ and $h8$ in the standard chess terminology). Is it possible to cover this pruned chessboard of 62 cells using 31 non-overlapping dominoes?

7.4 There are 12 coins on a table and A and B do the following. A and B both know the number of coins with heads up and tails up, to start with. A does not look at the board and B chooses a coin and turns it upside down. This operation is performed a number of times. Every time B performs such an operation, he rings a bell (he may choose any coin any number of times and perform the operation). At the end of the game, A is shown all the coins except one. Can A guess if this coin is heads or tails?

7.5 Show that for all $n \geq 4$, $1! + 2! + \cdots + n!$ is not a perfect square.

7.6 On the set $S = [20]$, perform the following operation successively. Choose any two distinct numbers say a and b from S, remove both a and b from S and add a (new) number $a + b + ab$ to S so that a new set say S_1 is obtained. Repeat the same operation on the set S_1 to obtain a new set S_2 (with 18 numbers) and so on till we are left with just one number say α. Show that independent of the manner in which the numbers are picked up, the value of α is the same. Also determine this value.

7.7 Prove Theorem 7.4.1.

7.8 Refer to Section 7.4. From the graph D_n, draw a new graph ${D_n}^*$ as follows. The vertices of ${D_n}^*$ are the 2^n edges of D_n. For two edges e and e' of D_n, define (directed) adjacency from e to e' provided the head of e equals the tail of e'. Show that the graph ${D_n}^*$ is just the graph D_{n+1}. This is exploited to obtain recursively a construction of an Eulerian circuit in D_{n+1} (and hence a de Bruijn sequence in M_{n+1}) from a given Eulerian circuit in D_n. Refer to van Lint and Wilson [57].

7.9 Suppose we have 5 red beads and 4 blue beads arranged regularly on a circle. For every pair of adjacent beads, make a new bead of color red (place it between the two existing beads under consideration) if the two beads are of the same color and make a new bead of color blue if the two beads are of different colors. Perform this operation for every pair of adjacent beads. Finally, erase all the old beads so that we still have 9 beads of two colours (but arranged differently and possibly also with different numbers of beads of each colour). Is it possible to obtain all the beads of the same colour by repeating the above operation several times?

7.10 In exercise 7.9, suppose we had 4 red beads and 5 blue beads. What would be the answer to the same question?

7.11 The product of 22 integers is 1. Can their sum be 0?

7.12 Numbers 1 to 10 are written on a line (with gaps to put the symbols $+$ and $-$). Can we put pluses and minuses between them so that the sum is 0?

7.13 25 boys and 25 girls are to be seated on a round table. Is it possible to arrange this so that no person has only boys to his/her left and right?

7.14 Let G be a simple graph and suppose that the degree of each vertex is an even number. Prove that we can assign directions to the edges of G such that the resulting digraph has the property that $d^+(x) = d^-(x)$ holds for every vertex x of G.

7.15 Let X be a set of order $n \geq 2$ and let C be a collection of subsets of X with the following property: $\emptyset \notin C$ and every proper subset Y of X has a non-empty intersection with an *even number* of sets in C. Prove that C is the collection of all the subsets of X.

7.16 On the standard 8×8 chessboard with the usual alternate black and white colouring of the 64 cells, we remove one white and one black cell *anywhere on the chessboard.* Can the resulting board of 62 cells be covered by 31 non-overlapping dominoes? Compare this with exercise 7.3.

7.17 Show that product of two even or two odd permutations is an even permutation while product of an even and an odd permutation is an odd permutation. Hence show that the set of all even permutations is a group under multiplication (composition) of permutations. By actually setting up an explicit bijection, show that the number of even permutations is $\frac{n!}{2}$. *The subgroup of all even permutations is called the alternating group and is an index 2 subgroup of the group of all the permutations.*

7.18 A Hadamard matrix of order n is a square matrix $H = [h_{i,j}]$ such that each entry of H is ± 1 and $HH^t = nI_n$. Thus every two distinct rows of H are orthogonal. Prove the following.

(a) Interchange of rows/columns of a Hadamard matrix results in a Hadamard matrix of the same order.

(b) Multiplying each entry of a fixed row (or a fixed column) of a Hadamard matrix results in a Hadamard matrix of the same order.

(c) By using the previous two methods one can bring a Hadamard matrix H into a normalized form for which the first row consists of all the entries equal to 1.

(d) Write down the equation giving inner products among the first, second and the third row and hence show that if $n > 2$, then n must be a multiple of 4.

Comment: A long standing conjecture asserts that there is a Hadamard matrix of order $4t$ for any natural number t; see [57].

7.19 Show that a 10×10 chessboard cannot be covered by nonoverlapping T-shaped tetraminoes of the following kind.

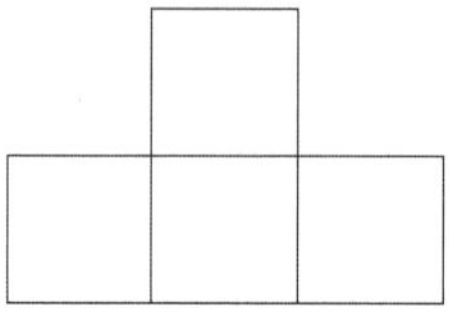

Figure 7.23: T-shaped Tetramino

7.20 A rectangular floor is covered by a certain number of non overlapping tiles each of which is either a 1×4 tile or a 2×2 tile. One tile got smashed and has to be replaced but a tile of the other type is the only tile available. Can we rearrange the tiles so that the floor can still be covered by removing the smashed tile and use the new available tile of the other type?

7.21 Each one of the numbers $a_1, a_2, \ldots, a_{50}$ are written around a circle regularly where each $a_i = \pm 1$. One is allowed to ask questions concerning the product $a_i a_{i+1} a_{i+2}$, a product of three consecutive integers. What is the least number of questions required in order to assert that the product of all the a_i's can be correctly guessed?

7.22 Solve the following congruences:

(a) $x^2 \equiv 5(mod\, 227)$

(b) $x^2 \equiv -5(mod\, 227)$

(c) $x^2 \equiv 7(mod\, 1009)$

(d) $x^2 \equiv 51(mod\, 229)$.

7.23 Compute the following Legendre symbols.

$$\left(\frac{127}{397}\right), \left(\frac{449}{569}\right), \left(\frac{198}{547}\right), \left(\frac{300}{1081}\right), \left(\frac{373}{733}\right), \left(\frac{-1}{1533}\right), \left(\frac{59}{2017}\right)$$

$$\left(\frac{89}{4651}\right), \left(\frac{131}{881}\right), \left(\frac{223}{271}\right), \left(\frac{2}{1613}\right), \left(\frac{307}{631}\right), \left(\frac{593}{769}\right), \left(\frac{37}{2029}\right)$$

7.24 Let p be an odd prime. For which values of p is $\left(\frac{-2}{p}\right) = 1$? If $p > 3$, then for which values of p is $\left(\frac{3}{p}\right) = 1$?

7.25 A Knight's move on a chessboard takes it to a nearest square of opposite colour from a given square the new square is not the nearest diagonally opposite square. A knight's closed tour on a chessboard begins at some cell, visits every cell and returns back to the same cell. Euler proved that a Knight's closed tour is possible on the standard 8×8 chessboard. Consider a $4 \times n$ chessboard with 4 rows and n columns.

(a) Show that there is no Knight's move from the first row to the last (fourth) row and hence the set of the $2n$ cells from the first and the fourth row is *an independent* set A in the Knight's graph.

(b) Prove that if we have an Eulerian circuit then the Knight must visit cells in A and not in A alternately and hence must visit cells of only one colour in A.

(c) Conclude that we do not have a knight's tour on a $4 \times n$ chessboard.

Comment: Refer to Watkins [58].

7.26 This exercise is borrowed from R. Thomas [53]. Define an (infinite) graph with vertex set the set of all real numbers and two vertices x and y adjacent if $|y-x| = 3^k$ for some integer k.

(a) Show that this graph has no odd circuits and hence it is a bipartite graph with bipartition (X, Y) say.

(b) Use the following result from analysis: Let A be a subset of $\mathbb{R}$ such that A has a positive Lebesgue measure. Then there is a $\delta > 0$ such that $\forall\ \alpha$ with $|\alpha| < \delta$, $A \cap (A + \alpha) \neq \emptyset$.

(c) Conclude that both X and Y are not Lebesgue measurable.

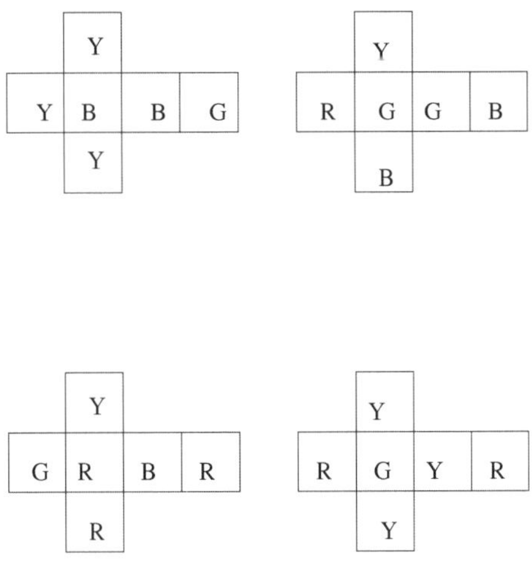

Figure 7.24: Cubes with faces coloured

7.27 Consider the set of 4 cubes with faces coloured in red, blue green and yellow as shown in Figure 7.24. Show that the instant insanity puzzle has a solution for this colouring.

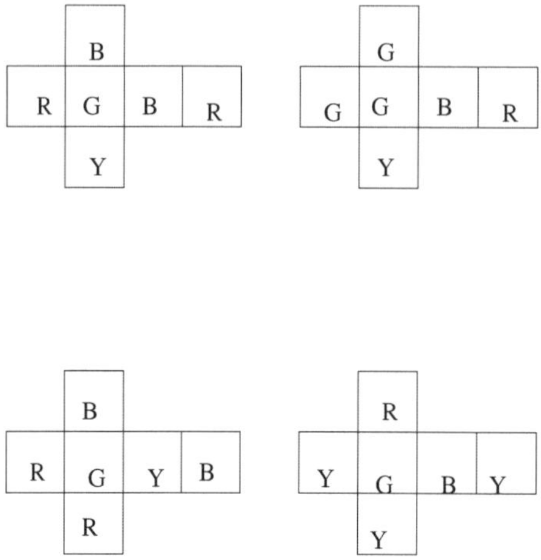

Figure 7.25: Cubes with faces coloured

7.28 Consider the set of 4 cubes with faces coloured in red, blue green and yellow as shown in Figure 7.25. Show that the instant insanity puzzle has a solution for this colouring.

7.29 Consider the set of 4 cubes with faces coloured in red, blue green and yellow as shown in Figure 7.26. Show that the instant insanity puzzle has no solution for this colouring.

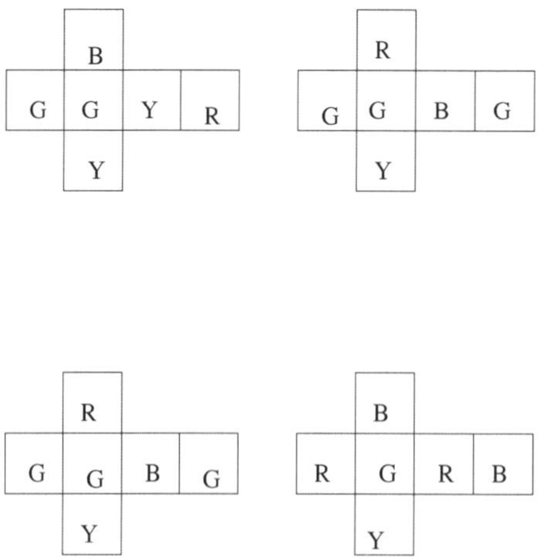

Figure 7.26: Cubes with faces coloured

7.30 Show that the following 4×4 matrix is non-singular.

$$\begin{pmatrix} 1423 & 3214 & 2134 & 3412 \\ 4132 & 1243 & 2314 & 2134 \\ 1342 & 2134 & 2431 & 3124 \\ 3142 & 4312 & 3214 & 3421 \end{pmatrix}.$$

7.31 Determine the sign of the following permutations on the set $[n] = [9]$:

$$\begin{pmatrix} 1,2,3,4,5,6,7,8,9 \\ 2,7,8,1,5,4,9,3,6 \end{pmatrix}$$

$$\begin{pmatrix} 1,2,3,4,5,6,7,8,9 \\ 9,8,7,6,5,4,3,2,1 \end{pmatrix}$$

and the permutation $(1568)(3427)$.

7.32 Start with the positive integers $1, 2, \ldots, 4n-1$. In one move, you can choose two (existing) numbers and replace them by their difference. Show that after $4n-2$ moves you will be left with an even integer.

7.33 Start with the standard 8×8 chessboard with the usual (alternate) white and black coloring of the cells of the chessboard. In one move, you can either choose a row or a column or a 2×2 sub chessboard and change the colors of all the cells on that row or column of the sub chessboard. The aim is to achieve a chessboard with exactly one black cell after repeatedly making the moves. Can that be done?

7.34 We have 13 white, 15 black and 17 red chips on a table. In one move, we can choose two chips of different colors, throw them out and add two chips of the third colour. Is it possible, through a sequence of such moves, to obtain all the chips of the same colour on the table?

7.35 A $3 \times 3 \times 3$ butter cube (like a Rubik's cube) consists of 27 small unit cubes. A mouse starts on some outer small cube and eats one small cube per day in such a manner that the small cube eaten on any day shares a face with the small cube eaten on the previous day. Is it possible for the mouse to arrange his eating pattern over 27 days so that the center cube is the one the mouse will eat on the last day?

7.36 Numbers 0 and 1 are arbitrarily written on the six faces of a cube. In one move, we erase the number on (each) face and replace it by the average of the four neighboring faces. After 26 moves, it was observed that we are back to the original numbering we started with. Show that this is possible only if the initial numbering had all the faces numbered 1 or all the faces numbered 0.

7.37 Work out the previous problem (Exercise 7.36) with numbers 0 and 1 arbitrarily written on the eight vertices of a cube. In one move, we replace a number on a vertex by the average of the three neighboring vertices.

7.38 Consider the following three Nim games with 6 piles given by the three sequences. In each case, find a good move or assert (with reason) that there is no good move.

(a) $(9, 14, 6, 7, 3, 7)$.

(b) $(39, 57, 64, 98, 34, 86)$.

(c) $(1, 4, 9, 16, 25, 36)$.

7.39 Consider the Nim game with n piles in which the i-th pile has $a_i = i$ chips. For which values of n does the first player A have a winning move and for which values of n is A in a losing position?

7.40 Given a 10×10 chess board, define a new piece called "lion" with the following property: The lion can move one square down, or one square to the right or one square diagonally upward and to the left. That is, it can only move to three possible squares on its next move. Starting from any original square, can the lion visit all the 100 squares exactly once and return back to its original square?

7.41 We are given an array (a matrix) of $m \times n$ cells in which initially numbers ± 1 are written on every cell such that exactly one cell has -1 written on it. During a move, we choose some cell on which -1 is written, replace this -1 by a 0 and simultaneously multiply all the numbers on the neighboring cells by -1 (we say that two cells are neighbors if they have a common side). We proceed through a succession of such moves hoping to covert the array into one that has only 0's written on each cell. Find out the values of m and n for which this can be done.

7.42 Beginning at the point $(0, 0)$ a snail crawls one unit moving either parallel to the x-axis or y-axis in 15 seconds. At the end of every 15 seconds interval, it makes either a right or left turn (that is, turns 90^0 in some direction). Show that when it eventually reaches its starting position, it would do so in full minutes.

7.43 Let (a_i) be a sequence of numbers each equal to ± 1 such that for some n, we have

$$a_1 a_2 + a_2 a_3 + \cdots + a_{n-1} a_n + a_n a_1 = 0$$

Show that 4 divides n.

7.44 A die (natural) is put on the corner square of a 50×50 chessboard. It rolls on the board and reaches the opposite corner. The sum of the numbers on the bottom of the die (the face touching the board) is S. Show that S ranges between 342 and 351.

7.45 An ant moves on the usual 8×8 chessboard beginning in some square (cell) and facing some direction. Each of the 64 cells is initially labeled L or R and then ant moves only horizontally or vertically. The ant moves according to the following set procedure. It takes a step forward and looks at the label of new cell (it has just entered). If the cell is labeled L, then it turns left 90° and if the cell is labeled R, then it turns right 90°. While exiting the cell the ant changes the label of the cell (from L to R and R to L), takes its next step forward and repeats this procedure over and over again. Given these rules, is it possible to devise an initial labeling procedure so that the ant will remain confined to the chessboard (without ever forcing it to leave the chessboard)? Also answer the question for any $n \times n$ chessboard.

7.46 45 cups are placed on a table in face up position. Can one change the position of 6 of them at a time (this will have to be done several times and the same cup may change position several times) so that finally all the cups are in face down position?

7.47 n identical balls are initially separated into a certain number m of piles where $m \geq 2$. At each step, we are allowed to perform the following operation. Pick up any two non-empty piles, the first containing p balls and the second containing q balls where $p \geq q$. Then remove q balls from the first pile and put them in the second pile. We continue performing this operation repeatedly with the idea of eventually to have all the n balls in a single pile. Prove that regardless of the initial distribution of balls into piles this can be achieved iff n is a power of 2.

7.48 On some cell (square) of a standard 8×8 chessboard, the number 2007 is written. On the remaining 63 cells numbers from 1 to 63 are written with one number on each cell but in a completely arbitrary manner. A move consists of choosing any row or any column, replacing each even number in that row (respectively column) by *any* odd number and each odd number in that row (respectively column) by *any* even number. We repeat these moves any number of times each time choosing a row or column and finally wish to obtain the same number on all the cells of the chessboard. Can this be done?

7.49 Initially, we have 1001 chips on a table and the chips are in a single heap. At every stage, we choose a heap, throw away one chip from the heap and then divide the heap into two heaps of chips (not necessarily of equal size). Thus, in the first stage, we get two heaps with the total number of chips equal to 1000. We then choose one of these two heaps and so on. The aim is to end with a certain number of heaps each containing exactly 3 chips. Can this be achieved?

7.50 Every cell on the standard 8×8 chessboard is colored white or black. In one move, we are allowed to choose some hook (3 cells such that one cell shares a common side with each of the other two and the cells do not form a 3×1 rectangle) and interchange colours on the 3 cells in that hook. Is it possible to colour all the entries in white after a certain number of moves regardless of the initial coloring?

7.51 Two players A and B take turns removing chips from a pile that initially contains n chips. The first player A cannot remove all the n chips at the beginning. A player, on his turn, must remove at least one chip. However he cannot remove more chips than those his opponent removed on his previous turn. The player who removes the last chip is the winner. What is the winning strategy for player A? (*Note: The problem is considerably more difficult if we allow any player to remove up to two times the number of chips removed by his opponent on the previous turn; this is called the Fibonacci Nim and is discussed in Chapter 11*).

7.52 n children are sitting forming a circle and each child has a certain number of chocolates to start with. On every move each child gives half of his chocolates to the child on his right. In case the number of chocolates with him is an odd number, the child borrows one chocolate from a large reservoir of chocolates. This process continues with the next round and so on. Does the process eventually stabilize with all children having the same number of chocolates or does it continue for ever?

7.53 Work out the previous problem with the following modification. If a child has an odd number of chocolates, then he eats one chocolate.

7.54 Work out the same problem with the following modification. A child gives half of his chocolates to his right and half to his left.

7.55 Numbers from 1 to 15 are written on the 15 cells of a 4×4 chessboard leaving the cell at position $(4, 4)$ with no number on it and remaining 15 cells filled with one number on each cell. A move consists of exchanging a gap (cell with no number on it) with an adjacent cell. Thus the gap moves to an adjacent cell and the number on the adjacent cell is written on the cell where we had a gap. We have to construct a sequence of moves, from a given position with numbers written on the 15 cells except the cell at $(4, 4)$ so that at the end we reach

1	2	3	4
5	6	7	8
9	10	11	12
13	14	15	

For which initial configurations can we achieve this?

7.56 Let n be a natural number. We have a set of $2n + 1$ people each with weight in full kilograms with the following property. If we remove any person then the remaining $2n$ persons can be divided into two sets of n persons each such that the sum of the weights in both the sets is the same. Show that all the people have the same weight.

7.57 Show that the assertion in the previous exercise cannot be made under the weaker hypothesis that given any person, the remaining $2n$ persons can be divided into some two subsets (not necessarily with n persons each that add to equal weights.

7.58 To every vertex of a regular pentagon, an integer is initially assigned in such a way that the sum of all the five numbers is positive. The following operation is allowed. We choose three consecutive vertices on which numbers x, y, z are written (in that order) such that $y < 0$. We then replace the three numbers by

$$x' = x + y, \; y' = -y, \; z' = y + z$$

respectively. We repeatedly apply this operation choosing a vertex on which a negative number is written. Can the operation be continued for ever?

7.59 Consider the (undirected) simple graph G constructed on the set of all the subsets of $[n]$ as follows. A subset A and a subset B are adjacent if $A \subset B$ and $|B| = |A| + 1$. Show that the graph G (called the Boolean lattice) is n-regular and is isomorphic to the hypercube graph Q_n.

Chapter 8

Pigeonhole principle

8.1 Introduction

If the average weight of students in a school classroom is 41 kg. (kilograms) then there is some student who weighs at least 41 kg. and some student who weighs at the most 41 kg. In fact, if there is one student with weight larger than 41 kg., then there must be some other student whose weight is less than 41 kg. There are four suits (spades, hearts, diamonds and clubs in the hierarchy) in a pack of 52 cards where every bridge player receives 13 cards in his hand. I was told of a story of a mathematician from a prestigious research institute who was taken out to play bridge by his bridge loving friend. No matter how the thirteen cards are dealt to you, said our friend to the mathematician, in some suit, you must receive at least 4 cards. The mathematician could not see this for quite some time and when he was convinced, he exclaimed, "Oh, that follows from the Pigeonhole principle!" An average bridge player, who loves the game of bridge, is hardly aware that this fact is a consequence of the Pigeonhole principle. In some sense, the Pigeonhole principle is intuitively too obvious and too commonsense (perhaps like the game of bridge) to be given the status of a principle. In its most general form, following are statements that go under the name "Pigeonhole Principle".

(a) *Simplest form*: When $n+1$ pigeons are to be put in n boxes, there is at least one box that receives two (or more) pigeons.

(b) *Slightly more profound*: If we put $kn+1$ objects in n boxes, then some box must receive $k+1$ or more objects.

(c) *Equivalently*: If n non-negative integer quantities have an average strictly greater than α then there must be at least one quantity whose value is at least $\lceil \alpha \rceil$, where $\lceil x \rceil$ denotes the smallest integer greater than or equal to x.

All these statements have twins that are other sides of the coin. Thus the first statement has a twin which reads like: When $n-1$ pigeons are to be put in n boxes, there is at least one box that remains empty (and receives no pigeons). The reader should similarly find the statements corresponding to the other two. Finally, all such kinds of statements are called *Pigeonhole principle or the Box principle.* It seems the *Box principle* also called the *Dirichlet drawer principle* was used by Dirichlet and we will soon see the use of this principle as a non-trivial application to Diophantine approximations. *Since the topic at an elementary level has much less by way of theory, we begin by giving a number of interesting examples.*

Example 8.1.1. A student has 6 weeks (that is, 42 days) to prepare for her examination and she has decided that during this period she will put in a total of 70 hours towards her preparation for the examination. She decides to study in full hours every day, studying at least one hour on each day. We have to prove that no matter how she schedules her studying pattern, she will study for exactly 13 hours during some consecutive days.
If a_i denotes the number of hours she studies on the i-th day, then we are given that each a_i is a positive integer and the sum $a_1 + a_2 + \cdots + a_{42} = 70$. To get the required succession of days, we must find some m and j such that $m \leq j$ and $a_m + \cdots + a_j$ equals 13. Trying out all the possibilities, is close to impossible as well as dumb. Let b_i denote the partial sum $a_1 + \cdots + a_i$ which is the number of hours she studies upto the i-th day. Our given data then translates into

$$1 \leq b_1 < b_2 < \cdots < b_{41} < b_{42} = 70 \tag{8.1}$$

and we have to find some $i < j$ such that $b_i + 13 = b_j$ that is, $b_j - b_i = a_{i+1} + \cdots + a_j = 13$ (clearly then $i < j$). Hence besides the 42 numbers in $B = \{b_1, b_2, \cdots, b_{42}\}$ we also look at 42 more numbers in $B' = \{b_1 + 13, b_2 + 13, \cdots, b_{42} + 13\}$ which are also 42 different numbers and the largest among them is $b_{42} + 13 = 70 + 13 = 83$. Hence the $2 \times 42 = 84$ numbers in $B \cup B'$ are actually among the positive integers from 1 upto 83 and hence by the pigeonhole principle, we see that two numbers in $B \cup B'$ must be equal. As we already saw the numbers in B are all distinct and so are the numbers in B'. Hence, we must have for some i and j, $b_i + 13 = b_j$ giving the required succession of days when she studied for exactly 13 hours.

Example 8.1.2. Take the same set up as in Example 8.1.1. The logic also applies in case one wants to prove that there are consecutive days during which the student studied for exactly r hours where r is a positive integer less than 13. Extending this logic further, assume that there is no succession of days during which the student studied for 55 hours and let k be a positive integer, $20 \leq k \leq 27$. We wish to show that there is a succession of days during which the student studied for exactly k hours or a succession of days during which she studied for $55 - k$ hours.

To solve this problem, we take the set B as before and let $C = \{b_i + k : i = 1, 2, \cdots, 42\}$ and let $C' = \{b_i + 55 : i = 1, 2, \cdots, 42\}$. Then in these three sets we have a total of $3 \times 42 = 126$ positive integers of which the largest is $b_{42} + 55 = 70 + 55 = 125$ and hence by pigeonhole principle, some two numbers must be equal. The numbers in each of the three sets B, C, C' are all distinct as we just saw. If some number in B equals a number in C', then we must have for some i and j, $b_j = b_i + 55$ giving a succession of days during which she studies for exactly 55 hours which is disallowed by the assumption. If a number in B equals a number in

C, then we have a succession of days during which the student studies for k hours. In the last alternative, two numbers from C and C' are equal giving $b_j + k = b_i + 55$ which implies $b_j - b_i = 55 - k$ completing the proof of the assertion.

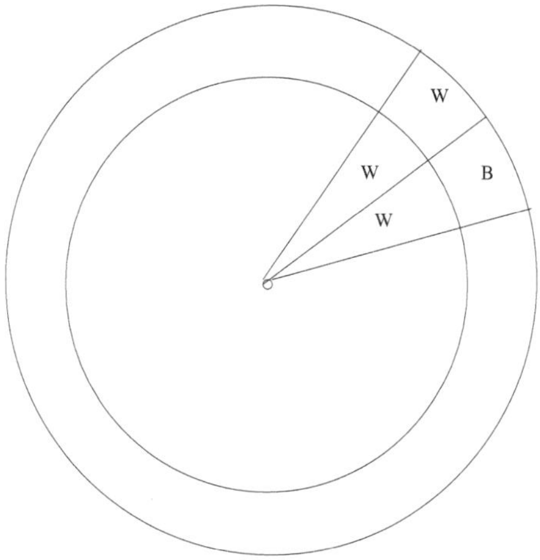

Figure 8.1: Two concentric discs and sectors

Example 8.1.3. Consider two concentric discs of different radii the inner smaller and the outer larger such that each disc is divided into 200 equal (central or radial) sectors (each making an angle of $\frac{360^o}{200}$ degrees at the center). The sectors on the inner disc are colored in white and black and so are all the sectors on the outer disc (see Figure 8.1). The only stipulation that applies is that there are exactly 100 sectors of each color on the inner disc. The sectors on the outer disc are colored in any manner (using the two colors). We wish to show that it is possible to suitably rotate the outer disc so that we get color agreements on at least 100 sectors.
To solve this problem, we make a two-way counting as follows. Clearly, we may keep the outer disc fixed and label the sectors from 0 through 199 in an anticlockwise manner. We now rotate the inner disc through an angle of R_j degrees where $R_j = j \times \alpha$ and α is the angle $\frac{360^o}{200}$ (in degrees). Let $S = \{(R_j, t) : j = 0, 1, \cdots, 199\}$ be the set of ordered pairs (R, t) where R refers to a rotation (through a certain angle R_j) and t is a sector where the two discs have the same colour. Hold a certain sector t on the outer disc fixed and count the number of rotations R that produce colour alignment *on that sector.* Since the inner disc has exactly 100 sectors of each colour, no matter what the colour of the segment on the disc is, we get alignment for exactly 100 rotations R. Adding this over all sectors t on the outer disc shows that $|S| = 100 \times 200$. This two-way counting argument shows that (since there are exactly 200 rotations), we get alignment of colours for 100 segments per rotation, on an average and hence by pigeonhole principle there must be some rotation, for which we get an alignment of colours for ≥ 100 sectors proving the assertion.

Example 8.1.4. Suppose we are given two concentric regular octagons one smaller and the other larger. Numbers from 1 to 8 are (already) written on the corners of the smaller octagon (but in any order) and we have to write numbers from 1 to 8 on the corners of the larger octagon so that the following stipulation is met. There are 8 possible rotations of either octagon through angles that are multiples of 45°. For every such rotation, we find out as to how many corners the numbers (on the corresponding corners) are matching. For every possible rotation, we must have at least one agreement of numbers. Can this be done?
To put the problem in the set up of the earlier problem, we have two concentric discs, each divided into 8 equal sectors that subtend an angle of 45° degrees at the center. Numbers from 1 through 8 are written on the sectors of the smaller disc and we have to write numbers from 1 through 8 on the sectors of the larger disc so that for every rotation, we get agreement of numbers on at least one sector. In order to simplify understanding of the problem, we may assume that the numbers are from 0 to 7 (instead of 1 to 8) since the solution cannot depend on the choice of 8 numbers. Also, if we apply a permutation σ to the set $\{0, 1, \cdots, 7\}$ then the problem has an affirmative answer if and only if the same problem with σ applied to numbers written on each sector has an affirmative answer. *Hence, without loss of generality, we may assume that the numbers on the inner disc are from 0 to 7 written in an anticlockwise manner.* Thus we have a (cyclic) permutation $a_0, a_1, \cdots, a_7$ on the inner disc where $a_i = i$ and we have to choose a (cyclic) permutation $b_0, b_1, \cdots, b_7$ on the outer disc so that no matter which rotation we choose, we always get an agreement of numbers on some sector. Notice that initially (which corresponds to zero rotation), we have numbers $a_i = i$ and b_i on the i-th sector. Since rotations are relative movements of the two discs, we may hold the outer disc fixed (in position) and rotate only the inner disc (through angles that are multiples of $\alpha = 45^o$) and thus a rotation through 3α will move the sector i on the inner disc to the sector $i + 3$ (with everything read modulo 8). Holding the outer disc fixed, we find out which rotation will produce an agreement of numbers on the i-th sector. For this to happen, we must get the number b_i on the i-th sector of the inner disc. Clearly, this can happen only for one rotation. Hence, if we let S denote the set of all ordered pairs of the form (rotation, sector) where we count for each rotation, on which sectors we get agreement of numbers (exactly as in the previous example), then $|S| = 8$. Since we have only 8 rotations, assuming that the question has an affirmative answer, we get, using the pigeonhole principle, exactly one agreement of numbers per rotation or there is a one to one correspondence between rotations and the sectors that produce agreement of numbers.
Suppose that we have to rotate through $r\alpha$ in order to get an agreement of numbers on the j-th sector. On this sector the outer disc has b_j and on the inner disc, the sector that was at location $j-r$ gets rotated to the j-th sector and hence, on this sector, the number must be b_j. This produces the equation $b_j = a_{j-r}$. From the preceding paragraph, we see that j and r uniquely determine each other. But $a_{j-r} \equiv j - r \ (mod\ 8)$ and hence with everything read modulo 8, we get $b_j + r \equiv j$. Summing this over all r (and hence over all j and all b_j), we obtain

$$\sum_{i=0}^{7} i + \sum_{i=0}^{7} i \equiv \sum_{i=0}^{7} i$$

Since this congruence has no solution (because 8 does not divide $\binom{8}{2}$), it follows that the answer to the question is that *it cannot be done.*

Example 8.1.5. Our last example is the following commonly occurring problem. A physical instructor at a school arranged mn children in a school in an $m \times n$ array $A = [a_{i,j}]$ such that the heights are monotonically increasing from left to right. That is, we have $a_{i,1} \leq a_{i,2} \leq \cdots \leq a_{i,n}$ for all the rows i. Now suppose that the instructor decides to shuffle the children in each column so that the heights are increasing from the front to the back (thus in the changed array $B = [b_{i,j}]$, we have $b_{1,j} \leq b_{2,j} \leq \cdots \leq b_{m,j}$ in each column j). Is it still true that the original arrangement $b_{i,1} \leq b_{i,2} \leq \cdots \leq b_{i,n}$ remains intact for every row i? We wish to prove that that is the case.

In fact, if that were not true, arranging children so that the heights are (monotonically) increasing from left to right as well as from the front to the back simultaneously, would have been a very difficult task in the absence of such an assertion. To that end, we have the following.

Lemma 8.1.6. *Let $A = (a_1, a_2, \cdots, a_n)$ and $B = (b_1, b_2, \cdots, b_n)$ be two sequences of real numbers such that $a_i \leq b_i \; \forall \; i = 1, 2, \cdots, n$. Rearrange A and B into two (new) sequences C and D respectively where $C = (c_1, c_2, \cdots, c_n)$ and $D = (d_1, d_2, \cdots, d_n)$ such that C is a rearrangement of A and D is a rearrangement of B and C and D are both monotonically increasing: $c_1 \leq c_2 \leq \cdots \leq c_n$ and $d_1 \leq d_2 \leq \cdots \leq d_n$. Then $c_i \leq d_i \; \forall \; i = 1, 2, \cdots, n$.*

Proof We have $d_1 = b_i$ for some i and hence $d_1 = b_i \geq a_i \geq c_1$ since c_1 is the smallest among all the a_i's. Inductively, we can assume that $c_1 \leq d_1, \cdots, c_{k-1} \leq d_{k-1}$ and prove that $c_k \leq d_k$. Let $S = \{i : i = 1, 2, \cdots, n \text{ and } a_i \geq c_k\}$ and $T = \{j : j = 1, 2, \cdots, n \text{ and } b_j \leq d_k\}$. Then $c_k, c_{k+1} \cdots, c_n$ are all $\geq c_k$ implies that $|S| \geq n - k + 1$ and $d_1, d_2, \cdots, d_k$ are all $\leq d_k$ implies that $|T| \geq k$. By pigeonhole principle, we must have $|S \cap T| \geq 1$ (since $|S \cup T| \leq n$). So, there is some $r \in S \cap T$. Then we have $c_k \leq a_r \leq b_r \leq d_k$ and we are done. □

Lemma 8.1.6 implies the assertion in example 8.1.5 since we can shuffle members of each column independently.

8.2 Some more interesting applications

Historically, the pigeonhole principle (*called the Dirichlet drawer principle*) was first used by *Dirichlet* to prove the following approximation of an irrational number by a sequence of rational numbers.

Theorem 8.2.1. *(Dirichlet approximation) Let α be an irrational number. Then there exist infinitely many rational numbers $\frac{p}{q}$ such that $\left|\alpha - \frac{p}{q}\right| < \frac{1}{q^2}$.*

Proof To simplify the proof, observe first that it is enough to prove the assertion when α is positive, which we assume to be the case. We claim that for every positive integer n, there are positive integers p and q such that $q \leq n$ and $\left|\alpha - \frac{p}{q}\right| < \frac{1}{nq} \leq \frac{1}{q^2}$. If the claim is true, then indeed, we can find infinitely many such rational numbers $\frac{p}{q}$: Having found $\frac{p_1}{q_1}, \frac{p_2}{q_2}, \cdots, \frac{p_k}{q_k}$ such that $|\alpha - \frac{p_i}{q_i}| = \alpha_i < \frac{1}{q_i{}^2}$, let $\beta = min\{\alpha_1, \alpha_2, \cdots, \alpha_k\}$. Then $\beta > 0$ is an irrational number and we may choose n large that $n > \frac{1}{\beta}$. We then obtain a corresponding $\frac{p}{q}$ such that

$$\left|\alpha - \frac{p}{q}\right| < \frac{1}{nq} \leq \frac{\beta}{q} \leq \beta$$

and hence $\frac{p}{q} \neq \frac{p_i}{q_i}\ \forall\, i = 1, 2, \cdots, n$.

To prove the claim, let, for a positive integer j, $\{j\alpha\}$ denote the fractional part of α. That is, $j\alpha - \lfloor j\alpha \rfloor = \{j\alpha\}$. Then $\{j\alpha\}$ is an irrational number which is in the open interval $(0, 1)$. Consider the boxes (open intervals): $(\frac{i-1}{n}, \frac{i}{n})$ where $i = 1, 2, \cdots, n$. Then, for $j = 1, 2, \cdots, n, n+1$, we have $n + 1$ numbers $\{j\alpha\}$ which lie in these intervals. Hence, by pigeonhole principle, for some $j \neq k$, we must have both $\{j\alpha\}$ and $\{k\alpha\}$ in the same interval $(\frac{i-1}{n}, \frac{i}{n})$ and hence $|\{j\alpha\} - \{k\alpha\}| < \frac{1}{n}$. Let $\lfloor j\alpha \rfloor$ equal I_1 and $\lfloor k\alpha \rfloor$ equal I_2. Let, w.l.o.g. $j > k$. then $(j-k)\alpha = (I_1 - I_2) + (\{j\alpha\} - \{k\alpha\})$. Writing q and p for the integers $j - k$ and $I_1 - I_2$, we see that $|q\alpha - p| < \frac{1}{n}$ and hence $\left|\alpha - \frac{p}{q}\right| < \frac{1}{nq}$ as desired. □

Dirichlet approximation (Theorem 8.2.1) can be improved to an extent by having $\sqrt{5}$ in the denominator. This latter result is called *Hurwitz theorem* (see [44], or [14]). But, in general, there is no way of improving this result beyond that. Thus, we can not have q^3 in the denominator, instead of q^2 for *every irrational number* α, though this can be asserted for special irrational numbers called *the algebraic numbers.* Also, it is trivial to check that Theorem 8.2.1 is false for rational numbers α.

At this point, we are tempted to quote the following from a review of book on Problems in real analysis. The review appeared in the Mathematical Intelligencer and it was written by Jet Wimp [62]. "The proof (of Dirichlet approximation), which proceeds via the pigeonhole principle, is very simple. The pigeonhole principle has proved to be uncommonly powerful in the rapidly expanding field of combinatorics, and articles examining its philosophy and methodology have recently appeared. I want to discuss the principle a little here, as it seems to me *this technique shuffles uncomfortably back and forth between the domains of methods (used to prove theorems) and tricks (used to solve problems).* ······ I think of *the principle as a trick- not because of its conceptual transparency, but because those situations for which it is appropriate do not announce themselves.* In analysis, it is often clear just when one should reach for a Helly convergence theorem or for the Fourier inversion formula. ······ Many problems in combinatorics seem intuitively amenable to the pigeonhole principle. ··· Doesn't its frequent use qualify for it as a bona fide method? Perhaps, but prepping any given problems so that the principle can be used may invoke imaginative trickery."

The reviewer (Jet Wimp) then narrates an interesting anecdote that involves both the pigeonhole principle and a mathematician winning the Fields medal (equivalent of the Nobel prize): "A combinatoricist of my acquaintance recently visited a prestigious Eastern institution. He encountered in the hall, an eminent Fields medalist, not known for his humility. After a de riguer declaration of research interests, the conversation took a combative turn. "Combinatorics!", jeered the medalist, who himself worked in the far reaches of differentiable manifolds. "So much fuss over it, and it's all just counting. It's all trivial." "I don't think that's fair" was the retort. "But it is true! Let's bet a dinner. Tell me something in combinatorics I can't do in one hour and I'll buy dinner." "Let's see. *There's the Erdös-Szekers result.*"

Needless to say that the story ended with the Fields medalist buying a fabulous dinner for the combinatoricist and *here is the Theorem the Fields medalist could not prove in one hour.*

Theorem 8.2.2. *(Erdös-Szekeres theorem on Monotone subsequences) Let* $S = (a_1, a_2, \cdots, a_{mn+1})$ *be a sequence of* $mn+1$ *real numbers. Then either* S *has a monotonically increasing subsequence with* $m+1$ *terms or a strictly decreasing subsequence with* $n+1$ *terms.*

Proof By the length of a sequence, we mean the number of terms in that sequence. Suppose S does not have a monotonically increasing subsequence of length $m+1$. For $i \in [mn+1]$, let d_i denote the length of a longest monotonically increasing subsequence that begins at a_i (i.e. the first term is a_i). Then the assumption implies that $d_i \leq m$. We assert that if $i < j$, then $d_i = d_j \Rightarrow a_i > a_j$, for, if not, then to a monotonically increasing subsequence of length $d = d_i = d_j$ that begins at a_j, we may attach a_i on the left to get a monotonically increasing subsequence of length $d+1$ that begins at a_i, which is a contradiction. So, $a_i > a_j$. Using the pigeonhole principle, we conclude that for some t with $1 \leq t \leq m$, there must be at least $\lceil \frac{mn+1}{m} \rceil = n+1$ values i for which $d_i = t$. Let these values be $d_{i_1} = d_{i_2} = \cdots = d_{i_{n+1}} = t$ where we assume, w.l.o.g. that, $i_1 < i_2 < \cdots < i_{n+1}$. Then from what we just proved, $a_{i_1} > a_{i_2} > \cdots > a_{i_n} > a_{i_{n+1}}$ and we have a strictly decreasing subsequence of length $n+1$. $\square$

Theorem 8.2.3. *Let the vertices of a regular* n*-gon* $\mathbf{P}$ *be coloured in two colors red and blue in an arbitrary manner. If* $n \neq 3, 4, 6, 8$*, then there exist three vertices of* $\mathbf{P}$ *such that the three vertices are corners of an isosceles triangle and have the same colour.*

Proof When $n = 3, 4, 6, 8$, we can easily find a colouring that does not have the stipulated property. So, let $n \neq 3, 4, 6, 8$. First let $n = 2m+1$ be an odd number where, we number the vertices from 0 to $n-1 = 2m$ in an anticlockwise manner. Since the number of vertices is odd, we cannot color the vertices alternately in red and blue and hence there must be two adjacent vertices say 0 and 1 that receive the same color red. Then the color of 2 must be blue since $(0, 1, 2)$ is an isosceles triangle and for the same reason, the color of $-1 = 2m$ must be blue (to avoid isosceles triangle with vertices of the same colour). Now look at $m+1$. This vertex is at distance m

from both 0 and 1 and hence must have color blue. But then $(2m, m+1, 2)$ is an isosceles triangle with all the corners blue. Now let $n = 2m$ be an even number where $m \geq 5$. Then the colors cannot be alternately red and blue, for, in that case $(0, 2, 4)$ is an isosceles triangle with vertices of the same color. Hence, two adjacent vertices must have the same color and let w.lo.g., the color of both 0 and 1 be blue. Then we must give color red to both 2 and $-1 = 2m - 1$. This forces 5 to have color blue (because $(-1, 2, 5)$ is an isosceles triangle). Since 1 and 5 are both blue, 3 must be red and since 2 and 3 are red, 4 must be blue. Now -1 and 3 are red and hence 7 must be blue. But then $(1, 4, 7)$ is an isosceles triangle with all corners blue. □

We now work out a masterpiece problem of Erdös, which may look like the Erdös and Szekeres Theorem (Theorem 8.2.2) (it is tempting to think that it follows from Theorem 8.2.2) but is a deep statement with a considerably difficult proof.

Theorem 8.2.4. *Let n be a natural number and let $N_n = \{1, 2, \cdots, 2^{n-1}\}$. Let f be a function, $f : N_n \to N_n$ such that the following condition is satisfied: For every $i \in N_n$, we have $f(i) \leq i$ (thus, in particular, $f(1) = 1$). Then there exists a sequence of n distinct elements*

$$a_1 < a_2 < \cdots < a_n$$

in N_n satisfying

$$f(a_1) \leq f(a_2) \leq \cdots \leq f(a_n) \tag{8.2}$$

Proof It is easy to verify the assertion for small values of n upto 3. This forms a basis of the inductive proof. Let $n \geq 4$ and suppose the assertion holds for all the smaller numbers $m \leq n-1$. *For the sake of contradiction, we suppose that we do not have a sequence of length n satisfying* (8.2). Let $t(i)$ denote the length of the largest sequence satisfying $(*)$ that begins with $i \in N_n$ (thus $i = a_1$). By the assumption, $1 \leq t(i) \leq n-1$ for all i. We first claim that if $t(i) = t(j)$, then $f(i) \neq f(j)$. This is true because $f(i) = f(j)$ would enable us to adjoin i to the left of the sequence that begins with j (assuming w.l.o.g. that $i < j$) exactly as in the proof of Erdös-Szekeres Theorem (Theorem 8.2.2) and this will then give $t(i) \geq t(j) + 1$, a contradiction (*in fact, we must have $f(i) > f(j)$ if $i < j$*).
Here is the major claim: Let $t(j) = k$ (with $1 \leq k \leq n-1$). Let $m = n-k$. Then $f(j) \leq 2^{m-1}$. Suppose not. Then $f(j) > 2^{m-1}$ and the definition of f implies that $j > 2^{m-1}$. Here $m \leq n-1$. Let $g = f|N_m$, the restriction of g to N_m. Then g satisfies the same properties as f and hence by induction. we get a sequence $a_1 < a_2 < \cdots < a_m$ in N_m such that $g(a_1) \leq g(a_2) \leq \cdots \leq g(a_m)$. Since $t(j) = k$, we can form a sequence $b_1 < b_2 < \cdots < b_k$ such that $f(b_1) \leq f(b_2) \leq f(b_k)$. Here, $b_1 = j > 2^{m-1} \geq a_m$ as well as $f(b_1) = f(j) > 2^{m-1} \geq f(a_m)$ and hence the sequence $a_1 < a_2 < \cdots < a_m < b_1 < b_2 < \cdots < b_k$ satisfies (8.2), and also has length $m + k = n$, which is a contradiction to the assumption. This proves the claim. Now let T_k be the set of all those j for which $t(j) = k$. Then we have shown that $f(j) \leq 2^{m-1}$ and if $j \neq j'$ are in T_k, then $f(j) \neq f(j')$. It now follows that $|T_k| \leq 2^{m-1}$ (where $m = n-k$). Hence summing over all k, we get $N_n = \cup_{k=1}^{n-1} T_k$ and since this is a disjoint union, we get $2^{n-1} = |N_n| = \sum_{k=1}^{n-1} |T_k| \leq \sum_{k=1}^{n-1} 2^{n-1-k} = 1 + 2 + 2^2 + \cdots 2^{n-2} = 2^{n-1} - 1$.

This final contradiction completes the proof. □

8.3 Ramsey theory

In a gathering of six people, assume that every two persons are mutually friends or mutually enemies. We can then conclude that we have a subset of three people who are either mutually friends or mutually enemies. To prove this, let x be one of the six persons. Then using the pigeonhole principle, x has (at least) either $\lceil \frac{5}{2} \rceil = 3$ friends or 3 enemies and assume w.l.o.g. that x has three friends a, b and c. If a and b are mutually friends then the set $\{x, a, b\}$ has the property that every two persons in this set are mutually friends. The same assertion (having three persons who are mutually friends) holds if a and c are mutually friends or if b and c are friends. Hence we are left with the case that no pair in the set $\{a, b, c\}$ is a friendly pair. But then every two persons in the set $\{a, b, c\}$ are mutually enemies and we are done. To see that the same assertion does not necessarily hold if we have five people (instead of six), let the five people be $\{a, b, c, d, e\}$ with friendship given by the edges of the 5-cycle $(abcde)$ and enmity given by the remaining $5 = \binom{5}{2} - 5$ edges that form the 5-cycle $(acebd)$ as shown in Figure 8.2. Then we neither have a friendship triangle nor an enmity triangle.

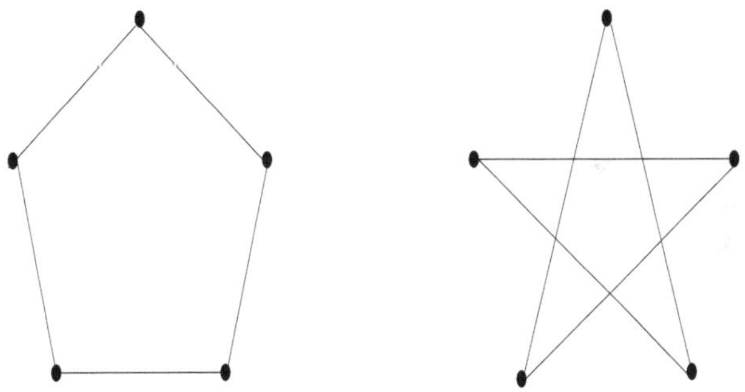

Figure 8.2: A colouring of K_5 that avoids a monochromatic triangle

The graphical language we used to describe friendship and enmity suggests the following definition.

Definition 8.3.1. Let p, q be integers ≥ 2. Then *the Ramsey number* $R(p, q)$ is *the smallest natural number* n with the following property: In *every coloring of the edges of the complete graph* K_n *in two colours red and blue, we either have a red* K_p *(that is, a complete subgraph on* p *vertices with all the edges in it colored red), or a blue* K_q.

Before embarking on the question of finding $R(p, q)$, let us understand what the definition is saying. We have already proved that $R(3, 3) = 6$: In any coloring of the edges of K_6 in two colors red and blue, we have either a red K_3 (*friendship triangle*) or a blue K_3 (*enmity triangle*). Hence $R(3, 3)$ exists and is ≤ 6. How do we know that

$R(3,3) \neq 5$? This is because of a coloring of the edges of K_5 in two colors that neither has a red K_3 nor a blue K_3 (given in Figure 8.2). In general, to show that $R(p,q) = n$, we must do two things: We must show that in *every coloring* of the edges of K_n in two colors we have a red K_p or a blue K_q and we must also show that $\exists$ *a coloring of* K_{n-1} *that has neither a red* K_p *nor a blue* K_q. If we cannot prove the second part but prove only the first part, then n gives an upper bound on $R(p,q)$ and $R(p,q)$ exists thanks to the well-ordering property of natural numbers (*the numbers for which every colouring of* K_n *must have a red* K_p *or a blue* K_q *includes* n *and hence is non-empty and therefore this set must have least element*). The upshot is that the existence of Ramsey number is guaranteed as soon as we manage to prove an upper bound on it.

Theorem 8.3.2. *Let $p, q \geq 2$. Then the following assertions hold.*

(a) $R(p,2) = p$.

(b) $R(2,q) = q$.

(c) *For $p, q \geq 3$ we have $R(p,q) \leq R(p-1,q) + R(p,q-1)$ and hence inductively $R(p,q)$ exists.*

(d) $R(p,q) \leq \binom{p+q-2}{p-1}$.

(e) $R(p,p) \leq \binom{2p-2}{p-1} < 4^p$.

Proof Consider a K_p colored in two colors red and blue. If all the edges are red, then we have a red K_p. Otherwise, we have a blue edge and we get a blue K_2. This property does not hold if we have a K_{p-1} with all the edges colored red. This proves (a) and (b) follows by symmetry. Consider (c). Let n denote the sum $R(p-1,q) + R(p,q-1)$. Let x be some vertex of K_n with edges that are colored in red and blue. Let A and B respectively denote the set of vertices that are joined to x by red and blue edges. If $|A| \leq R(p-1,q) - 1$ and $|B| \leq R(p,q-1) - 1$, then the degree of x is at the most $R(p-1,q) + R(p,q-1) - 2$ which is a contradiction since $n = R(p-1,q) + R(p,q-1)$ (we have used the pigeonhole principle here). Hence either $|A| \geq R(p-1,q)$ or $|B| \geq R(p,q-1)$ and w.lo.g. assume the former to be true. Then the complete graph on A has been colored in two colors and since $|A| \geq R(p-1,q)$, we have either a red K_{p-1} or a blue K_q. In the latter case, we are done. Consider the former. Then we have a subset A' of A consisting of $p-1$ vertices such that all the edges in A' are red. Then $A'' = \{x\} \cup A'$ has the property that all the edges in A'' are colored red and $|A''| \geq p$. (d) is proved using (c) and the use of induction and Pascal identity; this is left to the reader as an exercise. Finally consider (e) which is also proved by making induction on p. □

Definition 8.3.3. By a *monochromatic complete subgraph* K_k we mean a K_k all of whose edges are colored in the same color.

We depend on the following well-known approximation of the factorials; refer to Liu [36] for a proof.

Theorem 8.3.4. *(Stirling's formula) For all $k \geq 5$ we have*

$$\sqrt{2\pi k}(\frac{k}{e})^k \leq k!$$

Theorem 8.3.5. *Let k be a fixed integer, $k \geq 2$. If $\binom{n}{k} < 2^{\binom{k}{2}-1}$, then $\exists$ a coloring of K_n in red and blue such that we have neither a red K_k nor a blue K_k. In particular, if $n \leq 2^{k/2}$ then there is a coloring of K_n in red and blue such that we have neither a red K_k nor a blue K_k.*

Proof Assuming the first statement the second statement is a consequence of the Stirling inequality (Theorem 8.3.4): For $k = 2, 3, 4, 5, 6$ it is easy to see that $n \leq 2^{k/2}$ implies $\binom{n}{k} < 2^{\binom{k}{2}-1}$, Let $k \geq 7$. Then

$$\binom{n}{k} < \frac{n^k}{k!} \leq \frac{1}{\sqrt{2\pi}} \frac{2^{k^2/2}}{(k/e)^k} < \frac{1}{2} 2^{\frac{k^2-k}{2}}$$

and the assertion follows. Consider the first statement. Let A be a k-subset of the vertex set of K_n. In how many colorings of K_n does A become a red K_k? This allows only the freedom of coloring of all the edges not in A to be colored in one of the two colours and hence the number of such colorings is equal to $2^{\binom{n}{2}-\binom{k}{2}}$. Summing over all such A gives us an upper bound on the number of colorings in which we have a red K_k on some k-subset (these are not mutually exclusive). Exactly in the same way the number of colorings in which we have a blue K_k on some k-subset of vertices is equal to $\binom{n}{k}2^{\binom{n}{2}-\binom{k}{2}}$ and hence the total number of colorings in which there is at least one monochromatic K_k is at the most $2\binom{n}{k}2^{\binom{n}{2}-\binom{k}{2}}$. Since the total number of colorings is $2^{\binom{n}{2}}$, we see that the number of colorings that have no monochromatic K_k is at least

$$2^{\binom{n}{2}} - 2\binom{n}{k}2^{\binom{n}{2}-\binom{k}{2}} > 0$$

□

Theorem 8.3.6. *For all $k \geq 2$, we have:*

$$2^{k/2} < R(k,k) < 4^k$$

Remark 8.3.7. It is not known if the limit $lim_{k\to\infty} \sqrt[k]{R(k,k)}$ exists or not. If it does, it will lie between $\sqrt{2}$ and 4.

Instead of using only two colors, we may use three colors say red, blue and yellow to color the edges of K_n. Thus $R(3,3,3)$ is the smallest number n for which an arbitrary coloring of K_n in red, blue and yellow must give us a monochromatic K_3.

Lemma 8.3.8. $R(3,3,3) \leq 17$ *and* $R(3,3,3,3) \leq 66$.

Proof Consider an arbitrary coloring of K_{17} in red, blue and yellow and let x be some vertex. Let A, $A'A''$ respectively denote the subsets of vertices that are joined to x by red, blue, yellow edges. Then $|A| + |A'| + |A''| = 16$ and hence by pigeonhole principle, one of these sets, say A, w.l.o.g., must have size $\geq \lceil \frac{16}{3} \rceil = 6$. If we have some edge yz in A which is red, then (xyz) is a red K_3. If not, then all the edges in A are colored using only the colors blue and yellow and since $|A| \geq 6 = R(3,3)$, it follows that inside A, we have a blue or yellow K_3. The second assertion follows exactly in the same manner using the first assertion and the pigeonhole principle. □

It requires some more effort to prove that $K_3(3) = R(3,3,3) = 17$. It is among a few Ramsey numbers whose precise value is known. This requires one to show that $R(3,3,3) > 16$ by actually giving a coloring of K_{16} in three colors that avoids a monochromatic K_3. That was done by Greenwood and Gleason using a field with 16 elements (van Lint and Wilson [57]). We will content ourselves by defining $R_k(3) = R(3,3,\ldots,3)$ (with 3 occurring k times to indicate coloring in k colors) and proving the following upper bound (Theorem 8.3.9). $R_k(3)$ *is thus the least* n *that has the property that when the edges of* K_n *are coloured in* k *colors say* $a_1, a_2, \ldots, a_k$, *we must have a monochromatic* K_3 *in some color* a_i.

Theorem 8.3.9. $R_k(3) \leq 1 + \lfloor k!e \rfloor$ *for all* $k \geq 2$.

Proof We first claim that $\lfloor k!e \rfloor = \sum_{j=0}^{k} \frac{k!}{j!}$. To see this, we have $k!e$ equal to $[\sum_{j=0}^{k} \frac{k!}{j!}] + \delta$ where δ equals

$$\frac{1}{k+1}[1 + \frac{1}{k+2} + \frac{1}{(k+2)(k+3)} + \cdots] \;<\; \frac{1}{k+1}\{\sum_r (\frac{1}{(k+2)})^r\}$$
$$< \frac{1}{2}$$

and hence the claim is proved. Now consider the assertion of the Theorem which clearly holds when $k = 2$. So let $k \geq 3$. Let x be some vertex of K_n where $n = 1 + \lfloor k!e \rfloor$. Then there is a subset A with number of vertices $\geq \frac{1}{k}\lfloor k!e \rfloor = \sum_{j=0}^{k-1} \frac{(k-1)!}{j!} + \epsilon$ where ϵ is a positive number and all the vertices in A are joined to x by the same color say a_k. Thus $|A| \geq 1 + \lfloor (k-1)!e \rfloor$. We now argue exactly as before. Either there is an edge yz in A which is colored with color a_k and then we have a K_3 in color a_k on (xyz) or the edges in A are colored using only the colors $a_1, a_2, \ldots, a_{k-1}$. In that case, by induction $|A| \geq 1 + \lfloor (k-1)!e \rfloor \geq R_{k-1}(3)$ and hence A has a monochromatic K_3. □

We now look at a related problem. Suppose we color natural numbers in the set $[n]$ in two colors red and blue and we wish to avoid a monochromatic solution to $x + y = z$. *A monochromatic solution here means three numbers* x, y, z *satisfying the equation* $x + y = z$ *such that all the three numbers have the same colour.* We wish to avoid a monochromatic solution. Note that we do not insist that x, y, z are all distinct. Suppose, w.l.o.g, we give color red to 1. Then $1 + 1 = 2$ and hence we cannot given color red to 2 (else $1+1 = 2$ has a monochromatic solution) and hence 2 must be blue.

For the same reason, 4 cannot then be blue and hence must be red. This now forces 3 to be blue. But then 5 cannot be red (because $1+4=5$) nor can it be blue (because $2+3=5$). We thus see that $n=5$ is the least number with the property that in an an arbitrary coloring of the set $[n]$, we must have a monochromatic solution to $x+y=z$. This motivates the following definition.

Definition 8.3.10. Let $k \geq 2$ be a natural number. *A sum-free set* is a set S of natural numbers such that we do not have a solution to $x+y=z$ with $x,y,z \in S$. Also, *$N(k)$, the Schur number for k* is the smallest number n with the property that when the elements of $[n]$ are arbitrarily colored in k colors $a_1, a_2, \ldots, a_k$ we must have a monochromatic solution to $x+y=z$, that is no color class in $[n]$ is sum-free.

Theorem 8.3.11. *$N(2)=5$ and for $k \geq 3$, we have $N(k) \leq R_k(3) \leq 1+\lfloor k!e \rfloor$.*

Proof The first statement has already been proved. Let $k \geq 3$ and let $n = R_k(3)$. Given a coloring of the elements of $[n]$ in k colors, we obtain a coloring of the edges of K_n as follows. Let i, j be two distinct vertices of K_n where we assume, w.l.o.g. that $i < j$. Then $j - i \in [n]$. Give color a_t to the edge (ij) of K_n if the color of $j-i$ is a_t. Since $n = R_k(3)$, we have a monochromatic K_3 with vertices i_1, i_2, i_3 and we may assume, w.l.o.g. that $i_1 < i_2 < i_3$. Writing $x = i_2 - i_1,\ y = i_3 - i_2$ and $z = i_3 - i_1$ we see that x, y, z have the same color and the equation $x+y=z$ is satisfied. □

Theorem 8.3.12. *Let $k \geq 2$ be a natural number. If we have a coloring of the natural numbers in the set $[n]$ in k colors such that each color class is sum-free, then it is possible to give a coloring of the natural numbers in the set $[3n+1]$ in $k+1$ colors such that each color class is sum-free. In particular, $N(k) \geq 1 + \frac{3^k-1}{2}$.*

Proof Let N denote the set $[n]$ and N' the set $[3n+1]$. We are given a coloring of N in colors $1, 2, \ldots, k$ such that each color class is sum-free. If x in N has been colored i , then give the same color i to $3x$ and $3x-1$ in N'. This colors all the numbers in N' in the first k colors except the numbers that are $1 \ (mod\ 3)$. Give color $k+1$ to all these numbers. We claim that in this coloring every color class in N' is sum-free. Let $a+b=c$ in N' and suppose that a, b, c are in the same color class i. First let $i = 1, 2, \ldots, k$. Then $a = 3x$ or $3x-1$ and $b = 3y$ or $3y-1$. If $(a,b) = (3x, 3y)$, then $x+y=z$ (where $c = 3z$) and we have x, y, z all colored i (in the colouring of $[n]$), a contradiction. We get the same contradiction if $(a,b) = (3x, 3y-1)$ or $(a,b) = (3x-1, 3y)$. Finally, if $(a,b) = (3x-1, 3y-1)$, then $a+b \equiv 2 \ (mod\ 3)$ and hence $a+b$ must get the new color $k+1$. This leaves with the only possibility that a, b, c are all colored $k+1$. But that cannot happen since $a, b \equiv 1 \ (mod\ 3)$ implies that $c \equiv 2 \ (mod\ 3)$ and hence c cannot get color $k+1$. □

We close the discussion on Ramsey numbers with the following definition of (generalized) Ramsey numbers and their determination.

Definition 8.3.13. Let $t \geq 2$ be a fixed natural number. Let $a, b \geq t$ be natural numbers, Then *the Ramsey number $M(t; a, b)$* is *the smallest number n* such that given *any partition* of the set of all t-subsets of an n-set N into two (color) classes say red and blue, one of the following is true: Either there exists a subset A of order a such that all the t-subsets of A are red or there exists a subset B of order b such that all the t-subsets of B are blue.

Theorem 8.3.14. *Let $t \geq 2$ and let $a, b, \geq 2$. Then the Ramsey number $M(t; a, b)$ exists.*

Proof It is easy to check that $M(t; t, b) = b$ and $M(t; a, t) = a$. Also, we have already proved (Theorem 8.3.2) that the assertion is true when $t = 2$. We can therefore run an induction on all of t, a, b and hence assume that the statement holds for $t - 1$ (with any a, b) as also for $a - 1$ and $b - 1$ with the given t. So, let $t \geq 3$ and $a, b \geq t + 1$. By the induction hypothesis, $c = M(t; a - 1, b)$ and $d = M(t; a, b - 1)$ exist. Finally, let $n = M(t - 1; c, d) + 1$. We claim that n has the required property. Let N be a set of order n and suppose we are given a partition of the set of all the t-subsets of N into two classes red and blue. Let x any element of N and write $N' = N - \{x\}$. Given a $(t-1)$-subset T' of N' we look at the color of $T = T' \cup \{x\}$ and give the same color to T'. Thus all the $(t-1)$-subsets (as well as t-subsets) of N' have been given color red or blue. Since $|N'| = M(t - 1; c, d)$, we have some subset C of order c (of N') with the property that all the $(t-1)$-subsets of C are red or some subset D of order d (of N') with the property that all the $(t - 1)$-subsets of D are blue. Consider the first possibility. Since $c = M(t; a - 1, b)$, we see that C contains either a subset A' whose size is $a - 1$ and all the t-subsets of A' are colored red or a subset B whose size is b such that all the t-subsets of B are colored blue. In the latter situation, we are done. Consider the former and let $A = A' \cup \{x\}$. Then $|A| = a$ and if T is a t-subset of A, then either $x \in T$ in which case color of T is red because color of the $(t-1)$-set $T' = T - \{x\} \subset C$ is red or $x \notin T$ in which case, $T \subset A'$ and hence must have red color. The remaining part of the proof is left to the reader. □

Recall that *a convex subset* S of $\mathbb{R}^2$ (and $\mathbb{R}^n$, in general) is a subset with the property that the line segment $\{(1-\lambda)x+\lambda y : 0 \leq \lambda \leq 1\}$ lies completely in S if x and y are in S. Some examples of convex sets in $\mathbb{R}^2$ are a circular disk, any straight line segment, the upper half plane and a convex polygon (including its interior). A set S of points of $\mathbb{R}^2$ is said to be *in general position if no three points of S are collinear*. We wish to prove a Theorem of Erdös and Szekeres which asserts that a set with sufficiently many points in general position ensures a subset of points forming the vertices of a convex polygon with many vertices. To prepare ourselves for this result, we have two preliminary lemmas.

Lemma 8.3.15. *Let A be a set of five points in general position in $\mathbb{R}^2$. Then some four of these five points are vertices of a convex quadrilateral.*

Proof Look at the convex hull H of A. This is the smallest convex set of $\mathbb{R}^2$ that contains A and most importantly, it is a polygon whose vertices come from A. Hence if H is a pentagon or a quadrilateral, we are done. Suppose H is a triangle with vertices a, b, c. Then the other two points x, y of A are inside this triangle H. The line xy cannot intersect all the three sides (inside H) and we can assume that xy does not meet bc. Then b, c, y, x form vertices of a convex quadrilateral. □

Lemma 8.3.16. *Let $n \geq 4$ and let S be a set of n points in $\mathbb{R}^2$ such that any four points in S are vertices of a convex quadrilateral. Then the n points in S form vertex set of a convex n-gon.*

Proof This is evidently true if $n = 4$. So let $n \geq 5$ and we prove the assertion by making induction on n. If $x_1, x_2, \ldots, x_{n-1}$ are $n - 1$ points of S, then by induction, the convex hull of these points is a convex $(n - 1)$-gon P_{n-1} whose vertices can be assumed to be in the order $x_1, x_2, \ldots, x_{n-1}$ in a clockwise manner, by relabeling of the vertices. Let $y = x_n$ be the remaining n-th vertex. First suppose that y lies inside P_{n-1}. Then by looking at the triangles $x_i x_{i+1} x_{n-1}$ with $i = 1, 2, \ldots, n - 2$, we see that y must be in one of these triangles say $x_i x_{i+1} x_{n-1}$. But then, x_i, x_{i+1}, x_{n-1}, y form corners of a non-convex quadrilateral, which is contrary to the assumption. So y must be outside P_{n-1}. Now look at the straight lines yx_i and locate those for which the entire polygon P_{n-1} is on one side of this straight line. We get two such points x_i and x_j where we assume that $i < j$. If $j \geq i + 2$, then the quadrilateral on the vertices y, x_i, x_{i+1}, x_j is non-convex, which is a contradiction to the assumption. Hence $j = i + 1$ and the situation must be as on right part of Figure 8.4.
It is now clear that the vertices $x_1, x_2, \ldots, x_i, x_n, x_{i+1} \ldots, x_{n-1}$ form the corners of a convex n-gon and we are done. □

Let m be a natural number. By $K(m)$ we mean the smallest integer n such that given any set of n points in $\mathbb{R}^2$ that are in general position, some subset of m points is a vertex set of a convex m-gon. Lemma 8.3.15 shows that $K(4) = 5$ and it is also known that $K(5) = 9$. In fact, the following Theorem of Erdös and Szekeres asserts that $K(m)$ exists for all m.

Theorem 8.3.17. *(Erdös-Szekeres) $K(m) \leq M(4; 5, m)$.*

Proof Let $n = M(4; 5, m)$ and let S be a set of order n with the points of S in general position. Color all the 4-subsets of a set S of order n (the given set of n points in general position) into two colors blue and red as follows. If a 4-subset is a vertex set of a convex quadrilateral, then color it red and color the 4-subset blue otherwise. Since $|S| = n$, using Ramsey's Theorem, we either have a subset A of order 5 all of whose 4-subsets are blue or a subset B of order m all of whose 4-subsets are red. Using Lemma 8.3.15, the first possibility cannot arise. So, we have a subset B with $|B| = m$ such that every 4-subset of B is a vertex set of a convex quadrilateral. Using Lemma 8.3.16, we get a convex m-gon formed by the vertices of B. □

Remark 8.3.18. The problem of determining the least number of points in general position in $\mathbb{R}^2$ such that we are assured the existence of a convex m-gon among them (the number $K(m)$ in our notation) is a long cherished problem of Erdös who conjectured that $K(m)$ equals $2^{m-2} + 1$. He proved that $K(m) \geq 2^{m-2} + 1$ and also obtained an upper bound for $K(m)$. Erdös's conjecture is true when $m = 4, 5$ and 6 but this has not been proved for any other value of m so far.

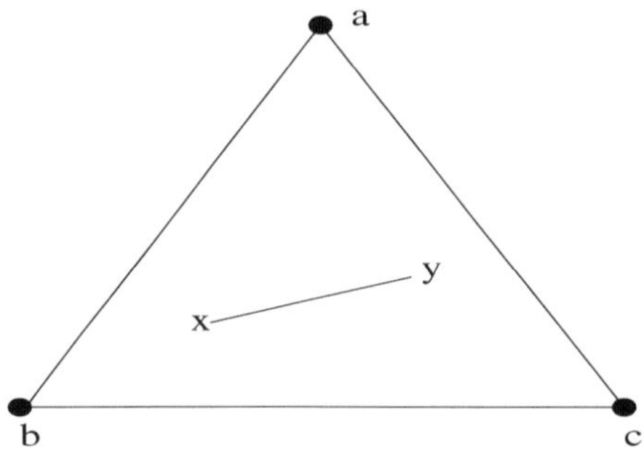

Figure 8.3: Five points in general position with convex hull a triangle

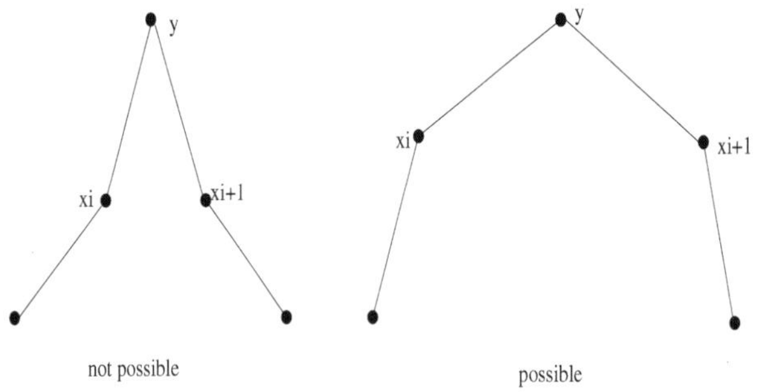

Figure 8.4: A part of the convex polygon

8.4 From finite to the infinite

Let $\mathbb{N}$ denote the set of natural numbers and consider the complete graph with vertex set $\mathbb{N}$ (this is a graph with countably infinite number of edges). Fix a natural number k and color the edges of this countably infinite graph in k colors. *Can we be sure that no matter how the coloring is done, there is an infinite subset A of $\mathbb{N}$ such that all the edges of A are colored in the same color?* The answer is yes. Just begin with vertex $x_1 = 1$ whose degree is infinite while the number of colors is k, which is finite. Hence, by pigeonhole principle, there must be some color i such that infinitely many numbers in $\mathbb{N} - \{1\}$ are joined to 1 by color i. Call this set A and look at the smallest number in this set (which exists, by well-ordering of natural numbers) and call it x_2. Then A has infinitely many elements and these are joined to x_2 by k colors. So, there is some color say j such that infinitely many elements of A forming the subset B of A are joined to x_2 by color j. Again choose the smallest element of B and continue in this manner. We then get a sequence of natural numbers:

$$x_1 < x_2 < \cdots < x_r < \cdots$$

and a function

$$f : \{x_1, x_2, \ldots, x_r, \ldots\} \to [k]$$

such that $f(x_r) = j$ means *infintely many* x_m *with* $m > r$ *are joined to* x_r *by color* j. Since we have an infinite set $x_1, x_2, \ldots, x_r, \ldots$ as a domain and a finite set as range, there must be some j such that j is the image of infinitely many x_r's. Let this subsequence be

$$x_{i_1} < x_{i_2} < \cdots < x_{i_r} < \cdots .$$

Then every pair of vertices in this set is joined by color j as desired. We have thus proved the following.

Theorem 8.4.1. *Let the edges of the complete graph on* $\mathbb{N}$ *be arbitrarily coloured in* k *colours. Then there exists a monochromatic complete subgraph* H *of* $\mathbb{N}$ *such that* H *is an infinite subset of* $\mathbb{N}$.

As an application of Theorem 8.4.1 we show the existence of Ramsey numbers for all natural numbers p, q.

Theorem 8.4.2. *(cf. Theorem 8.3.2) Let* $p, q \geq 2$. *Then* $R(p, q)$ *exists.*

Proof We will not give an upper bound as in the proof of Theorem 8.3.2 but will show that the negation of the assertion leads to a contradiction. So, assume that *for every* n, we have a coloring of K_n that has neither a red K_p nor a blue K_q. We thus have a sequence of colored graphs, one for each n,which has this property. We assume that the vertex set of K_n is $[n]$. Fix some natural number m and consider the restriction of each colored K_n with $n \geq m$ to $[m]$. Since number of K_n's is infinite, while the total number of colorings of K_m is no more than $2^{\binom{m}{2}}$ (which is finite), we see that we get an infinite sequence of K_n's that agree on $[m]$. Then consider the restriction of these to the set $[m+1]$ and draw the same conclusion. Proceeding in this manner, we get a coloring of all the edges of the complete graph on $\mathbb{N}$. This coloring cannot have a red K_p nor a blue K_q contradicting what we proved earlier (Theorem 8.4.1 that we have a red or blue complete subgraph which is infinite. □

Remark 8.4.3. The proof technique adopted in Theorem 8.4.2 is called *Compactness principle*; refer to [28] . The reader should draw a parallel between the argument made in the preceding paragraph and similar arguments in point-set topology. In fact, we have the Bolzano-Weierstrass property (or limit point compactness) and a Theorem that states that every compact space has Bolzano-Weierstrass property. The converse holds for a metric space and the proof of this assertion uses pigeonhole principle in the same form as is used in the preceding paragraph. One application of this theme appears in exercises.

8.5 Exercises for Chapter 8

8.1 Consider the problem of putting numbers on the corners of two concentric regular n-gons with numbers from 1 through n written in such a way that for every possible rotation, we get an agreement of numbers on at least one corner (Example 8.1.4). We saw in that example that the problem has no solution when $n = 8$. For which numbers n, does the problem have a solution?

8.2 We are given an $m \times n$ array $A = [a_{i,j}]$ of real numbers $a_{i,j}$ such that for every row i, the difference $|a_{i,j} - a_{i,k}| < d$, where $j, k = 1, 2, \ldots, n$ and d is some positive constant. Now suppose that the numbers in each column are arranged in monotonically increasing order to get a new array $B = [b_{i,j}]$ (so that $b_{1,j} \leq b_{2,j} \leq \cdots \leq b_{m,j}$). Show that this operation still preserves the original property: $|b_{i,j} - b_{i,k}| < d$.

8.3 Let $X \subset [99]$ such that $|X| = 10$. Show that it is possible to select two disjoint subsets Y and Z of X such that both are non-empty and

$$\sum_{y \in Y} y = \sum_{z \in Z} z$$

8.4 Let $\{a_1, a_2, \ldots, a_{1995}\}$ be a sequence of positive integers whose sum is 3989. Show that there are r consecutive terms in this sequence whose sum is 95.

8.5 Show that for an arbitrary integer N there exists a multiple of N that contains only the digits 0 and 7 (e.g. when $n = 3$ we have $259 \times 3 = 777$ and when $n = 4$ we have $1925 \times 4 = 7700$).

8.6 Show that in any given set A of 13 distinct real numbers, there are at least two numbers x and y such that

$$0 < \frac{x - y}{1 + xy} \leq 2 - \sqrt{3}$$

8.7 A point $(a_1, a_2, \ldots, a_n)$ with all a_i's real numbers is called a lattice point if every a_i is an integer. Show that if L is a subset of $2^n + 1$ lattice points then there are two points a and b in L such that the midpoint of the segment ab is also a lattice point.

8.8 There are 1958 computers that use one of the 6 languages to communicate with each other. Any two computers communicate with each other in exactly one of the six languages. Show that there are three computers that mutually communicate in the same language.

8.9 Let α be a positive irrational number and consider the straight line L in $\mathbb{R}^2$ whose equation is $y = \alpha x$. Let M be the set of all lattice points in $\mathbb{R}^2$ except the origin. Show that L contains no point of M but contains points that are arbitrarily close to M.

8.10 Let C be the set of all the points on a circle with unit circumference. A frog jumps along the circumference jumping α in one second, where α is a positive irrational number. let $\epsilon > 0$ be a real number and suppose that on the circumference of the circle, we have a ditch of width ϵ. Prove that no matter what ϵ is, the frog must eventually fall in the ditch.

8.11 Let A be a set of 20 distinct integers chosen from the arithmetic progression $1, 4, 7, \ldots, 100$. Prove that there are two integers in A whose sum is 104.

8.12 Let $n \geq 3$ be an odd number. Show that among the set of numbers

$$\{2^1 - 1, 2^2 - 1, \ldots, 2^{n-1} - 1\}$$

there is at least one number which is divisible by n.

8.13 For a finite set A of integers, denote by $s(A)$ the sum of all the elements of A. Let S be a subset of the set $[15]$ such that for every disjoint subsets B and C of S, we have $s(B) \neq s(C)$. Show that $|S| \leq 5$.

8.14 A chess player has 77 days to prepare for a serious tournament. He decides to practice by playing at least one game per day and a total of 132 games. Show that there is a succession of days during which he must have played exactly 21 games.

8.15 We select 38 even positive integers all less than $1,000$. Show that there are two of them whose difference is at the most 26.

8.16 (a) Let a and b be natural numbers that are coprime to each other. By looking at the remainders $mod\ b$ of the numbers pa with $p = 1, 2, \ldots, b-1$, prove that $\exists$ natural numbers x and y such that $ax - by = 1$.

(b) Prove that two integers a and b are coprime to each other iff for some integers x and y, we have, $ax + by = 1$.

(a) Let a, b and c be given integers. Then prove that the linear equation $ax + by = c$ has a solution in integers iff the g.c.d. of a and b divides c.

(c) We note that what we have in fact proved, in the language of algebra, is that every ideal in the ring of integers $\mathbb{Z}$ is a principal ideal.

8.17 We are given that the infinite sum $\sum_p \frac{1}{p^2}$ is convergent and converges to a number which is strictly less than 0.46 where the sum is over all p that are prime numbers. An integer m is called a square-free integer if it has no proper divisors that are square numbers. Show that for any n the set $[n]$ has at least $(0.54) \times n$ squarè-free elements.

8.18 An integral domain is a commutative ring with 1 in which both the cancellation law holds. Show that a finite integral domain is a field.

8.19 A drawer contains 6 pairs of white, 5 pairs of black, 5 pairs of red and 4 pairs of green socks. How many socks must we choose to ensure that we have taken out two socks of the same color? How many socks must we choose to ensure that we have taken out two socks with different colors?

8.20 Prove that it is possible to color the vertices of a regular hexagon and vertices of a regular octagon in two colors red and blue in such a way that we do not have three vertices forming the vertices of an isosceles triangle.

8.21 Fifteen problems each with an answer "Yes" or "No" are posed to $1,600$ students. Suppose each student wrote answer to every question and no student answered two consecutive questions correctly. Is it true that some two students have an identical pattern of answers?

8.22 An international society has $1,978$ members each getting a distinct number from 1 to 1978. Suppose the members belong to six different countries. Show that there is at least one member whose number is the sum of the numbers of two members of the same country or twice as large as the number of a member from the same country.

8.23 Given any three distinct integers, show that there exist at least two say a and b such that the difference $a^3b - b^3a$ is a multiple of 10.

8.24 A graph G is formed by two disjoint cycles C_m and C_n where m and n are both odd and all the vertices of the first cycle are made adjacent to all the vertices of the second cycle. We thus have $m + n + mn$ edges. The edges are colored in two colours red and blue so that no triangle is monochromatic. Show that all the edges inside C_m and C_n must be colored with the same colour.

8.25 Show that if we are given 52 integers, then we can always select two of them such that their sum of difference is a multiple of 100. Is this also true if 52 is replaced by 51?

8.26 Let a be a natural number such that a is coprime to both 2 and 5. Show that for any natural number n, there is a power of a with the last $n + 1$ digits $00 \cdots 001$ where the block of zeros has n zeros.

8.27 Let p be a prime other than 2 and 5. Show that among the p integers

$$1, 11, 111, \ldots, \underbrace{111 \cdots 111}_{p\ times}$$

where the last number has p ones, at least one number is a multiple of p.

8.28 Consider the sequence (a_n) of natural numbers constructed as follows. a_1 is any number. Then a_2 is constructed by writing any digit b except 9 after a_1 and in general, a_n is obtained from a_{n-1} by attaching any $b \neq 9$. Show that the sequence (a_n) has infinitely many composite numbers.

8.29 Let $S = [20]$. Let all the subsets of S of order 9 be given some (arbitrary) label from the set S itself. Show that there exists a set T of size 10 such that no element of T is the label for the subset of the remaining 9 element subset of T.

8.30 In a room of area 5 square meters, we have to put 9 rugs (of arbitrary and possibly irregular shapes) each of area 1 square meter. Show that no matter how we arrange the rugs, there will be two rugs that overlap with each other over a portion of area at least $\frac{1}{9}$.

8.31 Fifty one small insects are placed inside a square of side 1. Prove that at any moment there are at least three insects that can be commonly covered by a a single disk of radius $\frac{1}{7}$.

8.32 Show that in any convex hexagon, there is a diagonal which cuts off an area not more than one sixth the area of the hexagon.

8.33 Show that if ten 2-digit numbers are given, then one can always choose two disjoint subsets among them that have the same sum.

8.34 Let k be a positive integer and $n = 2^{k-1}$. Prove that, from any given $2n - 1$ positive integers, one can always find n integers whose sum is divisible by n.

8.35 If 6 points are placed in a 3×4 rectangle, prove that some two points have distance $\leq \sqrt{5}$.

8.36 Thirty three rooks are placed on the standard 8×8 chessboard. Prove that we can choose some five among them that are non-attacking.

8.37 We are given a 5×41 chessboard with each cell colored white or black in an arbitrary manner. Prove that we can select three rows and three columns such that the board formed of these 9 cells has all the cells of the same color.

8.38 Each of the m cards is labeled one of the numbers $1, 2, \ldots, m$. Suppose that the sum of the labels of any subset of cards is not a multiple of $m + 1$. Prove that each card is labeled with the same number.

8.39 Prove that in any convex $2n$-gon, there is some diagonal that is not parallel to any side.

8.40 Let $a_1 a_2 \cdots a_n$ be a permutation of the set $[n]$. Show that if n is odd, then the product

$$(a_1 - 1)(a_2 - 2) \cdots (a_n - n)$$

is an even number. Is this also true if n is an even integer?

8.41 Show that if $n + 1$ integers are chosen from the set $[2n]$, then there are two integers that are coprime to each other.

8.42 Show that if $n + 1$ integers are chosen from the set $[2n]$, then there are two integers a and b among the chosen ones such that a divides b.

8.43 For every integer in the set $\{n+1, n+2, \ldots, 2n\}$ look at its largest odd divisor and add all such largest odd divisors of all the integers. Show that this sum is equal to n^2.

8.44 Prove that if 5 points are chosen in a square of side 2, then there are two points whose distance from each other is at the most $\sqrt{2}$.

8.45 We are given $n + 1$ weights each in full kilograms so that the weights add to a total of m kg. and $m \leq 2n$, Prove that it is possible to obtain a sum of n kg. by taking some of the weights.

8.46 Show that the conclusion of the previous exercise also holds if we have n weights (in full kilograms) adding to a total weight of $2n$ kilograms except when n is odd and each weight is 2 or $n = 2$ and the two weights are 1 and 3.

8.47 Twenty five points are given in the plane. It is known that given any three of them, one can always choose two that are at distance less than one centimeter. Prove that there are thirteen points among the given points that can be covered by a disk of radius one.

8.48 A student has 37 days to prepare for her examination. On each of these 37 days, she decides to study in multiples of one hour and also estimates that a total of 60 hours of study will suffice for the preparation. Show that there is a succession of days in which she must have studied exactly 13 hours.

8.49 Ten players take part in a round robin (every player plays every other player exactly once) tournament. The scoring rules are: one point for a win, -1 for a loss and no point for a draw. It was observed that more than 70 percent of the games were drawn. Prove that there are two players that ended up with identical scores.

8.50 In a gathering of n people every two individuals either know each other or do not know each other. Show that there must exist two individuals who know the same number of people.

8.51 Show that if x is any non-negative real number, then among the numbers $x, 2x, \ldots, (n-1)x$, there is at least one that differs from an integer by at most $\frac{1}{n}$.

8.52 By a k-sum in a given sequence $(a_1, a_2, \ldots)$, we mean any sum of the form $a_i + a_{i+1} + \cdots a_{i+k-1}$ of k consecutive terms (beginning with any a_i). We are given that in a sequence, every 7-sum is negative while every 11-sum is positive. What is the largest number of terms in such a sequence?

8.53 50 integers are chosen among the first 100 integers (i.e. from the set $[100]$) in such a manner that no two of the chosen integers have sum equal to 100. Show that there is at least one perfect square among the numbers chosen.

8.54 Recall that a positive real number between 0 and 1 is a rational number if and only if its decimal expansion has a repeated block after finitely many decimal places. Let α be a real number whose decimal expansion is

$$0.123456789101112\cdots$$

where we keep writing natural numbers one after the other in that order. Prove that α is irrational.

8.55 We wish to give an alternative proof of Lemma 8.1.6 through the following steps.

(a) Prove the assertion in Lemma 8.1.6 for $n = 2$ as follows. Let a 2×2 array be given by

$$\begin{matrix} a & b \\ c & d \end{matrix}$$

where $a \leq b$ and $c \leq d$ holds. If the numbers in each column are permuted so that we have smaller element in the first row, then show that in the resulting array, the numbers are still monotonically increasing in each row.

(b) Consider the set-up of the hypothesis of Lemma 8.1.6. Compare the last two members a_{n-1}, a_n and b_{n-1}, b_n of both the sequence and use the first part to show that if we rearrange them, then the conclusion $c_i \leq d_i$ remains valid (here we move at the most two members of each of the sequences and keep the remaining members of both the sequences fixed).

(c) Beginning with $j = n$ and then $j = n-1$ and so on, use the previous step to arrive at a rearrangement in which (after $n-1$ comparisons) the members of both the sequences that are at the front are the smallest members and the conclusion $c_i \leq d_i$ still remains valid.

(d) Repeat the above step (with the last $n-1$ members of both the sequences) to obtain sequences where, at the second position, we have the smallest but one member.

(e) How many times does one have to carry out the operation described to above to arrange the members of both the sequences in ascending order and arrive at the conclusion of the Lemma 8.1.6?

8.56 *Here is a different proof of the Erdös-Szekeres Theorem on Monotone Subsequences (Theorem 8.2.2).* Let d_i (respectively r_i) denote the length of the longest monotone increasing (respectively strictly decreasing) subsequence that begins at the i-th term a_i of the given sequence and *assume both* $d_i \leq m$ *and* $r_i \leq n$. Use pigeonhole principle to show that this leads to a contradiction.

8.57 Let for $n = 3, 4 \ldots, N$ denote the set $\{1, 2, \ldots, 2^{n-1} - 1\}$. Show that there is a function, $f : N \to N$ such that the following condition is satisfied: For every $i \in N$, we have $f(i) \leq i$ but f does not satisfy the property given by equation (8.2) and therefore, the assertion in Theorem 8.2.4 is sharp (the size of the domain 2^{n-1} cannot be made smaller and still derive the same conclusion).

8.58 In a football tournament of 28 teams each team plays one match against every other team. The team gets 2 points for a win, 1 for a draw and 0 points for a loss. It was observed that more than 75 percent matches were drawn. Prove that there are two teams that finished with the same score.

8.59 Prove that $R(p, q) \leq \binom{p+q-2}{p-1}$.

8.60 Use the probability set up of Chapter 6 to prove Theorem 8.3.5 as follows. Let, for a k-subset A of the vertex set, X_A denote the indicator r.v. of A. That is, the X_A takes value 1 iff we have a red K_k on A. Find the expectation of X_A. Let X denote the sum of all X_A's, and Y the corresponding r.v. for the occurrence of a blue K_k. Letting Z denote the sum $X + Y$, conclude from the given data that $E(Z) < 1$ and hence there is some coloring that has neither a red K_k nor a blue K_k.

8.61 Suppose we have a coloring of the edges of K_n in colors $a_1, a_2, \ldots, a_k$ such that we do not have a monochromatic K_3. We wish to produce a coloring of K_{2n} in colors $a_1, a_2, \ldots, a_k$ and a new color a_{k+1} that has no monochromatic K_3 as follows. Let A and B two disjoint sets each of order n. Inside both A and B use the (old) colors $a_1, a_2, \ldots, a_k$ and the coloring of K_n avoiding a monochromatic K_3. The new color a_{k+1} is used only on all the edges in $A \times B$. Show that this coloring has no monochromatic K_3. Conclude that $R_{k+1}(3) \geq 2R_k(3)$ and hence $R_k(3) \geq 2^k$.

8.62 This exercise is borrowed from Knuth, Art of Computer Programming [34]. We have already defined a tree. In this exercise, we consider an infinite tree T (one that has infinitely many vertices). Given any tree T, we can (conveniently) declare any one of its vertices say x_0 as its roots and draw the tree from top to bottom (with x_0 on top) with neighbors of x_0 at level 1, their neighbors at level 2 and so on. By the definition of a tree, all the vertices of T get exhausted in this manner. The tree looks upside down (opposite to the way in which a natural tree looks like). The vertices at level 1 are the children of x_0, those at level 2 are her grandchildren and so on. Since T is infinite and we have a book on combinatorics, we define T to be *locally finite* if every vertex has a finite degree or equivalently, each vertex has only finitely many children. *Prove that a locally finite infinite tree has an infinite path.* Also show that in this assertion, the assumption of local finiteness cannot be dropped. *This assertion is called Konig's' infinity Lemma. A word of caution: The inductive proof requires a judicious choice of a descendant at every stage starting from the root (refer to the proof of Theorem 8.4.1. In fact, Knuth has some interesting piece of information. A randomly chosen path in the progeny tree of a certain community (as empirically verified by a British statistician) ends only after finitely many steps with the last person having no children and this happens with a high probability!*

Chapter 9

Some geometry

9.1 Regular polytopes and tessellations of the plane

In this chapter our aim is to introduce the reader to some results from geometry that have connections with combinatorics. This is by no means exhaustive. However, we believe that the present chapter very much fits in the theme of this book. In particular we go on to describe regular polytopes in 3-dimensions, tessellations or tilings of the plane, partitioning a rectangle into squares and some other interesting topics.
A graph that can be drawn on the plane $\mathbb{R}^2$ without edges crossing each other is called a *planar graph.* For example, all trees are planar and so are all cycles. The complete graph K_5 on five vertices a, b, c, d, e *is not planar.* This can be verified as follows. We attempt to draw it on the plane without edges crossing. Such a drawing must have the 5-cycle C with the given vertices as the corners of C as in Figures 7.2 or 8.2. We still have to drawn five edges $(ac, (ad), (bd), (be)$ and (ce). Of these, only two can be drawn inside C without crossings and only two can be drawn outside without edges crossing leaving us with one edge which cannot be drawn without a crossing. Under stereographic projection, a graph drawn on a sphere with edges represented by arcs on the sphere (and not crossing each other) can be projected on the plane to give a planar drawing of a graph and conversely. In this correspondence, the unbounded region of the graph drawn on the plane is mapped to that region on the sphere which encloses the north pole. Note also that every planar drawing of a graph must have exactly one unbounded region. We begin by proving the following formula that gives an equation connecting the number of vertices, edges and faces (regions) of a planar graph.

Theorem 9.1.1. *(Euler's formula) Let G be a connected planar graph. Let n denote the number of vertices, e the number of edges and f the number of faces (regions) of G. Then the following equation holds.*

$$n - e + f = 2 \tag{9.1}$$

Proof We make induction on n. When $n = 1$, we have only one vertex x with loops around that vertex as shown Figure 9.1. If we have e loops, then including the outside unbounded region, we have $e + 1$ regions (or faces) in all. This is seen as follows.

Adding an extra edge (loop) divides an existing region into two regions (at a rigorous level, this is a consequence of the Jordan curve theorem). Hence, inductively when we have e loops there are $e+1$ regions. So, $n=1$ and $f=e+1$ proving the assertion.

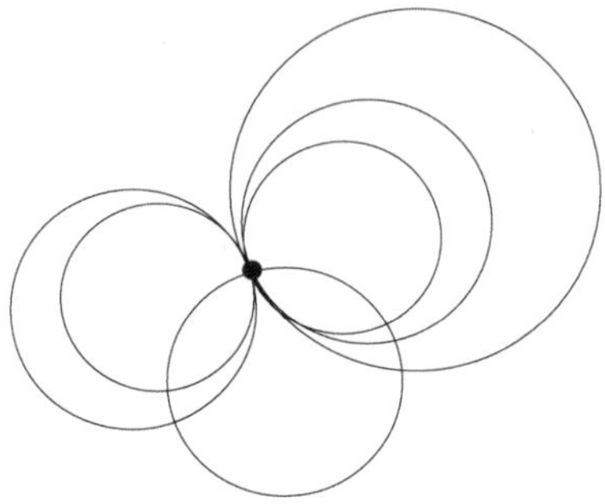

Figure 9.1: a bouquet of loops at a vertex

Now let $n \geq 2$ and then we have an edge $f=(xy)$ where $x \neq y$. We contract the edge f. Thus we identify the two vertices x and y and call the new vertex a, Each edge incident at x or y is also incident at a. This contraction does not reduce the number of regions.

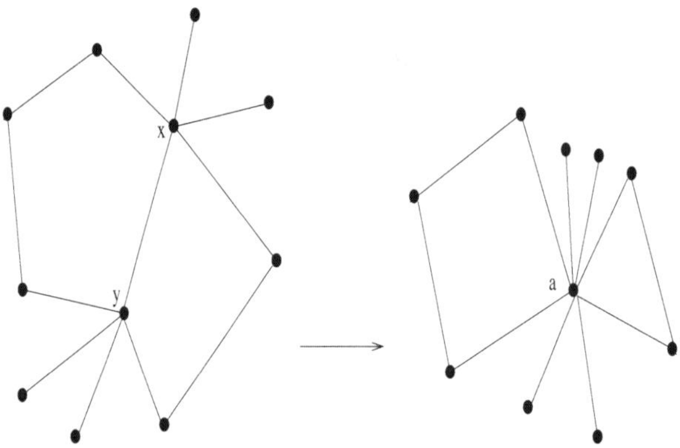

Figure 9.2: contraction of a planar graph at an edge

Thus the new graph G' has $n'=n-1,\ e'=e-1$ and $f'=f$ and by induction, $n'-e'+f'=2$ which implies that $n-e+f=2$. □

Corollary 9.1.2. *Let G be a simple connected planar graph. Then* $e \leq 3n-6$. *Further, if G is also bipartite, then* $e \leq 2n-4$.

Proof In a planar drawing of G, each edge is on the boundary of exactly two faces and since G is simple each face has at least three edges which gives the inequality: $2e \geq 3f$ and hence $f \leq \frac{2e}{3}$. Substitute this in the Euler equation (equation (9.1)), to get $e \leq 3n - 6$. Further, if G is bipartite, then each face is bounded by at least 4 edges and the inequality is improved to $f \leq \frac{e}{2}$. Euler equation again yields the desired conclusion. □

Example 9.1.3. The complete graph K_5 on five vertices is not planar because if it were, then $10 = e \leq 3n - 6 = 15 - 6 = 9$, a contradiction.

Definition 9.1.4. Let $m \geq 3$. *A regular m-gon* (in the two dimensional space $\mathbb{R}^2$) is a polygon with m sides (and also m vertices) in which all the sides are equal and all the angles are also equal.

Every angle of a regular m-gon has measure $(\frac{m-2}{m})\pi$. Also the weaker stipulation that all the sides be equal is sufficient only when $m = 3$.

Definition 9.1.5. *A regular polytope* **P** *(in $\mathbb{R}^3$)* is a convex object bounded by faces that are planes and with the following two properties: Each face is a congruent regular m-gon with $m \geq 3$ (and for a fixed m) and the number of faces meeting at a vertex is k (where $k, m \geq 3$ are fixed numbers).

A regular tetrahedron (shown in Figure 9.3) is a regular polytope with $m = k = 3$. In general, for a regular polytope n, e and f will respectively denote the number of vertices, edges and faces. For a regular tetrahedron, these numbers are $4, 6$ and 4 respectively. The other familiar example of a regular polytope is that of a cube where the triple (n, e, f) equals $(8, 12, 6)$ (see Figure 7.13 or 9.4).

Theorem 9.1.6. *(Classification of regular polytopes) There are exactly five regular polytopes (upto a certain notion of equivalence in the euclidean 3-space). These are listed below.*

	n	e	f	m	k
Regular Tetrahedron	4	6	4	3	3
Cube	8	12	6	4	3
Regular Octahedron	6	12	8	3	4
Regular Dodecahedron	20	30	12	5	3
Regular Icosahedron	12	30	20	3	5

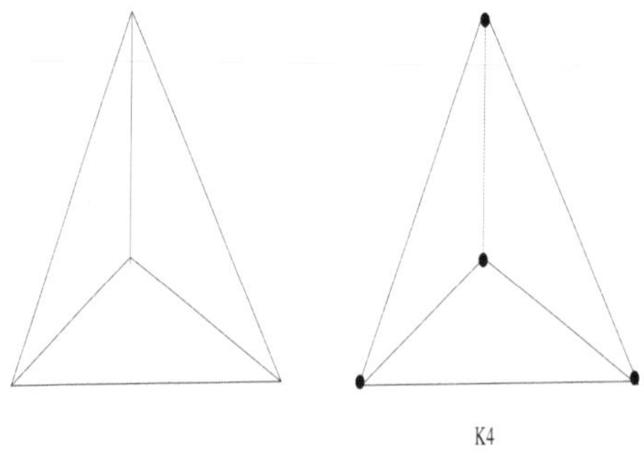

Figure 9.3: Tetrahedron and its graph

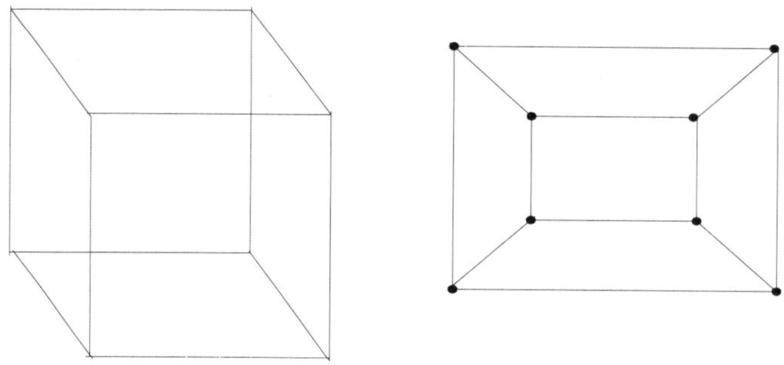

Figure 9.4: The graph of a cube

The equivalence indicated in the statement of the Theorem is that caused by rotations, reflections and translations as also contractions. The first are in fact isometries of R^3 and the first two constitute the orthogonal group $O_3(\mathbb{R})$. But we postpone the discussion of groups to the later Chapter 14.

Proof Given a regular polytope **P**, we circumscribe it by a sphere and make a stereographic projection on the real plane $\mathbb{R}^2$. This is done in such a way that the north pole of the sphere is in the interior of some region. This region then gets mapped onto the infinite region of the planar graph G of **P** (and the south pole is mapped onto the origin). In the case of a regular tetrahedron, the graph G looks like the second drawing adjacent to the regular tetrahedron which is a complete graph K_4 on 4 vertices. That graph is drawn in Figure 9.3 while the graph of the cube is drawn in Figure 9.4.

Since the graph G of a regular polytope **P** is planar, we can now invoke the Euler equation (Theorem 9.1.1). Assume that each face of **P** is a regular m-gon with k faces meeting at each vertex v. Since every edge accounts for two vertices and also bounds exactly two faces, a two-way counting, gives: $nk = fm = 2e$ and hence $n = \frac{2e}{k}$ and $f = \frac{2e}{m}$. These values can be substituted in the Euler equation (9.1) to get the equation:

$$\frac{1}{k} + \frac{1}{m} = \frac{1}{2} + \frac{1}{e} \tag{9.2}$$

which gives the inequality:

$$\frac{1}{k} + \frac{1}{m} > \frac{1}{2} \tag{9.3}$$

Notice that if k and m were both ≥ 4, then the L.H.S. (which is a decreasing function of k and m) is $\leq \frac{1}{2}$ which contradicts (9.3). Hence one of k or m must equal 3. Let $k = 3$. Then, for $m \geq 6$, we see that the L.H.S. is again $\leq \frac{1}{2}$. Using the symmetry of the inequality (in k and m), we see that the only possible pairs are:

$$(k, m) = (3, 3), (3, 4), (3, 5), (4, 3), (5, 3)$$

The values of k and m can be used in equation (9.2) to obtain the value of e and hence n and f are determined. These are given in the table. □

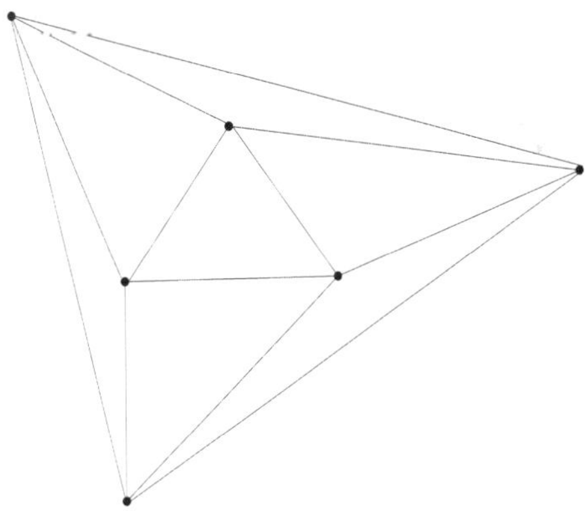

Figure 9.5: The graph of a regular octahdron

The graphs of a regular octahedron, a regular dodecahedron, and a regular icosahedron are drawn in Figure 9.5, Figure 9.6 and Figure 9.7 respectively. We note, in passing that the inequality (9.3) known to the Greeks could also be obtained using the following consideration. Since each face is a regular m-gon, the angle at any vertex v on any face is $\frac{m-2}{m}\pi = (1 - \frac{2}{m})\pi$. Now look at all the angles on all the faces incident at v. If we add all these angles, we get the sum $k(1 - \frac{2}{m})\pi$, which must be strictly less than $360^o = 2\pi$ (why?). Hence $k(1 - \frac{2}{m})\pi < 2\pi$ which yields the inequality (9.3). This argument has the advantage of proving the next interesting theorem (Theorem 9.1.8).

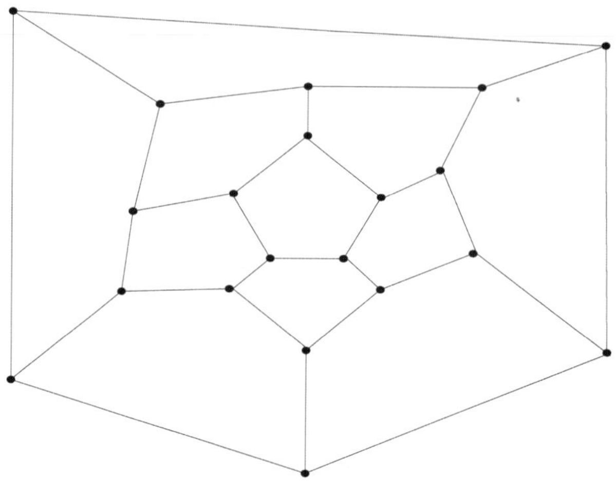

Figure 9.6: The graph of a regular dodecaahedron

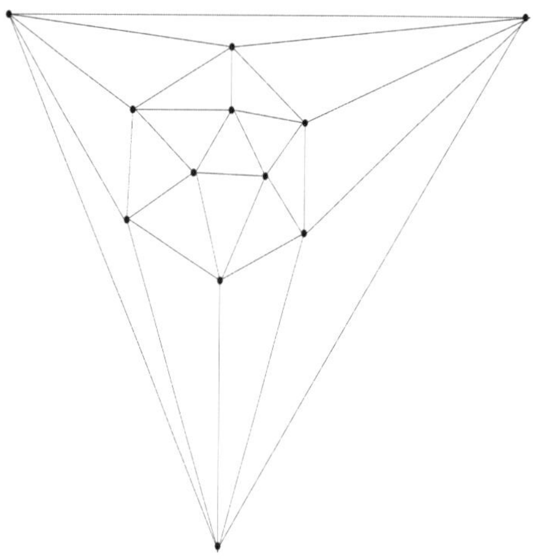

Figure 9.7: The graph of a regular icosahedron

Definition 9.1.7. Let $k, m \geq 3$ be fixed integers. *A regular tessellation (or tiling)* of the plane $\mathbb{R}^2$ is a covering of the plane $\mathbb{R}^2$ by *congruent* non-overlapping regular m-gons such that the number of faces (tiles) meeting at any vertex v is equal to k.

An example of a regular tessellation is the tiling of the plane by tiles each one of which is a square (here, both k and m are equal to 4). In fact, the Greek argument yields the following theorem whose proof is left to the reader as an exercise.

Figure 9.8: The square tiling of the plane

Theorem 9.1.8. *Regular tessellations of the plane exist only when the regular polygon is either an equilateral triangle, a square or a regular hexagon. Here, the pair* (k, m) *takes the values* $(6, 3), (4, 4), (3, 6)$ *and these are the only possible values.*

In the first case, we have a tiling of the plane by congruent equilateral tiles with six tiles meeting at a vertex. In the second case, we have a tiling by congruent squares with four tiles at a vertex and in the third, we have a tiling of the plane by congruent regular hexagons with three hexagons meeting at a vertex. Observe that in some sense a regular tessellation of the plane is a polytope with infinitely many faces. This theme was explored by Coxeter [19].

Definition 9.1.9. Let $k \geq 3$ and let $a_1, a_2, \ldots, a_k$ be integers each at least 3 and also assume, w.l.o.g. that $a_1 \leq a_2 \leq \cdots \leq a_k$. A *homogeneous tessellation* of $\mathbb{R}^2$ is a covering of the plane $\mathbb{R}^2$ by *congruent* non-overlapping regular polygons with each polygon an a_i-gon for some i and such that the all tiles that are a_i-gons are congruent and locally, at every vertex, each tile occurs exactly once with an identical situation at every vertex.

In order to classify all the homogeneous tessellations, we use the Greek idea to get the equation: $\sum_{i=1}^{k}(1 - \frac{2}{a_i})\pi = 2\pi$ which simplifies to

$$\sum_{i=1}^{k} \frac{1}{a_i} = \frac{k}{2} - 1 \tag{9.4}$$

This equation, can indeed be exploited to *classify all the homogeneous tessellations.* The L.H.S. of the equation is a decreasing function of the variables. Since each $a_i \geq 3$, the L.H.S. is $\leq \frac{k}{3}$ and hence $\frac{k}{2} - 1 \leq \frac{k}{3}$ which gives the inequality $k \leq 6$ with equality iff each $a_i = 3$. In that case, we have $a_1 = \cdots = a_6 = 3$ and we, in fact, have a regular tessellation of the plane by equilateral triangles. This case is already covered in Theorem 9.1.8. So, let $k \leq 5$. *First, let* $k = 5$. If $a_1 \geq 4$, then $\frac{5}{2} - 1 = \frac{3}{2} \leq \frac{5}{4}$ which is false. So, $a_1 = 3$. Again, if $a_2 \geq 4$, then we get $\frac{3}{2} - \frac{1}{3} = \frac{7}{6} \leq 1$ which is false. So, $a_2 = 3$. Finally, if $a_3 \geq 4$, then $\frac{3}{2} - \frac{2}{3} = \frac{5}{6} \leq \frac{3}{4}$ which is also false and hence we get $(a_1, a_2, a_3) = (3, 3, 3)$ and we have only a finite number of possibilities for the pair (a_4, a_5). *Now let* $k = 4$. If $a_1 \geq 4$, then we are forced to get the equality

$a_1 = a_2 = a_3 = a_4 = 4$ and we have the regular tessellation by squares that has already been discussed. Let $a_1 = 3$. Then we get $a_2 \leq 4$ and the substitution $a_2 = 3$ or $a_2 = 4$ in the equation (9.4) gives only a finite number of possibilities for the quadruple (a_1, a_2, a_3, a_4). *Finally, let* $k = 3$. Then $a_1 \leq 6$ with equality iff $a_1 = a_2 = a_3 = 6$ and in that case, we get a regular tessellation of the plane by regular hexagons (which has already been covered in theorem 9.1.8). Thus we are left with $a_1 = 3$ and $a_1 = 4$ and we get only a finite number of possibilities in these cases. In fact, all the 11 homogeneous tessellations are listed in the theorem below, whose proof is left to the reader as an exercise. Notice also that as we have just indicated there are possibilities that are numerically feasible but no homogeneous tessellations exist in those cases (such as $a_1 = 5, a_2 = 5, a_3 = 10$).

Theorem 9.1.10. *There are exactly* 11 *homogeneous tessellations of the plane. The values of* k *(the number of polygons at a vertex) and the sequence* $(a_1, a_2, \ldots, a_k)$ *are listed below.*

(1) $k = 3$ *and* $(a_1, a_2, a_3) = (3, 12, 12)$.

(2) $k = 3$ *and* $(a_1, a_2, a_3) = (4, 6, 12)$.

(3) $k = 3$ *and* $(a_1, a_2, a_3) = (4, 8, 8)$.

(4) $k = 3$ *and* $(a_1, a_2, a_3) = (6, 6, 6)$*; this is a regular tessellation by regular hexagons.*

(5) $k = 4$ *and* $(a_1, a_2, a_3, a_4) = (3, 3, 6, 6)$.

(6) $k = 4$ *and* $(a_1, a_2, a_3, a_4) = (3, 4, 4, 6)$.

(7) $k = 4$ *and* $(a_1, a_2, a_3, a_4) = (4, 4, 4, 4)$ *and this is a regular tessellation by squares.*

(8) $k = 5$ *and* $(a_1, a_2, a_3, a_4, a_5) = (3, 3, 3, 3, 6)$.

(9) $k = 5$ *and* $(a_1, a_2, a_3, a_4, a_5) = (3, 3, 3, 4, 4)$. *Here we have rows of equilateral triangles and squares arranged alternately.*

(10) $k = 5$ *and* $(a_1, a_2, a_3, a_4, a_5) = (3, 3, 3, 4, 4)$. *Here we have an equilateral triangles on each side of a square.*

(11) $k = 6$ *and* $(a_1, a_2, a_3, a_4, a_5, a_6) = (3, 3, 3, 3, 3, 3)$ *and this is a regular tessellation by equilateral triangles.*

We wish to draw the attention of the reader to the fact that the possibilities listed in Theorem 9.1.10 actually exist and are not just parametrically feasible. For example, equation (9.4) permits the possibility $(a_1, a_2, a_3) = (5, 5, 10)$ which does not exist since an attempt to construct it will result in overlapping regions. Also the possibilities Theorem 9.1.10 (9) and (10) are different. The former is drawn in Figure 9.9 while in the latter case, we have four equilateral triangles on four sides of a square as well as three squares on the three sides of each equilateral triangle. The reader should make his own drawing for this as well as all the other cases in Theorem 9.1.10. Alternatively, these drawings are available in Berman and Fryer [6].

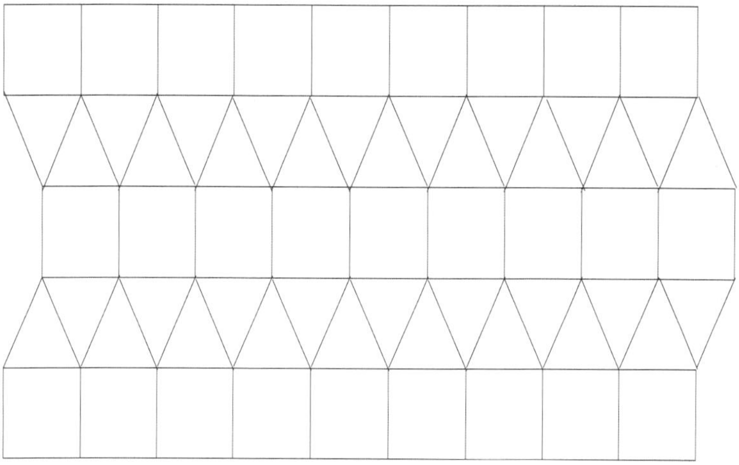

Figure 9.9: A tiling of the plane by squares and equilateral triangles

9.2 Some more geometry

We now collect some results pertaining to portioning a rectangle into squares.

Theorem 9.2.1. *Let R be an $a \times b$ rectangle. Then R can be partitioned into congruent squares iff $\frac{a}{b}$ is a rational number.*

Proof Let R be partitioned into congruent squares each of side λ. Then, for some integers m and n, we must have $a = m\lambda$ and $b = n\lambda$ and therefore $\frac{a}{b} = \frac{m}{n}$ which is a rational number. Conversely if $\frac{a}{b} = \frac{m}{n}$ where m, n are integers, then we let $\lambda = \frac{a}{m} = \frac{b}{n}$ to see that R can be partitioned into mn congruent squares each with side λ. □

Definition 9.2.2. A partition of a rectangle R into *incongruent squares* consists of a partition of R into a finite number of square such that *every two squares are mutually incongruent.* Such a partition of R is called an *incongruent partition.*

Figure 9.10 is an example of an incongruent partition of a 32×33 rectangle. Note that Figure 9.10 *is only a schematic representation (where squares are represented by rectangle).*

Lemma 9.2.3. *Let R be a rectangle partitioned into incongruent squares. Let A be the smallest square in the partition. Then A is placed in the middle of R with exactly four squares surrounding it (upto rotation) (as shown in Figure 9.11).*

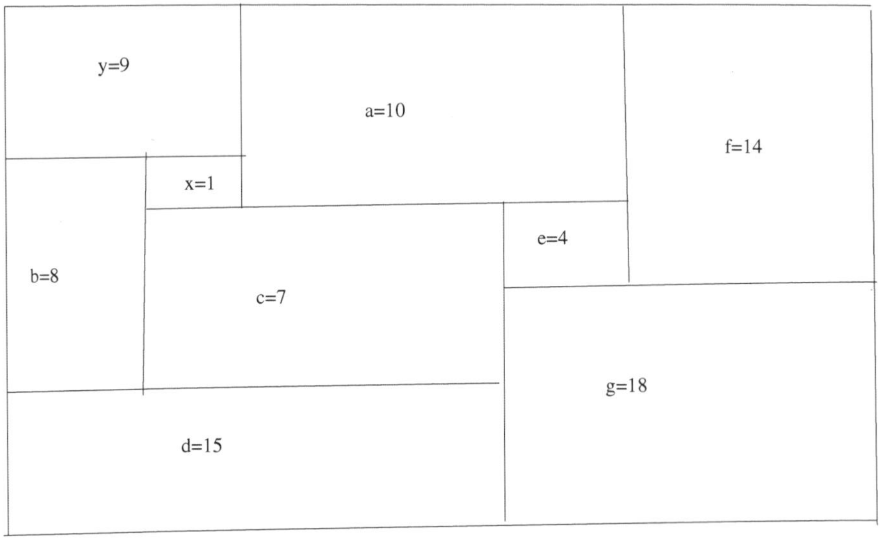

Figure 9.10: A 32×33 rectangle partitioned into 9 incongruent squares

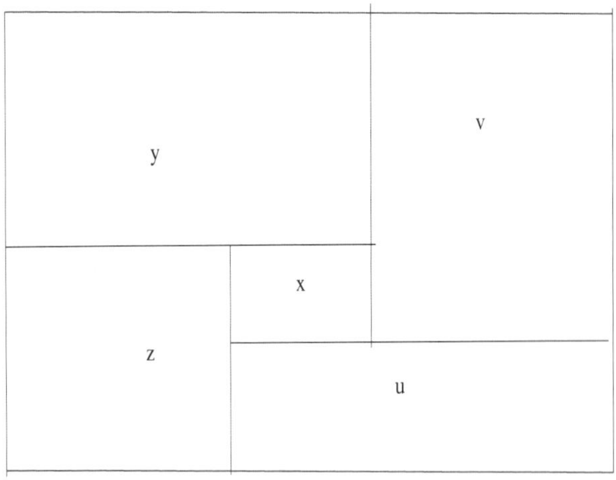

Figure 9.11: Small square surrounded by four larger incongruent squares

Proof If A occurs at a corner of R, then the two squares attached to A on its internal sides have larger sides and hence must overlap each other, a contradiction. If A occurs on some side of R say the left wall, then look at the two squares one immediately below and one immediately above A. Both these squares must protrude beyond A creating a horizontal tunnel which will have to be covered by squares with side $\leq$ the side of A, a contradiction. Hence A must occur in the middle of R. Suppose on the top of A, we have two squares that partially cover the top of A (and hence necessarily protrude to the left and right of the top of A), then we must have exactly one square each covering the left and right wall of A that must protrude below the bottom of A (since their sizes are bigger than that of A). This again creates a tunnel like situation at the bottom of A which must be covered by squares of size smaller than that of A, a contradiction. This argument proves that at the most and hence exactly one square is on each of the four sides of A and the picture (after rotation) must look like that in Figure 9.11. □

Lemma 9.2.3 proves that we require at least five squares for an incongruent partition of a rectangle R. Suppose we have such a partition into five squares with the smallest square (of side) x and the four others surrounding it denoted by y, z, u, v as shown in Figure 9.11. We immediately see that

$$x = y - z = z - u = u - v = v - y$$

and hence by adding all the four numbers on the R.H.S. we get $4x = 0$ which is a contradiction. We thus conclude that it is not possible to partition a rectangle into five incongruent squares. We now show that a partition of a rectangle R into 9 incongruent squares (as shown in Figure 9.10) is possible by solving linear equations.

$$\begin{aligned}
a &= x + y \\
b &= y - x \\
c &= b - x = y - 2x \\
d &= b + c = 2y - 3x \\
e &= (x + a) - c = (2x + y) - (y - 2x) = 4x \\
f &= a + e = 5x + y \\
g &= f + e = 9x + y.
\end{aligned}$$

Equating the left side length of R with its right side length, we get

$$y + b + d = 4y - 4x = f + g = 14x + 2y$$

from which we get $y = 9x$ and hence by taking $x = 1$ we get a partition of a 32×33 rectangle. Since x is arbitrary, we have proved the following theorem.

Theorem 9.2.4. *There exists a partition of a rectangle into incongruent squares.*

Remark 9.2.5. There are many incongruent partitions of a rectangle into squares; refer to Bollobas [10], for details.

Theorem 9.2.6. *There is no incongruent cube partition of a rectangular parallelepiped.*

Proof Suppose there is a an incongruent partition of a rectangular parallelepiped P which is an $a \times b \times c$. Look at the cubes on the bottom of P. These give a partition of the bottom rectangle R (which is an $a \times b$ rectangle) into incongruent squares and hence by Lemma 9.2.3, the smallest among these say C must be surrounded by four larger ones. The cubes that give rise to these squares *edge above the smallest cube creating a vertical tunnel.* This creates a rectangular parallelepiped (whose base is the top of the smallest cube C at the bottom level) which is $a' \times b' \times c'$ with $a' = b'$ is the top of C at the bottom level. This has to be covered by incongruent cubes and hence repeating the argument, we get another smaller tunnel inside the original tunnel which also needs to be covered by incongruent cubes. Continuing in this manner, we get an infinite sequence of tunnels (and cubes) which contradicts the assumption that we have only a finite number of cubes. □

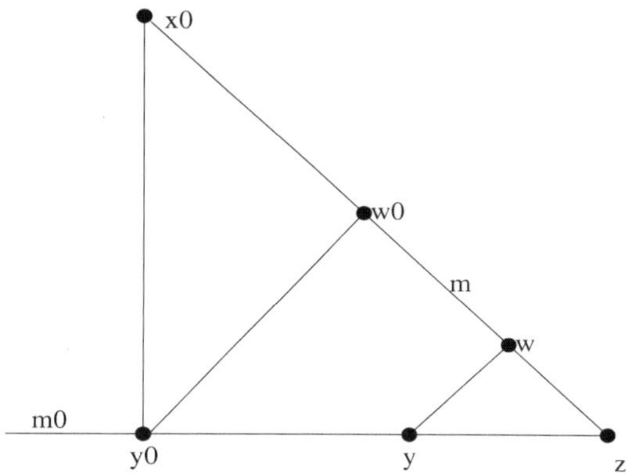

Figure 9.12: Proving the Sylvester-Gallai theorem

Theorem 9.2.7. *(Sylvester-Gallai Theorem) Let S be a finite set of points in $\mathbb{R}^2$ with the following property. Any (straight) line l that contains (joins) two points of S must contain at least three points of S. Then all the points of S are on a single line.*

Proof Suppose there is no such line that contains all the points of S. Let R be the set of all the lines that contain at least two (and hence by the stipulation, at least three) points of S. Then R is a finite set and for every point-line pair (p, l) with $p \in S$ and $l \in R$ such that p *is not* on l, we denote by $d(p, l)$, the distance between p and l. Let d_0 be the minimum of all such distances. Then d_0 is a positive real number. Let $d_0 = d(x_0, m_0)$. We show that there is a pair (y, m) for which $d(y, m) < d_0$ *which leads to the desired contradiction.* Let y_0 be the foot of the perpendicular from x_0 to m_0. Then at least two (of the three) points on m_0 are on the same side of y_0 on m_0 and let these be y and z in that order and as shown in Figure 9.12. Drop perpendiculars y_0w_0 and yw from y_0 and y onto the line m joining z and x_0. By assumption, $m \in R$. We have $x_0y_0 > y_0w_0$ since the first line is the hypotenuse and in the right angled triangle $x_0y_0w_0$. The line yw is in the triangle y_0w_0z and is parallel to the side y_0w_0. Hence we have

$$d_0 = d(x_0, m_0) = x_0y_0 > y_0w_0 \geq yw = d(y, m)$$

which is a contradiction. □

The remaining results in this section are from Bollobas [10].

Theorem 9.2.8. *Let R be a rectangle partitioned into a finite number of subrectangles $R_1, R_2, \ldots, R_m$. such that each R_i has at least one side an integer (thus either the length or the breadth of every R_i is an integer). Then R has the same property. That is at least one side of R (the length or breadth) is an integer.*

Proof Situate R with vertices $(0,0), (a,0), (0,b), (a,b)$ in the first quadrant of the real plane. We have to show that either a or b is an integer. We are given a division of R into R_i' s and R_i has either its vertical side with integer length or its horizontal side with integer length or both. Make a bipartite graph G with bipartition (X, Y) such that X consists of integer lattice points $\{(x,y) : x, y \in \mathbb{Z}^+\}$ in the first quadrant and Y consists of all the m rectangles R_i partitioning R. We draw an edge from a lattice point (r,s) to a rectangle R_i if (r,s) is one of the (four) vertices of R_i. Consider an R_i with vertices

$$v_{1,1} = (x_1, y_1), v_{2,1} = (x_2, y_1), v_{1,2} = (x_1, y_2), v_{2,2} = (x_2, y_2)$$

Let w.l.o.g. $x_2 - x_1$ be an integer. Then either both $v_{1,1}$ and $v_{2,1}$ are in X or none of these points is in X and the same also holds for $v_{1,2}$ and $v_{2,2}$. It thus follows R_i has degree $0, 2$ or 4 and hence summing over all R_i, the total number of edges of G (which is just the number of edges from Y to X) is an even number. Now suppose that neither a nor b is an integer. Then no vertex of R except the origin is in X and its degree is clearly 1. Let v be any other lattice point which is a vertex of some subrectangle R_i. Then v is a vertex of either two or four subrectangles. This shows that the total number of edges of G (which is the number of edges from X to Y) is an odd number, a contradiction. □

We give another proof since it can be generalized to higher dimensions. Fix an ϵ, a small positive real number. Define $f : \mathbb{R} \to \mathbb{R}$ by $f(x) = x + \epsilon$ if x is not an integer and $f(x) = x$ if x is an integer. Extend this to $\mathbb{R}^2$ by letting $g(x,y) = (f(x), f(y))$. The number ϵ is chosen as follows. Let (x,y) be a vertex of some subrectangle R_i and let $\epsilon(x,y) = min\{\lceil x \rceil - x, \lceil y \rceil - y\}$ and let ϵ_0 denote the minimum over all $\epsilon(x,y)$ where (x,y) is a vertex of some R_i and $\epsilon(x,y) > 0$. Assuming, for the sake of contradiction that a and b are both not integers, it follows that ϵ_0 is a positive real number between 0 and 1. Restrict ϵ to the open interval $(0, \epsilon_0)$. Applying g to the vertices of R_i obtains a shifted rectangle S_i and a shifted rectangle S from R. Finally the areas of S and S_i are calculated easily. Denoting these by $|S|$ and $|S_i|$ respectively, we have $|S| = \sum_i |S_i|$. Here, $|S| = (a+\epsilon)(b+\epsilon)$ *is a quadratic in* ϵ. Consider S_i. By the choice of ϵ, if one side of R_i is an integer, then the same side of S_i is also the same integer (and hence depends only on the coordinates of the vertices of R_i, which are constants). Therefore, summing over all i, we see that $|S|$ has the form $r\epsilon + t$ where r and t are constants that depend only on the given partition. We thus have a situation where a quadratic in ϵ equals a linear expression in ϵ. Since this holds for infinitely many ϵ, we arrive at a contradiction. The proof technique can be generalized to prove the n-dimensional version. *By a box in n dimensions we mean a rectangular parallelepiped B (thus the edges are at right angles).*

Theorem 9.2.9. *Let a box B in $\mathbb{R}^n$ be partitioned into subboxes $B_i, i = 1, 2, \ldots, m$. Let k be a fixed integer between 1 and n. Assume that for each subbox B_i at least k of its sides are integers. Then at least k sides of B are integers.*

The proof, which is left to the reader is essentially the same. On one side we get a polynomial whose degree is at the most $n - k$. For this to be true on the other side, at least k sides of B must be integers.

1	1	1	2
2	2	1	2

not in the same way

1	1	1	1
2	2	2	2

in the same way

Figure 9.13: Tiles not in the same way and in the same way

Let an $A \times B$ rectangle be tiled (partitioned) by congruent $a \times b$ rectangles. The partition (or the tiles arranged) is said to be "in the same way" if the tiles are parallel to each other. For this to happen, after interchanging A and B, if required we must have a divides A and b divides B. In the two tilings in Figure 9.13 where the cells in each 1×2 tile are numbered by its domino (or box) number, *the first tiling does not have tiles in the same way while the second does.* This is because in the first tiling, two tiles are horizontal and two are vertical while in the second tiling all the four tiles are horizontal. We can thus make a general definition for the n-dimensional space.

Definition 9.2.10. Let B be a rectangular box $B = A_1 \times A_2 \times \cdots \times A_n$ in n-dimensions. Let B be partitioned into congruent subboxes B_i with $i = 1, 2, \ldots, m$ where each B_i is an $a_1 \times a_2 \times \cdots \times a_n$ box (here we assume that all the numbers involved are integers). Then the partition is said to be *in the same way* if the subboxes are arranged parallel to each other.

For this to happen, as we just observed, we should be able to permute $A_1, A_2, \ldots, A_n$ (if required) so that a_i divides A_i for every i.

Theorem 9.2.11. *Let a_i and A_i be positive integers where $i = 1, 2, \ldots, n$. Let an n dimensional box B with sides $A_1 \times A_2 \times \cdots \times A_n$ be partitioned into sub boxes each of which has sides $a_1 \times a_2 \times \cdots \times a_n$. where a_1 divides a_2, a_2 divides a_3, $\cdots$, a_{n-1} divides a_n. Then it is possible to obtain a partition of B into the same number of subboxes with the subboxes arranged in the same way.*

Proof As was observed it suffices to prove that after relabeling (permuting), we should be able to show that a_i divides A_i for every i. Consider a_n. In Theorem 9.2.9, replace the word integer (i.e. divisible by 1) by a multiple of a_n. We have thus placed the box in the first positive (out of 2^n) part of the space $\mathbb{R}^n$ and are looking at the lattice points with lattice given by multiples of a_n. Since each subbox has at least one side divisible by a_n, it follows from the Theorem, that the same holds for the bigger box B say a_n divides A_j. By relabeling, we may assume that a_n divides A_n. Now consider a_{n-1}. Since a_{n-1} divides a_n, each subbox has at least two sides that are multiples of a_{n-1} Application of Theorem 9.2.9 (with lattice points that are multiples of a_{n-1}) shows that at least two sides of B are multiples of a_{n-1}, one of which is A_n and by relabeling, we may assume that the other is A_{n-1}. Continue in this manner and repeatedly use Theorem 9.2.9 to conclude that a_k divides A_k for all k. □

Corollary 9.2.12. *Let a brick B in three dimensions be partitioned into subbricks each of which is a $1 \times 2 \times 4$. Then it is possible to partition B in such a way that all the bricks are arranged in the same way.*

9.3 Triangulations and Sperner's lemma

We now look at the following question. Start with a (big enough) triangle T (in $\mathbb{R}^2$) with vertices A, B and C labeled $1, 2, 3$ in an anticlockwise manner.

Definition 9.3.1. *A triangulation of T* is a division of the closed triangle T into regions each of which is a triangle. Such a division is called *proper* if every two smaller triangles T' and T'' (considered as closed triangles) have either a vertex of each or a side of each in common or have an empty intersection. We also assume that if a vertex of a baby (i.e. smaller) triangle T' lies on some side of another baby triangle T'' then it must also be a vertex of T''. The vertices of all the baby triangles are called *the vertices of the triangulation.*

Definition 9.3.2. Label the new vertices of the triangulation (vertices other than A, B, C that have already been labeled) using one of the labels $1, 2, 3$. Such a labeling is called a *Sperner labeling* if any vertex that lies on the side of T whose end-points are labeled with i and j receives either the label i or the label j (but cannot receive the third possible label), there being no restriction on the labels of vertices that are in the interior of T. Also Call a baby triangle of the triangulation with a given Sperner labeling *a complete triangle* if the three vertices of this triangle receive three distinct labels.

In Figure 9.14, we have a Sperner labeling with 7 complete baby triangles. These are marked with small circles drawn inside them. The basic assertion of Sperner's Lemma that is proved in this section is that a triangulation with a Sperner labeling always *has at least one complete triangle.*

Lemma 9.3.3. *Let $I = [a, b]$ be a closed interval on the real line with $a < b$. Divide I into closed subintervals $I_i = [a_{i-1}, a_i]$ where*

$$a = a_0 < a_1 < \cdots < a_i < \cdots < a_{n-1} < a_n = b$$

Label the vertices the end points a and b of I with labels 0 and 1 respectively. Label all the intermediate points $a_1, a_2, \ldots, a_{n-1}$ with one of the labels 0 or 1 in any arbitrary manner. Call a subinterval I_i complete if its end points a_{i-1} and a_i receive labels 0 and 1 respectively or 1 and 0 respectively. Then the number of complete subintervals is an odd number.

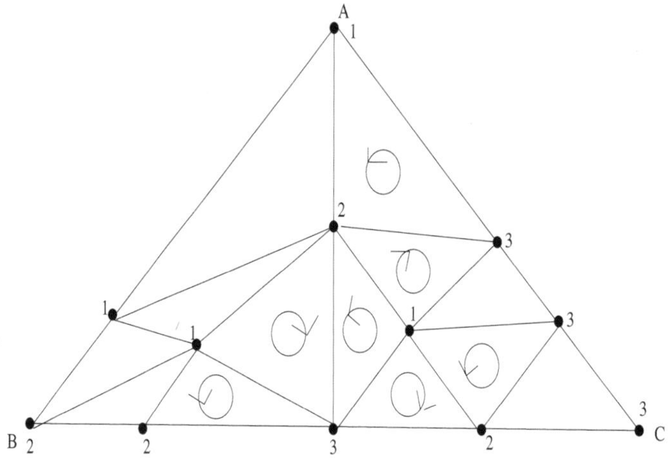

Figure 9.14: Triangulations and Sperner Labeling

Proof Use induction on n with the result being true when $n = 1$ since a and b receive different labels and hence $[a, b]$ is complete. Let the result hold for n and let γ be the number of complete subintervals when we have n subintervals in all. Subdivide the j-th subinterval I_j by inserting an extra point x in the open interval (a_{j-1}, a_j) so that we now have $n+1$ subintervals $I_1, I_2, \ldots, I_{j-1}, I_j', I_j'', I_{j+1}, \ldots, I_n$ where I_j' and I_j'' denote the subintervals $[a_{j-1}, x]$ and $[x, a_j]$ respectively. Label x by 0 or 1 and let γ' be the number of complete subintervals. To prove the assertion, it will be sufficient to show that $t = \gamma' - \gamma$ equals 0 or 2. Let the labels of both a_{j-1} and a_j be 0. Then $t = 0$ if x is also labeled 0 and $t = 2$ if x is labeled 1 (in the latter case, both I_j', I_j'' are complete while I_j is not). The same is also true in case both a_{j-1} and a_j are labeled 1. Now let a_{j-1} be labeled 0 and let a_j be labeled 1. Then regardless of the label of x, one of I_j', I_j'' is complete and the other is not while I_j is complete and hence $t = 0$. The same holds if a_{j-1} is labeled 1 and a_j is labeled 0. This shows that $t = 0$ or $t = 2$. □

Lemma 9.3.4. *(Sperner's Lemma) Let T be a triangle with vertices A, B, C labeled $1, 2, 3$ in an anticlockwise manner. Let T have a proper triangulation with a Sperner labeling of vertices of the triangulation. Then the number of complete baby triangles is odd.*

Proof Define a graph G with each vertex corresponding to a baby triangle and an extra vertex ∞ that corresponds to the outside of T. Let R, R' be two baby triangles or R a baby triangle and $R' = \infty$. We make an edge (R, R') in G if R and R' share an edge (xy) such that the vertices x and y get labels 1 and 2 (either x is labeled 1 and y is labeled 2 or vice versa). Let R be a baby triangle with a positive degree. Then on one edge of R, we have labels 1 and 2. If the third label of R is 3, then we have a complete triangle whose degree in G is 1. It is also clear that no matter what the label on the third vertex is, the degree of R cannot be 3. Hence a baby triangle R has an odd degree iff it is complete. Let a baby triangle R with positive degree be incomplete. Then its labels are in the set $\{1, 2\}$ where both 1 and 2 occur and one of these labels repeats. Hence, in this case the degree of R is 2. We conclude that an an incomplete R has an even degree. Now consider the vertex ∞. Lemma 9.3.3 shows that the number of complete subintervals on the line AB is an odd number and hence ∞ has an odd degree. Since the number of vertices of odd degree in any graph is even (Chapter 1), it follows that there is an odd number of complete baby triangles. □

Let a triangle T be properly triangulated with a Sperner labeling of the vertices of the triangulation. Consider the number of complete triangles with labels $1, 2, 3$ arranged in an anticlockwise manner (which coincides with the original complete and anticlockwise labeling of the vertices A, B, C of T) as shown on the left hand side complete triangle in Figure 9.15. Let the number of such anticlockwise complete triangles be α. Let the number of the other type of complete triangles (shown on the right hand side complete triangle in Figure 9.15) be denoted by β. Sperner's lemma (Lemma 9.3.4) just proved says that $\alpha + \beta$ is odd. Can we say something more? In Figure 9.14, for example, $\alpha = 4$ and $\beta = 3$. A complete baby triangle of anticlockwise type and a complete baby triangle of clockwise type are indicated with proper orientations of the circles in the interior. Note that $\alpha - \beta = 1$ which, *we want to assert to be true for all triangulations with Sperner labeling.* In particular, $\alpha - \beta = 1$ implies that $\alpha + \beta$ is odd proving Sperner's Lemma (Lemma 9.3.4).

Theorem 9.3.5. *(Stronger form of Sperner's Lemma) With the notations and terminology as before,* $\alpha - \beta = 1$.

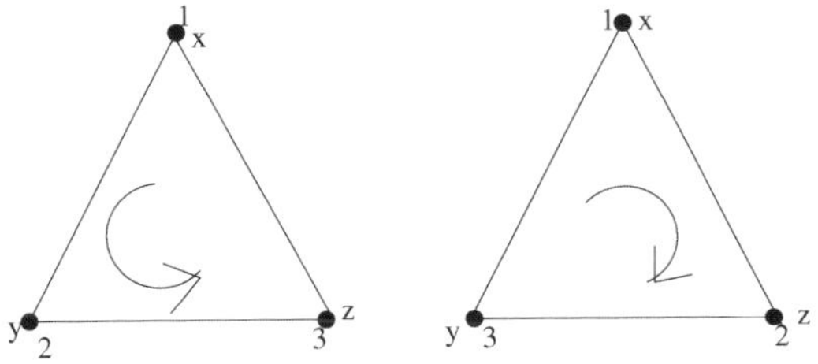

Figure 9.15: Complete oriented triangles of different types

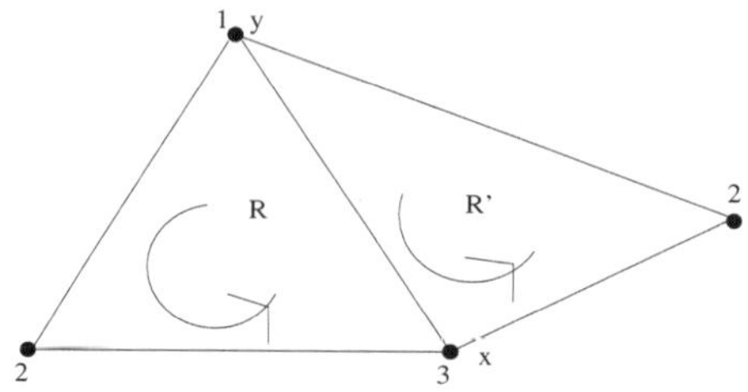

Figure 9.16: Two triangles with a common boundary edge xy

Proof Define a function f on the set of all (directed) edges $\overrightarrow{(xy)}$ of the triangulation as follows. f takes values in $\mathbb{Z}_3$ and make all the computations in this additive group. Orient each baby triangle R in an anticlockwise direction. Then every edge of the baby triangle R also gets directed. Suppose $e = \overrightarrow{(xy)}$ is an edge directed from x to y. Then $f\overrightarrow{(xy)} \in \{0, 1, -1\}$. Let x be labeled i and let y be labeled j. Then $f\overrightarrow{(xy)} = t$ where $t \equiv j - i$. Then f is well-defined and if an edge $e = (xy)$ bounds two baby triangles, then $f\overrightarrow{(yx)} = -f\overrightarrow{(xy)}$ because in one triangle the edge will get directed from x to y and in the other, it will get directed from y to x as shown in Figure 9.16.
Now define $f(R)$ for each baby triangle as follows. Let R be oriented in an anticlockwise manner as per the convention with vertices x, y, z in that order. Then

$$f(R) = f\overrightarrow{(xy)} + f\overrightarrow{(yz)} + f\overrightarrow{(zx)}$$

For example for the first complete triangle R shown in Figure 9.16, we see that the f-value for each side is 1 and hence $f(R) = 3$ while for the second complete triangle the f-values are -1 for each side and hence $f(R)$ is -3. Therefore, if a complete triangle R is of anticlockwise type then $f(R) = 3$ and if it is of clockwise type, then $f(R) = -3$. Now consider a baby triangle R which is incomplete. If all the the three vertices of R are labeled with the same number i, then each side of R has f value 0 and hence $f(R) = 0$. Suppose only two labels say i and j occur. Then one label, say j, must occur twice and the situation is as shown in Figure 9.17.

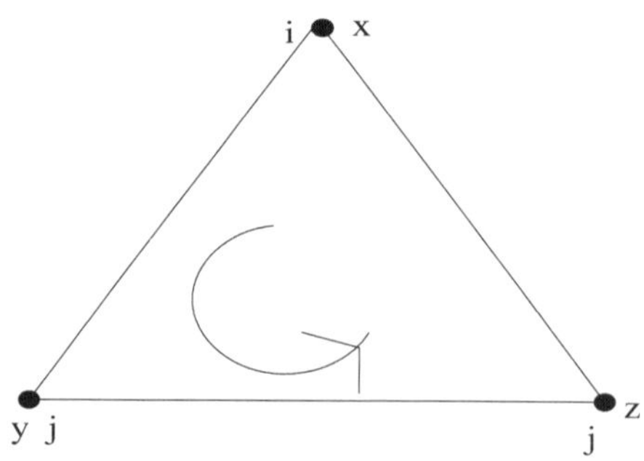

Figure 9.17: An incomplete triangle contributes 0

Clearly $f(\overrightarrow{yz}) = 0$ while $f(\overrightarrow{zx}) = -f(\overrightarrow{xy})$ and hence $f(R) = 0$. We thus conclude that $f(R) = 0$ for any incomplete baby triangle. Recall that the number of complete anticlockwise (respectively clockwise) baby triangles is α (respectively β) and hence we have

$$S = \sum_R f(R) = 3\alpha - 3\beta = 3(\alpha - \beta)$$

where the sum is over all the baby triangles R of the triangulation. We now evaluate S in a different manner. Observe that any edge $e = (xy)$ which is not on the sides AB, BC or CA of the original triangle T is an edge in two triangles say R and R' where its f-value in one triangle is negative of its f-value in the other triangle as shown in Figure 9.16. Thus each internal edge (xy) has a net contribution of 0 to S. Hence S equals $S_{AB} + S_{BC} + S_{CA}$ where $S_{AB} = \sum f(\overrightarrow{xy})$ where the sum is over all the edges on the boundary AB and S_{BC} and S_{CA} are similarly defined.

Consider an edge $\overrightarrow{(xy)}$ lying on the side AB of T. If x is labeled 1 and y is labeled 2, then $f(\overrightarrow{xy}) = 2 - 1 = 1$ while if x is labeled 2 and y is labeled 1, then $f(\overrightarrow{xy}) = -1$. In all the other cases (when x and y receive the same label 1 or 2), we have $f(\overrightarrow{xy}) = 0$. We thus see that the S_{AB} is the difference between the number of complete subintervals on AB with left vertex labeled 1 and the right vertex labeled 2 and the number of complete subintervals on AB with left vertex labeled 2 and the right vertex labeled 1. As an exercise, the reader is asked to prove that this number is 1. Therefore, $S_{AB} = 1$. Similarly S_{BC} and S_{CA} are 1 each and hence $S = 3$. This shows that $3(\alpha - \beta) = 3$ proving the assertion. □

Theorem 9.3.6. *(Brouwer's fixed point theorem) Let T be triangle with vertices A, B, C (where T is treated as a closed triangle) and let $f : T \to T$ be a continuous function. Then there exists a point $\overline{x}$ in T such that $f(\overline{x}) = \overline{x}$. That is f has a fixed point.*

Proof Use the Barycentric division w.r.t. the convex basis A, B, C of T and represent each point of T as a convex combination $x_1 A + x_2 B + x_3 C$ where each $x_i \geq 0$ and $x_1 + x_2 + x_3 = 1$. Since such a representation is unique, we may identify this point with the triple (x_1, x_2, x_3) of non-negative real numbers whose sum is 1. If $f(x_1, x_2, x_3) = (y_1, y_2, y_3)$, then the function $f_i : (x_1, x_2, x_3) \to y_i$ is a continuous function. Now divide the region T into subregions T_i defined as follows:

$$T_i = \{(x_1, x_2, x_3) : f(x_1, x_2, x_3) = (y_1, y_2, y_3) \text{ and } y_i \leq x_i\}$$

Each point $(x_1, x_2, x_3) \in T_i$ for some i, for, if not, then $x_i < y_i \;\; \forall\, i$ which implies that $1 = \sum_i y_i > \sum_i x_i = 1$, a contradiction. Hence T is a union of the closed and bounded (and hence compact) sets $T_i, i = 1, 2, 3$. Also, if we have some $(x_1, x_2, x_3) \in T_i \;\forall\, i$, then $1 = \sum_i x_i \geq \sum_i y_i = 1$ and hence equality must hold everywhere so that $y_i = x_i \;\forall\, i$ and therefore $(y_1, y_2, y_3) = (x_1, x_2, x_3)$. We now make a triangulation of T into very small baby triangles and label the vertices according to which T_i they belong to. The vertices on the boundary get label of one of the labels of the end-vertices: Suppose $(x_1, x_2, 0)$ is on the boundary AB. Then this point belongs to either T_1 or T_2, for if not, then $y_1 > x_1$ and $y_2 > x_2$ and hence $1 = \sum_i y_i \geq y_1 + y_2 > x_1 + x_2 = 1$ a contradiction. Hence the vertex labeling of the triangulation is a Sperner labeling. Using Sperner's Lemma (Lemma 9.3.4), we have a baby triangle which is complete. We can, in fact successively divide T into baby triangles with lengths of sides tending to 0. This gives us three sequences, one in each T_i. Since each T_i is compact, we can find common subsequences (using the Bolzano-Weierstrass property) of triangles each of which is convergent and since the lengths of the sides of the triangles tend to 0, all the the three subsequences in T_1, T_2, T_3 tend to the same point (x_1, x_2, x_3). Clearly, this point is in all the T_i's and hence is a fixed point of f. □

Brouwer's fixed point theorem, which is more general than what we just proved states that every continuous function from the unit disc D_2 in the plane to itself has a fixed point. This follows using a homeomorphism between D_2 and T. Finally analogues of Sperner's Lemma also exist in higher dimensions. This is treated in the exercises (Exercise 9.18).

9.4 Introduction to Euclidean Ramsey theory

Though conceptually part of the general frame of Ramsey theory, the problems of Euclidean Ramsey theory are of a different nature. *Here we color all the points of the real plane $\mathbb{R}^2$ (the Euclidean space) and ask various questions on finding monochromatic configurations.* For example, if we 2-color $\mathbb{R}^2$ in red and blue, are we sure, we get either a red equilateral triangle (that is an equilateral triangle with all vertices red) or a blue equilateral triangle (commonly to be called a monochromatic triangle)? There is thus a paradigm shift from the classical graph Ramsey theory which we dealt with in Chapter 8.

Lemma 9.4.1. *Let $\mathbb{R}^2$ be 2-colored in red and blue such that both the colors red and blue occur. Then there exist two points x and y at distance 2 such that x and y have different colors.*

Proof If a and b are two points such that a is red and b is blue and distance between them is ≤ 4, then we can find a point c which is at distance 2 from both of them. Depending on the color of c, we can then find the required pair. If a and b are at distance > 4 then we can construct a path of piecewise straight lines say $a = a_0, a_1, a_2, \ldots, a_n = b$ such that for every $i = 0, 1, \ldots, n-1$ the segment $a_i a_{i+1}$ has length 2. Since a_0 is red and a_n is blue, there is some i for which a_i is red and a_{i+1} is blue. □

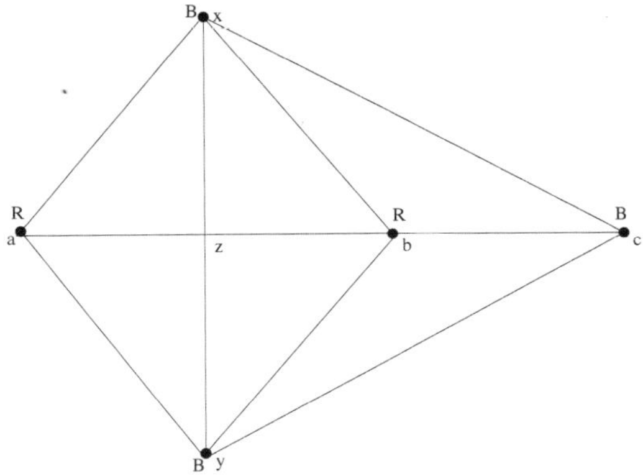

Figure 9.18: A monochromatic equilateral triangle with side 1 or $\sqrt{3}$

Theorem 9.4.2. *Let $\mathbb{R}^2$ be 2-colored. Then either there is an a monochromatic equilateral triangle with side 1 or a monochromatic equilateral triangle with side $\sqrt{3}$.*

Proof We may assume that points of both the colors red and blue exist. Then using Lemma 9.4.1, there are two points a and c at distance 2 such that a is red and c is blue. Let b denote the mid-point of ac and let w.l.o.g. b be red. Construct two unit equilateral triangles axb and ayb, on base ab, one above and the other below as shown in Figure 9.18.
If x is red, then axb is the required unit red triangle. Similarly, we are also done if y is red. Hence we must assume that both x and y are blue. If z is the mid-point of ab (which is also mid-point of xy), then $zc = \frac{3}{2}$ and $xz = \frac{\sqrt{3}}{2}$ implies $xc = yc = \sqrt{3}$ showing that we have a blue equilateral triangle with each side of length $\sqrt{3}$. □

Theorem 9.4.3. *There exists a coloring of $\mathbb{R}^2$ in two colors that does not have a monochromatic unit length equilateral triangle.*

Proof In $\mathbb{R}^2$, consider the horizontal lines $y = k\frac{\sqrt{3}}{2}$ where $k \in z$. Color the semiopen (closed below and open above) horizontal bands between the parallel lines $y = k\frac{\sqrt{3}}{2}$ and $y = (k+1)\frac{\sqrt{3}}{2}$ alternately in red and blue. This coloring does not have the unit equilateral monochromatic triangle. □

Theorem 9.4.4. *In any coloring of $\mathbb{R}^2$ in 2 colors, we have a monochromatic triangle T which is congruent to the triangle with sides $1, \sqrt{3}$ and 2 and with angles $30^o, 60^o$ and 90°.*

Proof Using Theorem 9.4.2, we may assume that we have a monochromatic equilateral triangle of unit length (the proof in the other case is left to the reader). So, let ABC be a red triangle with all of its sides equal to 1. Let T be the given triangle XYZ with $xy = \sqrt{3}, yz = 1$ and XZ (which is the hypotenuse) of length 2 as shown in Figure 9.19.

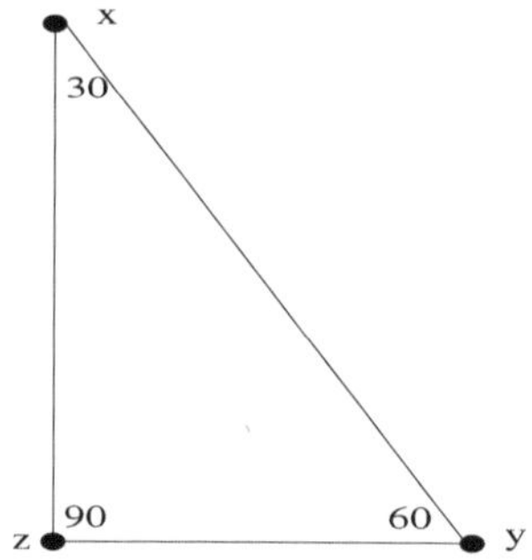

Figure 9.19: A $30^o, 90^o, 60^o$ triangle xyz

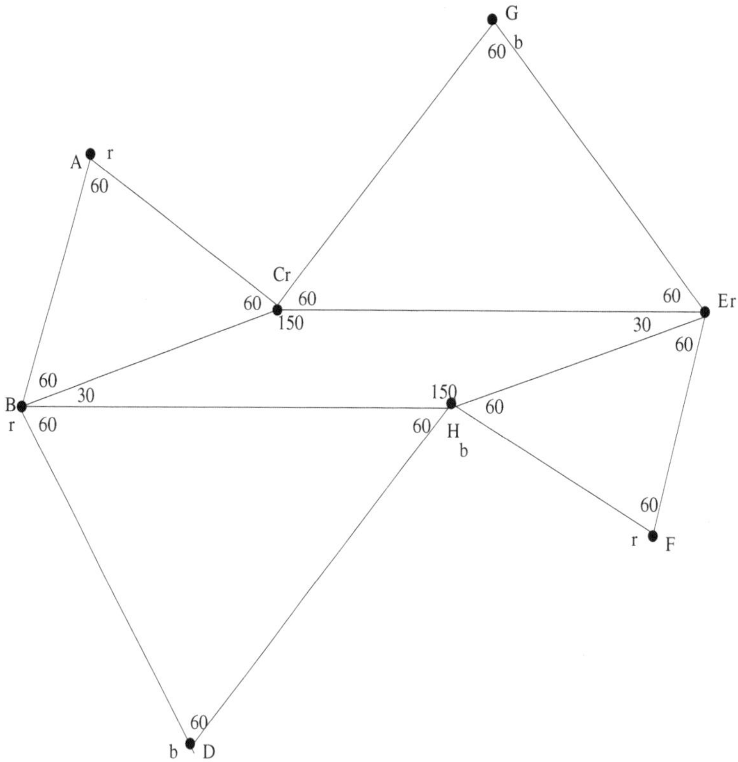

Figure 9.20: Monochromatic triangle

Draw the parallelogram $BCEH$ with one of its sides BC and the other side $CE = \sqrt{3}$. Thus $BH = \sqrt{3}$ and $EH = 1$ and also such that the $\angle CBH = 30^O$ (hence $\angle BCE = 150^o = \angle BHE$ and $\angle CEH = 30^o$). Draw equilateral triangles BHD (with BH as the base), EHF (with EH as the base) and CEG (with CE as the base). We note that all of the following triangles have angles $30^o, 60^o$ and

90° and hence are congruent to T. Use small letters r and b to respectively denote red and blue colours (of vertices). If we wish to avoid a monochromatic triangle congruent to T, the following triangles cannot have all vertices of the same color: $DBC, HBA, DHF, GCA, GHE, CEF$. Therefore D is blue, H is blue, F is red and G is blue. Finally since G and H are blue and GHE is not monochromatic, we see that E must be red. But C and F are already red and hence CEF is red. This contradiction completes the proof. □

We wind up the discussion in this section by noting that the results included here are just a tip of an iceberg! One of the questions concerning Theorem 9.4.2 and Theorem 9.4.3 is the following. Let T be a triangle *which is not an equilateral triangle.* Then is it true that in any coloring of the plane in two colors, we must have a monochromatic copy of T? Theorem 9.4.4 answers this question in the affirmative if T is *any right angled triangle with angles* $30^o, 60^o$ *and* 90°. It is a long standing conjecture of Erdös that we can obtain a monochromatic copy of T in any colouring of the plane *provided* T *is not an equilateral triangle.* A long research paper of Erdös and others gives several results in this area, many of which concern not just coloring of points in the plane, but even coloring of points in $\mathbb{R}^3$. We refer the reader to that article [23] and [28] for a wealth of knowledge in this area.

9.5 Exercises for Chapter 9

9.1 Show that the utility graph (Figure 7.5 in Chapter 7) is not planar.

9.2 Show that if G is a simple planar graph, then G has a vertex of degree less than or equal to 5.

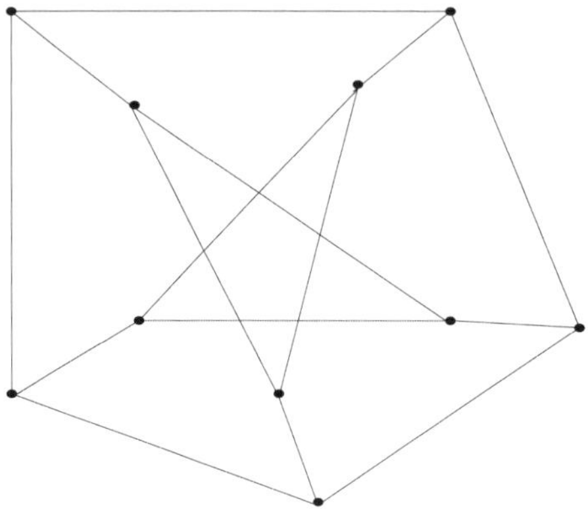

Figure 9.21: Petersen graph

9.3 Show that if we remove a single edge from K_5 or $K_{3,3}$, then these graphs can be drawn on the plane.

9.4 Both K_5 and $K_{3,3}$ can be drawn on a torus without crossings. Can you see this?

9.5 Show that the graph in Figure 9.21 (called the *Petersen graph*) is not planar (hint: smallest cycle is of length 5).

9.6 Let a simple graph G be drawn on the plane (with a planar drawing). G is said to be a triangulation of the plane if each face is a triangle. Show that for a connected simple graph, we have $e(G) \leq 3n - 6$ with equality iff G is a triangulation of the plane.

9.7 Show that every angle of a regular m-gon has measure $(\frac{m-2}{m})\pi$. Also show that all the angles of a triangle are equal iff all the sides are equal (and in that case, we have an equilateral triangle) but this assertion is false when $m \geq 4$.

9.8 In the case of a regular tessellation of $\mathbb{R}^2$, how can we modify (9.3)? Use this modification to prove Theorem 9.1.8.

9.8 Prove Theorem 9.1.10 (classification of homogeneous tessellations).

9.10 Where is the crucial condition, "any line picking up two points of S must also contain a third point of S" in the statement of Sylvester-Gallai Theorem (Theorem 9.2.7) used in the proof of that theorem? Exhibit a finite configuration of points in $\mathbb{R}^2$ in which not all points are on a line (but we do have lines that pick up only two points of S).

9.11 Show that the statement of Sylvester-Gallai Theorem (Theorem 9.2.7) is not true if the set S is discrete (in topological sense) but is allowed to have infinitely many points.

9.12 Show that a rectangle cannot be partitioned into incongruent squares where the number of incongruent squares is $6, 7$ or 8.

9.13 Consider the following proof of Theorem 9.2.8. We tile $\mathbb{R}^2$ in two colors white and black using half-squares as follows. The color of (x, y) is white iff $\lfloor x \rfloor \leq x < \lfloor x \rfloor + \frac{1}{2}$ and $\lfloor y \rfloor \leq y < \lfloor y \rfloor + \frac{1}{2}$ or $\lfloor x \rfloor + \frac{1}{2} \leq x < \lceil x \rceil$ and $\lfloor y \rfloor + \frac{1}{2} \leq y < \lceil y \rceil$. Thus the entire plane is alternately coloured by white and black half-squares. Let T be a (smaller) rectangle with vertices $(x_1, y_1), (x_2, y_1), (x_1, y_2), (x_2, y_2)$ such that $x_2 = x_1 + d$ where d is a positive integer and let the rectangle T' obtained from T by a shifting the vertical lines of T but maintaining the same horizontal lines. Thus T' is a rectangle with left wall given by the line $x = \lfloor x_1 \rfloor$ and the right wall given by $x = \lfloor x_2 \rfloor$. Let w (respectively b) denotes the white area (respectively black area).

(a) Prove that $w(T) = w(T')$. Use this to show that $w(T) = b(T)$ (where $b(T)$ denotes the black area in T).

(b) Conclude that $w(R_i) = b(R_i)$ for each i and hence $w(R) = b(R)$ for the given rectangle R.

(c) Show that this is possible only when a or b is an integer.

9.14 Consider yet another proof of Theorem 9.2.8. Let $f(x, y) = \sin 2\pi x \sin 2\pi y$. Integrate this function over R by summing it over all the integrals obtained from R_i's and show that this integral is zero. Conclude that a or b must be an integer.

9.15 Prove Theorem 9.2.9.

9.16 Give an alternative proof of Corollary 9.2.12 using the following steps. Let an $a \times b \times c$ brick (in three dimensions) be partitioned into $1 \times 2 \times 4$ subbricks. Prove all of the following.

(a) Show that each one of ab, bc and ca is an even number.

(b) Show that 8 divides abc and hence if one of a, b, c is odd, then by relabeling we get 2 divides b and 4 divides c and the desired assertion follows.

(c) Let all of a, b, c be even. Let S be the set of all lattice points

$$S = \{(i, j, k) : i, j, k \in z,\ 1 \leq i \leq a,\ 1 \leq j \leq b,\ 1 \leq k \leq c\}$$

For a sub(brick) B', we denote by $s(B')$, the sum of all the co-ordinates of all the points of S in B'. Here we treat B' to be open on one side and closed on the other parallel side. Thus treat B itself as $0 < x \leq a,\ 0 < y \leq b,\ 0 < z \leq c$. Show that $s(B')$ is a multiple of 8 for any subbrick B'.

(d) Show that

$$s(B) = \frac{abc(a + b + c + 3)}{2}$$

(e) Use this to show that one of a, b, c is a multiple of 4 and hence complete the proof.

9.17 Prove the following analogue of Sperner's Lemma (lemma 9.3.4) for $\mathbb{R}^3$. Let $\mathbf{P}$ denote a tetrahedron in $\mathbb{R}^3$ with vertices labeled $1, 2, 3, 4$. A subdivision of $\mathbf{P}$ into subtetrahedron is said to be proper if every two subtetrahedron either share a triangular face, or an edge or a vertex in common. Label all the vertices of the subdivision in such a way that the vertices that are on the boundary face i, j, k must receive one of the labels i, j or k and those on an edge ij must get one of the labels i and j. Such a labeling is called a Sperner labeling. A subtetrahedron is complete if it receives all the four labels. Show that in any Sperner labeling we must have at least one complete subtetrahedron.

9.18 State and prove a higher dimensional analogue of Exercise 9.17.

9.19 Prove that when $\mathbb{R}^2$ is 2-colored with strips alternately colored as in Theorem 9.4.3, we do not have a monochromatic unit triangle.

9.20 Complete the proof of Theorem 9.4.4 under the assumption that we have an equilateral monochromatic triangle with side $\sqrt{3}$ (instead of 1).

Chapter 10

Advanced counting numbers

10.1 Stirling numbers

In the earlier chapters, we have looked at many enumeration parameters such as the number of combinations (binomial coefficients), number of various types of permutations as well as counting the number of permutations with specified properties. In the present chapter, we are mainly interested in two other important enumeration parameters called the Stirling and Catalan numbers. This section deals with Stirling numbers and in Section 10.2, we deal with Catalan numbers.

Definition 10.1.1. Let n, k be non-negative integers. The *Stirling number of second kind* $S(n, k)$ is the total number of partitions of an n-set into k disjoint, non-empty and unordered subsets.

The term *unordered* here refers to the subsets (as members of a partition) being unordered. That is as good as putting n distinguishable objects (elements of a set) into k boxes, *where there is no way of distinguishing one box from the other and such that each box contains at least one object.* Here are some examples.

Example 10.1.2. We take the n-set to be $[n] = \{1, 2, \ldots, n\}$. Let $(n, k) = (5, 2)$. Since we have two non-empty subsets say of orders n_1 and n_2, we may assume that $n_1 \leq n_2$ and hence $n_1 = 1$ or $n_1 = 2$. Since the two subsets are complements of each other, it follows that the first (smaller) subset determines the larger uniquely. The first subset can be chosen in $\binom{5}{1} + \binom{5}{2} = 15$ ways and therefore $S(5, 2) = 15$. As another example, let $(n, k) = (6, 2)$. Proceeding exactly as before, we see that $n_1 \leq 3$. Does that mean, by the same argument that the required Stirling number equals

$$\binom{6}{1} + \binom{6}{2} + \binom{6}{3}?$$

Not really, because, though it is true that a subset of order 1 (and order 2) determines its complement, which is a subset of order 5 (or order 4) uniquely, the subsets of order three occur in complementary pairs and hence should be counted only once (they have been counted two times in the earlier count). Hence the correct answer is

$$S(6,2) = \binom{6}{1} + \binom{6}{2} + \frac{1}{2}\binom{6}{3} = 6 + 15 + 10 = 31$$

These two examples illustrate the level of difficulty in computation of $S(n,k)$ to us. Consider $S(6,3)$. Here, we have 3 non-empty subsets to choose and hence we can make three cases: One subset of order 4 (and the other two of order one each), One subset of order 3, the other two of orders 2 and 1 and all the three subsets of order 2. The first case can arise in $\binom{6}{4} = \binom{6}{2} = 15$ ways. The second case can arise in $\binom{6}{3} \times 3 = 60$ ways. Finally, the third case, where we have to put 6 objects in three boxes with each box receiving 2 objects each. We get the multinomial coefficient $\binom{6}{2,2,2} = 90$. However, this is the required number in case the boxes are labeled (or ordered). To get the number of unordered partitions, we must divide this number by $3! = 6$ giving the number $90/6 = 15$. Hence, $S(6,3) = 15 + 60 + 15 = 90$. We note in passing that, by convention, $S(0,0) = 1$ while $S(n,0) = 0$ for every positive integer n.

Lemma 10.1.3. *Let n and k be positive integers with $n \geq k$. Then*

$$S(n,k) = S(n-1,k-1) + kS(n-1,k)$$

Proof Let X be an n-set with $x \in X$ and let $Y = X - \{x\}$. Every partition of $X = \cup_{i=1}^{k} X_i$ into disjoint non-empty subsets X_i gives a natural partition of Y as $Y = \cup_{i=1}^{k} Y_i$ into disjoint subsets Y_i where $Y_i = X_i - \{x\}$. Given such a partition $\{X_1, X_2, \cdots, X_k\}$ of X into non-empty subsets, we can assume, w.l.o.g that $x \in X_k$. Depending then on whether X_k consists of x alone (and then Y_k is empty) or X_k has at least two elements (and then Y_k is non-empty), we have two distinct cases. In the first case, we get a partition of Y into $k-1$ mutually disjoint non-empty subsets $\{Y_1, Y_2, \ldots, Y_{k-1}\}$. This process is also reversible. Given any such partition $\{Y_1, Y_2, \ldots, Y_{k-1}\}$, we take $X_i = Y_i \; \forall \; i = 1, 2, \ldots, k-1$ and $X_k = \{x\}$. Clearly this case can arise in $S(n-1, k-1)$ ways and the number of partitions of X of this type (where the subset X_k is a singleton) equals $S(n-1, k-1)$. Consider the second case. Here, we have a partition $\{Y_1, Y_2, \ldots, Y_k\}$ of Y with each Y_i non-empty. This can be clearly achieved in $S(n-1,k)$ ways. However, given any such partition $\{Y_1, Y_2, \ldots, Y_k\}$, we can choose to insert x into *any one of the* Y_i and depending on which Y_i we choose, we get a different partition of X. Thus the correspondence between the set of partitions of Y into k non-empty unordered disjoint subsets and the set of partitions of X into k non-empty unordered disjoint subsets with the subset containing x having at least two elements, is $1 : k$. This shows that the number of partitions of the second type is equal to $kS(n-1,k)$ proving the assertion. □

Example 10.1.4. As an application, we have $S(7,3) = S(6,2) + 3S(6,3) = 31 + (3 \times 90) = 301$ using our earlier calculation.

Proposition 10.1.5. *We have the following collection of small Stirling numbers with $n \leq 6$.*

(a) $S(1,1) = 1$.

(b) $S(2,1) = S(2,2) = 1$.

(c) $S(3,1) = S(3,3) = 1$ *and* $S(3,2) = 3$.

(d) $S(4,1) = S(4,4) = 1,\ S(4,3) = \binom{4}{2} = 6$ *and* $S(4,2) = 7$.

(e) $S(5,1) = S(5,5) = 1,\ S(5,4) = \binom{5}{2} = 10,\ S(5,2) = 15$ *and* $S(5,3) = 25$.

(e) $S(6,1) = S(6,6) = 1,\ S(6,2) = 31,\ S(6,3) = 90,\ S(6,4) = 65$ *and* $S(6,5) = 15$.

Proof To find $S(4,3)$,observe that the three non-empty subsets must have sizes $1,1$ and 2. Hence, we just have to choose two elements out of 4 which can be done in $\binom{4}{2} = 6$ ways. To compute $S(4,2)$, we could choose the smaller subset of size 1 and that can be done in 4 ways. We could also choose the smaller subset and hence both the subsets to have order 2 and since the subsets are unordered, we get $S(4,2) = 4 + \frac{1}{2}\binom{4}{2} = 7$. Easier here is to use Lemma 10.1.3 and get $S(4,2) = S(3,1) + 2S(3,2) = 1 + 6 = 7$. Similarly, $S(5,2) = S(4,1) + 2S(4,2) = 1 + 14 = 15$ and $S(5,3) = S(4,2) + 3S(4,3) = 7 + (3 \times 6) = 25$. Finally, $S(6,3) = S(5,2) + 3S(5,3) = 15 + (3 \times 25) = 90$ and $S(6,4) = S(5,3) + 4S(5,4) = 25 + (4 \times 10) = 65$. □

Definition 10.1.6. A sequence $(a_1, a_2, \ldots, a_n)$ is called a *unimodal sequence* if for some r between 1 and n, we have $a_1 \leq a_2 \leq \cdots \leq a_r \geq a_{r+1} \geq \cdots \geq a_{n-1} \geq a_n$.

An example of a unimodal sequence is the sequence $(\binom{m}{k})$ of binomial coefficients where m is a fixed positive integer and $k = 0, 1, \ldots, m$.

Theorem 10.1.7. *Let n be a positive integer. The sequence*

$$(S(n,1), S(n,2), \ldots, S(n,n-1), S(n,n))$$

is unimodal and in fact, there is some $M(n)$ such that either

$$1 = S(n,1) < S(n,2) < \cdots < S(n,M(n)) > S(n,M(n)+1) > \cdots > S(n,n) = 1$$

or

$$1 = S(n,1) < S(n,2) < \cdots < S(n,M(n)-1) = S(n,M(n)) > \cdots > S(n,n) = 1$$

Proof Using the table of $S(n,k)$ (see Exercise 10.5), we observe that the assertion is true for small values of n. Let the assertion hold for all values of $m \leq n$ and we prove the assertion for $n+1$. First let $2 \leq k \leq M(n)$. Then $S(n+1,k) = S(n,k-1) + kS(n,k)$ and $S(n+1,k-1) = S(n,k-2) + (k-1)S(n,k-1)$ and hence we get

$$\begin{aligned} S(n+1,k) - S(n+1,k-1) &= [S(n,k-1) - S(n,k-2)] \\ &+ k[S(n,k) - S(n,k-1)] + S(n,k-1) \end{aligned}$$

By making induction (on n), each term on the R.H.S. is positive and hence $S(n+1,k) > S(n+1,k-1)$. Now let $M(n)+2 \le k \le n$. Then using Exercise 10.4, we have

$$S(n+1,k) = \sum_{j=1}^{n} \binom{n}{j} S(j,k-1)$$

which implies that

$$S(n+1,k) - S(n+1,k-1) = \sum_{j=1}^{n} \binom{n}{j} [S(j,k-1) - S(j,k-2)]$$

Here $j \le n$ on the R.H.S. and since $M(j) \le M(n)$ we get $S(j,k-1) < S(j,k-2)$ on the R.H.S and thus the R.H.S. is negative proving $S(n+1,k) < S(n+1,k-1)$. Therefore,

$$S(n+1,1) < S(n+1,2) < \cdots < S(n+1,M(n))$$

and

$$S(n+1,M(n)+1) > S(n+1,M(n)+2) > \cdots > S(n+1,n+1)$$

Thus the sequence $(S(n+1,1), S(n+1,2), \cdots S(n+1,n), S(n+1,n+1))$ is indeed unimodal with $M(n+1) = M(n)$ or $M(n+1) = M(n)+1$. □

We note in passing that there is no example of $n \ge 3$ for which the second situation in the statement of the Theorem holds. *That is, there is no known example where two equal maxima at $M(n)$ and $M(n)+1$ occur.*

Theorem 10.1.8. *Let n and k be positive integers. Then the following assertions hold.*

(a) The number of surjective (i.e. onto) functions from an n-set to a k-set is equal to $k!S(n,k)$.

(b) Hence for positive integers n and m. we have

$$\begin{aligned} m^n &= \sum_{k=1}^{m} \binom{m}{k} k!S(n,k) \\ &= \sum_{k=1}^{m} [m]_k S(n,k) \end{aligned}$$

Proof Using (a), (b) follows easily: Take A and B to be two sets of orders n and m respectively. The L.H.S. counts the number of all the functions f from A to B. If C is a subset of order k of B, then the number of surjective functions from A to C is, by (a), equals $k!S(n,k)$. Since every function f from A to B has a unique range C which is some k-subset of B where $1 \le k \le m$ and since C can thus be chosen in $\binom{m}{k}$ ways, (b) follows by summing over all k; (a) the is left to the reader (Exercise 10.12). □

Theorem 10.1.9. *Let n be a positive integer. Then we have the following (formal) polynomial identity.*

$$x^n = \sum_{k\geq 1} S(n,k)[x]_k \tag{10.1}$$

Proof Observe that since n is fixed, both the sides of (10.1) are polynomials of degree $\leq n$. Since two distinct polynomials of degree $\leq n$ cannot agree on more than n points, it will now suffice to show that we have at least $n+1$ points x where (10.1) holds. But this is certainly true for every value of x which is a positive integer m, using Theorem 10.1.8. Hence (10.1) is true for infinitely many values of x which proves the assertion. □

Since $S(n,0) = 0$ for n a positive integer, (10.1) is true with the sum on the R.H.S. running from 0 (instead of 1). Let n be a fixed positive integer. Then for every non-negative integer $r \leq n$, we have a linear equation

$$x^r = \sum_{k=0}^{r} S(r,k)[x]_k \tag{10.2}$$

We thus get a system of $n+1$ linear equations that express the L.H.S. $\{x^0, x^1, x^2, \ldots, x^n\}$ in terms of $\{[x]_0 = 1, [x]_1 = x, [x]_2, \ldots, [x]_n\}$ with coefficients that are Stirling numbers of second kind. For example, with $n = 6$, we have the matrix equation

$$\begin{bmatrix} x^0 \\ x^1 \\ x^2 \\ x^3 \\ x^4 \\ x^5 \\ x^6 \end{bmatrix} = \begin{bmatrix} 1 & 0 & 0 & 0 & 0 & 0 & 0 \\ 0 & 1 & 0 & 0 & 0 & 0 & 0 \\ 0 & 1 & 1 & 0 & 0 & 0 & 0 \\ 0 & 1 & 3 & 1 & 0 & 0 & 0 \\ 0 & 1 & 7 & 6 & 1 & 0 & 0 \\ 0 & 1 & 15 & 25 & 10 & 1 & 0 \\ 0 & 1 & 31 & 90 & 65 & 15 & 1 \end{bmatrix} \begin{bmatrix} [x]_0 \\ [x]_1 \\ [x]_2 \\ [x]_3 \\ [x]_4 \\ [x]_5 \\ [x]_6 \end{bmatrix}$$

Notice that the coefficient matrix A on the right hand side is a lower triangular integer matrix with 1 on the diagonal. Such a matrix A is invertible and in fact, the inverse B is also an integer matrix with 1 on the diagonal. Formally, we have a change of basis formula. Let V denote the vector space of all real polynomials whose degree is at the most n. Then V is an $(n+1)$-dimensional vector space with bases $\{x^0, x^1, x^2, \ldots, x^n\}$ and $\{[x]_0, [x]_1, \ldots, [x]_n\}$. Here the matrix A gives change of basis from the second basis to the first basis and the matrix B gives change of basis from the first basis to the second. The matrix A can either be considered as an infinite matrix when we think of it as the change of basis matrix from the space of *all the polynomials over* $\mathbb{R}$ or we may truncate it and look at only the first few rows and corresponding columns. The 7×7 matrix above consists of the first 7 rows and 7 columns of that infinite matrix. This prompts the following definition.

Definition 10.1.10. Let n and k be positive integers with $k \leq n$. *The Stirling number of first kind* $s(n,k)$ is defined by the formula

$$[x]_n = \sum_{k=1}^{n} s(n,k)x^k \tag{10.3}$$

By convention (and also consistent with Definition 10.1.10) we let $s(n,k) = 0$ if $n < k$, $s(n,0) = 0$ if n is a positive integer and $s(0,0) = 1$.

As an example, consider $n = 4$. To find $s(r,k)$ for $1 \leq r \leq 4$, we can either invert the matrix A or alternatively use (10.3). Thus we have $s(1,1) = 1$. Since $[x]_2 = x^2 - x$, we get $s(2,1) = -1$ and $s(2,2) = 1$. Similarly,

$$[x]_3 = x(x-1)(x-2) = x^3 - 3x^2 + 2x$$

gives us $s(3,1) = 2$, $s(3,2) = -3$ and $s(3,3) = 1$. Finally,

$$[x]_4 = x(x-1)(x-2)(x-3) = x^3 - 6x^2 + 11x - 6x$$

gives us $s(4,1) = -6$, $s(4,2) = 11$, $s(4,3) = -6$ and, of course, $s(4,4) = 1$. We therefore, see that the first five rows and columns of the inverse of the matrix A is B where B equals

$$\begin{bmatrix} 1 & 0 & 0 & 0 & 0 \\ 0 & 1 & 0 & 0 & 0 \\ 0 & -1 & 1 & 0 & 0 \\ 0 & 2 & -3 & 1 & 0 \\ 0 & -6 & 11 & -6 & 1 \end{bmatrix}$$

Recall that we defined the numbers $c(n,k)$ (Chapter 3) as the number of permutations of an n-set that have exactly k distinct cycles. Using Theorem 3.1.8, it is easy to see that $c(n,k) = |s(n,k)|$ and in fact, this is the reason for calling the numbers $c(n,k)$ *signless Stirling numbers of first kind.*

Definition 10.1.11. Let n be a non-negative integer. The *Bell number* B_n is defined to be the number of all (unordered) partitions of an n-set where by convention, we take $B_0 = 1$.

Clearly, we have $B_1 = 1, B_2 = 2$ (the two partitions of the set $\{1,2\}$ are $\{\{1,2\}\}$ and $\{\{1\},\{2\}\}$) and $B_3 = 5$. In general, for a positive integer n, we have $B_n = \sum_{k=1}^{n} S(n,k)$ which gives an expression for the Bell number in terms of Stirling numbers of second kind.

Lemma 10.1.12. *Let n be a positive integer. Then*

$$B_{n+1} = \sum_{k=0}^{n} \binom{n}{k} B_k$$

Proof Let $|X| = n+1$ and let $x \in X$. Let $Y = X - \{x\}$. In any partition of X, the set A containing x is uniquely determined. If we let $B = A - \{x\}$, then $C = Y - B$ is also uniquely determined as a subset of Y. If $|B| = r$ then we are partitioning the set C of order $n-r$ (into any number of subsets) and this can be done in B_{n-r} ways. Since B (and hence A) can be chosen in $\binom{n}{r}$ ways, we obtain the sum $B_{n+1} = \sum_{r=0}^{n} \binom{n}{r} B_{n-r}$ which equals $\sum_{k=0}^{n} \binom{n}{k} B_k$ using the standard binomial identity $\binom{n}{r} = \binom{n}{n-r}$. □

Theorem 10.1.13. *The Bell numbers B_n are given by the following expression.*

$$B_n = \sum \frac{n!}{\prod_{k=1}^{n} (k!)^{a_k} a_k!}$$

where the sum is taken over all the n-tuples $(a_1, a_2, \ldots, a_n)$ with $a_1 + 2a_2 + \cdots + na_n = n$.

Proof Consider a partition of an n-set X in which we have a_k subsets of size k. The multinomial coefficient $\binom{n}{a_1, 2a_2, \ldots, na_n}$ first gives us the number of ways in which the set gets divided into boxes with k-th box containing ka_k elements. This needs to be further partitioned by dividing the k-th subset of size ka_k into a_k subsets each of size k and that can be done in $\prod_{k=1}^{n} \binom{ka_k}{k, k, \ldots, k}$ ways. Since all the a_k subsets of size k should be treated as unordered, we must divide the product by $a_k!$ for each k to get the required number:

$$\binom{n}{a_1, 2a_2, \ldots, na_n} \times \prod_{k=1}^{n} \binom{ka_k}{k, k, \ldots, k} \times \prod_{k=1}^{n} \frac{1}{a_k!}$$

which gives the expression on the R.H.S. □

Theorem 10.1.14. *The following identity is satisfied by the Bell number B_n.*

$$B_n = \frac{1}{e} \sum_{k=1}^{\infty} \frac{k^n}{k!}$$

Proof We make use of the identity

$$\begin{aligned} S(n,k) &= \frac{1}{k!} \sum_{j=0}^{k-1} (-1)^j (k-j)^n \binom{k}{j} \\ &= \sum_{j=0}^{k-1} \frac{(-1)^j}{j!} \frac{(k-j)^n}{(k-j)!} \end{aligned}$$

given by Exercise 10.20 which is also valid for $k > n$ by the next Exercise 10.21. Therefore,

$$\begin{aligned} B_n &= \sum_{k=1}^{\infty} S(n,k) \\ &= \sum_{k=1}^{\infty}\sum_{j=0}^{k-1} \frac{(-1)^j}{j!}\frac{(k-j)^n}{(k-j)!} \\ &= \sum_{j\geq 0}\left\{\sum_{k\geq j}\frac{(k-j)^n}{(k-j)!}\right\}\frac{(-1)^j}{j!} \\ &= \sum_{j\geq 0}\left\{\sum_{r=0}^{\infty}\frac{(r)^n}{r!}\right\}\frac{(-1)^j}{j!} \\ &= \left\{\sum_{j\geq 0}\frac{(-1)^j}{j!}\right\}\left\{\sum_{r=0}^{\infty}\frac{(r)^n}{r!}\right\} \\ &= \frac{1}{e}\sum_{k=0}^{\infty}\frac{k^n}{k!}. \end{aligned}$$

□

10.2 Catalan numbers

Let R be a set with a binary operation $\times$ which is not necessarily associative or commutative. Thus the product between x and y takes two different forms xy and yx and these could be distinct elements of R. For three elements x, y, z, we could first arrange the elements in a particular order say x, y, z (this can be done in $3! = 6$ ways since we are just looking at all the permutations). We then have two possible products $(x \times y) \times z = (xy)z$ and $x \times (y \times z) = x(yz)$ and thus we have $6 \times 2 = 12$ possibilities. In general, let g_n denote the total number of products (in any order) among n elements $x_1, x_2, \ldots, x_n$ and let f_n denote the total number of products among the same n elements but in a specified order. Then the relationship between f_n and g_n is $g_n = n! \times f_n$ and we have already seen that $g_2 = 2$ and $g_3 = 12$. We wish to find an expression for g_n directly as a function of n. For this purpose we set up a recursive relationship between g_n and g_{n-1}.

In order to achieve this, we locate the position where x_n occurs, in a long product involving all of $x_1, x_2, \ldots, x_n$. If we delete x_n from this expression, we get a unique product among the elements $x_1, x_2, \ldots, x_{n-1}$. For example, in

$$(((x_2x_3)x_6)((x_1x_9)x_8))((x_4x_7)x_5)$$

the term x_9 is in the product $(x_1x_9)x_8$ and if we just delete x_9, what we get is

$$(((x_2x_3)x_6)(x_1x_8))((x_4x_7)x_5)$$

Similarly, in $((x_1x_2)x_3)(x_7((x_4x_5)x_6))$ we have $n = 7$ and the term x_7 is in the product $A \times (x_7 \times B)$ where $A = ((x_1x_2)x_3)$ and $B = (x_4x_5)x_6$. Deleting x_7 gives us $((x_1x_2)x_3)((x_4x_5)x_6)$. There are two exceptions to this. Either x_n may occur at the very end as in the case $(((x_1x_2)(x_3x_4))(x_5x_6))x_7$ in which case we can still delete x_7 and get $((x_1x_2)(x_3x_4))(x_5x_6)$ or x_n may occur at the beginning like in $x_5((x_1x_2)(x_3x_4))$, where again we can delete it to get $(x_1x_2)(x_3x_4)$. It thus follows that we obtain a unique expression involving $x_1, x_2, \ldots, x_{n-1}$ from a given expression involving $x_1, x_2, \ldots, x_n$ by deleting x_n. Conversely, let an expression involving $x_1, x_2, \ldots, x_{n-1}$ be given. Let us have a closer look at a product of m elements. In any such product, we actually perform $m - 1$ multiplication operations (this is clear since multiplication is a binary process). Any element x_i participates (as x_i) in exactly one multiplication (after which it does not remain x_i but is actually a product of at least two elements). Hence a required expression involving x_n can be obtained from a given expression that involves $x_1, x_2, \ldots, x_{n-1}$ in the following ways. Let the given expression involving $x_1, x_2, \ldots, x_{n-1}$ be E. We see that E has been obtained by performing $n - 2$ multiplication operations. We choose one of these $n - 2$ multiplications say $A \times B$ and *insert x_n in that expression.* For example, let $n = 7$ and let $E = ((x_1x_3)x_2) \times ((x_5x_6)x_4)$ which has the form $A \times B$ with $A = (x_1x_3)x_2$ and $B = (x_5x_6)x_4$, we can insert x_7 in the product $A \times B$ in the following four ways.

$$\begin{aligned}
(x_7 \times A) \times B &= (x_7((x_1x_3)x_2))((x_5x_6)x_4) \\
(A \times x_7) \times B &= (((x_1x_3)x_2)x_7)((x_5x_6)x_4) \\
A \times (x_7 \times B) &= ((x_1x_3)x_2)(x_7((x_5x_6)x_4)) \\
A \times (B \times x_7) &= ((x_1x_3)x_2)(((x_5x_6)x_4)x_7)
\end{aligned}$$

while in the same expression E, if we were to insert x_7 in the multiplication $x_1 \times x_3$, we get the following four expressions.

$$(((x_7 \times x_1)x_3)x_2)((x_5x_6)x_4), (((x_1 \times x_7)x_3)x_2)((x_5x_6)x_4)),$$
$$((x_1 \times (x_7 \times x_3))x_2)((x_5x_6)x_4), ((x_1 \times (x_3 \times x_7))x_2)((x_5x_6)x_4)$$

Since a given product E involving $x_1, x_2, \ldots, x_{n-1}$ has $n - 2$ multiplications, we get $4 \times (n - 2) = 4n - 8$ expressions involving $x_1, x_2, \ldots, x_n$ in this manner. Thus x_n can be *inserted* in $4n - 8$ ways in the multiplications in E. In these insertions, x_n multiplies some smaller expression and hence the expression E *does not remain intact after insertion of* x_n. Finally the two expressions that we did not count so far are pre and post multiplication (left and right multiplication) of E by x_n giving us $x_n \times E$ and $E \times x_n$. Clearly, in these we are not inserting x_n in E and hence these are not covered in the earlier counting. To sum up, a single expression E that contributes to g_{n-1} gives rise to $(4n - 8) + 2 = 4n - 6$ expressions that contribute to g_n and we get the formula

$$g_n = (4n-6)g_{n-1} \quad \forall\, n \geq 2$$

This also means that $g_{n-1} = [4(n-1)-6]g_{n-2} = (4n-10)g_{n-2}$ and hence iteration gives

$$\begin{aligned} g_n &= (4n-6)g_{n-1} \\ &= (4n-6)(4n-10)g_{n-2} \\ &= \cdots\cdots \\ &= (4n-6)(4n-10)\cdots \times 6 \times 2 \end{aligned}$$

since $g_2 = 2$. We may also use induction on n and arrive at the same formula.

Definition 10.2.1. Let n be a natural number. The *n-th Catalan number C_n* is defined by $C_n = \frac{1}{n}\binom{2n-2}{n-1}$. Here we treat $\binom{0}{0}$ as 1 and hence $C_1 = 1$.

Theorem 10.2.2. *Let the product of n elements $x_1, x_2, \ldots, x_n$ be formed in the specified order $x_1, x_2, \ldots, x_n$. Then the number of ways of doing this (the total number of ways of parenthesizing the product with elements prescribed in a given order) is equal to the n-th Catalan number C_n.*

Proof We just have to prove that $f_n = \frac{1}{n!}g_n = C_n$. We have,

$$\begin{aligned} f_n &= \frac{1}{n!}(4n-6)(4n-10) \times \cdots \times 10 \times 6 \times 2 \\ &= \frac{1}{n!}2^{n-1}(2n-3)(2n-5) \times \cdots \times 5 \times 3 \times 1 \\ &= \frac{1}{n!}2^{n-1}\frac{(2n-2)!}{(2n-2)(2n-4) \times \cdots \times \times 6 \times 4 \times 2} \\ &= \frac{1}{n!}2^{n-1}\frac{(2n-2)!}{2^{n-1}(n-1)!} \\ &= \frac{(2n-2)!}{n!(n-1)!} \\ &= \frac{1}{n}\binom{2n-2}{n-1} \end{aligned}$$

□

Let $\mathbf{P}$ be a convex n-gon. We have already encountered the notion of a triangulation of $\mathbf{P}$ in Chapter 9. This consists of division of $\mathbf{P}$ into regions each of which is a triangle.

Definition 10.2.3. *A diagonal triangulation* T of a convex n-gon $\mathbf{P}$ is its triangulation which has the property that every vertex (of every triangular region) is a vertex of $\mathbf{P}$ (hence no new vertices are introduced) and every side of each triangular region is either a diagonal of $\mathbf{P}$ or is a side of $\mathbf{P}$.

The first example in Figure 10.1 is not a diagonal triangulation since two diagonals intersect at an internal vertex which is not a vertex of $\mathbf{P}$ while the second and third are both diagonal triangulations.

Definition 10.2.4. Two triangulations of a convex n-gon are considered *the same* if they use the same diagonals. *The total number of diagonal triangulations* of a convex n-gon is denoted *by* T_n.

The second and third examples in Figure 10.1 are both diagonal triangulations of a convex hexagon but are not the same since one uses the three diagonals v_1v_3, v_3v_5 and v_5v_1 while the other uses the diagonals v_1v_4, v_2v_4 and v_5v_1. However, we have exactly three diagonals in both the diagonal triangulations and Lemma 10.2.5 below proves this in general. The task before us is to find out the number T_n of diagonal triangulations of a convex n-gon. By definition, we take $T_2 = 1$ and clearly $T_3 = 1$ while $T_4 = 2$ as shown in Figure 10.2.

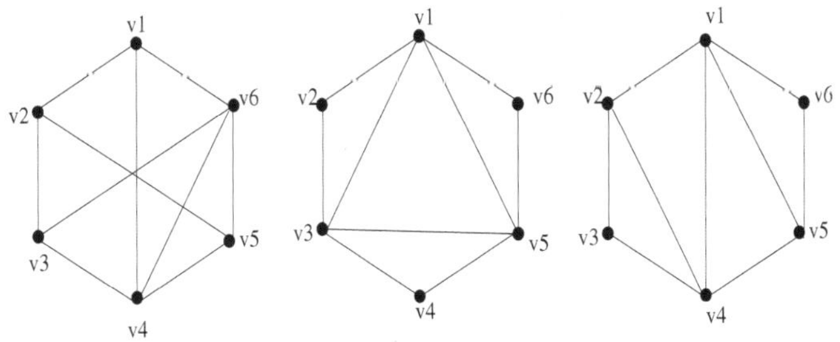

Figure 10.1: non-diagonal and diagonal triangulations of a convex hexagon

Lemma 10.2.5. *Consider a diagonal triangulation T of a convex n-gon $\mathbf{P}$. Then the number of triangles in which T divides $\mathbf{P}$ is $n-2$ and the number of diagonals used in T is $n-3$.*

Proof If t denotes the number of triangles, then sum of all the angles is equal to $t \times \pi$. But this sum is just the sum of all the angles of the convex n-gon $\mathbf{P}$ (since we have no other vertices in the triangulation T) which is $(n-2)\pi$ and hence $t = n-2$. If d denotes the number of diagonals used in T, then each diagonal borders two triangles of T and each side of $\mathbf{P}$ borders exactly one triangle of T which gives $2d+n = 3t = 3(n-2)$ by a two-way counting. Hence $d = n-3$. □

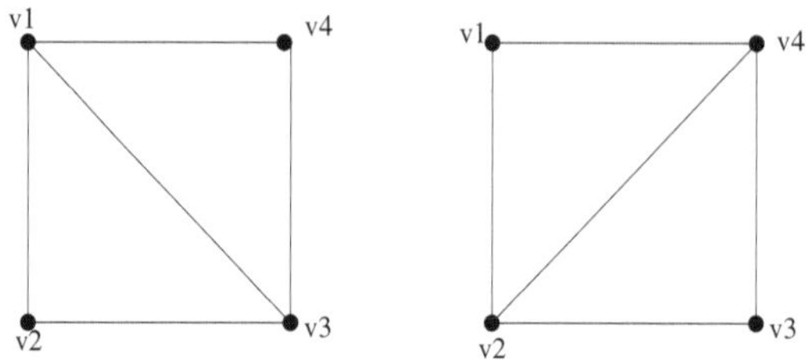

Figure 10.2: The two diagonal triangulations of a convex quadrilateral

Theorem 10.2.6. *Let $n \geq 2$. Then*

$$T_{n+1} = T_2 T_n + T_3 T_{n-1} + \cdots + T_{n-1} T_3 + T_n T_2$$

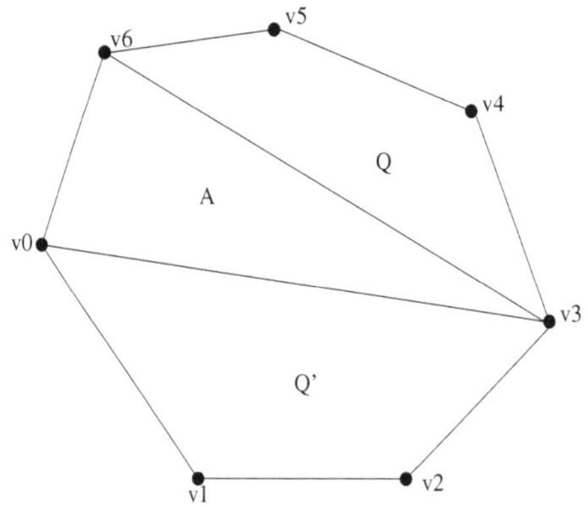

Figure 10.3: Division of a convex polygon into three parts

Proof Let the vertices a convex $(n+1)$-gon $\mathbf{P}$ be numbered $v_0, v_1, \ldots, v_n$ in an anticlockwise manner (as shown in Figure 10.3 for the case $n = 6$). Let T be a diagonal triangulation of $\mathbf{P}$. The side $v_0 v_n$ of $\mathbf{P}$ must belong to a unique triangle A with vertices v_0, v_n and some v_k where $k = 1, 2, \ldots, n-1$.
If we now cut out A from $\mathbf{P}$, we get two different convex polygons: Q below and Q' above A (we allow the possibility that Q or Q' is empty in the respective cases $k = 2$ and $k = n-1$). T thus gives rise to a diagonal triangulation of Q and a diagonal triangulation of Q'. Conversely, given a pair of diagonal triangulations of Q and Q', we get a unique diagonal triangulation of $\mathbf{P}$ that involves the triangle A. Here Q is a polygon with $k+1$ vertices $v_0, v_1, \ldots, v_k$ and Q' is a polygon with $n-k+1$ vertices $v_k, v_{k+1}, \ldots, v_n$. Since $k = 1, 2, \ldots, n-1$, we obtain, by summing over all k, $T_{n+1} = \sum_{k=1}^{n-1} T_{k+1} T_{n-k+1}$ which proves the result. □

Example 10.2.7. We have $T_4 = 2$ and hence $T_5 = T_2T_4 + T_3T_3 + T_3T_2 = (1 \times 2) + (1 \times 1) + (2 \times 1) = 5$ and these diagonal triangulations of a convex pentagon are shown in Figure 10.4.

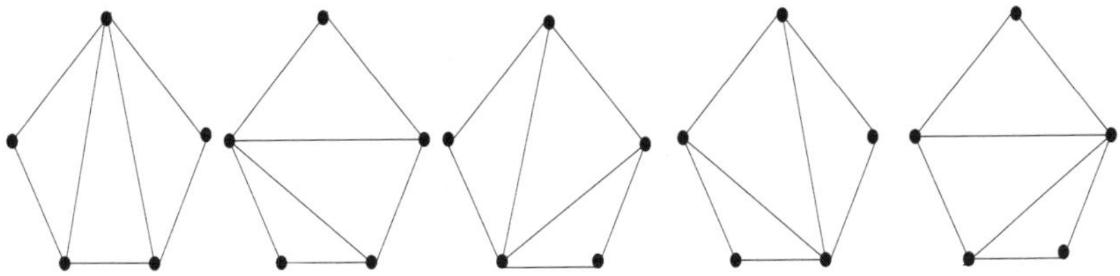

Figure 10.4: The five diagonal triangulations of a convex pentagon

We also have $T_6 = T_2T_5 + T_3T_4 + T_4T_3 + T_5T_2 = 5 + 2 + 2 + 5 = 14$ and these diagonal triangulations of a convex hexagon are shown on a later page.

Theorem 10.2.8. *For all $n \geq 4$, we have*

$$(n-3)T_n = \frac{n}{2}(T_3T_{n-1} + T_4T_{n-2} + \cdots + T_{n-2}T_4 + T_{n-1}T_3)$$

Proof We label the vertices of a convex n-gon $\mathbf{P}$ using the labels $v_0, v_1, \ldots, v_{n-1}$ in an anticlockwise manner. Suppose we wish to use the diagonal v_0v_k in the diagonal triangulation T where k is one of $2, 3, \ldots, n-2$. Arguing exactly as before, we see that we can cut along the diagonal v_0v_k and get two convex polygons Q and Q' that have been diagonally triangulated. Here Q has vertices $v_0, v_1, \ldots, v_k$ and is therefore a convex $(k+1)$-gon and Q' has vertices $v_k, v_{k+1}, \ldots, v_{n-1}, v_0$ and is therefore a convex $(n-k+1)$-gon. Thus the number of diagonal triangulations that use the diagonal v_0v_k for some k is obtained by summing over all k. This equals

$$\sum_{k=2}^{n-2} T_{k+1}T_{n-k+1} = T_3T_{n-1} + T_4T_{n-2} + \cdots + T_{n-2}T_4 + T_{n-1}T_3$$

Observe that this expression does not depend on v_0 and by symmetry, we can count the cardinality of the set $S = \{(T, D)\}$ where T is a diagonal triangulation and D is a diagonal used by T. Using Lemma 10.2.5, the number of diagonals used by any T is $n-3$ and summing over all vertices of $\mathbf{P}$ (notice that a diagonal v_iv_j gets counted twice, once for v_i and once for v_j) we get

$$|S| = \frac{n}{2}(T_3T_{n-1} + T_4T_{n-2} + \cdots + T_{n-2}T_4 + T_{n-1}T_3)$$

which gives the desired result. □

Theorem 10.2.9. *The number of diagonal triangulations T_{n+1} of a convex $(n+1)$-gon is given by $C_n = \frac{1}{n}\binom{2n-2}{n-1}$, the n-th Catalan number.*

Proof Using Theorem 10.2.6, we get

$$T_{n+1} - 2T_n = T_3T_{n-1} + \cdots + T_{n-1}T_3$$

and hence using Theorem 10.2.8, we get

$$\frac{n}{2}(T_{n+1} - 2T_n) = (n-3)T_n$$

which implies that $nT_{n+1} = (4n-6)T_n$. We now wish to exploit this recurrence relation. To that end, define $E_n = (n-1)T_n$. Then we get $E_{n+1} = \frac{4n-6}{n-1}E_n$ which gives

$$\frac{E_{n+1}}{E_n} = \frac{4n-6}{n-1} = \frac{2(2n-3)}{n-1} = \frac{(2n-2)(2n-3)}{(n-1)(n-1)}.$$

We then also get

$$\frac{E_n}{E_{n-1}} = \frac{(2n-4)(2n-5)}{(n-2)(n-2)}.$$

Observe that $E_3 = 2T_3 = 2$ and $E_2 = 1$. Therefore, iteration gives

$$\begin{aligned} E_{n+1} &= \frac{E_{n+1}}{E_n} \times \frac{E_n}{E_{n-1}} \times \cdots \times \frac{E_3}{E_2} \\ &= \frac{(2n-2)(2n-3)(2n-4)(2n-5) \times \cdots \times 2 \times 1}{(n-1)^2 \times (n-2)^2 \times \cdots \times 1} \\ &= \frac{(2n-2)!}{(n-1)!(n-1)!} \\ &= \binom{2n-2}{n-1} \end{aligned}$$

and the theorem is proved. □

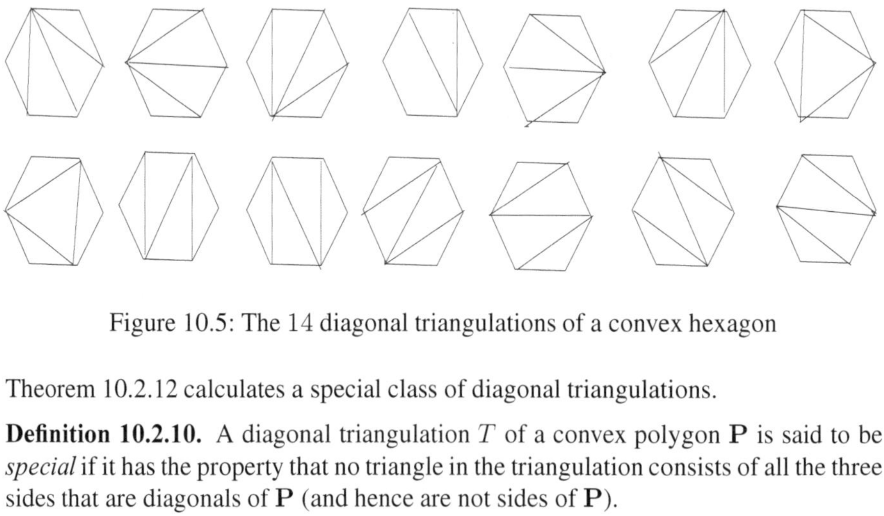

Figure 10.5: The 14 diagonal triangulations of a convex hexagon

Theorem 10.2.12 calculates a special class of diagonal triangulations.

Definition 10.2.10. A diagonal triangulation T of a convex polygon $\mathbf{P}$ is said to be *special* if it has the property that no triangle in the triangulation consists of all the three sides that are diagonals of $\mathbf{P}$ (and hence are not sides of $\mathbf{P}$).

One of the diagonal triangulations of a convex hexagon shown in Figure 10.1 is special and the other is not. In fact, in the enumeration of all the diagonal triangulations of a convex hexagon in Figure 10.5, only two are not special. Thus for a special diagonal triangulation T of a convex n-gon $\mathbf{P}$ with $n \geq 4$, each triangle in T either has one side which is a diagonal (and the other two sides are the sides of $\mathbf{P}$) or two sides that are diagonals (and the third side is a side of $\mathbf{P}$).

Lemma 10.2.11. *Let $n \geq 4$ and $\mathbf{P}$ a convex n-gon. Let T be a special diagonal triangulation of $\mathbf{P}$. Then out of the $n-2$ triangles of the triangulation, exactly 2 say A and B have the property that two of their sides are sides of $\mathbf{P}$. Also, given any diagonal d used in the triangulation T, we have A and B on either side of d.*

Proof Make a two-way counting. Let a and b refer to the number of triangles that use two (respectively one) side of $\mathbf{P}$. Then $2a+b=n$ and $a+b=n-2$ proving $a=2$. Let d be a diagonal used in T with x and y, the end vertices of d. The diagonal d divides $\mathbf{P}$ into two subpolygons. Let z and w denote vertices of $\mathbf{P}$ that are adjacent to x and y respectively and on the same side of d. By the definition of a special diagonal triangulation, we have a triangle with side d that must use one of xz or yw of say xz. Then a triangle containing d is xyz and uses the diagonal $d' = yz$. Proceeding in the same manner from d', the desired conclusion must follow. □

Theorem 10.2.12. *Let β_n denote the number of special diagonal triangulations of a convex n-gon $\mathbf{P}$ where $n \geq 4$. Then $\beta_n - n2^{n-5}$.*

Proof The assertion is easy to check for $n = 4$. Let $n \geq 5$. Let x be a vertex with vertices v and w adjacent to x in $\mathbf{P}$. How many special diagonal triangulations T use the triangle vxw? Here vx and xw are sides of $\mathbf{P}$ and vw is a diagonal say d. Let this number be γ. Let z be adjacent to v and u adjacent to w. Then z and u are distinct. Look at the other triangle containing the diagonal vw. Since $n \geq 5$ such a triangle must use one more diagonal which must be either vu or wz. This gives us 2 choices. Suppose $d' = (vu)$ is the diagonal used in T. Then arguing exactly in the same manner, we have two choices at d'. Continuing in this manner, we get a sequence of diagonals exhausting all but one diagonal and hence $\gamma = 2^{n-4}$ (as the number of diagonals is $n-3$). By symmetry, we must get the same number γ at every vertex. Since any given special diagonal triangulation has two triangles that use two sides of $\mathbf{P}$ and hence two distinct points like x, we see that $\beta_n = \frac{n}{2}2^{n-4} = n2^{n-5}$. □

We now discuss the famous box office problem as the next application of Catalan numbers. We have $2n$ people standing in a line (a queue) to purchase tickets for a movie at the box office. The price of each ticket is Rs. 50 and each person either has a Rs. 100 note or a Rs. 50 note with him. The cashier at the box office has no money at his disposal at the beginning. Half the persons, that is n people have Rs. 50 note each and the other half that is remaining n people have Rs. 100 note each. We thus have a sequence of length $2n$ consisting of n terms that are F each (F for a fifty rupees note) and n terms that are H each (H for a hundred rupees note). Such a sequence of n H's and n F's is called a *workable sequence* if, at every stage of processing of the sequence, the cashier has change left with him in case it is required. That is, we should not be in a situation where the next person has Rupees 100 and the cashier has no Rs. 50 note left to give back the change to that person. For example, with $n = 1$, the only workable sequence is FH while for $n = 2$, there are two workable sequences given by $FFHH$ and $FHFH$ (the sequence $FHHF$ is not workable since at the time when the third person goes with his Rs. 100 note the cashier has given away his Rs.50 note to the second person and has no change to give back to the third person). The requirement then is that at every stage, the $\#F$'s should be $\geq$ the $\#H$'s which is what we call a workable sequence. For $n = 3$, workable sequences are $FFFHHH, FFHFHH, FHFFHH, FFHHFH$ and $FHFHFH$ giving a total of 5 workable sequences.

Definition 10.2.13. A sequence of length $2n$ with each term equal to F or H such that the sequence has n terms equal to F and n terms equal to H is called a *workable sequence if at every stage k, the number of F's is greater than or equal to the number of H's.*

Theorem 10.2.14. *Let a_n denote the total number of workable sequences of length $2n$. Then $a_n = C_{n+1}$.*

Proof We have a total of $\binom{2n}{n}$ sequences. From this, if we subtract the total number of non-workable sequences, then we get the number of workable sequences. Consider a non-workable sequence. Look at the first stage at which this sequence became non-workable and call that stage a snag. What this means is that just before the snag, the $\#F$'s was $\geq$ the $\#H$'s and exactly at the snag, the $\#F$'s becomes strictly smaller than the $\#H$'s. This can happen only by having the $\#F$'s equal to the $\#H$'s just before the snag and having an H at the snag. In particular, the snag occurs at an odd stage say $2m + 1$, where $0 \leq m \leq n - 1$. Thus a non-workable sequence has the form XZ where X equals $x_1 x_2 \cdots x_{2m} x_{2m+1}$, the portion upto the snag and Z has length $2n - 2m - 1$. Notice that in X, the $\#H$'s is one more than that of F's. Change the subsequence X to $Y = y_1 y_2 \cdots y_{2m+1}$ where we interchange F's and H's at all positions including and before the snag. Specifically, we have $y_i = H$ if $x_i = F$ and $y_i = F$ if $x_i = H$. Consider the sequence YZ of length $2n$. In X, the number of H's is one more than that of F and hence in Y, the number of F's is one more than that of H and in Z, we have one F more than H. Thus, in the sequence YZ, the number of F's is $n+1$ (and the number of H's is $n-1$). Also, in Y, the number of H's is greater than or equal to the number of F's at all positions before the snag, and at the snag, the number of F's is just one more than that of the number of H's. It follows that the correspondence between the set of all the non-workable sequences (with n H's and n F's) and the set of all the sequences of length $2n$ with exactly $n + 1$ F's is a $1 - 1$ correspondence and as a consequence, the number of non-workable sequences is equal to $\binom{2n}{n+1}$. Therefore, the number of workable sequences is a_n where a_n is given by:

$$\begin{aligned} a_n &= \binom{2n}{n} - \binom{2n}{n+1} \\ &= \frac{2n!}{n!n!} - \frac{2n!}{(n+1)!(n-1)!} \\ &= \frac{2n!}{n!n!}\left[1 - \frac{n}{n+1}\right] \\ &= \frac{1}{n+1}\binom{2n}{n} \\ &= C_{n+1} \end{aligned}$$

□

As a final application, consider the following problem. We have a set of $2n$ children all of them of different heights. We wish to arrange them in two rows of n children each in such a way that children in each row are arranged in increasing order of their heights (from the left to the right) and each child in the first row is shorter than the corresponding child in the second row. In how many ways can this be done? Consider the case when $n = 2$ and let the order $1 < 2 < 3 < 4$ give the ordering on the four children (thus 1 is the shortest and 4 the tallest). Then the only two possible workable arrangements are:

$$\begin{pmatrix} 1 & 2 \\ 3 & 4 \end{pmatrix} \quad and \quad \begin{pmatrix} 1 & 3 \\ 2 & 4 \end{pmatrix}$$

Definition 10.2.15. An arrangement of the elements of $[2n]$ into two rows and n columns given below

$$\begin{pmatrix} a_1 & a_2 & \cdots & \cdots & a_n \\ b_1 & b_2 & \cdots & \cdots & b_n \end{pmatrix}$$

is called a *workable arrangement* if $a_1 < a_2 < \cdots < a_n$ and $b_1 < b_2 < \cdots < b_n$ and $a_i < b_i$ for all $i = 1, 2, \ldots, n$.

Let **A** denote the set of all the workable arrangements on $[2n]$ and let **B** denote the set of all the sequences of length $2n$ of F's and H's with n F's and n H's such that the sequence is a workable sequence. Let a_n denote the total number of workable arrangements on $[2n]$, that is $a_n = |\mathbf{A}|$. Using Theorem 10.2.13 we know that $|\mathbf{B}| = C_{n+1}$ and hence to prove the following theorem, it suffices to set a bijection between the sets **A** and **B**.

Theorem 10.2.16. *The number of workable arrangements on $[2n]$ is the $(n+1)$-th Catalan number C_{n+1}.*

Proof Given an arrangement in **A**, write F exactly at those numbers in the first row and H at the remaining (and write the numbers in a sequence of length $2n$. For example, with $n = 4$, the workable arrangement

$$\begin{pmatrix} 1 & 2 & 3 & 6 \\ 4 & 5 & 7 & 8 \end{pmatrix}$$

gives us the workable sequence $FFFHHFHH$. Such a sequence clearly has equal number of F's and H's. Further, at the stage of a k-th H, such an H, must have been preceded by at least k F's since this position (in the workable arrangement) is at k-th position in the second row and this number is larger than at least k numbers in the first row. Conversely, given a member of $\mathbf{B}$, label the terms of the sequence from 1 to $2n$ from the first to the last and then make an arrangement with all the Fs in the first row and all the Hs in the second row. For example, the workable sequence $FFHFHHFFHH$ corresponds to

$$\begin{pmatrix} 1 & 2 & 4 & 7 & 8 \\ 3 & 5 & 6 & 9 & 10 \end{pmatrix}$$

This bijection completes the proof of the theorem. □

Theorem 10.2.17. *The number of monotonically increasing functions $f : [n] \to [n]$ is equal to $\binom{2n-1}{n-1}$.*

Proof Let f be such a function. Draw a bar chart or bar graph of f by plotting the points $\{(i, f(i)) : i = 1, 2, \ldots, n\}$ (and a vertical line from the x-axis to the point $(i, f(i))$). Since $f(1) \geq 1$, we can now join these points as follows Draw a vertical line from $(1, 1)$ to $(1, f(1))$ then a horizontal line from $(1, f(1))$ to $(2, f(1))$ and then a vertical line from $(2, f(1))$ to $(2, f(2))$ and so on. In general, we have a horizontal line from $(i, f(i))$ to $(i+1, f(i))$ and then a vertical line from $(i+1, f(i))$ to $(i+1, f(i+1))$ and finally a horizontal move $(n, f(n))$ to $(n+1, f(n))$ followed by (if required) a vertical move from $(n+1, f(n))$ to $(n+1, n)$. We thus get an increasing lattice path from $(1, 1)$ to $(n+1, n)$. On the other hand, given an increasing lattice path from $(1, 1)$ to $(n+1, n)$, we can easily obtain the monotonically increasing function f. This $1-1$ correspondence establishes the equality of the number of such functions and the number of increasing lattice paths from $(1, 1)$ to $(n+1, n)$. This latter number can be found from the observation that *any such path must involve n moves to the right and $n-1$ moves up* and hence the number of such paths is equal to the number of sequences of U's and R's whose length is $2n-1$ in which exactly $n-1$ terms are U, which is the binomial coefficient $\binom{2n-1}{n-1}$. □

10.3 Exercises for Chapter 10

10.1 Prove that $S(n, 2) = 2^{n-1} - 1 \; \forall \, n \geq 2$.

10.2 Prove that $S(n, n-1) = \binom{n}{2}$.

10.3 Show that both the numbers $S(8n, 8n-2)$ and $s(8n, 8n-2)$ are even for all n.

10.4 Prove that

$$S(n+1, k) = \sum_{j=1}^{n} \binom{n}{j} S(j, k-1)$$

where $2 \leq k \leq n$.

10.5 Make a small table of Stirling numbers $S(n, k)$ for $n \leq 8$.

10.6 Let a coin be tossed n times repeatedly with probability of heads equal to p at each toss where $0 < p < 1$ is a fixed real number. Let α_k be the probability that we have exactly k heads. Show that the sequence $(\alpha_0, \alpha_1, \ldots, \alpha_{n-1}, \alpha_n)$ is unimodal.

10.7 Prove that

$$S(n, n-2) = \binom{n}{3} + 3\binom{n}{4} = \frac{1}{4}(3n-5)\binom{n}{3}$$

10.8 Prove that

$$S(n, n-3) = \frac{1}{2}(n^2 - 5n + 6)\binom{n}{4}$$

10.9 Prove the recurrence relation

$$s(n, k) = s(n-1, k-1) - (n-1)s(n-1, k)$$

for the Stirling numbers of the first kind.

10.10 Let $1 \leq k \leq n$. Show that $s(n, k) = (-1)^{n-k} c(n, k)$.

10.11 Show that $n! = \sum_{k=1}^{n} (-1)^{n-k} c(n, k) n^k$.

10.12 Show that the number of ways of putting n balls of distinct colors into k distinct boxes with each box containing at least one ball is equal to $k! S(n, k)$.

10.13 Prove that the sequence of signless Stirling numbers of first kind is unimodal.

10.14 (a) Find $S(5, k)$ for all k.

(b) Find $c(5, k)$ for all k.

(c) Use (b) to find $s(5, k)$ for all k from $c(5, k)$.

(d) Find $s(5, k)$ directly using (10.3).

(a) Find $s(5, k)$ using the inverse of the matrix whose coefficients are $S(5, k)$.

10.15 Show that
$$\sum \binom{n}{n_1, n_2, \ldots, n_k} = k!S(n,k)$$
where the sum is taken over all the k-tuples $(n_1, n_2, \ldots, n_k)$ with $n_1 + n_2 + \cdots + n_k = n$ and $n_i \geq 1 \; \forall \; i$.

10.16 Let $n > m$ Prove the following identities.
$$\sum_{r=0}^{n-m} (-1)^r c(n, m+r) S(m+r, m) = 0.$$
$$\sum_{r=0}^{n-m} (-1)^r S(n, m+r) s(m+r, m) = 0.$$

10.17 Show that the Stirling number of second kind $S(n,k)$ is equal to the sum of all the products of the kind $x_1 x_2 \cdots x_{n-k}$ of $n-k$ elements where $x_1 \leq x_2 \leq \cdots \leq x_{n-k}$ and $x_i \in [k]$. For example,
$$S(5,2) = (1 \times 1 \times 1) + (1 \times 1 \times 2) + (1 \times 2 \times 2) + (2 \times 2 \times 2) = 15$$

10.18 Show that the signless Stirling number of the first kind $c(n,k)$ is equal to the sum of all the products of the kind $x_1 x_2 \cdots x_{n-k}$ of $n-k$ elements where $x_1 < x_2 < \cdots < x_{n-k}$ and $x_i \in [n-1]$. For example,
$$c(5,3) = (1 \times 2) + (1 \times 3) + (1 \times 4) + (2 \times 3) + (2 \times 4) + (3 \times 4) = 35$$

10.19 Consider the following generalization of the Stirling numbers of second kind. Let $i \geq 1$ be a fixed integer and define $S_i(n,k)$ to be the number of unordered partitions of an n-set into k subsets such that each subset has at least i elements. Show that

(a) $S_i(n,k) = \binom{n-1}{i-1} S_i(n-i, k-i) + k S_i(n-1, k)$.

(b) $S_i(n,k) = \frac{1}{k!} \sum \frac{n!}{j_1! j_2! \cdots j_k!}$ where the sum is taken over all the k-tuples $(j_1, j_2, \ldots, j_k)$ with each $j_r \geq i$ and $j_1 + j_2 + \cdots + j_k = n$.

10.20 Prove the following identity where $1 \leq m \leq n$.
$$\begin{aligned} S(n,m) &= \frac{1}{m!} \sum_{k=0}^{m-1} (-1)^k (m-k)^n \binom{m}{k} \\ &= \frac{1}{m!} \sum_{j=1}^{m} (-1)^{m-j} j^n \binom{m}{j} \end{aligned}$$

10.21 Prove that the expression for the Stirling number $S(n,m)$ obtained in Exercise 10.20 also holds for $m > n$ (when the L.H.S. is zero).

10.22 Let A_n denote the number of distinct ordered sets $\{x_1, x_2, \dots, x_n\}$ such that each x_i is a non-negative integer and $\sum_{i=1}^{n} x_i = n$ and $\sum_{i=1}^{k} x_i \geq k \; \forall \; k = 1, 2, \dots, n$. Show that $A_n = C_{n+1}$.

10.23 Consider all the lattice paths that go from $(0, 0)$ to (n, n) that consists of moves that are horizontal to the right and vertical upwards (and hence n right and n up moves) such that the path remains on or above the diagonal line $y = x$. Show that the number of such lattice paths is a Catalan number.

10.24 What would be the answer to the previous question (in Exercise 10.23) if a path is to remain on or below the diagonal line?

10.25 Remove the variable x_8 from the following parenthesized product to get a unique parenthesized product in variables $x_1, x_2, \dots, x_7$.

(a) $((x_1x_2)((x_4x_5)x_8))((x_3x_6)x_7)$

(b) $(((x_1x_5)((x_2x_5)(x_8x_3)))x_4)x_7$

(c) $((x_1x_3)(x_2x_4))((x_6x_8)(x_5x_7))$

(d) $((x_1x_2)(((x_3x_4)x_5)x_7)x_8)x_6$

(e) $((x_1x_2)(((x_3x_4)x_5)x_7)x_6)x_8$

10.26 In the following expressions of parenthesized products involving the variables $x_1, x_2, \dots, x_7$, insert the variable x_8 in the multiplication indicated by $\overline{\times}$.

(a) $(((x_3x_5)\overline{\times}x_6)((x_1x_4)x_7))x_2$

(b) $((x_1x_2)((x_3x_4)\overline{\times}(x_5x_6)))x_7$

(c) $(((x_1x_2)(x_3x_4))(x_5x_6))\overline{\times}x_7$

(d) $(x_7(x_1x_2))((x_5x_6)\overline{\times}(x_3x_4))$

10.27 Let t_n be the number of sequences $(a_1, a_2, \dots, a_{2n})$ of $2n$ terms adding to zero such that each a_i is ± 1 and $\sum_{i=1}^{k} a_i \geq 0 \; \forall \; k = 1, 2, \dots, 2n$. Show that $t_n = C_{n+1}$.

10.28 Let u_n denote the number positive integer sequences $a_1, a_2, \dots, a_n$ such that $a_1 \leq a_2 \leq \dots \leq a_n$ and $a_i \leq i \; \forall \; i = 1, 2, \dots, n$. Show that $u_n = C_{n+1}$.

10.29 Suppose we have a set of $2n$ points forming vertices of a regular $2n$-gon and lying on the circumference of a unit circle. We wish to join the points in pairs so that we get n chords that are non-intersecting. In how many ways can this be done?

10.30 Find the number of monotonically increasing functions

$$f : [n] \to [n]$$

with the property that $f(j) \leq j \; \forall \; j = 1, 2, \dots, n$.

10.31 Show that the number of sequences $(a_0, a_1, a_2, \ldots, a_{2n})$ of non-negative integers with the property that $a_0 = a_{2n} = 0$ and $|a_i - a_{i+1}| = 1 \ \forall\ i = 0, 1, \ldots, 2n-1$ is equal to C_{n+1}.

10.32 Construct a $1-1$ correspondence between the set of all parenthesized products among the variables $x_1, x_2, \ldots, x_n$ in that order and the set of all increasing lattice paths from $(0,0)$ to $(n-1, n-1)$ that make either a right horizontal move or a vertical upward move through the following steps.

a Enclose a given parenthesized product making proper left and right parentheses. For example $(x_1x_2)x_3 = ((x_1x_2)x_3)$ and

$$((x_1x_2)((x_3x_4)x_5))(x_6x_7) = (((x_1x_2)((x_3x_4)x_5))(x_6x_7))$$

Check that, we get exactly $n-1$ open (or left) parentheses.

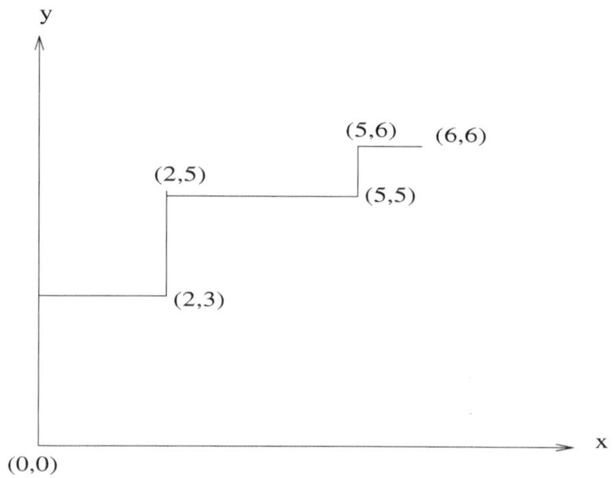

Figure 10.6: A lattice path

(b) For every open parenthesis make a vertical up U-move and for every variable (except the last variable x_n) make a horizontal right R-move Thus the parenthesized expression $(((x_1x_2)((x_3x_4)x_5))(x_6x_7))$ gives us the sequence of moves $UUURRUURRRUR$ and hence the path $(0,0) \to (0,1) \to (0,2) \to (0,3) \to (1,3) \to (2,3) \to (2,4) \to (2,5) \to (3,5) \to (4,5) \to (5,5) \to (5,6) \to (6,6)$ as shown in Figure 10.6.

(c) Check that every parenthesized product in the n variables and in the order $x_1, x_2, \ldots, x_n$ gives a well-defined increasing lattice path consisting of up and right moves that goes from $(0,0)$ to $(n-1, n-1)$.

(d) Show that the correspondence is a $1-1$ correspondence and hence the number of increasing lattice paths from $(0,0)$ to $(n-1, n-1)$ is equal to C_n.

10.33 The next few exercises (including this exercise) gives connections of oriented trees with Catalan numbers. The trees we deal with here are *unlabeled (unlike in Chapter 3) but are rooted and oriented.* Recall also from chapter 3 that a rooted and oriented binary tree is a tree with a root in which each vertex has at the most two children: the left child and the right child. Thus we have two rooted, oriented and binary trees with 2 vertices and five such trees on three vertices. These are drawn below.

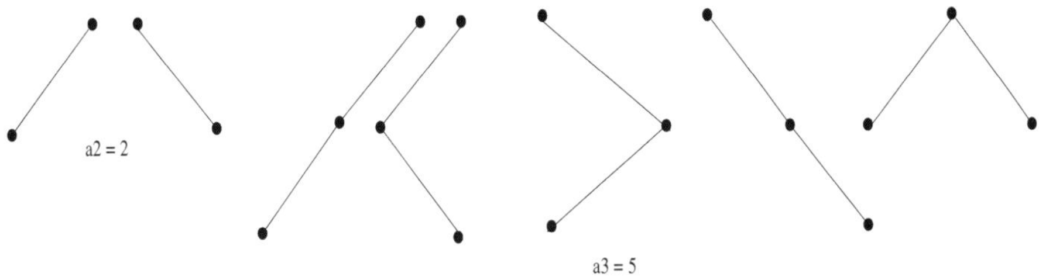

Figure 10.7: Binary, rooted, unlabeled, oriented trees on 2 and 3 vertices

Let a_n denote the number of such trees on n vertices. Then we have $a_2 = 2$ and $a_3 = 5$. Show that $a_n = C_{n+1}$, the $(n+1)$-th Catalan number.

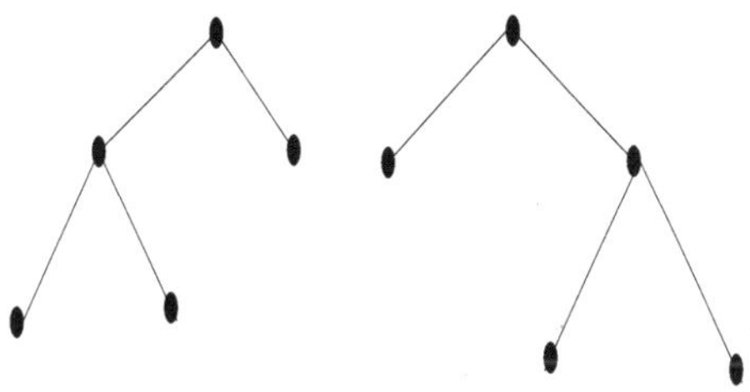

Figure 10.8: Binary, rooted, unlabeled, oriented full trees with 3 leaves

10.34 A rooted, oriented binary tree is called *full* if each vertex has either two children or no children. Thus, in figure 10.7 in Exercise 10.33, only the last tree on three vertices is full. Show that a full rooted tree with n leaves must have $n-1$ vertices that are not leaves (and thus has an odd number of vertices). The two rooted, binary, oriented and full trees with three leaves are drawn in Figure 10.8. Let b_n denote the number of rooted, unlabeled and full binary trees. Thus $b_3 = 2$. Show that $b_n = C_n$, the n-th Catalan number.

10.35 We now consider non-isomorphic, rooted, oriented (and unlabeled) trees. Thus, being oriented, the following two trees in Figure 10.9 are *not isomorphic.*

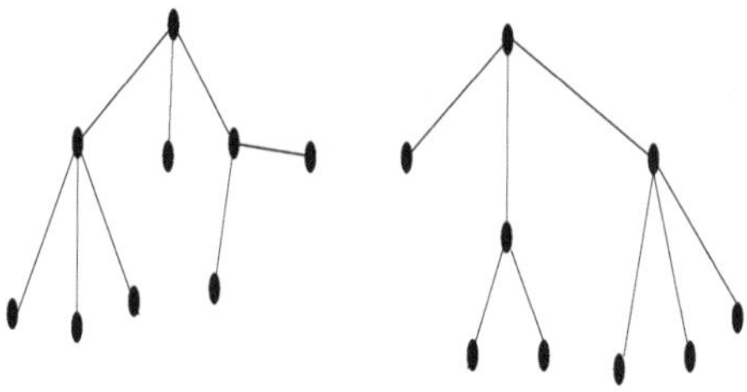

Figure 10.9: Non-isomorphic rooted and oriented trees

Let γ_n denote the number of non-isomorphic trees on $n+1$ vertices. Show that $\gamma_n = C_{n+1}$, the $(n+1)$-th Catalan number.

10.36 Let $\beta = a_1a_2\cdots a_n$ denote a permutation on $[n]$. We say that β *avoids* 132 *if we do not have* $i_1 < i_2 < i_3$ *such that* $a_{i_1} < a_{i_2} > a_{i_3}$. Show that the number of permutations on $[n]$ that avoid 132 is equal to C_{n+1}. There are similar notions of permutations avoiding other members of the symmetric group S_3 For the connections between them refer to Bona [11, 12].

Chapter 11

Recurrence relations

11.1 Introduction

In the earlier chapters, we have encountered a number of instances where a sequence $(H(n) : n = 0, 1, 2, \ldots)$ is given through a function that describes the n-th term $H(n)$ of the sequence in terms of $H(n-1), H(n-2), \ldots$. One of the early instances of this is the Pascal identity of the binomial coefficients. There are also many other examples such as the derangement number D_n, and the Stirling and Catalan numbers encountered in chapter 10 where recurrence relations were used. There are essentially two equivalent definitions of the number $n!$. We define $n! = 1 \times 2 \times \cdots \times n$. In this case, $n!$ is directly computed using a formula as a function of n. On the other hand, we may also define $n! = n \times (n-1)!$. This obtains $n!$ in terms of $(n-1)!$ and we say that $n!$ *is now recursively defined.* There is a conceptual difference between the two approaches. We do not need to know the entire sequence when it is recursively defined and hence are not interested in an explicit formula. It is the knowledge of the previous $H(r)$'s that helps us find $H(n)$. We give a number of examples.

Example 11.1.1. We begin by giving an elementary, almost trivial example. On an interest rate of rupees r per hundred per annum, let A_n denote the amount obtained after n years on a principal of rupees P. If the interest is a simple interest, then the amount after n years is $P + nP\frac{r}{100} = P(1 + \frac{nr}{100})$. However, if it is compound interest, which is generally the case, then the amount A_{n-1} at the end of $n-1$ becomes the principal for the (last) n-th year giving $A_n = A_{n-1}(1 + \frac{r}{100})\ \forall\, n \geq 1$ with $A_0 = P$ as the initial condition. By iteration, (we have $A_{n-1} = A_{n-2}(1 + \frac{r}{100})$ and hence $A_n = A_{n-2}(1 + \frac{r}{100})^2$ etc.) we get $A_n = P(1 + \frac{r}{100})^n$. This could also be seen by induction. Observe that with a simple interest on an interest rate of 5 percent per annum, it takes twenty years for the amount to double while it takes much less number of years if the interest is compounded annually.

Example 11.1.2. On the standard 8×8 chessboard, start by putting one grain on the first cell, two on the second, 4 on the third and, in general, put on the n-th cell twice the number of grains that were put on the $(n-1)$-th cell. We put 2^{n-1} grains on the n-the cell and if a_n is the number of grains put on the first n cells, then $a_n - a_{n-1} = 2^{n-1}$

and we also get $a_{n-1} - a_{n-2} = 2^{n-2}$ and so on. By adding all such equations, we get $a_n = 2^{n-1} + 2^{n-2} + \cdots + 2 + 1 = 2^n - 1$. In an ancient Indian story, a king agrees to give all the grains on all the sixty four squares of the chessboard to the person seeking alms. Thus the king would end up giving $2^{64} - 1$ grains. We leave it to the reader to figure out how many billion quintles. this works to and how difficult it would be for the king to meet the condition!

Example 11.1.3. This example, known as the *towers of Hanoi* recurrence relation is a must *in the paradigm of computer science.* We have a set n disks of different radii with a hole at the center of each disk so as to enable them to slide up and down three (vertical) pegs that are firmly fixed on the ground. The disks are put on the three pegs and can be shifted from any one peg to the other. A single move consists of shifting one disk from its existing peg to some other peg strictly adhering to the rule that a larger disk cannot be placed on top of a smaller disk. Initially all the n disks are arranged on the first peg say A in such a manner that the largest disk is at the bottom, the smallest at the top and such that the sizes are decreasing from the bottom to the top. Using one of the two other pegs say B and C as an auxiliary peg we have to transfer the entire mass of all the disks from A to C or B *with least number of moves.* Let $H(n)$ denote the number of moves when we have n disks. Clearly, $H(1) = 1$. When we have two disks D_1 on top of D_2, both on the peg A, we can first move D_1 to B (D_2 cannot be otherwise moved), then we move D_2 to C, the unoccupied disk and then shift D_1 on top of D_2 on the peg C, in the third move. Hence $H(2) = 3$. We let D_i denote the i-th disk where the size of D_i is smaller than that of D_{i+1} for $i = 1, 2, \cdots, n-1$. Thus D_1 is the smallest disk and D_n the largest disk. Consider three disks D_1, D_2, D_3. Since D_3 cannot move (but has to eventually move) unless both D_1 and D_2 have moved out of A, we must first shift both D_1 and D_2 making one peg available for D_3 to move out of A. Thus D_1 and D_2 must both be shifted to the same peg say C and as we just saw (this is the recursion) the number of moves required is $H(2) = 3$. In the fourth move, we shift D_3 to B. Now the peg A is vacant and we remember that it requires $H(2) = 3$ moves to shift the mass D_1 and D_2 from peg C to peg B. Hence we see that it is both necessary and sufficient that we have $3 + 1 + 3 = 7$ moves to shift three disks from the initial peg to some other peg and therefore, $H(3) = 7$. Note that we have also proved that this cannot be done in less than 7 moves: there is a middle stage when the largest disk D_3 gets moved. For that to happen, we must have a peg, say B that is vacant and that requires shifting of both D_1 and D_2 to C which needs 3 moves. After D_3 has been moved, we require 3 more moves to shift D_1 and D_2 on top of D_3.

Theorem 11.1.4. *Let $H(n)$ denote the least number of moves required when n disks are to be shifted from one peg to the other in the towers of Hanoi problem. Then*

$$H(n) = 2H(n-1) + 1 \quad \forall\, n \geq 2$$

Proof Let $n \geq 2$ and let the n disks be initially arranged in the order of decreasing radii from the bottom to the top. In order to shift the n disks from peg A to some other peg C, we can divide the entire procedure in three parts: the first part that consists of shifting the $n-1$ disks $D_1, D_2, \cdots, D_{n-1}$ from A to some other disk say C. Then the second part that consists of shifting the largest disk from A to the unoccupied peg B. The final part then consists of using the vacant peg A in order to shift the entire mass of the $n-1$ disks $D_1, D_2, \cdots, D_{n-1}$ from disk C to disk B. The three parts have to be followed exactly in that order if we wish to minimize the number of moves. The second part consists of shifting the largest disk D_n out of A and this cannot be done without completion of the first part that consists of moving all the smaller disks out of A and keeping one peg say B vacant at the end so that D_n can be moved there. Thus the first part is essential before the second part and the second part is essential before the third part which is required to achieve the final goal. By induction, the first part requires $H(n-1)$ moves and so does the third part. So, we get $H(n) \geq 2H(n-1)+1$. On the other hand, the first and third part each can be carried out in $H(n-1)$ moves (this is the recursion) and hence we require no more than $2H(n-1)+1$ moves giving $H(n) \leq 2H(n-1)+1$ proving the assertion. □

Having proved Theorem 11.1.4, how can we use it to solve the recurrence relation given in that theorem? *Solving* the recurrence relation here means obtaining an explicit formula for $H(n)$ as a function of n (which does not depend on the previous $H(j)$'s). This can be done in two ways. $H(n-1) = 2H(n-2)+1$ and hence if we continue in that manner, then we get

$$\begin{aligned} H(n) &= 2H(n-1)+1 \\ &= 2(2H(n-2)+1)+1 \\ &= 4H(n-2)+3 \\ &= 4(2H(n-3)+1)+1 \\ &= 8H(n-3)+7 \\ &= \cdots \\ &= 2^j H(n-j) + (2^j - 1) \\ &= 2^j(2H(n-j-1)+1)+1 \\ &= 2^{j+1}H(n-j-1) + (2^{j+1}-1) \\ &= \cdots \\ &= 2^{n-1}H(1) + (2^{n-1}-1) \\ &= 2^n - 1 \end{aligned}$$

This procedure is called *iteration.* We have used it in earlier chapters. Alternatively, we could also have made induction on n which requires a little guess work. Assuming that $H(n-1) = 2^{n-1} - 1$ and using the recurrence relation of Theorem 11.1.4, also obtains the proof of the following theorem.

Theorem 11.1.5. *$H(n) = 2^n - 1$ where $H(n)$ is the least number of moves required to solve the n-disk towers of Hanoi problem.*

The towers of Hanoi problem was first published in print form by Lucas in 1833. It appears to be ancient with similar stories of pegs and disks at the temple of Benares in north India where the priests make one move per second and the number of disks is 64. Some kind of apocalypse is supposed to occur (prophesy) when all the moves are completed. How long does that take?

11.2 Fibonacci recurrence relation

Though there are a number of ways of introducing Fibonacci numbers and the sequence, we take a more traditional approach and introduce them through the famous rabbit story. A pair of rabbits takes exactly two months to give birth to a new (baby) pair of rabbits. Thus, *as a rule each pair of rabbits gives rise to a new pair of rabbits after two months and then after the end of each subsequent month.* To begin with, that is, at time 0, there is only one pair of rabbits. Let a_n denote the number of pairs of rabbits after completion of n months. We have $a_0 = 1$ and since the pair takes two months to produce a new pair, we still have only one pair of rabbits at the end of the first month. So $a_1 = 1$. However, at $n = 2$, the initial pair of rabbits gives birth to a new pair and hence $a_2 = 2$. At the end of three months, the initial pair gives rise to yet another new pair of rabbits while the pair produced at the end of two months requires more time (two months) to give birth to a new pair. Hence, $a_3 = a_2 + 1 = 3$. What happens when $n = 4$ which is at the end of four months? The initial pair gives birth to a new pair and the pair that was born at the end of two months also gives rise to a new pair of rabbits. So, we get $a_4 = a_3 + 2 = 5$. How do we get a formula, in general? All the pairs that were around at the end of $n-1$ months are very much there at the end of n months and these are a_{n-1} in number while each new pair produced at the end of n months has parents that were alive at the end of $n-2$ months giving the number of new pairs equal to a_{n-2}. Hence we get the recurrence relation $a_n = a_{n-1} + a_{n-2}$. The sequence generated by this recurrence relation with the initial condition $a_0 = a_1 = 1$ is called a Fibonacci sequence named after Fibonacci whose real name was Leonardo of Pisa. He lived in the thirteenth century Europe and is credited with the introduction of modern place based numeration system (supposed to have originated in India) to the Western world.

Definition 11.2.1. *The Fibonacci sequence (and number)* (f_n) is defined by the recurrence relation $f_n = f_{n-1} + f_{n-2} \quad \forall\, n \geq 2$ where $f_0 = f_1 = 1$.

Here is a small table of the first 15 Fibonacci numbers.

n	0	1	2	3	4	5	6	7	8	9	10	11	12	13	14
$p(n)$	1	1	2	3	5	8	13	21	34	55	89	144	233	377	510

We are tempted to give another interesting occurrence of Fibonacci numbers and this is from Graham Knuth and Patashnik [27].

Example 11.2.2. A certain species of bees, the reproductive process is peculiar in the following way. The male of the species called the *drone* is produced asexually by the female of the species while the female of the species is produced by both the drone and the female. In plain terms, a drone has only mother but no father while a female has both the father and the mother. Thus it follows that a drone has only one grandmother and only one grandfather (both on his mother's side). In general, if x_n and y_n denote the number of n-th level grandparents of a drone, female and male respectively, then $x_{n+1} = x_n + y_n$ and $y_{n+1} = x_n$. We ask the reader as an exercise to show that these are Fibonacci numbers.

Lemma 11.2.3. *Let f_n denote the n-th Fibonacci number. Then*

$$\sum_{j=1}^{n} f_{2j} = f_{2n+1} - 1 \tag{11.1}$$

and

$$\sum_{j=1}^{n} f_{2j-1} = f_{2n} - 1 \tag{11.2}$$

Proof Both the statements have identical proofs and are clearly valid for $n = 1$ and hence we consider only the second statement. Assuming it to be true for n, we see that $\sum_{j=1}^{n+1} f_{2j-1} = \sum_{j=1}^{n} f_{2j-1} + f_{2n+1} = f_{2n} - 1 + f_{2n+1} = f_{2n+2} - 1$ as desired. □

Notation 11.2.4. For natural numbers r and s, we write $r << s$ to mean $r \leq s - 2$.

Definition 11.2.5. *A Fibonacci expression* is an expression of the form $f_{a_1} + f_{a_2} + \cdots + f_{a_r}$ where a_i's are natural numbers and $a_1 >> a_2 >> \cdots >> a_r$. Notice that an equivalent form for a Fibonacci expression is $\sum_{i=1}^{m} \epsilon_i f_i$ where each ϵ_i equals 1 or 0 and $\forall\, i = 1, 2, \ldots, m-1$ we have $\epsilon_i \epsilon_{i+1} = 0$ ensuring that *consecutive Fibonacci numbers do not appear in the Fibonacci expression.* A Fibonacci expression is called *full* if one of the following conditions holds: a_1 is even and then $a_1 = 2r$ so that we necessarily have $a_2 = 2r - 2,\ a_3 = 2r - 4, \ldots, a_r = 2$ or a_1 is odd and then $a_1 = 2r - 1$ so that we necessarily have $a_2 = 2r - 3,\ a_3 = 2r - 5, \ldots$ and then $a_r = 1$. Thus a full Fibonacci expression has the form $\sum_{j=0}^{m-1} f_{2m-2j}$ or $\sum_{j=0}^{m-1} f_{2m-1-2j}$.

Example 11.2.6. $f_7 + f_3 + f_2$ is not a Fibonacci expression (since both f_2 and f_3 occur) while $f_7 + f_4 + f_2$ is but it is not a full Fibonacci expression. This expression represents the number $f_7 + f_4 + f_2 = 21 + 5 + 2 = 28$. Finally, the expressions $f_8 + f_6 + f_4 + f_2$ and $f_5 + f_3 + f_1$ are both full Fibonacci expressions.

Lemma 11.2.3 just proved gives the value of a full Fibonacci expression. This is used to prove the following.

Lemma 11.2.7. *Let $f_{a_1} + f_{a_2} + \cdots + f_{a_r}$ be a Fibonacci expression. Then $f_{a_1} + f_{a_2} + \cdots + f_{a_r} \leq f_{a_1+1} - 1$ with equality iff the given Fibonacci expression is full.*

Proof Let a_1 be even say $a_1 = 2m$ (the other case is similar). Then the given condition implies that $a_2 \leq 2m - 2,\ a_3 \leq 2m - 4,\ \ldots, a_r \leq 2m - 2r + 2$ (we must have $m \geq r$) and hence we get

$$\begin{aligned} f_{a_1} + f_{a_2} + \cdots + f_{a_r} &\leq f_{2m} + f_{2m-2} + \cdots + f_{2m+2-2r} \\ &\leq f_{2m} + f_{2m-2} + \cdots + f_2 \\ &= f_{2m+1} - 1 \\ &= f_{a_1+1} - 1 \end{aligned}$$

and equality implies equality everywhere (note that every f_i is positive) and hence we must have $r = m$ so that the given Fibonacci expression has the form $f_{2m} + f_{2m-2} + \cdots + f_2$ which is full. $\square$

Theorem 11.2.8. *(Zeckendorf's theorem) For every natural number n there is a unique Fibonacci expression $n = f_{m_1} + f_{m_2} + \cdots + f_{m_r}$.*

Proof This is clearly true for $n = 1$ and we can make an induction on n. Let $n \geq 2$ and assume that the result holds for all natural numbers $n' < n$. Let k be the largest number such that $f_k \leq n$. Then $f_k \leq n < f_{k+1}$. We first observe that *any* Fibonacci expression for n must have its first summand f_k. For, if not, then let $n = f_{m_1} + f_{m_2} + \cdots + f_{m_r}$ be a Fibonacci expression. Clearly then $f_{m_1} \leq f_k$ and hence $m_1 \leq k$. If $m_1 \leq k - 1$, then Lemma 11.2.3 implies that $n = f_{m_1} + f_{m_2} + \cdots + f_{m_r} \leq f_{m_1+1} - 1 \leq f_k - 1 \leq n - 1$, a contradiction. Also, if $n = f_k$, then we cannot have any other Fibonacci expression for n (precisely for the same reason). Hence the proof is complete if $n = f_k$. Now let $n > f_k$ and let $n' = n - f_k$. By induction, we have a unique Fibonacci expression $n' = f_{m_1} + f_{m_2} + \cdots + f_{m_r}$ for n'. In this expression, if $m_1 \geq k - 1$, then we get $n = f_k + n' \geq f_k + f_{k-1} = f_{k+1} > n$, a contradiction. So, we must have $m_1 \leq k - 2$, i.e., $k >> m_1$ and hence $n = f_k + f_{m_1} + f_{m_2} + \cdots + f_{m_r}$ is the unique Fibonacci expression for n as desired. $\square$

Example 11.2.9. Here are a few examples of Fibonacci expressions.

(a) $50 = 34 + 13 + 3 = f_8 + f_6 + f_3$

(b) $504 = 377 + 89 + 34 + 3 + 1 = f_{13} + f_{10} + f_8 + f_3 + f_1$

(c) $604 = 377 + 144 + 55 + 21 + 5 + 2 = f_{13} + f_{11} + f_9 + f_7 + f_4 + f_2$

(d) $800 = 510 + 233 + 55 + 2 = f_{14} + f_{12} + f_9 + f_2$

The game *Fibonacci Nim* is played with two players A and B taking turns. Each player on his turn, takes away (removes) a certain number of chips from a given (only one) pile with a certain number of chips. Each player must remove at least one chip on his turn with A beginning the game. The player to remove the last chip is declared the winner. The stipulations are as follows. Initially the pile has ≥ 2 chips. *Player A on his very first turn cannot take away all the chips.* Further, *on each subsequent turn by any player, the player cannot remove more than two times the number of chips removed by his opponent on the previous turn.*

As an example, consider the pile that initially contains $n = 15$ chips. Player A removes 3 chips. If he removes 5 or more chips then the table would be left with at most 10 chips and the rule of the game permits B to take away all of them and winning the game. So assume that A removes only 3 chips. Then player B must remove between 1 and $2 \times 3 = 6$ chips. Notice that if B removes 4 or more chips, then the table would be left with at the most $15-(3+4) = 8$ chips and A can remove all of them to emerge the winner. So, it is not advisable for B to remove more than 3 chips. Suppose B removes only 2 chips. We are now left with $15 - (3 + 2) = 10$ chips on the table. Since B has removed 2 chips, A must remove between 1 and $2 \times 2 = 4$ chips on his turn. Suppose A takes away 2 chips. The table now has only 8 chips and B is allowed to (and must) remove between 1 and 4 chips. Hence any move on B's part where he takes away 3 or more chips is a win for A (he can just walk away with all the remaining chips). Suppose then that B takes away 2 chips and A on his turn takes away 1 chip. The table now has only 5 chip and B must remove either 1 or 2 chips. If B removes 2 chips, then A can remove the remaining 3 winning the game. If B removes only 1 chip, then A also removes only one chip and this is a win for A since A can ensure that he removes the last chip.

Though Fibonacci Nim is a progressively finite game, the rules of the game do not allow us to apply the theory of Chapter 7 to this situation (as a motivation, the reader may wish to look at Exercise 7.50 in chapter 7). That is because as to how many chips a player can remove is not decided right at the beginning of the game (as in the case of classical Nim game) but it is bounded by twice the number of chips removed by the opponent on the move immediately preceding the move under consideration. This number keeps changing at each turn since it depends on the removal at the previous turn. We thus have a somewhat dynamic situation. To solve the Fibonacci Nim, we use the Fibonacci expression of a non-negative integer given in Theorem 11.2.8.

Definition 11.2.10. Let n be a non-negative integer. Let

$$n = f_{a_1} + f_{a_2} + \cdots + f_{a_r}$$

be the Fibonacci expression for n. Then *the weight of* n denoted by $w(n)$ is r. Also, by definition, we take $w(0) = 0$.

Example 11.2.11. Since $12 = f_5 + f_3 + f_1$, we see that $w(12) = 3$. Since $100 = f_{10} + f_5 + f_3$ we have $w(100) = 3$. Finally, $f_{11} = 144$ implies that $w(144) = 1$.

The winning strategy for A is outlined in the next two paragraphs.

A's turn: The number of chips on the table is $n = f_{a_1} + f_{a_2} + \cdots + f_{a_k}$ where it is possible for A (due to the previous move of B, or otherwise, right at the beginning of the game) to remove f_{a_k} chips (the last summand in the Fibonacci expression). Then A actually removes f_{a_k} chips from the table leaving the table with n' chips with the Fibonacci expression $n' = f_{a_1} + f_{a_2} + \cdots + f_{a_{k-1}}$. Since $w(n) = k$ and $w(n') = k-1$, a move by A reduces the weight by 1.

B's turn: The number of chips on the table is some $n = f_{b_1} + f_{b_2} + \cdots + f_{b_m}$. Here, B is allowed to take away at the most M chips where $M < f_{b_m}$ (this ensures that B cannot follow the same strategy that A chooses to follow). Hence B removes x chips where $1 \leq x \leq M$ and the number of chips left on the table is some n' where

$$n' = f_{b_1} + f_{b_2} + \cdots + f_{b_{m-1}} + f_{c_1} + f_{c_2} + \cdots + f_{c_s}$$

and

$$b_1 >> b_2 >> \cdots >> b_{m-1} >> c_1 >> c_2 >> \cdots >> c_s$$

where $s \geq 1$ and hence $w(n') = m - 1 + s \geq m = w(n)$. Therefore, B's move either increases the weight of the table or keeps it unchanged *but cannot decrease it.*

Theorem 11.2.12. *The strategy described above ensures a win for player A except in the case where the initial number of chips n, is itself some Fibonacci number f_r (then A cannot remove all of n, by the stipulated rule) and in fact, in that case, B has an assured win by following the same strategy.*

Proof The second part of the statement is obvious: if the number of chips on the table at the start itself is $n = f_r$, then A cannot take away all of n and is forced to do what is written under B's turn. B can then follow A's strategy and force A to make the moves written for B and then B is assured a win by interchanging positions.
Suppose at some intermediate stage of the game the number of chips on the table is $n = f_{a_1} + f_{a_2} + \cdots + f_{a_k}$ where A can (by the stipulated rules) remove f_{a_k} chips from the table. Then A does remove those many chips from the table. Indeed, if $k = 1$, then the game is over and A has won. Then the table is now left with $n' = f_{a_1} + f_{a_2} + \cdots + f_{a_{k-1}}$ chips. Since A has removed f_{a_k} chips B can remove at the most $2f_{a_k}$ chips from n'. Here $a_{k-1} >> a_k$ and we have $f_{a_{k-1}} = f_{a_{k-1}-1} + f_{a_{k-1}-2}$ where the first summand on the R.H.S. is strictly larger than f_{a_k} while the second summand on the R.H.S. is greater than or equal to f_{a_k}. Hence it follows that B cannot take away $f_{a_{k-1}}$ chips from the table. Thus B is presented with a table of some n chips (not the same n) where

$$n = f_{b_1} + f_{b_2} + \cdots + f_{b_m}$$

and B *cannot (is not allowed to)* remove f_{b_m} (or more) chips from the table. Therefore B must take away some x chips from the table where $1 \leq x < f_{b_m}$. This ensures that B cannot mimic A's moves. Now suppose B removes x chips and $x + y = f_{b_m}$. Then both x and y are positive numbers and $y \leq f_{b_m} - 1$. Let the Fibonacci expression for y be $y = f_{c_1} + f_{c_2} + \cdots + f_{c_s}$. Since $b_{m-1} >> b_m > c_1$, the Fibonacci expression for the number $n' = n - x$ of chips remaining on the table is

$$n' = f_{b_1} + f_{b_2} + \cdots + f_{b_{m-1}} + f_{c_1} + f_{c_2} + \cdots + f_{c_s}$$

and $s \geq 1$ ensures that $w(n') \geq w(n)$. Hence B *cannot decrease the weight of the number on the table.* We just need to prove that at this stage, A is in a position to remove the smallest part f_{c_s} from the table which will complete the proof. Since B has removed x chips on his previous turn, this would be true if $f_{c_s} \leq 2x$. Suppose $f_{c_s} > 2x$ and let the Fibonacci expression for x be $f_{d_1} + f_{d_2} + \cdots + f_{d_r}$. Then $f_{c_s} > 2f_{d_1}$ implies that $c_s > d_1$ and if $c_s = d_1 + 1$ then $f_{c_s} = f_{d_1} + f_{d_1 - 1} > 2f_{d_1}$ implies that $f_{d_1-1} > f_{d_1}$ which is a contradiction. Therefore, $c_s >> d_1$ and thus $x + y = f_{b_m}$ has another Fibonacci expression

$$f_{c_1} + f_{c_2} + \cdots + f_{c_s} + f_{d_1} + f_{d_2} + \cdots + f_{d_r}$$

Since x and y are both positive, so are s and r and the expression above implies that $w(f_{b_m}) = w(x + y) \geq 2$ a contradiction. This contradiction proves that $f_{c_s} \leq 2x$ and hence A can remove f_{c_s} chips on his turn. □

Definition 11.2.13. Let $(a_0, a_1, \cdots, a_n)$ be a sequence of integers. The *continued fraction* associated with this sequence is the number

$$a_0 + \cfrac{1}{a_1 + \frac{1}{a_2} + \frac{1}{a_3 + \frac{\cdots}{\frac{\cdots}{a_n}}}}$$

Equivalently one may define this expression recursively to be $a_0 + b_1$ where b_1 equals the reciprocal of the continued fraction associated with the sequence $(a_1, a_2, \cdots, a_n)$.

For example, the continued fraction associated with the sequence $(2, 3, 4)$ is

$$2 + \frac{1}{3 + \frac{1}{4}} = 2 + \frac{1}{\frac{13}{4}} = 2 + \frac{4}{13} = \frac{30}{13}.$$

The continued fraction associated with a finite sequence of integers is a rational number. We can also associate a continued fraction with an infinite sequence of integers in the same manner.

Lemma 11.2.14. *The continued fraction associated with the integer sequence*

$$(1, 1, \ldots, 1, \ldots)$$

is

$$\tau = \frac{1 + \sqrt{5}}{2}$$

Proof τ equals

$$1 + \cfrac{1}{1 + \frac{1}{1 + \frac{1}{1 + \cdots}}}$$

where the denominator of the second summand is equal to τ giving us the equation

$$\tau = 1 + \frac{1}{\tau}$$

Since this $\tau > 1$, it is the positive root of $x^2 - x - 1 = 0$. So we get the desired value of τ. □

Definition 11.2.15. The number $\tau = \frac{1+\sqrt{5}}{2}$ is called the *golden ratio.*

Theorem 11.2.16. *Let $\tau_n = \frac{f_{n+1}}{f_n}$ where $n = 0, 1, 2, \ldots$ (and f_n is the n-th Fibonacci number). Then the sequence (τ_n) converges to τ.*

Proof Let β_n be the continued fraction associated with the sequence $(a_0, \ldots, a_n) = (1, 1, \ldots, 1)$. Then $\beta_0 = \tau_0$ and we may inductively assume that $\beta_n = \tau_n$. Then $\beta_{n+1} = 1 + \frac{1}{\beta_n} = 1 + \frac{1}{\tau_n}$. Writing $\tau_n = \frac{f_{n+1}}{f_n}$, we get

$$\beta_{n+1} = \frac{f_n + f_{n+1}}{f_{n+1}} = \frac{f_{n+2}}{f_{n+1}} = \tau_{n+1}$$

as desired. Now taking the limit as $n \to \infty$, we see that $\tau_n \to \tau$ since the continued fraction corresponding to the infinite continued fraction $(1, 1, \ldots, 1, \ldots)$ is τ. An alternative proof is suggested in Exercise 11.13. □

Theorem 11.2.17. *(Cassini's identity) Let f_n denote the n-th Fibonacci number. Then $\forall\, n \geq 1$, we have*

$$(f_n)^2 = f_{n+1}f_{n-1} + (-1)^n$$

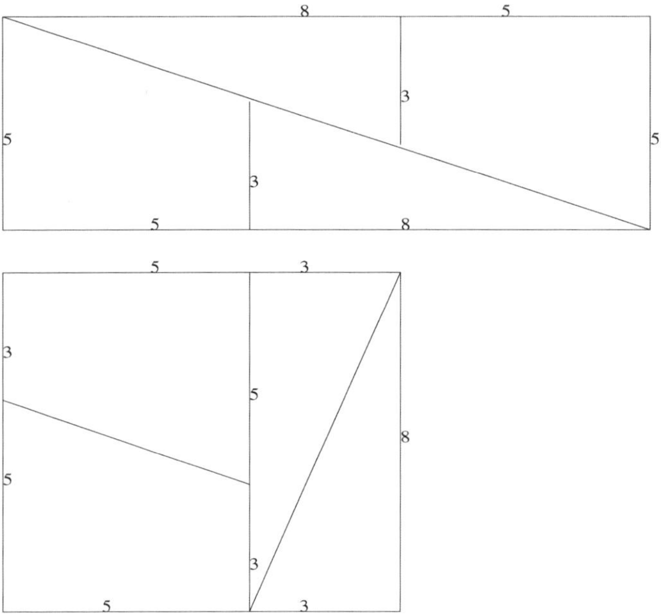

Figure 11.1: A paradox based on Cassini identity

Proof The statement is true for $n = 1$ and we use induction on n. Then

$$\begin{aligned} f_{n+2}f_n &= (f_{n+1} + f_n)f_n \\ &= f_{n+1}f_n + (f_n)^2 \\ &= f_{n+1}f_n + f_{n+1}f_{n-1} + (-1)^n \\ &= f_{n+1}(f_n + f_{n-1}) + (-1)^n \\ &= (f_{n+1})^2 - (-1)^{n+1} \end{aligned}$$

□

Cassini's identity gives rise to the following geometrical paradox. Assuming n to be odd say $n = 5$, we have $f_5 = 8$ and hence $(f_5)^2 = 64 = (f_6 f_4) - 1 = (13 \times 5) - 1$. We cut a 5×13 rectangle into four pieces, two congruent trapezoids and two congruent right angled triangles and fit them into an 8×8 square. If this were true, then we would arrive at the impossible $65 = 64$! As an exercise, the reader is asked to resolve this. Look at Figure 11.1 and see Exercise 11.5.

We now consider the question of the number of ways of writing a non-negative number n as an (ordered) sum in which every summand is either 1 or 2. For example, when $n = 4$, we have the following five ways of writing 4 as a sum in which each term is 1 or 2.

$$\begin{aligned} 4 &= 2+2 \\ &= 2+1+1 \\ &= 1+2+1 \\ &= 1+1+2 \\ &= 1+1+1+1 \end{aligned}$$

Let a_n denote the number of ways in which n can be expressed as a sum with each summand equal to 1 or 2 (and where the order in which the summands are written matters). Thus a_1, a_2 and a_3 equal $1, 2$ and 3 respectively while we just showed that $a_4 = 5$. We wish to show that a_n is the n-th Fibonacci number. Let $n = x_1 + x_2 + \cdots + x_r$ be such a sum. Since each x_i equals 2 or 1, we have two cases. In the first case, $n-2$ is written as a sum (of $r-1$ terms) with each term equal to 1 or 2 and in the second case, $n-1$ is written in the same fashion. We get the recurrence relation $a_n = a_{n-1} + a_{n-2}$ valid for all $n \geq 2$ with the provision that $a_0 = 1$. Since $a_1 = 1$, we see that $a_n = f_n$, the n-th Fibonacci number. In this counting, let k be the number of summands equal to 2. If r is the total number of summands, then $r - k$ is the number of summands equal to 1. Hence $r = n - k$. It follows that we have a total of $n - k$ summands of which k are 2's. Since the places of 2's determine the remaining places for 1's we see that the number of such sums in which 2 occurs k times is $\binom{n-k}{k}$. we obtain the following identity for Fibonacci numbers in terms of the binomial coefficients.

Theorem 11.2.18. *We have*

$$f_n = \sum_{k \geq 0} \binom{n-k}{k}$$

For many purposes, it is a good idea to define a Fibonacci sequence beginning with 0 (instead of 1). Thus we have a (new) sequence (F_n) where $F_0 = 0, F_1 = 1, F_2 = 1$, and, in general, $F_n = F_{n-1} + F_{n-2}$. This is just a shifted Fibonacci sequence (with the same recurrence relation), where $F_n = f_{n-1} \;\; \forall\, n \geq 1$ (and $F_0 = 0$). We call this sequence *Knuth-Fibonacci sequence* since it is extensively explored in the book by Graham, Knuth and Patashnik [27].

Theorem 11.2.19. *We have the following identities for the Knuth-Fibonacci numbers.*

(a) $\forall\, k \geq 1, \;\; F_{n+k} = F_k F_{n+1} + F_{k-1} F_n$

(b) $\forall\, n \geq 1, \;\; F_{2n} = F_n F_{n+1} + F_{n-1} F_n$

(c) $\forall\, n \geq 1, \;\; F_{3n} = F_{2n} F_{n+1} + F_{2n-1} F_n$

(d) $\forall\, n, k \geq 1$, F_n *divides* F_{kn}

(e) $g.c.d.(F_m, F_n) = F_g$ *where* $g = g.c.d.(m, n)$

Proof (a) certainly holds for $k = 1$ and for $k = 2$, it is just the Fibonacci recurrence relation. Hence, by induction,

$$\begin{aligned} F_{n+k+2} &= F_{n+k} + F_{n+k+1} \\ &= (F_k F_{n+1} + F_{k-1} F_n) + (F_{k+1} F_{n+1} + F_k F_n) \\ &= (F_k + F_{k+1}) F_{n+1} + (F_{k-1} + F_k) F_n \\ &= F_{k+2} F_{n+1} + F_{k+1} F_n \end{aligned}$$

For (b) and (c), substitute $k = n$ and $k = 2n$ respectively in (a). Consider (d) and let $k \geq 2$. Then, using (a), we have

$$F_{kn} = F_{(k-1)n+n} = F_{(k-1)n} F_{n+1} + F_{(k-1)n-1} F_n$$

where the first term on the R.H.S. is a multiple of F_n by induction. So F_n divides F_{kn}. Finally consider (e) and observe that using (d), F_g certainly divides both F_m and F_n. Let w.l.o.g., $m > n$ and write $m' = m - n$. Then, using (a), we get

$$F_m = F_{m'+n} = F_{m'} F_{n+1} + F_{m'-1} F_n$$

and therefore $g.c.d.(F_m, F_n) = g.c.d.(F_{m'} F_{n+1}, F_n) = g.c.d.(F_{m'}, F_n)$ because F_{n+1} and F_n are coprime. We can thus lower the subscripts F_j's and use the Euclidean algorithm to conclude that $g.c.d.(F_m, F_n) = F_g$. □

11.3 Linear homogeneous recurrence relations with constant coefficients

Definition 11.3.1. Let $(F(n))$ be a sequence. A *recurrence relation* on $(F(n))$ has the form

$$H(n) = F(H(n-1), H(n-2), \ldots, H(n-k))$$

where k is a fixed positive integer.

Definition 11.3.2. A recurrence relation is called a *linear recurrence relation if F is a linear function of its variables.*

Thus

$$F(H(n-1), H(n-2), H(n-3)) = n^2 H(n-1) + 3H(n-2) - H(n-3) + 6n$$

is linear while

$$F(H(n-1), H(n-2), H(n-3)) = H(n-1)^3 + H(n-2) - 7$$

is not since it is a cubic function in the variable $H(n-1)$. Similarly

$$F(H(n-1), H(n-2), H(n-3)) = H(n-1)H(n-2) + H(n-3)$$

is not a linear recurrence relation (since the term $H(n-1)H(n-2)$ is a quadratic).

Definition 11.3.3. A recurrence relation is called a *homogeneous recurrence relation* if F is a homogeneous relation of its variables.

For example,

$$F(H(n-1), H(n-2), H(n-3)) = H(n-1)H(n-2) + 3(H(n-2))^2$$

is homogeneous (of degree 2, *though it is not linear*) and so is

$$F(H(n-1), H(n-2)) = H(n-1)^3 + H(n-2)^3$$

Finally,

$$F(H(n-1), H(n-2), H(n-3)) = n^2 H(n-1) + \cos nH(n-2) + H(n-3) + 4n$$

is linear but not homogeneous (since the term $4n$ has degree 0 while other terms have degree 1).

Definition 11.3.4. A recurrence relation is said to have *constant coefficients if in each summand in the function $F(H(n-1), H(n-2), \ldots, H(n-k))$, the coefficient is a constant (and does not depend on n).*

Thus

$$F(H(n-1), H(n-2), H(n-3)) = 3H(n-1)^2 + 7H(n-2)^2 + 8$$

has this property (constant coefficients) while

$$F(H(n-1), H(n-2), H(n-3)) = 5nH(n-1) + 6H(n-2) - H(n-3)$$

does not.

Our interest is in studying a *Linear homogeneous recurrence relation with constant coefficients.* We look at a few more examples just to see what is the purview of our study (and what is not).

Example 11.3.5. (a) $H(n) = 3H(n-1) + 7H(n-2) + 12$

(b) $H(n) = 4\cos(H(n-1))$

(c) $H(n) = 3H(n-1)^2 + 8H(n-2)$

(d) $H(n) = 4H(n-1)^3 + 3nH(n-2) + 12$

(e) $H(n) = 5H(n-1) + 6H(n-3)$

(a) is linear and also has constant coefficients but is not homogeneous due to the presence of the last term 12 whose degree is 0. (b) involves a transcendental function cosine and hence is certainly not in our purview. (c) is neither linear nor homogeneous (since the terms have degrees 2 and 1 respectively). The relation (d) is neither linear (due to the presence of cube in the first term) and nor is homogeneous. In this recurrence relation, the second summand has coefficient $3n$ which is not a constant. Hence this recurrence relation satisfies none of the three criteria we led down. Finally, (e) is linear and homogeneous and the coefficients are $5, 0$ and 6 which ensures that it is a linear homogeneous recurrence relation with constant coefficients and is an object of our study. In short, a prototypical linear homogeneous recurrence relation with constant coefficients has the following form.

Definition 11.3.6. A *linear homogeneous recurrence relation with constant coefficients* is a relation of the form

$$H(n) = a_1H(n-1) + a_2H(n-2) + \cdots + a_kH(n-k) \;\; \forall n \geq k \tag{11.3}$$

This recurrence relation is said to have *order* k if $a_k \neq 0$.

A solution of equation (11.3) *is an explicit sequence* $h(n)$ whose terms also satisfy the recurrence relation (11.3). In order to determine the entire sequence $H(n)$ from (11.3), we must be given information about the first k (initial) values (*and hence a linear homogeneous recurrence relation with constant coefficients* (11.3) *of order* k *has* k *degrees of freedom*) $H(0), H(1), \ldots, H(k-1)$ of the sequence. First observe that $h(n) \equiv 0$ is a trivial solution of the recurrence relation (11.3). Also, if $h(n)$ and $h'(n)$ are two solutions of (11.3), then for arbitrary constants c and c' (constants here mean complex numbers), the sequence $g(n) = ch(n) + c'h'(n)$ is also a solution of (11.3). This is seen very easily: $\forall n \geq k$, we have

$$\begin{aligned} g(n) &= ch(n) + c'h'(n) \\ &= c[a_1h(n-1) + a_2h(n-2) + \cdots + a_kh(n-k)] \\ &\quad +c'[a_1h'(n-1) + a_2h'(n-2) + \cdots a_kh'(n-k)] \\ &= a_1(ch(n-1) + c'h'(n-1)) + a_2(ch(n-2) + c'h'(n-2)) + \cdots \\ &\quad +a_k(ch(n-k) + c'h'(n-k)) \\ &= a_1g(n-1) + a_2g(n-2) + \cdots a_kg(n-k) \end{aligned}$$

This finding just says that we have *a vector space of all the solutions of* (11.3) *(over the field of complex numbers)*. In order to find a non-zero solution of (11.3), we must take recourse to the following device called the characteristic polynomial of (11.3).

Definition 11.3.7. If $H(n) = a_1H(n-1) + a_2H(n-2) + \cdots + a_kH(n-k) \; \forall n \geq k$ with $a_k \neq 0$ is a given linear homogeneous recurrence relation (of order k), then its *associated characteristic polynomial* is the degree k polynomial $p(x)$ given by

$$p(x) = x^k - a_1x^{k-1} - a_2x^{k-2} - \cdots - a_{k-1}x - a_k \tag{11.4}$$

Evidently (11.3) *determines* $p(x)$ *and conversely and hence we can make abuse of language to call a root of* $p(x)$ *as a root of the recurrence relation* (11.3). *Such a complex number (which is a root of* $p(x)$*) is called a characteristic root (of* (11.3)*). Note that the assumption ensures that a characteristic root cannot be* 0.

Lemma 11.3.8. *Let* $q \neq 0$ *be a complex number. If* q *is a characteristic root of* (11.3), *then* $h(n) = q^n \;\forall\; n$ *is a solution of* (11.3) *and conversely if* $h(n) = q^n \;\forall\; n$ *is a solution of* (11.3), *then* q *is a characteristic root of* (11.3).

Proof $h(n) = q^n \;\forall\; n$ is a solution of (11.3)

$$\begin{aligned}
\iff \quad q^n &= a_1q^{n-1} + a_2q^{n-2} + \cdots + a_{k-1}q^{n-k+1} + a_kq^{n-k} \;\forall\; n \geq k \\
\iff \quad q^k &= a_1q^{k-1} + a_2q^{k-2} + \cdots + a_{k-1}q + a_k \\
\iff \quad p(q) &= 0
\end{aligned}$$

and this just means that q is a characteristic root. Note that the third step in the proof is consequence of $q^{n-k} \neq 0$ (because $q \neq 0$). □

Lemma 11.3.8 now manufactures a large number of (non-zero) solutions to the recurrence relation (11.3): if q_1, q_2 are two characteristic roots, then $h(n) = c_1{q_1}^n + c_2{q_2}^n$ is a solution to the recurrence relation (11.3) as we already observed. Using the fundamental theorem of algebra, the characteristic polynomial has k characteristic roots say $q_1, q_2, \ldots, q_k$ (not necessarily distinct) and hence extending the argument (from 2 to k), we see that

$$h(n) = c_1{q_1}^n + c_2{q_2}^n + \cdots + c_k{q_k}^n \tag{11.5}$$

is a solution to the recurrence relation (11.3) where $c_1, c_2, \cdots, c_k$ are (arbitrary) complex numbers.

Definition 11.3.9. A solution to the recurrence relation (11.3) of the form (11.5) is called a *general solution* if given any set of initial conditions, that is, given k arbitrary complex numbers $b_0, b_1, \ldots, b_{k-1}$, we can find complex numbers $c_1, c_2, \ldots, c_k$ such that $h(n)$ as given in (11.5) is a solution to the recurrence relation (11.3) *and* we also have $h(0) = b_0, h(1) = b_1, \ldots, h(k-1) = b_{k-1}$.

Theorem 11.3.10. *Let the characteristic polynomial* (11.4) *(and hence also the recurrence relation* (11.3)*) have* k *distinct roots* $q_1, q_2, \ldots, q_k$. *Then* (11.5) *is a general solution of the recurrence relation* (11.3).

Proof Since we already know that (11.5) is a solution of (11.3), it now suffices to show that we can choose constants $c_1, c_2, \ldots, c_k$ such that $h(j) = b_j \;\forall\; j = 0, 1, \ldots, k-1$. Since $h(j) = \sum_{i=1}^{k} c_i{q_i}^j$, this is equivalent to solving a system of linear equations

$$\begin{array}{rcl} c_1 + c_2 + \cdots + c_k & = & b_0 \\ c_1 q_1 + c_2 q_2 + \cdots + c_k q_k & = & b_1 \\ c_1 {q_1}^2 + c_2 {q_2}^2 + \cdots + c_k {q_k}^2 & = & b_2 \\ \cdots\cdots\cdots\cdots\cdots & = & \cdots \\ \cdots\cdots\cdots\cdots\cdots & = & \cdots \\ c_1 {q_1}^{k-1} + c_2 {q_2}^{k-1} + \cdots + c_k {q_k}^{k-1} & = & b_{k-1} \end{array}$$

for the k unknowns $c_1, c_2, \ldots, c_k$. This is a matrix equation of the form $Q\overline{c} = \overline{b}$ where $\overline{b}$ is the column vector $\overline{b} = (b_0, b_1, \ldots, b_{k-1})^t$ and $\overline{c} = (c_1, c_2, \ldots, c_k)^t$ and has a unique solution for $\overline{c}$ provided the coefficient matrix Q is non-singular. Here, the coefficient matrix Q equals

$$\begin{bmatrix} 1 & 1 & \cdots & \cdots & 1 \\ q_1 & q_2 & \cdots & \cdots & q_k \\ {q_1}^2 & {q_2}^2 & \cdots & \cdots & {q_k}^2 \\ \cdots & \cdots & \cdots & \cdots & \cdots \\ \cdots & \cdots & \cdots & \cdots & \cdots \\ {q_1}^{k-1} & {q_2}^{k-1} & \cdots & \cdots & {q_k}^{k-1} \end{bmatrix}$$

The matrix Q is the well-known Vandermonde matrix. Since q_i's are distinct, this matrix is non-singular (we do not prove this here since a more general statement is proved later on and in the exercises, reader is also asked to find the determinant of Q explicitly). □

Example 11.3.11. We give a number of examples.

(a) Consider the recurrence relation:

$$H(n) = 5H(n-1) - 6H(n-2) \;\forall n \geq 2$$

subject to the initial conditions: $H(0) = 2, H(1) = 5$. The characteristic polynomial is: $p(x) = x^2 - 5x + 6$ which factors into $(x-2)(x-3)$ and hence the characteristic roots of the recurrence relation are 2 and 3. We can thus assume the general solution to be $h(n) = c_1 2^n + c_2 3^n$. In order to find the constants c_1 and c_2, we use the initial conditions. Substitution of $n = 0, 1$ in the general solution produces the two equations

$$\begin{array}{rcl} c_1 + c_2 & = & 2 \\ 2c_1 + 3c_2 & = & 5 \end{array}$$

respectively. Multiplying the first equation by 2 and subtracting it from the second gives $c_2 = 1$ and hence $c_1 = 1$, so that the unique solution is $h(n) = 2^n + 3^n$.

(b) Consider

$$H(n+2) = 2H(n+1) + 3H(n) \; \forall \, n \geq 0$$

subject to the initial conditions: $H(0) = 3, H(1) = 1$. The characteristic polynomial is: $p(x) = x^2 - 2x - 3$ and hence the characteristic roots are 3 and -1 and the general solution is $h(n) = c_1 3^n + c_2(-1)^n$. Substitution of $n = 0, 1$ in the general solution produces the two equations

$$\begin{aligned} c_1 + c_2 &= 3 \\ 3c_1 - c_2 &= 1 \end{aligned}$$

respectively. Addition of the two equations gives $c_1 = 1$ and hence $c_2 = 2$. So the solution is $h(n) = 3^n + 2(-1)^n$.

(c) Now consider

$$H(n) = H(n-1) + 4H(n-2) - 4H(n-3) \; \forall n \geq 3$$

subject to the initial conditions: $H(0) = 6, H(1) = 4, H(2) = 12$. The characteristic polynomial is: $p(x) = x^3 - x^2 - 4x + 4$ and hence the characteristic roots are $1, 2$ and -2 and the general solution is $h(n) = c_1 + c_2 2^n + c_3(-2)^n$. Substitution of $n = 0, 1, 2$ in the general solution produces the three equations

$$\begin{aligned} c_1 + c_2 + c_3 &= 6 \\ c_1 + 2c_2 - 2c_3 &= 4 \\ c_1 + 4c_2 + 4c_3 &= 12 \end{aligned}$$

respectively and the general solution is $h(n) = 4 + 2^n[1 + (-1)^n]$.

(d) Recall the Fibonacci recurrence relation (Section 11.2) given by $f_n = f_{n-1} + f_{n-2}$ subject to the initial conditions $f_0 = f_1 = 1$. The characteristic polynomial is: $p(x) = x^2 - x - 1$ and hence the characteristic roots are τ and β where τ is the golden ratio. Specifically,

$$\tau = \frac{1+\sqrt{5}}{2}, \; \beta = -\frac{1}{\tau} = \frac{1-\sqrt{5}}{2}$$

We have $f_n = c_1\tau^n + c_2\beta^n$ where, using the initial conditions, we get

$$\begin{aligned} c_1 + c_2 &= 1 \\ c_1\tau + c_2\beta &= 1 \end{aligned}$$

Since $\tau - \beta = \sqrt{5}$ and $1 - \beta = \tau$, we have

$$c_1 = \frac{\tau}{\sqrt{5}}, \quad c_2 = -\frac{\beta}{\sqrt{5}}$$

and hence we obtain an explicit formula for the n-th Fibonacci number:

$$f_n = \frac{\tau^{n+1}}{\sqrt{5}} - \frac{\beta^{n+1}}{\sqrt{5}}$$

The following example is an illustration to show that the general solution found in Theorem 11.3.10 turns out not to be a *general solution* in the case of repeated roots of (11.3).

Example 11.3.12. Consider $H(n) = 6H(n-1) - 9H(n-2)$ with the initial conditions: $H(0) = 5$ and $H(1) = 25$. The characteristic polynomial $p(x) = (x-3)^2$ with a single characteristic root 3. If we now try a solution such as $h(n) = c_1 3^n + c_2 3^n = (c_1 + c_2)3^n$, then using the initial conditions we get $A3^0 = 5$ and $A3^1 = 25$ where $A = c_1 + c_2$. This leads to $A = 5$ and $3A = 25$ which are clearly inconsistent. The theory needs to be modified in order to take care of the situation of repeated roots of (11.3). This is done in the (next) Section 11.4.

11.4 The case of repeated roots

Let the characteristic polynomial $p(x)$ of the recurrence relation (11.3) split (over the field $\mathbb{C}$ of complex numbers) into $p(x) = (x - q_1)^{r_1}(x-_2)^{r_2} \cdots (x - q_m)^{r_m}$ where $q_1, q_2, \ldots, q_m$ are distinct non-zero complex numbers and $r_1, r_2, \ldots, r_m$ are positive integers (whose sum is k). We begin with the following purely algebraic result.

Lemma 11.4.1. *Let q be a non-zero complex number and let $f(x)$ be a non-zero polynomial. Let r be a positive integer. Recursively define: $f_1(x) = f(x)$, $f_2(x) = x{f_1}'(x), \ldots, f_r(x) = x{f_{r-1}}'(x)$. Then the following statements are equivalent.*

(a) q is a root of $f(x)$ with multiplicity $\geq r$.

(b) $f_i(q) = 0 \ \forall\, i = 1, 2, \ldots, r$.

Proof Let (a) hold. If $r = 1$, then there is nothing to prove. Let $r \geq 2$. Then q is a zero of multiplicity $\geq r - 1$ of $f'(x)$ and of $f_2(x) = xf'(x)$. Since $f_3, f_4, \ldots, f_r$ are defined in the same manner from $f_2(x)$, it follows by induction that $f_i(q) = 0 \ \forall\ i = 1, 2, \ldots, r$. Let (b) hold with $r \geq 2$. Again, since $f_2(q) = 0$ and $f_3, \ldots, f_r$ are defined from f_2 (in the same manner), it follows by induction that q is a root of $f_2(x)$ with multiplicity $\geq r - 1$. But $q \neq 0$ and hence q is a root of $f'(x)$ with multiplicity $\geq r - 1$ and therefore that of $f(x)$ with multiplicity $\geq r$. □

Lemma 11.4.2. *Let q be a zero of (11.3) with multiplicity $r \geq 2$ and let $j = 0, 1, \ldots, r-1$. Then $H(n) = n^j q^n$ is a solution of (11.3).*

Proof When $j = 0$, this has already been proved in Section 11.3. Let $j \geq 1$. Let $f(x) = x^{n-k}p(x)$. Then q is a zero of $f(x)$ with multiplicity r. Defining $f_j(x)$ recursively in the same manner as in Lemma 11.4.1, we see that

$$\begin{aligned}
f(x) &= x^n - a_1 x^{n-1} - a_2 x^{n-2} - \cdots - a_{k-1}x^{n-k+1} - a_k x^{n-k} \\
f_j(x) &= n^j x^n - a_1(n-1)^j x^{n-1} - a_2(n-2)^j x^{n-2} - \cdots - \\
&\quad a_{k-1}(n-k+1)^j x^{n-k+1} - a_k(n-k)jx^{n-k}
\end{aligned}$$

Therefore, using Lemma 11.4.1, we get $f_j(q) = 0$ which implies that

$$n^j q^n = a_1(n-1)^j q^{n-1} + a_2(n-2)^j q^{n-2} + \cdots + a_{k-1}(n-k+1)^j q^{n-k+1} + a_k(n-k)^j q^{n-k}$$

showing that $H(n) = n^j q^n$ is a solution of (11.3). □

The proof of the following theorem is more involved than that of an earlier theorem (Theorem 11.3.10) where we handled the case of distinct simple roots. However, the proof of the general case we present here *avoids the computation of the generalized Vandermonde determinant which we believe, is more difficult than showing that associated generalized Vandermonde matrix is nonsingular.*

Theorem 11.4.3. *Let the characteristic polynomial* $p(x)$ *factorize into*

$$p(x) = (x-q_1)^{r_1}(x-q_2)^{r_2}\cdots(x-q_m)^{r_m}$$

where $q_1, q_2, \cdots, q_m$ *are distinct non-zero complex numbers and* $r_1, r_2, \ldots, r_m$ *are positive integers. Then*

$$h(n) = \sum_{i=1}^{m}\left\{\sum_{j=1}^{r_i} c_{i,j} n^{j-1}\right\} {q_i}^n \tag{11.6}$$

where $c_{i,j}$ *with* $i = 1, 2, \ldots, m$ *and* $j = 1, 2, \ldots, r_i$ *are* k *arbitrary constants is a general solution of the recurrence relation* (11.3).

Proof Since the solution space is closed under linear combinations (as shown in Section 3), we see that $h(n)$ is a solution of (11.3) using the previous lemma (Lemma 11.4.2). It is therefore sufficient to prove that given arbitrary numbers $b_0, b_1, \ldots, b_{k-1}$, we can find (unique) constants $c_{i,j}$'s such that the following k linear equations are satisfied. Consider the equation determined by b_0:

$$c_{1,1} + 0 + \cdots + 0 + c_{2,1} + 0 + \cdots + c_{m,1} + 0 + \cdots + 0 = b_0$$

For this equation, the coefficient of $c_{i,1}$ is 1 while that of $c_{i,j}$ is zero if $j \geq 2$ (since $0^j = 0$). Next, we have

$$c_{1,1}q_1 + \cdots + c_{1,r_1}q_1 + \cdots + c_{m,1}q_m + \cdots + c_{m,r_m}q_m = b_1$$

$$c_{1,1}{q_1}^2 + c_{1,2}2{q_1}^2 + \cdots + c_{1,r_1}2^{r_1-1}{q_1}^2 + \cdots + c_{m,1}q_m + \cdots + c_{m,r_m}2^{r_m-1}{q_m}^2 = b_2$$

and so on. Since $h(s) = b_s$, we have, in general,

$$\sum_{i=1}^{m}\sum_{j=1}^{r_i} c_{i,j} s^{j-1} {q_i}^s = b_s \tag{11.7}$$

Here $0^u = 0$ if $u \geq 1$ and $0^0 = 1$. We have a matrix equation $Q\bar{c} = \bar{b}$ where $\bar{b}$ is the column vector $\bar{b} = (b_0, b_1, \ldots, b_{k-1})^t$ and

$$\bar{c} = (c_{1,1}, \ldots, c_{1,r_1}, \ldots, c_{m,1} \ldots, c_{m,r_m})^t$$

and the coefficient matrix Q is a $k \times k$ matrix given by $Q = [a_{s,(i,j)}]$ where $s = 0, 1, \ldots, k-1$ and $i = 1, 2, \ldots, m$ and $j = 1, 2, \ldots, r_i$ for a given i. Specifically $a_{s,(i,j)} = s^{j-1}{q_i}^s$. To simplify things, we observe that $Q = [Q_1, Q_2, \ldots, Q_m]$ where Q_i is a $k \times r_i$ matrix corresponding to the i-th root q_i. For the sake of simplicity, if we write q for q_i, then the first column of the matrix Q_i (the one corresponding to $j = 1$) is $[1, q, q^2, \ldots, q^{k-1}]^t$ while for $j \geq 2$ the j-th column of this matrix is

$$[0,\ q,\ 2^{j-1}q^2,\ 3^{j-1}q^3, \ldots,\ (k-1)^{j-1}q^{k-1}]^t$$

Suppose we prove that Q is non-singular. Then clearly $\overline{c} = Q^{-1}\overline{b}$ gives a unique solution and proves that the constants are indeed uniquely determined. Note that the non-singularity of Q is equivalent to the assertion that the row null-space of Q has dimension 0. Hence the proof is complete if we can prove the following claim.

Claim Let $\overline{d} = [d_0, d_1, \ldots, d_{k-1}]$ be a row vector such that $\overline{d}Q = [0, 0, \ldots, 0]$, the zero vector. Then we must have $\overline{d} = [0, 0, \ldots, 0]$.

Proof of the claim: Let

$$g(x) = d_0 + d_1 x + \cdots + d_{k-1}x^{k-1}$$

If $g(x) \equiv 0$ is not true, then $g(x)$ is a non-zero polynomial with degree $\leq k-1$ and hence has at the most $k-1$ zeros. Therefore, if we show that (counting multiplicities) $g(x)$ has $\geq k$ zeros, then we arrive at a contradiction. Since $\overline{d}$ is orthogonal to Q_i, we get the first equation (for $j = 1$ and with the simplification $q = q_i$): $\sum_{s=0}^{k-1} d_s q^s = 0$ and hence q is a root of $g(x)$. Next, let $2 \leq j \leq r_i - 1$. Then we get $\sum_{s=1}^{k-1} d_s s^{j-1} q^s = 0$. Since $g(x) = \sum_{s=0}^{k-1} d_s x^s$, it is easily seen that for all $j \geq 2$, we have $f_j(x) = \sum_{s=1}^{k-1} d_s s^{j-1} x^s$ and hence we get $f_j(q) = 0$ for all $j = 1, 2, \ldots, r_i$. Using Lemma 11.4.1, it follows that $q = q_i$ is a root of $g(x)$ with multiplicity $\geq r_i$. Adding all the multiplicities, we see that $g(x)$ has at least $\sum_{i=1}^{m} r_i = k$ roots, which is a contradiction. This completes the proof of the claim and also the theorem. □

Example 11.4.4. Here are some examples.

(a) As the first example, we look at $H(n) = 6H(n-1) - 9H(n-2)$ with the initial conditions: $H(0) = 5$ and $H(1) = 25$ discussed earlier in example 11.3.12. With the characteristic polynomial $p(x) = x^2 - 6x + 9 = (x-3)^2$, we have the characteristic root 3 with multiplicity 2 and hence general solution has the form $h(n) = (c_{1,1} + nc_{1,2})3^n$. The initial conditions give $c_{1,1} = 5$ and $c_{1,1} + c_{1,2} = \frac{25}{3}$ and hence the solution is $h(n) = [5 + \frac{10n}{3}]3^n$.

(b) Consider the recurrence relation

$$H(n) = -5H(n-1) - 6H(n-2) + 4H(n-3) + 8H(n-4)\ \forall\, n \geq 4$$

subject to

$$H(0) = 1,\ H(1) = -5,\ H(2) = -1,\ H(3) = 31$$

Here the characteristic polynomial is $p(x) = x^4 + 5x^3 + +6x^2 - 4x - 8$ and it is easily seen that $x - 1$ is a factor of $p(x)$ to get

$$p(x) = (x-1)(x^3 + 6x^2 + 12x + 8) = (x+2)^3(x-1)$$

with roots -2 (multiplicity 3) and the simple root 1. Hence the general solution has the form

$$\begin{aligned} h(n) &= (c_{1,1} + c_{1,2}n + c_{1,3}n^2)(-2)^n + c_{2,1}(1)^n \\ &= (c_{1,1} + c_{1,2}n + c_{1,3}n^2)(-2)^n + c_{2,1} \end{aligned}$$

It remains to find the four constants $c_{1,1}, c_{1,2}, c_{1,3}$ and $c_{2,1}$. We have

$$\begin{aligned} h(0) = c_{1,1} + c_{2,1} &= 1 \\ h(1) = -2(c_{1,1} + c_{1,2} + c_{1,3}) + c_{2,1} &= -5 \\ h(2) = 4(c_{1,1} + 2c_{1,2} + 4c_{1,3}) + c_{2,1} &= -1 \\ h(3) = -8(c_{1,1} + 3c_{1,2} + 9c_{1,3}) + c_{2,1} &= 31 \end{aligned}$$

Subtraction of $h(0)$ from each one of $h(s)$ where $s = 1, 2, 3$ produces the following three equations (after change of sign, if required):

$$\begin{aligned} 3c_{1,1} + 2c_{1,2} + 2c_{1,3} &= 6 \\ 3c_{1,1} + 8c_{1,2} + 16c_{1,3} &= -2 \\ 3c_{1,1} + 8c_{1,2} + 24c_{1,3} &= -10 \end{aligned}$$

From the second and third equations, we get $c_{1,3} = -1$ and hence the first two equations now yield

$$\begin{aligned} 3c_{1,1} + 2c_{1,2} &= 8 \\ 3c_{1,1} + 8c_{1,2} &= 14 \end{aligned}$$

to finally give us: $c_{1,2} = 1$ and $c_{1,1} = 2$ and therefore $c_{2,1} = -1$. The solution is:

$$h(n) = (2 + n - n^2)(-2)^n - 1$$

(c) Now suppose, for the same recurrence relation, we change the initial conditions to

$$H(0) = 3,\ H(1) = 2,\ H(2) = -10,\ H(3) = 66$$

Elimination of $c_{2,1}$ (using the first equation) from each of the last three equations results in the following three linear equations in $c_{1,1}, c_{1,2}, c_{1,3}$:

$$\begin{aligned} 3c_{1,1} + c_{1,2} + 2c_{1,3} &= 1 \\ 3c_{1,1} + 8c_{1,2} + 16c_{1,3} &= -13 \\ 3c_{1,1} + 8c_{1,2} + 24c_{1,3} &= -21 \end{aligned}$$

and the last two equations now yield $c_{1,3} = -1$. Then the first two equations give

$$\begin{aligned} 3c_{1,1} + c_{1,2} &= 3 \\ 3c_{1,1} + 8c_{1,2} &= 3 \end{aligned}$$

so that $c_{1,2} = 0$ and $c_{1,1} = 1$. This finally gives $c_{2,1} = 2$ and hence the general solution is $h(n) = (-1)^{n+1}(n^2 - 1)2^n + 2$.

11.5 Difference tables and sums of polynomials

Definition 11.5.1. For a real valued function $f(x)$, defined on $\mathbb{R}$, we define $\triangle f(x) = f(x+1) - f(x)$, *called the (forward) difference of* $f(x)$.

Observe that $\triangle f(x)$ is also a real valued function defined on $\mathbb{R}$. Differences are used in numerical analysis (particularly for the purpose of an interpolation formula) because they approximate the 'differential' of a function, but we will not get into that part of mathematics here. As an example, if $f(x) = x^2 + 2x + 3$, then $\triangle f(x) = [(x+1)^2 + 2(x+1) + 3] - [x^2 + 2x + 3] = 2x + 3$. *All through this section, the real valued functions we consider are polynomials in the vector space* $\mathbb{R}[x]$.

Theorem 11.5.2. *The following properties hold.*

(a) $\triangle$ *is a linear operator on* $\mathbb{R}[x]$. *That is, if* $f(x)$ *and* $g(x)$ *are two polynomials and* r *is a real number, then we have* $\triangle(f(x) + g(x)) = \triangle f(x) + \triangle g(x)$ *and* $\triangle(rf(x)) = r\triangle f(x)$.

(b) If $f(x)$ *is a polynomial of degree* $n \geq 1$, *then* $\triangle f(x)$ *is a polynomial of degree* $n-1$ *and if* $f(x)$ *is a constant, then* $\triangle f(x) \equiv 0$.

Proof It is enough to prove (b) since (a) is an easy exercise. Again using (a), it is enough to show that $\triangle f(x)$ is a polynomial of degree $n-1$ when $f(x) = x^n$ since the general result then follows by linearity. Here

$$\triangle f(x) = (x+1)^n - x^n = \sum_{k=0}^{n} \binom{n}{k} x^k - x^n = nx^{n-1} + \sum_{k=2}^{n} \binom{n}{k} x^k$$

which has the form $nx^{n-1}+$ terms of lower degree and hence the proof follows. Note that if we ignore the lower degree terms, we are left with nx^{n-1}, the derivative of x^n, *which partly explains the use of the term difference.* □

We can now define

$$\begin{aligned} \triangle^2 f(x) &= \triangle(\triangle f(x)) \\ &= \triangle(f(x+1) - f(x)) \\ &= \triangle f(x+1) - \triangle f(x) \\ &= (f(x+2) - f(x+1)) - (f(x+1) - f(x)) \\ &= f(x+2) - 2f(x+1) + f(x) \end{aligned}$$

In general, we have $\triangle^m f(x) = \triangle(\triangle^{m-1} f(x))\ \forall\ m \geq 2$. This allows us to define the *difference table of the polynomial function* $f(x)$ as follows. The 0-th (horizontal) line of this table has $f(0), f(1), f(2), \ldots$ (written from left to right). On the next or the first line we write the differences $\triangle f(0), \triangle f(1), \triangle f(2), \ldots$. This line can be calculated using the 0-th line because $\triangle f(j) = f(j+1) - f(j)$. Then the second line consists of $\triangle^2 f(0), \triangle^2 f(1), \triangle^2 f(2), \ldots$ and so on. as shown below.

$$\begin{array}{ccccccccccccc}
f(0) & & & & f(1) & & & & f(2) & & & & f(3) & & & & f(4) \\
& \triangle f(0) & & & & \triangle f(1) & & & & \triangle f(2) & & & & \triangle f(3) & & & \\
& & \triangle^2 f(0) & & & & \triangle^2 f(1) & & & & \triangle^2 f(2) & & & & \triangle^2 f(3) & & \\
& & & \triangle^3 f(0) & & & & \triangle^3 f(1) & & & & \triangle^3 f(2) & & & & & \\
& & & & \triangle^4 f(0) & & & & \triangle^4 f(1) & & & & & & & & \\
& & & & & \triangle^5 f(0) & & & & & & & & & & &
\end{array}$$

The entries here are written in such a way that every number $\triangle^{j+1} f(m)$ is the difference between the two numbers $\triangle^j f(m+1)$ which is above and to the right and $\triangle^j f(m)$ which is above and to the left. Since $f(x)$ has degree n, the $n+1$ entries $f(0), f(1), f(2), \ldots, f(n)$ uniquely determine f and hence also the entire difference table. It is, therefore enough to write only a small portion of the difference table (as we have done) since the remaining is uniquely determined. We can also add two difference tables (position wise) and also multiply (all the entries in) a given difference table by a real number. This just means that each polynomial is uniquely determined by its difference table and conversely and we have an isomorphism from $\mathbb{R}[x]$, the vector space of all the polynomials to the space of difference tables.

Definition 11.5.3. The ordered set $(f(0), \triangle f(0), \triangle^2 f(0), \ldots)$ consisting of the left-most entries in the difference table of f is called *the left edge* of (the difference table of) f.

Theorem 11.5.4. *Let f be a polynomial of degree n. Then:*

(a) The left edge of f is $(f(0), \triangle f(0), \triangle^2 f(0), \ldots, \triangle^n f(0), 0, 0, \ldots)$ and we can conveniently write this as $(f(0), \triangle f(0), \triangle^2 f(0), \ldots, \triangle^n f(0))$ after deleting the 0s at the end.

(b) The left edge of f determines f uniquely.

(c) Let n be a positive integer and let $f_n(x) = \binom{x}{n}$. Then the left edge of $f_n(x)$ is the sequence $(0, 0, \ldots, 0, 1)$ where 1 is at the n-th place.

Proof The difference table of f is the same as that of $\triangle f$ with the top or the 0-th row (consisting of $f(0), f(1), f(2), \ldots$) added to it. In particular, the left edge of f is the same as that of $\triangle f$ with the entry $f(0)$ augmented to the left. Using Theorem 11.5.2, $\triangle f$ is a polynomial of degree $n-1$ with the left edge $\triangle f(0), \triangle^2 f(0), \ldots, \triangle^n f(0))$ (using induction) and hence the left edge of f is $(f(0), \triangle f(0), \triangle^2 f(0), \ldots, \triangle^n f(0))$ as desired. For (b), use induction again to see that the left edge $(\triangle f(0), \triangle^2 f(0), \ldots, \triangle^n f(0))$ of $\triangle f$ determines it uniquely and in particular, $\triangle f(1), \triangle f(2), \ldots,$ are all uniquely determined. Since $f(j+1) =$

$f(j) + \triangle f(j)$, we see that all of $f(0), f(1), f(2), \ldots$ are recursively and uniquely determined. For (c), use Pascal identity (Theorem 1.2.6) to get

$$\triangle f_n(x) = \binom{x+1}{n} - \binom{x}{n} = \binom{x}{n-1} = f_{n-1}(x)$$

By induction, the left edge of $f_{n-1}(x)$ is $(0, 0, \ldots, 0, 1)$ with 1 at the $(n-1)$-th place and hence augmenting $f(0) = 0$ to this on the left, we get $(0, 0, \ldots, 0, 1)$ with 1 at the n-th place. $\square$

Example 11.5.5. Let $p(x) = x^4 + 3x^3 - 7x^2 + 5x + 2$. The difference table is

$$\begin{array}{ccccccccccc}
2 & & 4 & & 24 & & 116 & & 358 & & 852 \\
 & 2 & & 20 & & 92 & & 242 & & 494 & \\
 & & 18 & & 72 & & 150 & & 252 & & \\
 & & & 54 & & 78 & & 102 & & & \\
 & & & & 24 & & 24 & & & & \\
 & & & & & 0 & & & & &
\end{array}$$

The last entry (at the bottom) was not really necessary since we already know that $\triangle^5 f \equiv 0$.

The key to finding a sum of the form $\sum_{x=0}^{m} p(x)$ is the following identity proved as a generalization of Pascal identity (Theorem 1.2.6) in Chapter 1.

$$\sum_{x=0}^{m} f_n(x) = \sum_{x=0}^{m} \binom{x}{n} = \binom{m+1}{n+1}$$

To find $\sum_{x=0}^{m} p(x)$ we express $p(x)$ of degree n as a linear combination of the polynomials $f_r(x)$ and obtain the required sum. To facilitate this, recall the identity on Stirling numbers of second kind proved in Chapter 10 (Theorem 10.1.9): $x^n = \sum_{k=1}^{n} S(n,k)[x]_k$ valid for all positive integers n. From this, we get $x^n = \sum_{k=1}^{n} S(n,k)k!\binom{x}{k}$ and this can be used to obtain $p(x)$ in terms of $f_k(x)$ and the use of the generalized Pascal identity will then obtain the sum $\sum_{x=0}^{m} p(x)$ as a closed polynomial expression in the variable m.

Example 11.5.6. We wish to find the sum $\sum_{x=0}^{m}(x^4 + 3x^3 - 7x^2 + 5x + 2)$. Make a small list of Stirling numbers of second kind that are required: $S(4,1) = S(4,4) = 1$, $S(4,2) = 7$, $S(4,3) = 6$ Also, $S(3,1) = S(3,3) = 1$ and $S(3,2) = 3$. Finally,

$S(2,1) = S(2,2) = 1$. Hence use of these values obtains:

$$\begin{aligned}
x^4 + 3x^3 - 7x^2 + 5x + 2 &= \sum_{k=1}^{4} S(4,k)k!\binom{x}{k} + 3\sum_{k=1}^{3} S(3,k)k!\binom{x}{k} \\
&\quad -7\sum_{k=1}^{2} S(2,k)k!\binom{x}{k} + 5\binom{x}{1} + 2 \\
&= 24\binom{x}{4} + (36+18)\binom{x}{3} + (14+18-14)\binom{x}{2} \\
&\quad +(1+3-7+5)\binom{x}{1} + 2 \\
&= 24\binom{x}{4} + 54\binom{x}{3} + 18\binom{x}{2} + 2\binom{x}{1} + 2
\end{aligned}$$

Therefore,

$$\begin{aligned}
\sum_{x=0}^{m}(x^4 + 3x^3 - 7x^2 + 5x + 2) &= 24\sum_{x=0}^{m}\binom{x}{4} + 54\sum_{x=0}^{m}\binom{x}{3} \\
&\quad +18\sum_{x=0}^{m}\binom{x}{2} + 2\sum_{x=0}^{m}\binom{x}{1} + 2(m+1) \\
&= 24\binom{m+1}{5} + 54\binom{m+1}{4} + 18\binom{m+1}{3} \\
&\quad +2\binom{m+1}{2} + 2(m+1)
\end{aligned}$$

Is it possible to solve this question without recourse to Stirling numbers of second kind? The answer is yes, if we use the following Lemma.

Lemma 11.5.7. *Let $p(x)$ be a polynomial of degree n with the left edge $(a_0, a_1, \ldots, a_n)$. Then $p(x) = \sum_{j=0}^{n} a_j \binom{x}{j}$.*

Proof Left edge determines the difference table uniquely and hence it suffices to show that the left edge of the polynomial on the R.H.S. is also $(a_0, a_1, a_2, \ldots, a_n)$. But this is clear since the left edge of $f_j(x) = \binom{x}{j}$ is $(0, 0, \ldots, 1)$ with 1 at the j-th place and since the left edges also form a vector space isomorphic to $\mathbb{R}[x]$. □

Example 11.5.8. Take the same problem with $p(x) = x^4 + 3x^3 - 7x^2 + 5x + 2$ discussed in Example 11.5.6. As we have already seen, the left edge of this polynomial is $(2, 2, 18, 54, 24)$ and hence using Lemma 11.5.8, we get

$$p(x) = 24\binom{x}{4} + 54\binom{x}{3} + 18\binom{x}{2} + 2\binom{x}{1} + 2$$

Using generalized Pascal identity we get the same formula for the sum $\sum_{x=0}^{m} p(x)$ found earlier.

As a final application of the method of differences, we discuss the following problem. Let $h_2(n)$ denote the number of regions in which the real plane $\mathbb{R}^2$ is divided by n lines in 'general position'. This means that no two lines are parallel (and hence they intersect in a single point) while no point is on the intersection of three lines (no three lines are concurrent). Evidently, $h_2(1) = 2$, $h_2(2) = 4$ and $h_2(3) = 7$ as Figure 11.2 shows. In general, suppose n lines $L_1, L_2, \ldots, L_n$ are given to us in general position and these partition the plane into $h_2(n)$ regions. The $(n+1)$-th line $L = L_{n+1}$ is drawn in such a way that it intersects each one of $L_1, L_2, \ldots, L_n$ at distinct points, say p_i is the point of intersection of L with L_i and by relabeling the points $p_1, p_2, \ldots, p_n$ occur on the line L in that order from left to right. We thus get a partition of L into line segments (intervals) $I_1, I_2, \ldots I_n, I_{n+1}$ where I_j is the line segment between p_{j-1} and p_j where $j = 2, \ldots, n$ and I_1 and I_{n+1} are the infinite line segments bounded by p_1 and p_n respectively. Each of these line segments partitions an existing region into two new regions. We thus get $n+1$ new regions effectively and hence the equation: $h_2(n+1) = h_2(n) + (n+1)$. Let $h_1(n)$ equal to the number of line segments formed by taking n distinct points on a line. Clearly $h_1(n) = n+1$. We thus have, $\triangle h_2(n) = h_2(n+1) - h_2(n) = h_1(n)$. By the theory of difference tables, the left edge of the polynomial $h_1(x)$ is $(1,1)$ and since $h_2(0) = 1$, the left edge of $h_2(x)$ is $(1,1,1)$. We have thus proved the following theorem.

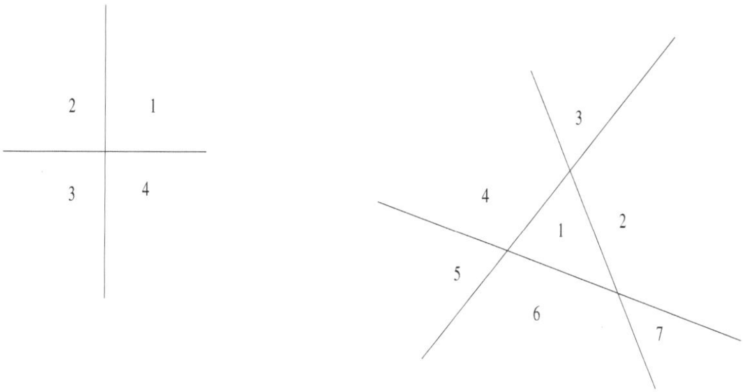

Figure 11.2: Lines in general position in the real plane

Theorem 11.5.9. *The number of regions into which the real plane $\mathbb{R}^2$ gets divided by n lines in general position is*

$$h_2(n) = \binom{n}{2} + \binom{n}{1} + \binom{n}{0} = \frac{n(n-1)}{2} + n + 1 = \binom{n+1}{2}$$

Generalization of this theorem to higher dimensions is left to the reader as Exercise 11.43. In three dimensions, a set of n planes is in general position if any two planes intersect in a line and no three planes pass through the same line, but three planes intersect in a point while no four planes are concurrent at a point. If $h_3(n)$ denotes the number of three dimensional regions we obtain, then arguing exactly in the same same manner, we get $\triangle h_3(n) = h_3(n+1) - h_3(n) = h_2(n)$ and hence

$$h_3(n) = \binom{n}{3} + \binom{n}{2} + \binom{n}{1} + 1$$

11.6 Other types of recurrence relations

No general method is known for solving a recurrence relation which is not of the type discussed in the earlier sections. In spite of that, we discuss some other recurrence relations that are *not linear, homogeneous with constant coefficients. An example of this is the towers of Hanoi problem considered in Section 1.* In what follows, we assume that the recurrence relation has the form

$$H(n) = F(H(n-1), H(n-2), \ldots, H(n-k)) + G(n)$$

where F is a linear and homogeneous function of its variables with constant coefficients and $G(n)$, the non-homogeneous part, is a function of the variable n. We also assume that $G(n)$ has some nice form such as a polynomial in n or an exponential function in n. In short, if $G(n) = 0$, then we are back to the situation discussed earlier. Thus we can solve the part $H(n) = F(H(n-1), \ldots, H(n-k))$ using the earlier theory upto determination of the coefficients. Then we set to find *particular solutions* for which we have the *thumb rules*. Assuming that $G(n)$ is a sum $G_1(n) + G_2(n) + \cdots + G_r(n)$, we try to obtain a particular solution in each case, $H(n) = F(H(n-1), \cdots H(n-k)) + G_i(n)$. For this, we equate the coefficients on both sides (treating the equation as an identity in n). That gives us r particular solutions. These are then combined with the general solution (obtained using the homogeneous part) and finally the general solution is found by determination of constant coefficients using the initial conditions. It remains to describe the thumb rules. Let $G_i(n) = f(n)$. Depending on what $f(n)$ is, the form of the particular solution is described in the following table. Here d and B are numbers that are to be determined by the given conditions.

$f(n)$	$Form\ of\ Particular\ solution$
d	B
db	$B_1n + B_0$
db^2	$B_2n^2 + B_1n + B_0$
d^n	Bd^n

We illustrate this procedure through the following examples.

Example 11.6.1. (a) Consider the recurrence relation $H(n) = 4H(n-1) - 3n + (2 \times 5^n)$ with the initial condition $H(0) = \frac{16}{3}$. We look at the particular solution in two parts:

$$H(n) = 4H(n-1) - 3n \ \ and \ \ H(n) = 4H(n-1) + (2 \times 5^n)$$

The first part has a solution (using the thumb rule): $h_1(n) = B_1 n + B_0$ and hence substitution gives:

$$\begin{aligned} B_1 n + B_0 &= 4B_1(n-1) + 4B_0 - 3n \\ &= (4B_1 - 3)n + (4B_0 - 4B_1) \end{aligned}$$

to get $B_1 = 4B_1 - 3 \Rightarrow B_1 = 1$ and $3B_0 = 4B_1 \Rightarrow B_0 = \frac{4}{3}$. Hence, we have $h_1(n) = n + \frac{4}{3}$. Next consider $H(n) = 4H(n-1) + (2 \times 5^n)$, the solution for which (using the thumb rule) has the form $h_2(n) = B \times 5^n$ to get

$$B \times 5^n = 4 \times (B \times 5^{n-1}) + (2 \times 5^n)$$

and hence $B = 10$ so that we have $h_2(n) = 2 \times 5^{n+1}$. Since the homogeneous part $H(n) = 4H(n-1)$ clearly has 4 as the characteristic root, we see that the general solution has the form

$$h(n) = (c \times 4^n) + 2 \times 5^{n+1} + \left(n + \frac{4}{3}\right)$$

Since $h(0) = c + 10 + \frac{4}{3} = \frac{16}{3}$, we get $c = -6$ and therefore the solution is

$$(-6 \times 4^n) + 2 \times 5^{n+1} + (n + \frac{4}{3})$$

(b) Consider $H(n) - 5H(n-1) + 6H(n-2) = 3n + 2$ subject to $H(0) = \frac{5}{4}$ and $H(1) = \frac{3}{2}$. Take $h_1(n) = B_1 n + B_0$ to get:

$$(B_1 n + B_0) - 5(B_1(n-1) + B_0) + 6(B_1(n-2) + B_0) = 3n + 2$$

and hence we get $B_1 = \frac{3}{2}$ and $B_0 = \frac{25}{4}$. So the general solution has the form

$$h(n) = c_1 2^n + c_2 3^n + \frac{3}{2}n + \frac{25}{4}$$

Using the initial conditions, we get $c_1 + c_2 + \frac{25}{4} = \frac{5}{4}$ while $2c_1 + 3c_2 + \frac{3}{2} + \frac{25}{4} = \frac{5}{4}$. This gives us the following two equations.

$$\begin{aligned} c_1 + c_2 &= -5 \\ 2c_1 + 3c_2 &= -9 \end{aligned}$$

and the solutions: $c_2 = 1$ and $c_1 = -6$ so that the final solution is

$$h(n) = (-6 \times 2^n) + 3^n + \frac{3}{2}n + \frac{25}{4}$$

(c) Consider $H(n) = H(n-1) + n$ subject to $H(0) = 0$. If we use the thumb rule to try $h_1(n) = B_1 n + B_0$, then we run into trouble as can be easily seen. The trouble here is that the homogeneous part is a recurrence relation with 1 as its characteristic root. In this case, we modify the thumb rule and merely iterate the recurrence relation to obtain

$$\begin{aligned} H(n) &= H(n-1) + n \\ &= H(n-2) + (n-1) \ + \ n \\ &= \cdots\cdots \\ &= 1 + 2 + \cdots + n \\ &= \binom{n}{2} \end{aligned}$$

We thus have *the modified thumb rule* that states that when 1 is a characteristic root, merely iterate the given recurrence relation.

(d) $H(n) = 6H(n-1) - 9H(n-2) + 4$ subject to $H(0) = 3$ and $H(1) = 10$. Here the particular solution is $h_1(n) = 1$ and the characteristic polynomial has 3 as a repeated root so that the general solution has the form

$$h(n) = (c_1 + c_2 n) \times 3^n + 1$$

The initial conditions give $c_1 = 2$ and $c_1 + c_2 = 3$ and hence the we have

$$h(n) = (2 + n) \times 3^n + 1$$

(e) Finally consider $H(n) = H(n-1) + (n^2 + n - 2)$ with $H(0) = -2$. Using the modified thumb rule, we see that $H(n) = \sum_{x=0}^{n}(x^2 + x - 2)$ and using the difference table method, we get the solution:

$$h(n) = 2\left\{\binom{n+1}{3} + \binom{n+1}{2} + \binom{n+1}{1}\right\}$$

11.7 Exercises for Chapter 11

Unless stated otherwise, f_n refers to the n-th Fibonacci number.

11.1 Consider the towers of Hanoi problem (Example 11.1.3) with n disks. Show that in an optimal (one that requires the least number of moves) scheme for shifting all the disks from peg A to some other peg, the disk D_k is moved 2^{n-k} times.

11.2 Refer to Example 11.2.2. Show that the number of n-th level grandmothers for a drone is f_{n-1} and the number of n-th level grandfathers is f_{n-2} while the same numbers for a female are f_n and f_{n-1} respectively. Hence conclude that the number of grandparents of a drone at the n-th level is f_n while the same for a female is f_{n+1}.

11.3 Show that consecutive Fibonacci numbers are coprime.

11.4 Show that $\forall\, k \geq 1$, F_n and F_{kn-1} are coprime (where F_n is the n-th Knuth-Fibonacci number).

11.5 (a) Prove that $f_n f_{n+1} = f_{n-1} f_{n+2} + (-1)^n$ for all $n \geq 1$.

(b) Use (a) (and/or Cassini's identity (Theorem 11.2.17)) to construct geometrical paradoxes similar to the one constructed in the chapter using any four consecutive Fibonacci numbers.

(c) Use (a) and (b) to resolve the geometrical paradox.

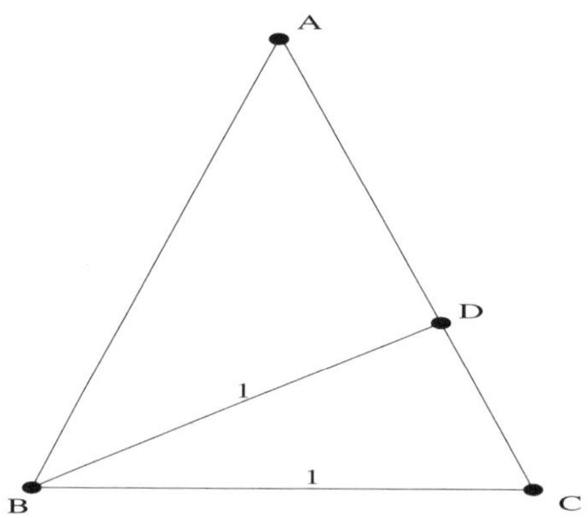

Figure 11.3: trigonometric ratios of 36°

11.6 A child climbs a staircase of n steps taking one or two steps at a time. Show that the total number of ways in which the child can climb the staircase is f_n.

11.7 Call a finite set A of positive integers 'fat' if every element of A is at least as large as the order of A. Thus $\{4, 7, 8\}$ is fat while $\{2, 5, 7\}$ is not fat. Treat the empty set $\emptyset$ as fat. Let a_n denote the number of fat subsets of the set $[n]$. Show that $a_n = f_{n+1}$, the $(n + 1)$-th Fibonacci number.

11.8 Draw an isosceles triangle ABC with the base angles B and C each equal to 72° and BC of unit length. Let BD be the angle bisector of B as shown in Figure 11.3.

(a) By considering similarity of triangles show that CD has length $\sigma = \frac{\sqrt{5}-1}{2}$.

(b) Show that $\cos 36^o = \frac{\tau}{2}$ where τ is the golden ratio.

(c) Use (b) to derive sine and cosine of the angles $54^o, 72^o$ and 18°.

11.9 Let **P** be a regular pentagon with each side of length 1 and let **Q** be the pentagon formed by the five diagonals of **P**. Show that **Q** is a regular pentagon and find the length of a side of **Q**.

11.10 Let R be a rectangle with length τ (the golden ratio) and breadth 1, as shown in Figure 11.4.

(a) Cut off (with a line parallel to the breadth of R) so as to form a unit square S_1 of side 1 and a rectangle R_1 whose length is 1 and breadth is $\tau - 1$. Show that R_1 is similar to R.

(b) By repeating (a) for R_1, obtain a square S_2 (of length $\frac{1}{\tau}$) and a rectangle R_2 of length $\frac{1}{\tau}$ and breadth $\frac{1}{\tau^2}$ and thus R_2 is also similar to R.

(c) Repeat the process to get a sequence of disjoint squares $S_1, S_2, \ldots$ such that the length of a side of S_n is $(\tau)^{-n+1}$ and a rectangle R_n similar to R.

(d) By summing areas of all the squares S_n, check that we get the area of the original rectangle R.

11.11 Prove the identity $F_n = 5F_{n-4} + 3F_{n-5} \ \forall\ n \geq 6$. Use this to prove that every fifth Knuth-Fibonacci number is a multiple of 5.

11.12 Show that a real number is a rational number iff it can be expressed as a finite continued fraction.

11.13 Here is an alternative way (Theorem 11.2.16) of proving that the sequence (τ_n) converges to τ.

(a) Use Exercise 11.4 to show that the subsequence (τ_{2m}) is a strictly increasing sequence and the subsequence (τ_{2m+1}) is strictly decreasing. Conclude that since both the subsequences are bounded, they are convergent.

(b) Use Cassini identity (Theorem 11.2.17) to show that $|\tau_{n+1} - \tau_n| \to 0$ as $n \to \infty$ and hence conclude using the first part that the point of convergence of both the subsequence is the same number τ which is also the point of convergence of the original sequence.

(c) Use the Fibonacci recurrence relation to get

$$\tau_{n+1} = 1 + \frac{1}{\tau_n}$$

and hence taking limits on both sides the point of convergence is a positive root of the quadratic $x^2 - x - 1 = 0$. Conclude that (τ_n) converges to τ.

Figure 11.4: A rectangle partition and golden ratio

11.14 Denote by T_n the set of all the permutations $\pi = a_1 a_2 \cdots a_n$ on the set $[n]$ with property that $\forall\, i = 1, 2, \ldots, n$ the number a_i is in the i-th column of the following array consisting of three rows and n columns:

	1	2	3	...	$i-1$	$\cdots$	$n-2$	$n-1$
1	2	3	4	...	i	$\cdots$	$n-1$	n
2	3	4	5	...	$i+1$	$\cdots$	n	

Let t_n denote the size of the set T_n. Obtain a recurrence relation and find t_n.

11.15 Let a_n denote the total number of binary sequences of length n in which every every 1 is adjacent to another 1. Find a recurrence relation for (a_n) and then solve it.

11.16 Solve the following recurrence relations.

(a) $a_n = 3a_{n-1} + 3a_{n-2} - a_{n-3}$ subject to $a_0 = a_1 = 1$ and $a_2 = 2$.

(b) $a_n = a_{n-1} + 9a_{n-2} - 9a_{n-3}$ subject to $a_0 = 0, a_1 = 1$ and $a_2 = 2$.

11.17 Show that $f_0 + f_1 + \cdots + f_{n-1} = f_{n+1} - 1$.

11.18 If the recurrence relation $a_n = ca_{n-1} + da_{n-2}$ has a general solution $a_n = A3^n + B6^n$, find c and d.

11.19 Solve the recurrence relation

$$a_n = 5a_{n-1} - 6a_{n-2} - 4a_{n-3} + 8a_{n-4}$$

for $n = 4, 5, 6, \ldots$ subject to the initial conditions $a_0 = 0,\ a_1 = 1,\ a_2 = 1$ and $a_3 = 2$.

11.20 Solve the following recurrence relation:

$$a_n - 5a_{n-1} + 6a_{n-2} = 1$$

subject to the conditions : $a_0 = 1,\ a_1 = 2$.

11.21 Solve the recurrence relation $H(n) = 8H(n-1) - 16H(n-2)$ for $n \geq 2$ subject to the initial conditions $H(0) = -1$ and $H(1) = 0$.

11.22 Solve the following recurrence relations.

(a) $a_n = 5a_{n-1} - 6a_{n-2} - 4a_{n-3} + 8a_{n-4}\ \forall\, n \geq 4$, subject to the initial conditions : $a_0 = 0, a_1 = 1, a_2 = 1, a_3 = 2$.

(b) $H(n) = 3H(n-2) - 2H(n-3)\ \forall\, n = 3, 4, 5, \ldots$ subject to the initial conditions $H(0) = 1, H(1) = 0, H(2) = 0$.

(c) $H(n) = 20H(n-1) - 17H(n-2) + 4H(n-3)\ \forall\, n \geq 3$, subject to the initial conditions $H(0) = 0, H(1) = -5/2, H(2) = -59/4$.

(d) $H(n) = -6H(n-1) + 12H(n-2) - 8H(n-3)\ \forall\, n \geq 3$, subject to the initial conditions $H(0) = 1, H(1) = -6, H(2) = 28$.

(e) $H(n) = 7H(n-1) - 10H(n-2)\ \forall\, n \geq 2$, subject to the initial conditions $H(0) = 0, H(1) = 3$.

(f) $H(n) = 3H(n-1) + 4H(n-2)\ \forall\, n \geq 2$, subject to the initial conditions $H(0) = H(1) = 1$.

(g) $H(n) - 7H(n-1) + 15H(n-2) - 9H(n-3) = 0\ \forall\, n \geq 3$, subject to the initial conditions $H(0) = 0, H(1) = 2, H(2) = 3$.

(h) $H(n) = 10H(n-1) - 25H(n-2)\ \ \forall\, n \geq 2$, subject to the initial conditions $H(0) = 1, H(1) = 2$.

(i) $H(n) = 10H(n-1) - 16H(n-2)\ \ \forall\, n \geq 2$, subject to the initial conditions $H(0) = 0, H(1) = 1$.

(j) $H(n) = 9H(n-1) - 15H(n-2) + 7H(n-3)\ \forall\, n \geq 3$, subject to the initial conditions $H(0) = 0, H(1) = 1, H(2) = 2$.

(k) $H(n) - 13H(n-1) - 40H(n-2) + 36H(n-3) = 0\ \forall\, n \geq 3$, subject to the initial conditions $H(0) = 1, H(1) = 1, H(2) = 0$.

11.23 Find a recurrence relation for the sequence $(H(n))$ where the n-th term is given by $H(n) = 5 \times 2^n + n \times 2^{n+1} - 4 \times 3^n$.

11.24 If the recurrence relation $H(n) = a_1H(n-1) + a_2H(n-2)$ has a general solution $H(n) = A2^n + B3^n$, what are the values of a_1 and a_2?

11.25 Solve the following recurrence relations.

(a) $H(n) = 3H(n-1) + (n^2 - 3)\ \forall\, n \geq 1$, subject to $H(0) = 1$.

(b) $H(n) = 3H(n-1) - 2H(n-2) + 3\ \forall\, n \geq 2$, subject to $H(0) = H(1) = 1$.

(c) $H(n) = 2(H(n-1))^2 + 1 \ \forall\, n \geq 1$, subject to $H(0) = 1$.

(d) $H(n+2) + 2H(n+1) - 8H(n) = 27 \times 5^n$ subject to $H(1) = -9$ and $H(2) = 45$.

(e) $(a_n)^2 = 2(a_{n-1})^2 + 1$, subject to $a_0 = 1$.

11.26 Let (a_n) denote the sequence satisfying the recurrence relation $a_n = 2a_{n-1} + a_{n-2}$ subject to $a_0 = 0$ and $a_1 = 1$.

(a) Show that $a_n = \frac{\alpha^n - \beta^n}{\alpha - \beta}$ where and $\alpha = 1 + \sqrt{2}$ and $\beta = 1 - \sqrt{2}$.

(b) Let (b_n) be a sequence defined by $b_0 = 2$ and $b_n = a_{n-1} + a_{n+1} \ \forall\, n \geq 1$. Then show that b_n's satisfy the same recurrence relation as the a_n's.

(c) Show that $b_n = \alpha^n - \beta^n \ \forall\, n$.

(d) Show that $a_{2n} = a_n b_n$.

(e) Show that $b_n \equiv \ 2 \ (mod\ 4) \ \forall\, n$.

(f) Show that 2^k divides a_n iff 2^k divides n.

Reference: From Erdös to Kiev by Honsberger [32], page 79.

11.27 Let $m \geq 2$ be a fixed natural number. We divide the unit circle into n congruent radial segments by drawing n radii that make an angle which is a multiple of $\frac{2\pi}{n}$. We wish to colour these segments in m colours such that each segment receives one colour and adjacent segments receive different colours. Let t_n denote the number of distinct colourings.

(a) Prove that $t_2 = m(m-1)$ and $t_3 = m(m-1)(m-2)$.

(b) Show that for all $n \geq 2$, we have

$$t_n = m(m-1)^{n-1} - t_{n-1}$$

(hint: when does a colouring become invalid?)

(c) By making induction on n otherwise, prove that

$$t_n = (m-1)^n + (-1)^n(m-1)$$

The numbers $t_n = t_n(m)$ are connected with the chromatic polynomial of the cycle C_n in graph theory; refer to West [60].

11.28 n points are chosen on the circumference of a circle and all possible chords joining every two points are drawn. Assume that no three chords intersect in the same point. Show that the number of regions thus formed is

$$a_n = \binom{n}{4} + \binom{n}{2} + 1$$

11.29 Solve the recurrence relation

$$H(n) = -2H(n-1) + 3H(n-2) - 4H(n-3) + 12H(n-4) - 8H(n-5)$$

subject to the initial conditions:

$$H(0) = 3, \ \ H(1) = -5, \ \ H(2) = 40, \ \ H(3) = -123, \ \ H(4) = 406$$

11.30 Prove that

$$f_0{}^2 + f_1{}^2 + \cdots + f_{n-1}{}^2 = f_{n-1}f_n$$

11.31 The 'regula falsi' method for solving $g(x) = 0$ begins with two approximate solutions x_0 and x_1 and finds x_n successively using the formula:

$$x_{n+1} = \frac{x_{n-1}g_n - x_n g_{n-1}}{g_n - g_{n-1}}$$

where g_i denotes $g(x_i)$. Show that if we begin with the initial approximations $x_0 = 1, x_1 = 1/2$ to solve $g(x) = x^2 = 0$, then x_n equals $\frac{1}{f_{n+1}}$ where f_n is the n-th Fibonacci number.

11.32 The Josephus problem is described as follows: n people numbered $1, 2, \ldots, n$ are standing equidistantly on a circle. Beginning at 1, we cross off every second remaining integer and continue crossing off the numbers until only one number is left who is declared winner of the game. For example, if n is 10, then the numbers crossed off, in that order, are $2, 4, 6, 8, 10, 3, 7, 1, 9$ and 5 survives. Let $J(n)$ denote the surviving integer.
Verify the entries in the following table :

n	1	2	3	4	5	6	7	8	9
$J(n)$	1	1	3	1	3	5	7	1	3

Show that $J(2n) = 2J(n) - 1$ and $J(2n+1) = 2J(n) + 1$.

11.33 Find $f_0 - f_1 + f_2 - \cdots + (-1)^n f_n$.

11.34 Prove all of the following (recall that F_n refers to the n-th Knuth-Fibonacci number).

(a) $F_{2n} \equiv 2F_nF_{n+1} \ (mod \ (F_n)^2)$

(b) $F_{2n+1} \equiv (F_{n+1})^2 \ (mod \ (F_n)^2)$

(c) $F_{3n} \equiv 3F_n(F_{n+1})^2 \ (mod \ (F_n)^2)$

(d) $F_{3n+1} \equiv (F_{n+1})^3 \ (mod \ (F_n)^2)$

(e) $F_{kn} \equiv kF_n(F_{n+1})^{k-1}\ (mod\ (F_n)^2)$

(f) $F_{kn+1} \equiv (F_{n+1})^k\ (mod\ (F_n)^2)$

(g) Show that $(F_n)^2$ divides F_{kn} iff F_n dives k

(h) Use (g) to conclude (Matijasevich Lemma, Graham-Knuth-Patashnik [27]) that $(F_n)^2$ divides F_m iff nF_n divides m.

11.35 Consider the sequence (a_n) of numbers

$$1, 9, 7, 7, 4, 7, 5, 3, 9, 4, 1, \ldots$$

where each each a_n for $n \geq 5$ is obtained from the four previous ones by the formula:

$$a_n = a_{n-1} + a_{n-2} + a_{n-3} + a_{n-4}\ \ (mod\ 10)$$

Note that each a_n is a single digit number. A word is a sequence of consecutive a_ns. Which of the following words occur in this sequence, after the 12-th term?

$$1234, \quad 3269, \quad 1977, \quad 0197, \quad 7019$$

(*Hint*: Work modulo 2).

11.36 Show that if $H(n) = 2(H(n-1) - H(n-2))\ \forall\, n \geq 2$ subject to $H(0) = 1$ and $H(1) = 2$, then the general solution is:

$$h(n) = (\sqrt{2})^n(\cos\frac{n\pi}{4} - \sin\frac{n\pi}{4})$$

11.37 Use of the identity $x^n = \sum_{k=1}^{n} S(n,k)k!\binom{x}{k}$ can actually evaluate Stirling numbers of second kind for us using the difference table of x^n. Do this for the case $n = 5$ and 6 and find Stirling numbers of second kind $S(5,k)$ and $S(6,k)$ for all k.

11.38 Find the sums:

(a) $1 + 2^2 + 3^2 + \cdots + m^2$

(b) $1 + 2^3 + 3^3 + \cdots + m^3$

(c) $\sum_{k=0}^{m} k(k+1)(k+2)$

In each case, observe that we have a polynomial of degree, one higher. Use this to make a guess for the sum

$$\sum_{k=0}^{m} k^n$$

11.39 Consider the set of all the 4^n sequences with each digit (entry) equal to $0, 1, 2, 3$. How many of these sequences have an even number of 0's?

11.40 Let b_n denote the determinant of the $n \times n$ matrix $A = [a_{i,j}]$ where

(a) $a_{i,i} = 2 \ \forall \ i$

(b) $a_{i-1,i} = 1 \ \forall \ i \geq 2$

(c) $a_{i,i+1} = 1 \ \forall \ i \leq n-1$

(d) $a_{i,j} = 0$ for other values of (i, j)

Obtain a recurrence relation for b_n and then solve it.

11.41 Consider the Vandermonde matrix Q in the proof of Theorem 11.3.10 (the case of distinct roots of a homogeneous recurrence relation with constant coefficients).

(a) By a suitable subtraction of columns, prove that $q_i - q_j$ is a factor of the $\det(Q)$, the determinant of Q, for all $i < j$.

(b) Using the definition of determinant, show that $\det(Q)$ is a homogeneous polynomial in the variables $q_1, q_2, \ldots, q_k$ whose degree is $\frac{k(k-1)}{2}$.

(c) Show that

$$\det(Q) = \prod_{1 \leq i < j \leq k} (q_j - q_i)$$

(d) Show that if q_i's are all distinct, then Q is non-singular.

11.42 Show that the n-th Fibonacci number is the integer nearest to $\frac{\tau^{n+1}}{\sqrt{5}}$ where τ is the golden ratio.

11.43 Clearly defining 'hyperplanes in general position', state and prove the m-dimensional version of Theorem 11.5.9.

11.44 *Conway's soldier game* is a solitaire (played with only one player) which is played on an infinite chessboard with a finite number of *pieces* arranged below the x-axis (with each piece occupying some cell below the x-axis and no two pieces occupying the same cell). A *move* consists jumping of one piece over another piece occupying an adjacent cell provided the next cell is vacant. Adjacent here means the cell that is immediately adjacent to the cell either horizontally or vertically and thus for each cell, there are four adjacent cells. The piece that is jumped over is removed (after the move) and the piece that jumps over now occupies a new cell (that was previously vacant) as shown in Figure 11.5.

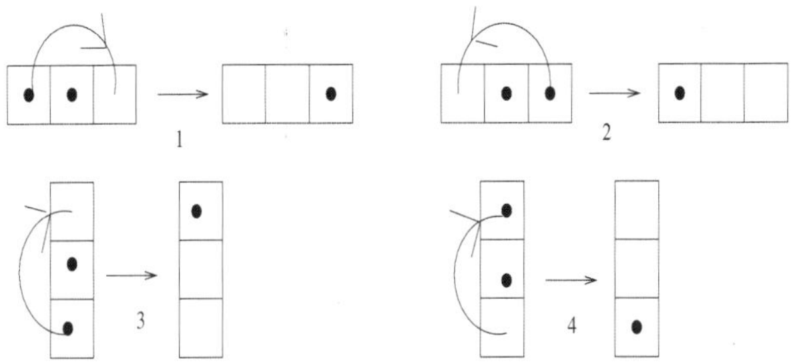

Figure 11.5: Possible jump moves

Through a sequence of moves, we have to reach a level above the x-axis, which is as high as possible.

(a) Show that to reach level 1, that is to reach $y = 1$, two pieces are both necessary and sufficient and the corresponding arrangement (upto isomorphism) is as shown in Figure 11.6.

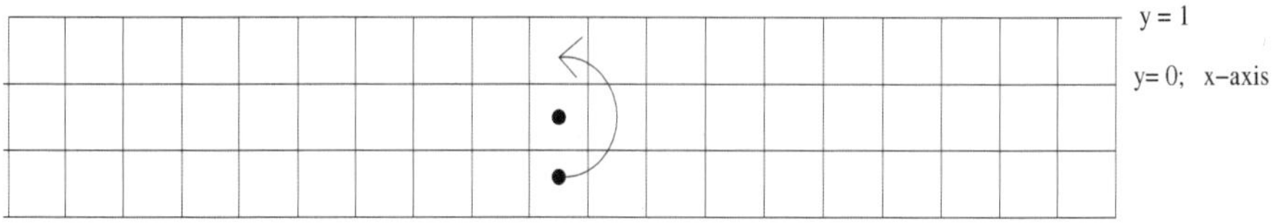

Figure 11.6: Reach level 1

(b) Show that to reach level 2, that is $y = 2$, four pieces are both necessary and sufficient and the corresponding arrangement (upto isomorphism) is as shown in Figure 11.7.

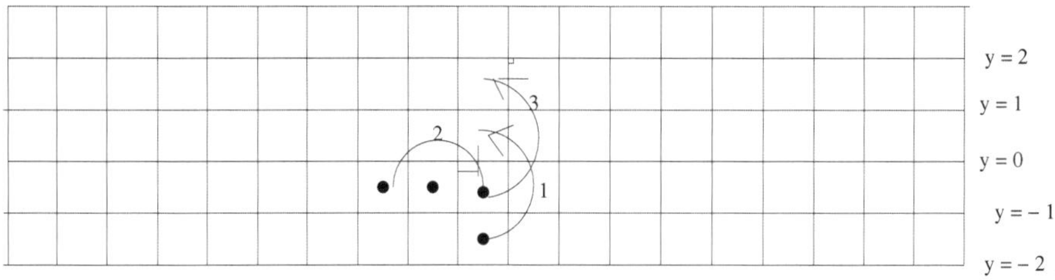

Figure 11.7: Reach level 2

(c) Show that to reach level 3, that is $y = 3$, eight pieces are sufficient with the corresponding arrangement shown in Figure 11.8.

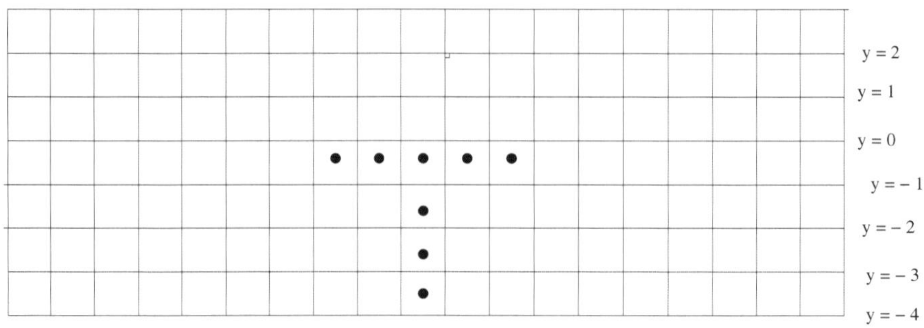

Figure 11.8: Reach level 3

(d) It is slightly harder to show that 20 pieces suffice to reach level 4. You may try this.

11.45 (Continuation of Exercise 11.44) We now wish to show that *no (finite) configuration S (arranged below the x-axis) can reach level* 5. Assume that S is such a configuration and arrive at a contradiction through the steps indicated below.

(a) Show that the given problem is *equivalent to arranging the pieces on the integer lattice points (i.e. $\mathbb{Z}^2$) in the real plane* so that the positions of the pieces are lattice points (x, y) with $y \leq 0$ and with the aim of reaching the point $(0, 5)$ using jump moves (here we assume that $(0, 5)$ is the first point on level 5 that is reached after a sequence of jumps).

(b) Let

$$\sigma = \frac{1}{\tau} = (\frac{1+\sqrt{5}}{2})^{-1} = \frac{\sqrt{5}-1}{2} \approx 0.618$$

denote the reciprocal of the golden ratio. For every lattice point (x, y) with $y \leq 5$, let

$$w(x, y) = \sigma^{|x|+5-y}$$

denote the weight of the lattice point (x, y). Notice that all the weights are positive numbers ≤ 1 and the weight of a point is just its distance from $(0, 5)$ where the distance is counted as the shortest lattice distance (path that uses only horizontal or vertical segments). For a given configuration S of pieces placed below the x-axis, let $w(S) = \sum w(x, y)$ where the sum is over the weights of all the lattice points (x, y) that are in S.

(c) Show that $\sigma^2 + \sigma = 1$ and hence, in general, we have $\sigma^n + \sigma^{n-1} = \sigma^{n-2}$ $\forall\, n \geq 2$.

(a) Show that for a horizontal move of type 1 or type 2, if the move is towards $(0, 5)$, then two pieces (cells) with total weight $\sigma^n + \sigma^{n-1}$ get replaced by a piece (cell) with weight σ^{n-2} and hence the total weight $w(S')$ of the new configuration remains unchanged.

(d) Show that this is also true for a move of type 3 (which is a vertical move towards the center).

(e) Finally show that for a horizontal move of either of the two types or a vertical move of type 4 (which is away from $(0, 5)$), the total weight $\sigma^n + \sigma^{n+1}$ gets replaced by σ^{n+2} and hence $w(S') < w(S)$. Conclude that in any one of the possibilities, the total weight $w(S)$ of a configuration, as the jump moves are applied, is a monotonically decreasing positive sequence of numbers.

(f) Show that, if a sequence of jump moves is to take us to $(0, 5)$, then the initial configuration S must have weight which is at least as large as 1.

(b) Let L denote the lower half plane. That is, $L = \{(x, y) : x, y \in \mathbb{Z}, y \leq 0\}$. Then show that $L = L_0 \cup (\cup_{k=1}^{\infty} L_k)$ where $L_0 = \{(0, y) : y \leq 0\}$ and for a positive integer k we define $L_k = \{(\pm k, y) : y \leq 0\}$.

(g) Show that $w(L_0) = \sigma^5(1-\sigma)^{-1}$.

(h) Show that $w(L_k) = 2\sigma^{k+5}(1-\sigma)^{-1}\ \forall\, k \geq 1$.

(i) Hence show that $w(L) = 1$ and conclude that $w(S) < 1$ to arrive at a contradiction.

Chapter 12

Generating functions

12.1 Introduction and examples

From "Generatingfunctionololgy" by H.S. Wilf [61]:

"Generating functions are a bridge between discrete mathematics, on one hand and continuous analysis (particularly complex variable theory) on the other. It is possible to study them solely as tools for solving discrete problems."

That is the aim of the present chapter.

"...... A generating function is a clothesline on which we hang up a sequence of numbers for display."

This chapter discusses a very important area that is at the heart of both the classical as well as applicable combinatorics. The origins of medieval time combinatorics were the mathematical questions on gambling that led to the rich branch of discrete probability as we already saw in Chapters 5 and 6 of this book. However, around the same time a gadget was being developed that could understand and also solve a large number of counting questions and that gadget is the concept of a generating function. The powerful tool of generating functions was first used by de Moivre in 1720 but was really championed by Euler in the second half of the eighteenth century. The name "generating function" was coined by Laplace in the late eighteenth century and was among highly used tools of Euler, the father figure of combinatorics. At a conceptual level, generating function is a device that embeds a combinatorial problem in the framework of algebra and most of the time also solves it. As we will see in the present chapter, *this is a paradigm shift from all the earlier methods of counting.* The paradigm shift permits use of sophisticated techniques of *algebraic combinatorics* (the branch that is now called with the same name) in the service of difficult combinatorial problems.

Suppose we have one object which we denote by x. Then the function $1+x = x^0+x^1$ enlists the two choices in front of us: x^0 means we do not choose the object x and x^1 means we choose it. How do we modify this if we have two distinct objects say x and

y but we are not allowed to choose both (at the same time). Then the function $1+x+y$ enlists all the three possibilities for us as before with 1 referring to choosing no object at all and the summands x and y referring to the choices of x and y respectively. However, if there is no restriction on choosing both x and y, then the function $(1+x)(1+y) = 1+x+y+xy$ enlists all the four possibilities. In this function y means choose y (but do not choose x) while xy means choose both x and y. We have a clear understanding here that $xy = yx$. If we now make the identification $y = x$, then we get the function $(1+x)^2 = 1+2x+x^2$. The objects are still different. How do we interpret the middle term $2x$ then? That just means that there are two ways in which only one object (out of two) is chosen while the term x^2 tells us that there is exactly one way of choosing both the objects. Extending the logic to n objects say $x_1, x_2, \ldots, x_n$, we can perceive the situation to be the following. Suppose the i-th object x_i is in the i-th box B_i. An agent is appointed to make the choice. He opens B_i and either selects x_i or rejects it. The generating function merely for the choice of x_i is $1 + x_i$. If the agent performs this act for each box, then the generating function is

$$(1+x_1)(1+x_2)\cdots(1+x_n)$$

and *if the interest is only in knowing how many objects are picked up, and not in who is picked up and who is not*, then the generating function we are looking for is obtained by the identification $x = x_1 = x_2 = \cdots = x_n$. This gives us the generating function $f(x) = (1+x)^n$. Since we already know that k objects out of n can be selected in $\binom{n}{k}$ ways (Chapter 1), we have

$$f(x) = (1+x)^n = \sum_{k=0}^{n} \binom{n}{k} x^k$$

giving an n-th proof of the binomial theorem.

In short, the notion of a generating function is too general. We can define it depending on what purpose we have in mind. Think of the situation when an object called x is available in copies, say upto ten copies of x are available. To make the example more juicy, just think of a box containing 10 identical chocolates. The term x^6 then represents the choice of a person who puts his hand in the box and grabs 6 chocolates. Thus

$$1+x+x^2+x^3+x^4+x^5+x^6+x^7+x^8+x^9+x^{10}$$

represents the function that lists all the possibilities. Suppose the box has 10 chocolates but we can choose only $2, 3, 5$ or 7 chocolates. Then the generating function is $x^2 + x^3 + x^5 + x^7$.

We begin with some simple examples.

Example 12.1.1. Fix a positive integer m and consider the generating function of the sequence (a_k) where $a_k = \binom{m}{k}$. Here $a_k = 0$ for all $k \geq m+1$ and hence the required generating function (g.f.) is $f(x)$ given by

$$\sum_{k=0}^{m} a_k x^k = \sum_{k=0}^{m} \binom{m}{k} x^k = (1+x)^m$$

Notice that the equality of the L.H.S. and the R.H.S. is the statement of the binomial theorem (Theorem 1.2.5).

Example 12.1.2. Suppose we have a box of chocolates, all identical and let the number of chocolates be infinite. Let a_k denote the number of ways in which we can pick up k chocolates. Then clearly a_k equals 1 for all k and the required g.f. is

$$1 + x + x^2 + x^3 + \cdots + x^k + \cdots = \frac{1}{1-x} = (1-x)^{-1}$$

How do we know that the L.H.S. expression equals the compact expression $\frac{1}{1-x} = (1-x)^{-1}$ on the R.H.S.? One way of answering this question is to observe that the series on the L.H.S. is a geometric series which converges to $(1-x)^{-1}$ if $|x| < 1$. However, we wish to avoid convergence questions and then a way out is the following

$$(1 + x + x^2 + \cdots)(1-x) = (1 + x + x^2 + \cdots) - (x + x^2 + \cdots)$$

is obtained by actually multiplying out and the expression on the R.H.S. equals 1.

Example 12.1.3. What is the g.f. $f(x)$ of the sequence $(0, 0, 0, 1, 1, 1, 1, \ldots)$?

$$\begin{aligned} f(x) &= x^3 + x^4 + x^5 + \cdots \\ &= (1 + x + x^2 + \cdots + x^n + \cdots) - (1 + x + x^2) \\ &= (1-x)^{-1} - (1 + x + x^2) \end{aligned}$$

Alternatively,

$$\begin{aligned} f(x) &= x^3 + x^4 + x^5 + \cdots \\ &= x^3(1 + x + x^2 + \cdots + x^n + \cdots) \\ &= x^3(1-x)^{-1} \end{aligned}$$

Check that the two expressions for $f(x)$ are (algebraically) the same.

Example 12.1.4. What is the g.f. $f(x)$ of the sequence $(1, 0, 1, 0, 1, 0, .., 1, 0, \ldots)$? We have

$$f(x) = 1 + x^2 + x^4 + \cdots = \sum_{k=0}^{\infty} x^{2k}$$

and after noting that the R.H.S. is an infinite geometric series (also called *a free monoid*) in x^2, we see that $f(x) = \frac{1}{1-x^2} = (1-x^2)^{-1}$.

Example 12.1.5. The g.f. of the sequence $(4, -3, 0, 1, 5, 2)$ (*by convention, the remaining terms of the sequence are assumed to to zero*) is the polynomial $g(x) = 4 - 3x + 0x^2 + 1x^3 + 5x^4 + 2x^5 = 4 - 3x + x^3 + 5x^4 + 2x^5$.

Example 12.1.6. The g.f. of the sequence $(1, -1, 1, -1, \ldots)$ is $f(x) = 1 - x + x^2 - x^3 + x^4 - x^5 + \cdots$ which is the geometric series $\sum_{m \geq 0}(-x)^m$ and hence equals $\frac{1}{1-(-x)} = (1+x)^{-1}$.

Example 12.1.7. As a final example, consider the sequence (a_n) where $a_n = \frac{1}{n!}$. Then the g.f. is

$$h(x) = \sum_{n\geq 0} \frac{x^n}{n!} = e^x$$

Definition 12.1.8. *A generating function (g.f.)* of a sequence (a_n) of (real or complex) numbers is the (formal) power series $f(x) = \sum_n a_n x^n$.

Note that the term *formal* here refers to not being interested in the question of convergence; in any case, every generating function clearly converges at 0 but we will not bother about that here. A formal power series thus means only an algebraic expression for us and hence there is a one-to-one correspondence

$$(a_n) \longleftrightarrow \sum_n a_n x^n$$

between a sequence and its generating function. We can also carry out *routine* algebraic operations on g.f.s. Thus if $f(x) = \sum_n a_n x^n$ and $g(x) = \sum_n b_n x^n$ are two g.f.s then their sum is the g.f. $h(x) = f(x) + g(x)$ corresponding to the sequence (c_n) where $c_n = a_n + b_n$ for all n. Also, if λ is a real or complex number, then $(\lambda f)(x)$ is the g.f. corresponding to the sequence whose n-th term is λa_n. These properties show that g.f.s form a vector space. Finally, if $f(x)$ and $g(x)$ are generating functions as above, what do we mean by the product $h(x) = f(x)g(x)$? If we actually multiply the power series on the right, we see that $h(x) = \sum_n e_n x^n$ where e_n is obtained by the convolution formula:

$$e_n = \sum_{k+m=n} a_k b_m = \sum_{k=0}^{n} a_k b_{n-k}$$

Here, each e_n is a finite sum and is therefore defined. This shows that g.f.s form *a ring and hence an algebra over the given field.* Lastly, with

$$f(x) = a_0 + a_1 + a_2 x^2 + \cdots$$

when can we find inverse (reciprocal) $t(x)$ of $f(x)$? If $t(x)$ equals

$$t_0 + t_1 + t_2 x^2 + \cdots$$

then $f(x)t(x) = 1$ forces $a_0 t_0 = 1$. Clearly then we must have $a_0 \neq 0$ and this condition also turns out to be sufficient for $f(x)$ to have an inverse (Exercise 12.1). We have already seen this in Example 12.1.2 where we noted that the inverse of the g.f. $1 + x + x^2 + \cdots$ is $1 - x$ (which is a polynomial).
We now attempt a solution of the towers of Hanoi problem (Example 11.1.3 and Theorem 11.1.4) with a recurrence relation given by $a_n = 2a_{n-1} + 1 \; \forall \; n \geq 2$ and with the initial condition $a_1 = 1$. Let

$$f(x) = a_1 x + a_2 x^2 + \cdots + a_n x^n + \cdots = \sum_{n\geq 1} a_n x^n$$

denote the g.f. of the sequence (a_n) (note the absence of the constant term in $f(x)$). We rewrite the recurrence relation as follows: $a_n - 2a_{n-1} = 1 \; \forall \; n \geq 2$ and hence (multiplying throughout by x^n and then summing over n), we get

$$\sum_{n=2}^{\infty} a_n x^n - 2\sum_{n=2}^{\infty} a_{n-1}x^n = \sum_{n=2}^{\infty} x^n$$

Since $a_1 = 1$, this gives us

$$(f(x) - x) - 2x\sum_{m=1}^{\infty} a_m x^m = x^2(1-x)^{-1}$$

which works to

$$(1-2x)f(x) = x + \frac{x^2}{1-x} = \frac{x}{1-x}$$

and hence

$$f(x) = \frac{x}{(1-2x)(1-x)}$$

Use of partial fractions gives

$$\begin{aligned} f(x) &= \frac{1}{1-2x} + \frac{1}{1-x} \\ &= \sum_{n=0}^{\infty} 2^n x^n - \sum_{n=0}^{\infty} x^n \\ &= \sum_{n=1}^{\infty} (2^n - 1)x^n \end{aligned}$$

Thus we get $a_n = 2^n - 1$ as expected.

The procedure outlined above, should, *in principle*, be able to handle *any* recurrence relation. However, obtaining explicit expression for a_n may not be very easy. In case we are dealing with linear homogeneous recurrence relation with constant coefficients, we can solve the recurrence relation (upto the ability of factoring polynomials over the field of complex numbers). We illustrate this through two examples.

Example 12.1.9. We wish to solve the recurrence relation

$$a_n = 3a_{n-1} + 10a_{n-2} \; \forall \; n \geq 2$$

subject to the initial condition : $a_0 = 4$ and $a_1 = 13$. We first write the given linear recurrence relation in the form $a_n - 3a_{n-1} - 10a_{n-2} = 0$ which is valid for all the values of $n \geq 2$. Let $f(x) = \sum_{n \geq 0} a_n x^n$ be the g.f. Now multiply by x^n and sum over all $n \geq 2$ to get

$$
\begin{aligned}
& \sum_{n\geq 2} a_n x^n - 3\sum_{n\geq 2} a_{n-1}x^n - 10\sum_{n\geq 2} a_{n-2}x^n = 0 \\
\Rightarrow & \sum_{n\geq 0}(a_n x^n - a_0 - a_1 x) - 3x\sum_{m\geq 0}(a_m x^m - a_0) - 10x^2\sum_{m\geq 0} a_m x^m = 0 \\
\Rightarrow & \quad (f(x) - 4 - 13x) - 3x(f(x) - 4) - 10x^2 f(x) = 0 \\
\Rightarrow & \quad (1 - 3x - 10x^2)f(x) = 4 + x
\end{aligned}
$$

and therefore,

$$f(x) = \frac{4+x}{1-3x-10x^2}$$

We can now apply a long division to the R.H.S.

$$
\begin{aligned}
4 + x &= 4(1-3x-10x^2) + 13x + 40x^2 \\
&= 4(1-3x-10x^2) + 13x(1-3x-10x^2) + 79x^2 + 130x^3 \\
&= 4(1-3x-10x^2) + 13x(1-3x-10x^2) + 79x^2(1-3x-10x^2) \\
&\quad +367x^3 + 790x^4 \\
&= (1-3x-10x^2)[4 + 13x + 79x^2 + 367x^3 + \cdots]
\end{aligned}
$$

showing that $f(x) = 4 + 13x + 79x^2 + 367x^3 + \cdots$. Hence we obtain a first few terms as:

$$a_0 = 4,\ a_1 = 13,\ a_2 = 79,\ a_3 = 367$$

However, to obtain a_n as a function of n, we factorize

$$1 - 3x - 10x^2 = (1-5x)(1+2x)$$

and write

$$\frac{4+x}{1-3x-10x^2} = \frac{A}{1-5x} + \frac{B}{1+2x}$$

Hence $A(1+2x) + B(1-5x) = 4 + x$. Substituting $x = \frac{1}{5}$ gives $\frac{7}{5}A = \frac{21}{5}$ and we have $A = 3$ and since $A + B = 4$ we get $B = 1$. Thus the solution to the given recurrence relation is

$$a_n = 3 \times 5^n + (-2)^n = 3 \times 5^n + (-1)^n 2^n$$

Example 12.1.10. Consider the following recurrence relation valid for all $n \geq 4$:

$$a_n = 3a_{n-1} + 9a_{n-2} - 23a_{n-3} + 12a_{n-4}$$

with the initial conditions $a_0 = a_1 = 4$, $a_2 = 29$ and $a_3 = 42$. Let $f(x) = \sum_{n\geq 0} a_n x^n$ be the g.f. Exactly as in the previous example, we get

$$(f(x) - 4 - 4x - 29x^2 - 42x^3) - 3x(f(x) - 4 - 4x - 29x^2)$$

$$-9x^2(f(x) - 4 - 4x) + 23x^3(f(x) - 4) - 12x^4 f(x) = 0$$

Therefore,

$$(1 - 3x - 9x^2 + 23x^3 - 12x^4)f(x) = 4 - 8x - 19x^2 + 11x^3$$

We have

$$1 - 3x - 9x^2 + 23x^3 - 12x^4 = (1 - x)^2(1 - 4x)(1 + 3x)$$

and hence

$$\begin{aligned} f(x) &= \frac{4 - 8x - 19x^2 + 11x^3}{(1 - x)^2(1 - 4x)(1 + 3x)} \\ &= \frac{A}{1 - x} + \frac{B}{(1 - x)^2} + \frac{C}{1 - 4x} + \frac{D}{1 + 3x} \end{aligned}$$

the second line using partial fractions. So we get

$$A(1 - x)(1 - 4x)(1 + 3x) + B(1 - 4x)(1 + 3x) + C(1 - x)^2(1 + 3x)$$
$$+D(1 - x)^2(1 - 4x) = 4 - 8x - 19x^2 + 11x^3$$

Both the sides are polynomials of degree 3 and hence we get (by equating coefficients), 4 linear equations in 4 unknowns A, B, C and D which can be uniquely solved to get the values of these constants. Alternatively, we note that this equation is an identity that must hold for *all values of* x. Letting $x = 1$ gives $-12B = -12$ and hence $B = 1$. Next let $x = \frac{1}{4}$. This gives $D \times \frac{7}{4} \times \frac{9}{16} = \frac{63}{64}$ and hence $D = 1$. Similarly letting $x = -\frac{1}{3}$ gives $C = 1$. Finally $A + B + C + D = 4$ gives $A = 1$. So we have

$$\begin{aligned} f(x) &= \frac{1}{1 - x} + \frac{1}{(1 - x)^2} + \frac{1}{1 - 4x} + \frac{1}{1 + 3x} \\ &= \sum_{n\geq 0} x^n + + \sum_{n\geq 0} \binom{n+1}{1} x^n + \sum_{n\geq 0} 4^n x^n + \sum_{n\geq 0} (-3)^n x^n \end{aligned}$$

and hence we get

$$a_n = (n + 2) + 4^n + (-3)^n = (n + 2) + 4^n + (-1)^n 3^n$$

In Chapter 10, we have already solved the problem of the number of diagonal triangulations of a convex polygon with solution in the form of a Catalan number. Here, we solve the same problem in a completely different setting of generating functions. Let then h_n denote the number of diagonal triangulations of *a convex* $(n + 1)$*-gon* P *where by convention we also let* $h_1 = 1$. Let the vertices of P be $x_0, x_1, \ldots, x_n$ (in an anticlockwise order). Fixing the side $x_n x_0$ as the base b we see that any diagonal triangulation must have a unique triangle T with base b and the third vertex some x_k. Removal of the triangle T (with vertices x_0, x_k, x_n) from P leaves us with two regions (to the left and right of T say R and S where R has vertex set $x_0, x_1, \ldots, x_k$

and S has vertex set $x_k, x_{k+1}, \ldots, x_n$ showing that R is a convex $(k+1)$-gon and S is a convex $(n-k+1)$-gon. We thus get the following recurrence relation

$$h_n = \sum_{\substack{k+m=n \\ k,m\geq 1}} h_k h_m$$

Let $h(x)$ denote the g.f. :

$$\begin{aligned} h(x) &= \sum_{n\geq 1} h_n x^n \\ &= x + x^2 + 2x^3 + \cdots \end{aligned}$$

Since we have summands on the R.H.S. that are products of the members of the sequence we find $(h(x))^2$:

$$(h(x))^2 = \sum_{n=2}^{\infty} \left\{ \sum_{\substack{k+m=n \\ k,m\geq 1}} h_k h_m \right\} x^n$$

Using the given recurrence relation, we obtain:

$$\begin{aligned} (h(x))^2 &= \sum_{n\geq 2} h_n x^n \\ &= h(x) - h_1 x \\ &= h(x) - x \end{aligned}$$

Now let $y = h(x)$ and treat the equation obtained $y^2 - y + x = 0$ as a quadratic equation in the unknown y. The two solutions of this quadratic are given by:

$$y = \frac{1 \pm \sqrt{(1-4x)}}{2}$$

Since $h(x)$ has its constant term equal to 0, we have $y(0) = h(0) = 0$ and we are therefore required to choose the negative sign in the above solution. This gives us

$$
\begin{aligned}
h(x) = y &= \frac{1}{2} - \frac{1}{2}(1-4x)^{\frac{1}{2}} \\
&= \frac{1}{2} - \frac{1}{2}\left\{\sum_{m=0}^{\infty}\binom{\frac{1}{2}}{m}](-4)^m x^m\right\} \\
&= -\frac{1}{2}\left\{\sum_{m=1}^{\infty}\binom{\frac{1}{2}}{m}(-1)^m 4^m x^m\right\} \\
&= -\frac{1}{2}\sum_{m=1}^{\infty}(-1)^m \frac{(\frac{1}{2})(\frac{1}{2}-1)\cdots(\frac{1}{2}-(m-1))}{m!}4^m x^m \\
&= x - \frac{1}{2}\sum_{m=2}^{\infty}(-1)^m \frac{(1)\times(-1)\times(-3)\cdots\times(-(2m-3))}{2^m m!}4^m x^m \\
&= x + \sum_{m=2}^{\infty}\frac{1\times 3\times 5\times\cdots\times(2m-3)}{2^{m+1}m!}\times 4^m x^m \\
&= x + \sum_{m=2}^{\infty}\frac{1\times 2\times 3\times\cdots\times(2m-3)(2m-2)}{2^{m+1}2^{m-1}m!(m-1)!}\times 4^m x^m \\
&= x + \sum_{m=2}^{\infty}\frac{(2m-2)!}{m!(m-1)!}x^m \\
&= x + \sum_{m=2}^{\infty}\frac{1}{m}\binom{2m-2}{m-1}x^m \\
&= \sum_{m=1}^{\infty}\frac{1}{m}\binom{2m-2}{m-1}x^m
\end{aligned}
$$

where we treat $\binom{0}{0} = 1$. *We have, therefore, reproved Theorem 10.2.9.*

We now wish to solve the following interesting question. Let $\alpha(n,k)$ denote *the sum of the smallest elements of all the k-subsets* A of the n-set $[n]$ (where $1 \le k \le n$). Though the problem of finding $\alpha(n,k)$ can be solved by other methods, we wish to solve it using generating functions. The idea here is to fix the natural number k and let $n \ge k$. To that end, fix k and let $b_n = \alpha(n,k)$. Thus $b_n = 0$ if $n < k$ and $b_k = 1$. Let $n \ge k$ and let $A = \{a_1, a_2, \ldots, a_k\}$ be a subset of $[n]$, where, we assume, w.l.o.g. that $1 \le a_1 < a_2 < \cdots < a_k \le n$. Construct a new sequence $(c_1, c_2, \cdots, c_k, c_{k+1})$ from A as follows:

$$c_1 = a_1,\ c_2 = a_2 - a_1,\ c_3 = a_3 - a_2, \ldots, c_k = a_k - a_{k-1},\ c_{k+1} = n - a_k$$

Observe that given $a_1, a_2, \ldots, a_k$ and n we can uniquely determine $c_1, c_2, \ldots, c_k, c_{k+1}$ and given $c_1, c_2, \ldots, c_k, c_{k+1}$, we can uniquely determine $a_1, a_2, \ldots, a_k$ and n because:

$$a_1 = c_1,\ a_2 = c_1 + c_2, \ldots, a_k = c_1 + \cdots + c_k,\ n = c_1 + \cdots + c_k + c_{k+1}$$

This one-to-one correspondence allows us to shift our attention to the new sequence $(c_1, c_2, \ldots, c_k, c_{k+1})$. The interest here is really in two things: c_1 and $c_1 + c_2 + \cdots + c_k + c_{k+1}$ because the former gives us the smallest element $a_1 = c_1$ while the latter is the sum equal to n. Let the g.f. for c_i be f_i for $i = 1, 2, \ldots, k+1$. Note the constraints. Each c_i is positive if $i = 1, 2, \ldots, k$ while c_{k+1} is non-negative. So,

$$f_i(x) = x^1 + \cdots + x^m + \cdots = x(1-x)^{-1}$$

for all $i = 1, 2, \ldots, k$ while

$$f_{k+1}(x) = x^0 + x^1 + \cdots + x^m + \cdots = (1-x)^{-1}$$

We can multiply to get $f(x) = f_1(x)f_2(x)\cdots f_k(x)f_{k+1}(x)$ and find the coefficient of x^n in $f(x)$. However, that does not solve the problem. What does it give us then? It just gives us the total number of k-subsets of the n-set $[n]$. What do we need to do? We are quite alright with all the f_i except the g.f. for c_1 which should count c_1 *not just in the exponent but also as coefficient (thus what we are looking for is not x^{c_1} but $c_1x^{c_1}$).* This is done by looking at $g_1(x) = x{f_1}'(x)$. Since $f_1(x) = x(1-x)^{-1}$, we have

$$g_1(x) = x[(1-x)^{-1} + (1-x)^{-2}] = x(1-x)^{-2}$$

Thus the correct g.f.to look for is

$$\begin{aligned} g(x) &= g_1(x)f_2(x)\cdots f_k(x)f_{k+1}(x) \\ &= x(1-x)^{-2}[x(1-x)^{-1}]^{k-1}(1-x)^{-1} \\ &= x^k(1-x)^{-(k+2)} \end{aligned}$$

Since the coefficient of x^n in $g(x)$ is the coefficient of x^{n-k} in $(1-x)^{-(k+2)}$. and

$$(1-x)^{-(k+2)} = \sum_{j=0}^{\infty} \binom{k+j+1}{j} x^j$$

The required number b_n is

$$\binom{k+(n-k)+1}{n-k} = \binom{n+1}{k+1}$$

Having used g.f.s to solve the problem, we now give a different solution that uses a remarkable amount of ingenuity; I am indebted to R.C.Cowsik for this solution. Given a subset $A = \{a_1, a_2, \ldots, a_k\}$ of $[n]$ with $1 \leq a_1 < a_2 < \cdots < a_k \leq n$, we construct subsets B of $\{0, 1, 2, \ldots, n\}$ where $B = \{a_0, a_1, a_2, \ldots a_k\}$. Here a_0 is any number less than a_1 and hence B can be chosen in a_1 ways. Evidently, A can be uniquely determined from B by deleting its smallest element. Since the interest is in summing up all a_1's we can do that by finding the total number of B's in $\{0, 1, 2, \ldots, n\}$. The required number b_n is just the number of $(k+1)$-subsets of the $(n+1)$-set $\{0, 1, 2, \ldots, n\}$ which is $\binom{n+1}{k+1}$. We have thus proved the following.

Theorem 12.1.11. *For a k-subset A of $[n]$, let* $\min A$ *denote the smallest element of A. Then*

$$\sum_A \min A = \binom{n+1}{k+1}$$

12.2 Money exchange problem

The famous problem we now discuss is the *money exchange problem.* We initially discuss this problem with coins of only two denominations say a rupees and b rupees where a and b are natural numbers. These coins are available in unlimited supply. We wish to find as many numbers n that can be obtained as non-negative linear combinations of a and b. Assume w.l.o.g that $a \leq b$. Take the specific case $a = 5$ and $b = 7$. We cannot evidently construct non-negative integer linear combinations that add to $8, 9, 11$ or 13 but can certainly cover numbers 17 and 19 since $17 = (2 \times 5) + 7$ and $19 = (1 \times 5) + (2 \times 7)$. *Call non-negative number n good if $n = xa + yb$ where $x \geq 0$ and $y \geq 0$.* Thus (for the given pair $(a, b) = (5, 7)$), the numbers 17 and 19 are good. Also, we have $24 = (2 \times 5) + (2 \times 7),\ \ 25 = 5 \times 5,\ \ 26 = 5 + (3 \times 7),\ \ 27 = (4 \times 5) + 7,\ \ 28 = 4 \times 7$. We can now argue that for every number $n \geq 29$, we can subtract a suitable multiple of 5 to obtain a number m that is one of the numbers $24, 25, 26, 27, 28$ and hence every number $n \geq 24$ is good. On the other hand 23 is bad (not good) as can be checked easily. *The kind of result we are seeking here is the existence of a number N such that n is good for all $n \geq N$.* First note that if a and b are not coprime, then their g.c.d. g is larger than 1. Then any integer combination of a and b must be a multiple of g which we do not desire to have. For example, if $a = 6$ and $b = 10$, then any integer linear combination of a and b can only be even and thus no odd number can be good. *We must thus assume that a and b are coprime.*

Definition 12.2.1. Let a and b be two coprime natural numbers such that both $a, b \geq 2$. Let n be a natural number such that $n = xa + yb$ where $x, y \geq 0$. Then n is called *good.* And n is called *bad* if it is not good.

Theorem 12.2.2. *Let a, b be two coprime natural numbers. Let $N = N(a, b) = (a - 1)(b - 1)$. Then the following assertions hold.*

(a) n is good $\forall\ n \geq N$.

(b) $N - 1$ is bad.

(c) Let $S = \{0, 1, \ldots, N - 1\}$ be the set of N numbers. Let $n \in S$ and let $n + n' = N - 1$. Then n is good iff n' is bad. Hence exactly half of the elements of S are good (respectively bad).

Proof From Exercise 8.16, we know that every non-negative integer n can be expressed in the form $n = xa + yb$, though the pair (x, y) is not unique. However, we can write n in the form $xa + yb$ where $0 \leq x \leq b - 1$ by suitably adding and subtracting multiples of ab. An expression for n of the form $xa + yb$ will be called a *valid expression*, when $0 \leq x \leq b - 1$. Not only every n has a valid expression but it is also unique. For, if $x'a + y'b$ is yet another valid expression, we may assume w.l.o.g. that $0 \leq x < x' \leq b - 1$. But then $xa + yb = x'a + y'b$ implies $(y - y')b = (x - x')a$ and since the R.H.S. is positive and a, b are coprime, we must have b dividing $x - x'$ which is a contradiction. Thus $x' = x$ and hence $y' = y$. Now let $n \geq N$ and let $n = xa + yb$ be a valid expression for n. If $y < 0$, then $y \leq -1$ and $n = xa + yb \leq (b - 1)a - b = ab - a - b = N - 1$ and hence $n \leq N - 1$, a contradiction. So $y \geq 0$ and the valid expression $xa + yb$ for n is also a good expression and hence n is good. Therefore, for all $n \geq N$, n is good proving (a). Consider $N-1 = ab - a - b = (b-1)a - b$ which is certainly a valid expression. If $N - 1 = xa + yb$ is a good expression for n, then $x \geq b$ will imply (since $y \geq 0$) that $N - 1 \geq ba$, a contradiction. So, $0 \leq x \leq b-1$ showing that $xa + yb$ is also a valid expression for n. Uniqueness of a valid expression then forces $x = b - 1$ and $y = -1$ showing that we cannot have a good expression for $N - 1$. This proves (b). Let $n \in S$ and let $n' := (N-1) - n$. Let $n = xa + yb$ be a valid expression for n. Let $x' = (b-1) - x$ and $y' = -1 - y$. Observe that $n = xa + yb$ iff $n' = x'a + y'b$ and the pair (x, y) determines the pair (x', y') and conversely. Also if n has a good expression, then it must be $n = xa + yb$ (since $n \leq N - 1 < ab$). It now follows that n is good iff n' is bad proving (c). □

Now we turn to the general problem. We are given coins of r different denominations say $a_1, a_2, \ldots, a_r$ (where a_i's are natural numbers). We have an unlimited supply of coins of each denomination. Given a non-negative integer n, we ask the question, is it possible to have $n = x_1a_1 + x_2a_2 + \cdots + x_ra_r$ with each $x_i \geq 0$ (and thus we have x_i coins of the i-th type so that the sum n is obtained)? Stated in this manner, it is a yes/no type of question. We are obviously looking for an assertion similar to the one proved in Theorem 12.2.1. Thus we wish to prove that there is an N such that $\forall\, n \geq N$ we can express n as a non-negative integer linear combination of $a_1, a_2, \ldots, a_r$. A necessary condition for this to happen is clearly that the g.c.d of $a_1, a_2, \ldots, a_r$ is 1. For example, if each a_i is an even number, then no odd number can be written as an integer linear combination of the a_i's. It turns out that the g.c.d. of a_i' is equal to 1 is also a sufficient condition.

Theorem 12.2.3. *(Schur's Theorem) Let $(a_1, a_2, \ldots, a_r)$ be an r-tuple of natural numbers with the property that $gcd(a_1, a_2, \ldots, a_r) = 1$. Then there is a natural number $N = N(a_1, a_2, \ldots, a_r)$ such that $\forall\, n \geq N$, we have $n = x_1a_1 + x_2a_2 + \cdots + x_ra_r$ and $x_i \geq 0$ for all $i = 1, 2, \ldots, r$.*

Proof Let h_n denote the number of ways in which n can be written as a sum $n = x_1a_1 + x_2a_2 + \cdots + x_ra_r$ and $x_i \geq 0$ for all $i = 1, 2 \ldots, r$. We wish to prove that h_n is positive for every $n \geq N$ (where N depends on $(a_1, a_2, \ldots, a_r)$). Let

$$H(x) = \sum_{n \geq 1} h_n x^n$$

denote the g.f. of the non-negative numbers h_n. Suppose first that the coins of only one type say a_i were available. Then we could only cover every n that is a multiple of a_i and that too only once. Hence the g.f. for the i-th type of coin is

$$\begin{aligned} H_i(x) &= x^0 + x^{a_i} + x^{2a_i} + \cdots + x^{ma_i} + \cdots \\ &= \sum_{j=0}^{\infty} (x^{a_i})^j \\ &= (1 - x^{a_i})^{-1} \end{aligned}$$

Hence we have

$$H(x) = \prod_{i=1}^{r} H_i(x) = \prod_{i=1}^{r} (1 - x^{a_i})^{-1}$$

What are the zeros of the polynomial $P(x) = \prod_{i=1}^{r} P_i(x) = \prod_{i=1}^{r}(1 - x^{a_i})$ in the denominator of $H(x)$? These must be ω where, for some $j = 1, 2, \ldots r$, we have $\omega^{a_j} = 1$. Since the zeros of $P(x)$ are roots of unity, they all lie on the unit circle and any zero of $P(x)$ has order m where m divides a_j for some j. Clearly, the factor $(1 - x)$ occurs in each term $(1 - x^{a_i})$ and hence the multiplicity of $1 - x$ in $P(x)$ is r. Since the a_i roots of the polynomial $1 - x^{a_i}$ are all distinct a_i-th roots of 1, we see that if there is a root ω that has multiplicity r in $P(x)$, then it must be a root of each one of the polynomials $1 - x^{a_i}$. If ω has order m then m must divide each one of $a_1, a_2, \ldots, a_r$ and hence m must divide the g.c.d. of these numbers which is equal to 1 forcing $m = 1$ and hence $\omega = 1$. *Thus every zero of $P(x)$ other than 1 has multiplicity strictly less than r.* We can now make a partial fraction expansion of the R.H.S. giving $H(x)$ to obtain

$$\frac{1}{(1 - x^{a_1})(1 - x^{a_2}) \cdots (1 - x^{a_r})} = U + V$$

where

$$U = \frac{A_r}{(1-x)^r} + \frac{A_{r-1}}{(1-x)^{r-1}} + \cdots + \frac{A_1}{1-x}$$

are the partial fractions corresponding to the root 1 of $P(x)$ and V is a sum of terms of the form

$$\frac{B}{(1 - \frac{x}{\omega})^s}$$

where ω is a zero of $P(x)$ (and hence also a root of unity) other than 1 and from what we just proved the exponent s must be strictly less than r. In the formula

$$\frac{1}{(1-y)^k} = \sum_{n=0}^{\infty} \binom{k+n-1}{k-1} y^n$$

we see that the n-th term equals

$$\frac{1}{(k-1)!}(n + k - 1) \cdots (n - 1)$$

which has the form (as a polynomial in n with a fixed k) $Dn^{k-1}+$ terms whose degrees are strictly less than $k - 1$ (here D is a constant). Hence each polynomial involved in V has degree strictly less than $r - 1$ and since the n-th term of $(1 - x)^{-r}$ has degree $r - 1$ as polynomial in n, we now see that h_n equals a polynomial in n whose degree is at the most $r - 1$. If this polynomial has degree strictly less than $r - 1$, then, in U, we must have $A_r = 0$. However, multiplying both the sides by $(1 - x)^r$ and then substituting $x = 1$ gives the L.H.S. equal to $\frac{1}{a_1 a_2 \cdots a_r}$ while the R.H.S. equals A_r and hence $A_r \neq 0$. Therefore h_n is a polynomial whose degree is $r - 1$. If the leading coefficient in this polynomial was negative then for all very large values of n, we would have $h_n < 0$ which is absurd: $h_n \geq 1$ at least for every n which is a multiple of $a_1 + a_2 + \cdots + a_r$. Hence H_n is a degree $r-1$ polynomial in n and with a positive leading term. Therefore, there is N large that $\forall\, n \geq N$, we must have $h_n > 0$ completing the proof. □

Remark 12.2.4. We conclude by noting that Schur's Theorem is only a beginning of a larger game. The number $N = N(a_1, a_2, \ldots, a_r)$ such that h_n is positive for every $n \geq N$ has been precisely determined in Theorem 12.2.2 when $r = 2$ but no closed formula for N seems to be known for $r \geq 3$. Also the asymptotic growth of h_n as well as determining the number of ways of forming non-negative integer linear combinations of $\{a_1, a_2, \ldots, a_r\}$ when a_i's are available in limited supply (the postage stamp problem) are related questions.

12.3 The idea of an exponential generating function

While defining the generating function of a sequence (a_n) as $\sum_{n=0}^{\infty} a_n x^n$ we declared that we do not care what the sequence (a_n) is and in particular, bypass the question of convergence of the series $\sum_{n=0}^{\infty} a_n x^n$. If x is a complex number, then the expression of the series makes sense (as a complex number) if $x = 0$. This is asking for the radius of convergence of the series so that we can write the given series as a concise function of the variable x. That is necessary when we wish to carry out algebraic operations on such power series and express the result as a function (in a closed formula). For example, when $\{a_n\}$ is the sequence given by $a_n = n!$, we cannot really give a concise form to $\sum_{n=0}^{\infty} n! x^n$ (since the radius of convergence is zero). Roughly, the notion of an exponential generating function comes into being precisely to handle such situations (of large terms of the given sequence of numbers.)

Definition 12.3.1. Let (a_n) be a sequence of (complex) numbers. Then the exponential generating function *(exponential* g.f. or *e.g.f.* for short) is the function $g(x) = \sum_{n=0}^{\infty} \frac{a_n x^n}{n!}$.

Note that the g.f.s we dealt with in Section 12.1 will now be called *ordinary g.f.s* (in contrast to the e.g.f.s that we look at in this section). Here are some examples. In all these examples the given sequence is denoted by (a_n).

Example 12.3.2.
$a_0 = 3, a_1 = 4, a_2 = -5, a_3 = 0, a_4 = 2$ and $a_n = 0\ \forall\, n \geq 5$. Then the exponential g.f. is

$$\begin{aligned} f(x) &= a_0 + a_1 x + \frac{a_2}{2!}x^2 + \frac{a_3}{3!}x^3 + \frac{a_4}{4!}x^4 + 0 + \cdots \\ &= 3 + x + \frac{-5}{2}x^2 + 0 + \frac{2}{24}x^4 \\ &= 3 + x - \frac{5}{2}x^2 + \frac{1}{12}x^4 \end{aligned}$$

Note that the e.g.f. is a polynomial in this case.

Example 12.3.3. Fix a positive integer m and let $a_k = [m]_k$ be the number of k-permutations of an m-set. Then $a_k = [m]_k = \frac{m(m-1)\cdots(m-k+1)}{k!}$ (and in particular, $a_k = 0 \,\forall\, k \geq m+1$). So, the e.g.f. is

$$\begin{aligned} f(x) &= \sum_{k=0}^{\infty} \frac{[m]_k}{k!} x^k \\ &= \sum_{k=0}^{m} \frac{[m]_k}{k!} x^k \\ &= \sum_{k=0}^{m} \binom{m}{k} x^k \\ &= (1+x)^m \end{aligned}$$

As we have already seen, $(1+x)^m$ is also an ordinary generating function (o.g.f.) of the sequence (b_k) where $b_k = \binom{m}{k}$.

Example 12.3.4. Let $a_n = 1 \,\forall\, n \geq 0$. Then the e.g.f. is

$$\begin{aligned} f(x) &= \sum_{n=0}^{\infty} a_n \frac{x^n}{n!} \\ &= 1 + x + \frac{x^2}{2!} + \frac{x^3}{3!} + \cdots + \frac{x^n}{n!} + \cdots \\ &= \sum_{n=0}^{\infty} \frac{x^n}{n!} \\ &= e^x \end{aligned}$$

and we now see the point in calling this new function, the exponential generating function.

Example 12.3.5. Let $a_0 = a_1 = a_2 = 0$ and $a_n = 1 \,\forall\, n \geq 3$. Then the e.g.f. is

$$\sum_{n=3}^{\infty} \frac{x^n}{n!} = e^x - \left(1 + x + \frac{x^2}{2}\right)$$

Example 12.3.6. Let $a_n = (-1)^n$. Then the e.g.f.is

$$f(x) = \sum_{n \geq 0} \frac{(-1)^n}{n!} x^n = \sum_{n \geq 0} \frac{(-x)^n}{n!} = e^{-x}$$

Given a sequence (a_n) when should one use an ordinary generating function (g.f. or o.g.f.) and when should one use an exponential generating function (e.g.f.)? Answer to this question, of course, depends on what we want to do (*generating functions are used to solve problems*). A general idea is given by the examples we dealt with earlier. Observe that, for a fixed m, the function $(1+x)^m$ is o.g.f. of the sequence (a_k) with

$a_k = \binom{m}{k}$ (k-combinations of an m-set) as well as e.g.f. of the sequence (b_k) with $b_k = [m]_k$ (k-permutations of an m-set). Hence *as a thumb rule, one should use o.g.f. if the given problem involves unordered sets while one should use e.g.f. if the given problem involves questions of order.* This is illustrated in the following example.

Example 12.3.7. We wish to colour the n cells of an $n \times 1$ chessboard (or a paper strip) in three colours red, blue and green. Let a_n denote the number of ways of doing this. Let $F(x)$ be the e.g.f of a_n's; *we should use an e.g.f. since the question involves order, the cells of the chessboard have positions* (it is not merely a question of how many cells are red but also where the red cells are placed). Imagine for a moment that we had only the red colour available and let $F_R(x)$ denote the corresponding e.g.f. Now no matter what n is, there is only one way of coloring (since only one color is available). Thus $F_R(x)$ is the e.g.f. of the constant sequence all of whose terms are equal to 1 and hence $F_R(x) = \sum_{m=0}^{\infty} 1 \times \frac{x^n}{n!} = e^x$. If we now denote the e.g.f.'s with only the color blue (respectively green) available by $F_B(x)$ and $F_G(x)$ respectively, then for the same reason, the e.g.f.s are: $F_B(x) = F_G(x) = e^x$. Now comes the punch in the argument. Since we are using e.g.f.'s if we multiply the three e.g.f.s, the coefficient of $\frac{x^n}{n!}$ in the product will give us the number of ways of coloring an $n \times 1$ chessboard in three colors. Check this; the choices of the cells for three different colour are given by the product, which, because of the convolution formula has the form $\binom{n}{r_1, r_2, r_3}$ where r_i's are non-negative integers that add to n. This shows that

$$F(x) = F_R(x)F_B(x)F_G(x) = (e^x)^3 = e^{3x} = \sum_{n=0}^{\infty} \frac{(3x)^n}{n!} = \sum_{n=0}^{\infty} 3^n \frac{x^n}{n!}$$

showing that $a_n = 3^n$. *That is not a big deal; the same conclusion can be arrived at using simpler arguments.* First the obvious recurrence relation: $a_n = 3a_{n-1}$ (with $a_1 = 3$) which could be solved by the standard theory of recurrence relations, or iteration or use of g.f. to get $a_n = 3^n$. Easier than that is the direct approach as follows. Each cell has three choice for getting colored and since these are independent choices we see that $a_n = 3 \times 3 \times \cdots \times 3 = 3^n$.

Consider the following question. We still have 3 colors to color the cells of an $n \times 1$ chessboard but we wish to color the board in such a way that an even number of cells are colored red and we must also have at least two blue cells. The 'free choice argument' we made earlier does not hold any more. If I start coloring the cells freely, how do I know that I end up with an even number of red cells (and not an odd number) and also how do I know that I have ensured that there are at least two blue cells? E.g.f.s do the job. Let the e.g.f.s for the red, blue and green colors be denoted by $f_R(x)$, $f_B(x)$ and $f_G(x)$ respectively. $f_R(x)$ corresponds to the sequence whose n-th term is 1 if n is even and is zero if n is odd. Then

$$f_R(x) = \sum_{n \geq 0; n \ even} \frac{x^n}{n!} = \frac{e^x + e^{-x}}{2}$$

and

$$f_B(x) = \sum_{n \geq 2} \frac{x^n}{n!} = e^x - 1 - x$$

Hence the required e.g.f. is

$$f(x) = \left(\frac{e^x + e^{-x}}{2}\right) \times (e^x - 1 - x) \times e^x = \frac{1}{2}\left[e^{3x} + e^x - e^{2x} - 1 - xe^{2x} - x\right]$$

Thus $a_n = 0$ if $n = 0, 1$ and for all $n \geq 2$ we have

$$a_n = \frac{3^n - 2^n - n2^{n-1} + 1}{2} = \frac{3^n - (n+2)2^{n-1} + 1}{2}$$

Just to double-check that we are doing okay, consider a_5. Since we must have $j \geq 2$ cells that are blue, there are four cases depending on what j is. When $j = 2$ we either have no red cells or exactly two red cells where the former can be done in only one way while the latter can be done in $\binom{3}{2} = 3$ ways giving a total of $\binom{5}{2}(1 + \binom{3}{2}) = 40$ possibilities. Next consider 3 blue cells. Then we either have no or two red cells and this gives us a total of $\binom{5}{3} \times 2 = 20$ possibilities. If we have 4 or 5 blue cells then we cannot have a red cell at all. Hence the number of possibilities is $\binom{5}{4} + 1 = 6$. That gives us $a_5 = 40 + 20 + 6 = 66$ which can be obtained from the general formula we have. *The strength here, is, of course that the formula works for every n and we do not have to carry out such messy computations separately for every n.*

12.4 E.g.f. of the sequence of Bell numbers

We give five proofs of the following well-known result that computes the exponential generating function of the Bell numbers B_n. Recall that (10.1.11 in Chapter 10) B_n is the total number of partitions of an n-set where the parts are unordered and by convention we have $B_0 = 1$.

Theorem 12.4.1. *The exponential generating function of the sequence of Bell numbers is given by the following closed formula.*

$$\sum_{n=0}^{\infty} B_n \frac{x^n}{n!} = e^{e^x - 1}$$

In order to give the first proof (due to Rota), which is sophisticated and not very easy, we need the following lemma, which is also of an independent interest. Recall that $[t]_n$ denotes the falling factorial $t(t-1)\cdots(t-n+1)$. If a power series is given in the form

$$f(t) = \sum_{n=0}^{\infty} a_n [t]_n$$

where a_n's are real numbers, then we define the linear operator L by $L(f(t)) = \sum_{n=0}^{\infty} a_n$ if the sum on the R.H.S. exists (that is the series with real terms is convergent). It is easily seen that L is a linear operator (whenever it is defined).

Lemma 12.4.2. *Let $g(t) = \sum_{n=0}^{\infty} b_n t^n$. Then $L(g(t)) = \sum_{n=0}^{\infty} b_n B_n$.*

Proof *(of the Lemma)* Fix a positive integer m and let $s_m = \sum_{n=0}^{m} b_n t^n$. Then we have:

$$s_m = \sum_{n=0}^{m}\left\{\sum_{k=0}^{n} S(n,k)[t]_k\right\} b_n = \sum_{k=0}^{m}\left(\sum_{n=0}^{m} S(n,k)b_n\right)[t]_k$$

Hence,

$$\begin{aligned} L(s_m) &= \sum_{k=0}^{m}\sum_{n=0}^{m} S(n,k)b_n \\ &= \sum_{n=0}^{m}\left\{\sum_{k=0}^{n} S(n,k)\right\} b_n \\ &= \sum_{n=0}^{m} b_n B_n \end{aligned}$$

Taking the limit as $m \to \infty$ we see that $L(g(t)) = \sum_{n=0}^{\infty} b_n B_n$ as desired. □

First Proof of Theorem 12.4.1 Let $e^x - 1 = y$. The proof evaluates $L(e^{tx})$ (as a function of t) in two ways and makes use of Lemma 12.4.2. We have $e^{tx} = \sum_{n=0}^{\infty} \frac{x^n}{n!} t^n$ and using this function as $g(t)$ in the Lemma gives $L(e^{tx}) = \sum_{n=0}^{\infty} B_n \frac{x^n}{n!}$ which is the e.g.f. we are looking for. Also, $e^x = 1 + y$ and hence

$$\begin{aligned} e^{tx} &= (1+y)^t \\ &= \sum_{n=0}^{\infty}\binom{t}{n} y^n \\ &= \sum_{n=0}^{\infty}\frac{y^n}{n!}[t]_n \end{aligned}$$

Therefore,

$$\begin{aligned} L(e^{tx}) &= \sum_{n=0}^{\infty}\frac{y^n}{n!} \\ &= e^y \\ &= e^{e^x - 1} \end{aligned}$$

The assertion is proved by equating the two expressions for $L(e^{tx})$. □

Second Proof $e^x - 1$ is the e.g.f. when we have a non-empty set of order n divided (trivially) into only one part. Similarly, $\frac{(e^x-1)^2}{2!}$ is e.g.f. of the situation when a non-empty set of order n (this is given by the coefficient of $\frac{x^n}{n!}$ in the e.g.f.) is divided into two non-empty subsets. Likewise, for a fixed k (with k positive), the e.g.f. $(e^x - 1)^k$ is the e.g.f. of the situation when n objects are put in k boxes with no box empty. Dividing this by $k!$ makes the boxes unordered. This is just the Stirling number $S(n,k)$.

Therefore,

$$\frac{(e^x-1)^k}{k!} = \sum_n S(n,k)\frac{x^n}{n!}$$

Now summing over all k, we get

$$\begin{aligned}\sum_k \sum_n S(n,k)\frac{x^n}{n!} &= \sum_n \left\{\sum_k S(n,k)\right\}\frac{x^n}{n!} \\ &= \sum_{n=1}^{\infty} B_n \frac{x^n}{n!}\end{aligned}$$

On the other hand,

$$\sum_{k=0}^{\infty} \frac{(e^x-1)^k}{k!} = e^{e^x-1}$$

So,

$$e^{e^x-1} = \sum_{n=0}^{\infty} B_n \frac{x^n}{n!}$$

□

Third Proof This proof is very similar to the second proof. Let $n \geq 1$. Then Exercise 10.15 tells us that

$$B_n = \sum_k \frac{1}{k!} \sum_{\substack{j_1+j_2+\cdots+j_k=n \\ j_i \geq 1 \ \forall\, i}} \binom{n}{j_1, j_2, \ldots, j_k}$$

This implies that

$$\sum_{n\geq 1} B_n \frac{x^n}{n!} = \sum_{k\geq 1} \frac{1}{k!} \sum_{\substack{j_1+j_2+\cdots+j_k=n \\ j_i \geq 1 \ \forall\, i}} \prod_{i=1}^{k} \frac{x^{j_i}}{j_i!}$$

Since the expression under the product is just $e^x - 1$ we get

$$\sum_{n\geq 1} B_n \frac{x^n}{n!} = \sum_{k\geq 1} \frac{(e^x-1)^k}{k!}$$

and therefore we have

$$\sum_{n\geq 0} B_n \frac{x^n}{n!} = \sum_{k\geq 0} \frac{(e^x-1)^k}{k!} = e^{e^x-1}$$

Fourth proof We use the identity

$$B_n = \sum_{k=0}^{n-1} \binom{n-1}{k} B_k$$

proved in Chapter 10 (Lemma 10.1.12). Let $f(x) = \sum_{n\geq 0} B_n \frac{x^n}{n!}$. Then

$$\begin{aligned}
f'(x) &= \sum_{n\geq 1} B_n \frac{x^{n-1}}{(n-1)!} \\
&= \sum_{n\geq 1} \left\{ \sum_{k=0}^{n-1} \binom{n-1}{k} B_k \right\} \frac{x^{n-1}}{(n-1)!} \\
&= \sum_{n\geq 1} \sum_{k=0}^{n-1} B_k \frac{x^k}{k!} \frac{x^{n-1-k}}{(n-1-k)!} \\
&= \left\{ \sum_{k=0}^{\infty} B_k \frac{x^k}{k!} \right\} \times \left\{ \sum_{m=0}^{\infty} \frac{x^m}{m!} \right\} \\
&= f(x)e^x
\end{aligned}$$

Now let $y = f(x)$. Then we have $\frac{dy}{dx} = ye^x$ and a solution to this differential equation is given by $\log y = e^x + c$ i.e., $y = e^{e^x + c}$. But $y(0) = f(0) = 1$ and hence we must have $c = -1$. Therefore $f(x) = e^{e^x - 1}$. □

Fifth proof In Chapter 10 (Theorem 10.1.14), we have shown that

$$B_n = \frac{1}{e} \sum_{k=0}^{\infty} \frac{k^n}{k!}$$

Hence

$$\begin{aligned}
\sum_{n=}^{\infty} B_n \frac{x^n}{n!} &= e^{-1} \sum_{n=0}^{\infty} \sum_{k=0}^{\infty} \frac{1}{n!} \frac{k^n x^n}{k!} \\
&= e^{-1} \sum_{k=0}^{\infty} \left\{ \sum_{n=0}^{\infty} \frac{(kx)^n}{n!} \right\} \frac{1}{k!} \\
&= e^{-1} \sum_{k=0}^{\infty} \frac{e^{kx}}{k!} \\
&= e^{-1} \sum_{k=0}^{\infty} \frac{(e^x)^k}{k!} \\
&= e^{-1} e^{e^x} \\
&= e^{e^x - 1}
\end{aligned}$$

□

12.5 Bernoulli numbers

In this section, study an interesting sequence of real numbers called the *Bernoulli numbers.* We will use them to find a formula for the sum

$$\sum_{j=0}^{n} j^m$$

as a closed function of n and show, in particular that this sum is a polynomial of degree $m+1$ in the variable n. We regret the use of the same notation B_n for Bernoulli numbers (this has been used in Section 12.4 as well as in the earlier part of the book to denote Bell numbers); this use is necessitated by the non-availability of a better symbol.

Definition 12.5.1. *Bernoulli numbers* B_n are defined by their exponential generating function:

$$\frac{z}{e^z - 1} = \sum_{n \geq 0} B_n \frac{z^n}{n!}$$

One way of seeing that this makes sense is to see that $\frac{e^z-1}{z}$ has constant term equal to 1 and the e.g.f. we are looking at just the reciprocal of that e.g.f. If we wish to use some analysis then we can apply L'Hospital's rule (at 0) to see that $\frac{z}{e^z-1}$ has a Maclaurin series which is given on the right hand side. The first two values are $B_0 = 1$ and $B_1 = -\frac{1}{2}$. To find the other values of B_n cross multiply to get the equation

$$\left(\sum_{j=0}^{\infty} B_j \frac{z^j}{j!}\right) \times \left(\sum_{k=0}^{\infty} \frac{z^k}{(k+1)!}\right) = 1$$

and hence $B_0 = 1$ and $\forall\, m \geq 0$ we have

$$\sum_{j=0}^{m} \binom{m+1}{j} B_j = 0 \tag{12.1}$$

This gives $\binom{2}{0}B_0 + \binom{2}{1}B_1 = 0$ and hence $B_1 = -\frac{1}{2}$. Also,

$$\binom{3}{0}B_0 + \binom{3}{1}B_1 + \binom{3}{2}B_2 = 0 \Rightarrow B_2 = \frac{1}{6}$$

Equation (12.1) can in fact be used to find B_m using the values $B_0, B_1, \ldots, B_{m-1}$ that are already obtained. Here is a small table of Bernoulli numbers.

n	0	1	2	3	4	5	6	7	8	9	10	11	12
B_n	1	$-\frac{1}{2}$	$\frac{1}{6}$	0	$-\frac{1}{30}$	0	$\frac{1}{42}$	0	$-\frac{1}{30}$	0	$\frac{5}{56}$	0	$-\frac{691}{2730}$

A look at the table indicates that B_n is zero if n is an odd number other than 1. Before we prove that in Theorem 12.5.2, here are some warm ups from trigonometry and hyperbolic functions. The trigonometric functions $\cos z$ and $\sin z$ are used to construct $\tan z = \frac{\sin z}{\cos z}$ and $\cot z = \frac{\cos z}{\sin z}$. Since sine is an odd function and cosine is an even function, it is clear that both $\tan z$ and $\cot z$ are odd functions. The corresponding hyperbolic functions are more easily described in terms of the exponential:

$$\cosh z = \frac{e^z + e^{-z}}{2} \quad \sinh z = \frac{e^z - e^{-z}}{2}$$

Expressed in this form, we also have

$$\cos z = \frac{e^{iz} + e^{-iz}}{2} \quad \sin z = \frac{e^{iz} - e^{-iz}}{2i}$$

These formulas readily give us: $\cos z = \cosh iz$ and $\sin z = -i \sinh z$. We also see that $\cosh z$ is even and $\sinh z$ is odd and hence the ratios $\tanh z = \frac{\sinh z}{\cosh z}$ and $\coth z = \frac{\cosh z}{\sinh z}$ are both odd functions. We also have $\cot z = i \coth iz$ and hence $z\cot z = iz\coth iz$.

Theorem 12.5.2. *The following assertions hold.*

(a) $\frac{z}{e^z-1} + \frac{z}{2} = \frac{z}{2}\coth \frac{z}{2}$

(b) $B_{2m+1} = 0 \; \forall \; m \geq 1$

(c) $z\coth z = \sum_{n\geq 0} 4^n B_{2n} \frac{z^{2n}}{(2n)!}$

(d) $z\cot z = \sum_{n\geq 0} (-4)^n B_{2n} \frac{z^{2n}}{(2n)!}$

Proof Consider (a) where the L.H.S. equals

$$\begin{aligned} \frac{z}{2}\left\{\frac{2}{e^z-1} + 1\right\} &= \frac{z}{2} \times \frac{e^z+1}{e^z-1} \\ &= \frac{z}{2} \times \frac{e^{\frac{z}{2}} + e^{\frac{-z}{2}}}{e^{\frac{z}{2}} - e^{\frac{-z}{2}}} \\ &= \frac{z}{2} \coth \frac{z}{2} \end{aligned}$$

The R.H.S. in (a) is an even function and therefore so is the L.H.S. This proves (b). From (a) again,

$$\frac{z}{2} \coth \frac{z}{2} = \sum_{n\geq 0} B_{2n} \frac{z^{2n}}{(2n)!}$$

Replacing $\frac{z}{2}$ by z proves (c). Finally (d) follows from (c) when z is replaced by iz. □

Definition 12.5.3. Let m be a non-negative and n a positive integers. By $S_m(n)$ we mean the sum:

$$S_m(n) = \sum_{k=0}^{n-1} k^m$$

By convention, $0^0 = 1$ so that $S_0(n) = n$. Thus $S_1(n)$ is the sum of the first $n-1$ natural numbers which is $\frac{n^2-n}{2}$. The following theorem computes the e.g.f. of the numbers $S_m(n)$ in terms of the Bernoulli numbers and hence furnishes an answer to the question of finding the sum $S_m(n)$.

Theorem 12.5.4. *The following identities hold.*

$$S_m(n) = m!\left\{\sum_{j+k=m} \frac{B_j}{j!} \times \frac{n^{k+1}}{(k+1)!}\right\} \tag{12.2}$$

$$S_{m-1}(n) = \frac{1}{m}\sum_{j=0}^{m-1} \binom{m}{j} B_j n^{m-j} \tag{12.3}$$

In particular, the sum $S_m(n)$ is a polynomial in n whose degree is $m+1$.

Proof

$$\begin{aligned}
\sum_{m\geq 0} S_m(n)\frac{z^m}{m!} &= \sum_{m\geq 0}\sum_{k=0}^{n-1} \frac{k^m z^m}{m!} \\
&= \sum_{k=0}^{n-1}\left\{\sum_{m\geq 0} \frac{(kz)^m}{m!}\right\} \\
&= \sum_{k=0}^{n-1} e^{kz} \\
&= \sum_{k=0}^{n-1} (e^z)^k \\
&= \frac{e^{nz}-1}{e^z-1} \\
&= \frac{e^{nz}-1}{z} \times \frac{z}{e^z-1} \\
&= \left\{\sum_{j=0}^{\infty} B_j \frac{z^j}{j!}\right\}\left\{\sum_{r\geq 1}(\frac{n^r z^{r-1}}{r!}\right\} \\
&= \left\{\sum_{j=0}^{\infty} B_j \frac{z^j}{j!}\right\} \times \left\{\sum_{k\geq 0} \frac{n^{k+1} z^k}{(k+1)!}\right\}
\end{aligned}$$

Equating coefficients, we get

$$S_m(n) = m!\left\{\sum_{j+k=m} \frac{B_j}{j!} \times \frac{n^{k+1}}{(k+1)!}\right\}$$

as desired proving (12.2). Replacing m by $m-1$, we have

$$\begin{aligned}
S_{m-1}(n) &= (m-1)!\left\{\sum_{j+k=m-1} \frac{B_j}{j!} \times \frac{n^{k+1}}{k+1}\right\} \\
&= \sum_{j+k=m-1} \frac{(m-1)!}{j!k!} \times \frac{n^{k+1}}{k+1} B_j \\
&= \sum_{j+k=m-1} \frac{1}{k+1}\binom{m-1}{k} \times n^{k+1} B_j \\
&= \frac{1}{m} \sum_{j+k=m-1} \binom{m}{k+1} \times n^{k+1} B_j \\
&= \frac{1}{m} \sum_{j=0}^{m-1} \binom{m}{j} B_j n^{m-j}
\end{aligned}$$

which proves equation (12.3). □

Example 12.5.5. Here are some examples that use Theorem 12.5.2.

$$\begin{aligned}
S_1(n) &= B_0 \frac{n^2}{2!} + B_1 \frac{n}{1!} \\
&= \frac{n^2}{2} - \frac{n}{2} = \frac{n(n-1)}{2}
\end{aligned}$$

Similarly,

$$\begin{aligned}
S_2(n) &= 2\left[B_0 \frac{n^3}{6} + B_1 \frac{n^2}{2} + B_2 \frac{n}{2}\right] \\
&= 2\left[\frac{n^3}{6} - \frac{n^2}{4} + \frac{n}{12}\right] \\
&= \frac{n(n-1)(2n-1)}{6}
\end{aligned}$$

Replacing $n-1$ by n gives the more familiar form:

$$1^2 + 2^2 + \cdots + n^2 = \frac{n(n+1)(2n+1)}{6}$$

12.6 Number theoretic functions

We now discuss a generating function which is useful in analytic, algebraic and combinatorial number theory and it is, in a sense, the correct generating function for the set-ups that are number theoretic in nature.

Definition 12.6.1. *A number theoretic function* f is a function whose domain is $\mathbb{N}$, the set of natural numbers (and whose range is the set of complex numbers). Let (a_n) be a sequence of complex numbers and let s be a formal variable. Then *the Dirichlet generating function (D.g.f.)* of this sequence is the formal series

$$F(s) = \sum_{n=1}^{\infty} \frac{a_n}{n^s}$$

Lemma 12.6.2. *There is a one-to-one correspondence between the following three sets: number theoretic functions, sequences of all complex numbers and the correspondence is given by:*

$$(a_n) \longleftrightarrow f(n) = a_n \longleftrightarrow F(s) = \sum_{n=1}^{\infty} a_n n^{-s}$$

Example 12.6.3. Here are some examples.

1. The number theoretic function I is defined by $I(1) = 1$ and $I(n) = 0 \; \forall \, n \neq 1$.
2. The number theoretic function u is defined by $u(n) = 1 \; \forall \, n$. Here, the corresponding D.g.f. is the famous Riemann Zeta function defined by

$$\zeta(s) = \sum_{n=1}^{\infty} \frac{1}{n^s}$$

3. The number theoretic function N is defined by $N(n) = n \; \forall \, n$.
4. Recall the definition (Definition 4.3.4) of the number theoretic function μ in Chapter 4. This takes values 0 and ± 1.

Lemma 12.6.4. *Let* $A(s) = \sum_{n=1}^{\infty} \frac{a_n}{n^s}$ *and* $B(s) = \sum_{n=1}^{\infty} \frac{b_n}{n^s}$ *be two D.g.f.s. Then their product* $C(s)$ *is a D.g.f. given by* $C(s) = \sum_{n=1}^{\infty} \frac{c_n}{n^s}$ *where* $c_n = \sum_{de=n} a_d b_e$.

Proof By way of example, what is c_{10}, which is the coefficient of 10^{-s} in the product: Each time we express 10 as a product of two natural numbers, we get a summand for c_{10} and hence $c_{10} = a_1 b_{10} + a_2 b_5 + a_5 b_2 + a_{10} b_1$. Clearly, this also holds in general and hence the proof. □

Lemma 12.6.4 tells us the way of multiplying two number theoretic functions f and g: if the corresponding D.g.f.'s are $A(s)$ and $B(s)$ respectively, then the D.g.f. of the product of f and g must be $A(s)B(s)$ prompting the following natural definition.

Definition 12.6.5. Let f and g be two number theoretic functions. Then *the product* $f * g$ of these functions is the number theoretic function h defined by

$$h(n) = \sum_{km=n} f(k)g(m).$$

Lemma 12.6.6. *Let f be a number theoretic function such that $f(1) \neq 0$. Then there exists a unique number theoretic function g such that $f * g = I$.*

Proof We define g recursively starting with $g(1) = \frac{1}{f(1)}$. Having found $g(k)$ for all $k < n$ uniquely, we have

$$g(n)f(1) = -\sum_{d|n, d\neq n} g(d)f\left(\frac{n}{d}\right)$$

since $I(n) = 0$. Hence $g(n)$ is uniquely determined. □

Lemma 12.6.6 shows that the set of all the number theoretic functions f for which $f(1) \neq 0$ is a multiplicative abelian group with I as the identity.

Definition 12.6.7. Let f be a number theoretic function which is not identically zero. Then f is called a *multiplicative function* if $f(mn) = f(m)f(n)$ holds whenever m and n are coprime. A multiplicative function is called a *completely multiplicative* function if $f(mn) = f(m)f(n)$ is true for all m and n.

An example of a completely multiplicative function is the function u (example 12.6.3) while the function ϕ (Euler's totient function) is multiplicative but not completely multiplicative since $\phi(12) = 4$ while $\phi(6)\phi(2) = 2 \times 1 = 2$. The familiar identity $\sum_{d|n} \mu(d) = 0$ if $n \neq 1$ can be expressed in the form $\mu * u = I$ and this tells us that the inverse of the Riemann zeta function is the D.g.f. given by $\sum_{n=1}^{\infty} \mu(n)n^{-s}$.

Theorem 12.6.8. *Let f and g be number theoretic functions. Then the following assertions hold.*

(a) If f is multiplicative, then $f(1) = 1$.

(b) Let $f(1) = 1$. Then f is multiplicative iff for all distinct primes $p_1, p_2, \ldots, p_r$ and for all natural numbers $a_1, a_2, \ldots, a_r$, we have

$$f({p_1}^{a_1}{p_2}^{a_2}\ldots, {p_r}^{a_r}) = \prod_{i=1}^{r} f({p_i}^{a_i})$$

(c) Let f be multiplicative. Then f is completely multiplicative iff for all primes p and for a natural number a, we have $f(p^a) = (f(p))^a$.

(d) If f and g are multiplicative, then so is $f * g$.

(e) Let g and $f * g$ be both multiplicative. Then so is f.

(f) If f is multiplicative, then so is its inverse (under the operation $*$).

Proof Let $f(n) \neq 0$. Since n and 1 are coprime, we have $f(n) = f(n \times 1) = f(n)f(1)$ and canceling the non-zero number $f(n)$, we get $f(1) = 1$ proving (a). For (b), let $q_i = {p_i}^{a_i}$. Since q_i and q_j are coprime, we can make induction on r to get$f(q_1q_2\cdots q_r) = f(q_1)f(q_2)\cdots f(q_r)$. For the converse, note that if a and b are coprime, then their prime factorization has no primes in common and we can use the given condition to show that $f(ab) = f(a)f(b)$. This proves (b). For (c), let f be completely multiplicative. Then $f(p^a) = (f(p))^a$ follows by making induction on a. Converse is trivial. Consider (d) and let a and b be coprime. Then $(f * g)(ab) = \sum_{de=ab} f(d)g(e)$. We can then write $d = \alpha\beta$ and $e = \alpha'\beta'$ where α and α' divide a and β and β' divide b to get (note that f and g are both multiplicative):

$$\begin{aligned}(f * g)(ab) &= \sum_{de=ab} f(d)g(e) \\ &= \sum_{\alpha\alpha'=a;\beta\beta'=b} f(\alpha)f(\beta)g(\alpha')g(\beta') \\ &= \{\sum_{\alpha\alpha'=a} f(\alpha)g(\alpha')\} \times \{\sum_{\beta\beta'=b} f(\beta)g(\beta')\} \\ &= (f * g)(a) \times (f * g)(b)\end{aligned}$$

Consider (e) and let $h = f * g$. Suppose there is a coprime pair (m, n) such that $f(mn) \neq f(m)f(n)$. Then we can find (m, n) such that mn is the smallest among all such pairs. Since both g and h are multiplicative, we have $h(1) = g(1) = 1$ (by (a)) which forces $f(1) = 1$. Then $f(1 \times 1) = f(1)$ and hence for the pair (m, n) we have $mn > 1$. Thus for every (a, b) such that $ab < mn$, and (a, b) a coprime pair, we have $f(ab) = f(a)f(b)$ while $f(mn) \neq f(m)f(n)$. Hence

$$\begin{aligned}h(mn) &= \sum_{a|m\ b|n} f(ab)g\left(\frac{m}{a}\right) \times g\left(\frac{n}{b}\right) \\ &= f(mn) + \sum_{a|m\ b|n\ ab<mn} f(a)f(b)g\left(\frac{m}{a}\right) \times g\left(\frac{n}{b}\right) \\ &= f(mn) + \left\{\sum_{a|m} f(a)g\left(\frac{m}{a}\right)\right\} \times \left\{\sum_{b|n} f(b)g\left(\frac{n}{b}\right)\right\} - f(m)f(n) \\ &= f(mn) + h(m)h(n) - f(m)f(n)\end{aligned}$$

Since h is multiplicative, we get $h(mn) = h(m)h(n)$ and hence $f(mn) = f(m)f(n)$ which is a contradiction. Finally (g) follows easily from (e) since $f * f^{-1} = I$ and both f and I are multiplicative. □

Theorem 12.6.9. *Let f be a multiplicative function. Then the following assertions hold.*

(a) f is completely multiplicative iff $g(n) = f^{-1}(n) = \mu(n)f(n)\ \forall\ n \geq 1$ (note that the product on the R.H.S. is an ordinary product).

(b) f satisfies

$$\sum_{d|n} \mu(d)f(d) = \prod_{p|n}(1 - f(p))$$

where the product on the R.H.S. is over all the prime divisors of n.

Proof Let $g = f^{-1}$ and first assume that $g(n) = \mu(n)f(n)$. To show that f is completely multiplicative, it suffices to show that $f(p^a) = (f(p))^a$ holds for every prime p (using the multiplicativeness of f) and for every natural number n. We have, for all $n > 1$,

$$\begin{aligned} \sum_{d|n} g(d) f\left(\frac{n}{d}\right) &= 0 \\ \Rightarrow \qquad \sum_{d|n} \mu(d) f(d) f\left(\frac{n}{d}\right) &= 0 \\ \Rightarrow \mu(1)f(1)f(p^a) + \mu(p)f(p)f(p^{a-1}) &= 0 \\ \Rightarrow \qquad f(p^a) &= f(p)f(p^{a-1}) \end{aligned}$$

and we are done by induction. Conversely, let f be completely multiplicative. Then

$$\begin{aligned} \sum_{d|n} g(d) f\left(\frac{n}{d}\right) &= \sum_{d|n} \mu(d) f(d) f\left(\frac{n}{d}\right) \\ &= \sum_{d|n} \mu(d) f(n) \\ &= \left(\sum_{d|n} \mu(d)\right) \times f(n) \\ &= I(n) \end{aligned}$$

and hence $g(n) = \mu(n)f(n)$ is the inverse of $f(n)$. Consider (b) and let $h(n) = \sum_{d|n} \mu(d)f(d)$. If we write $g(n) = \mu(n)f(n)$ then g is multiplicative because f is multiplicative. Then $h = g * u$ and hence h is also multiplicative. Let $a \geq 1$ and let p be any prime. Then

$$h(p^a) = \sum_{d|p^a} \mu(d) f(d) = f(1) - f(p) = 1 - f(p)$$

and hence by the multiplicativeness of h, we get $h(n) = \prod_{p|n}(1 - f(p))$ where the product is over all the primes p dividing n. □

Definition 12.6.10. *Liouville function* λ *is defined as follows.* $\lambda(1) = 1$ *and if* $n = p_1{}^{a_1} p_2{}^{a_2} \cdots p_r{}^{a_r}$ *is the prime power factorization of* $n > 1$, *then* $\lambda(n) = (-1)^{a_1+a_2+\cdots a_r}$. *Also define the function* $\sigma(n) = \sum_{d|n} d$.

Note that λ is completely multiplicative and σ is multiplicative because $\sigma = u * N$ and u and N are multiplicative. Also, if p is a prime and a any positive integer, then

$$\sigma(p^a) = 1 + p + p^2 + \cdots + p^a = \frac{p^{a+1} - 1}{p - 1}$$

Therefore, if $n = p_1{}^{a_1} p_2{}^{a_2} \cdots , p_r{}^{a_r}$ is a prime factorization of n, then

$$\sigma(n) = \prod_{i=1}^{r} \frac{p_i{}^{a_i+1} - 1}{p_i - 1}$$

As an elegant application of the theory of number theoretic functions, we derive a formula for the cyclotomic polynomial.

Definition 12.6.11. Let n be a natural number. Then the *cyclotomic polynomial* $\phi_n(x)$ (of degree $\phi(n)$) is defined by $\phi_n(x) = \prod(x - \alpha)$ where the product is over all the primitive n-th roots of unity.

Recall that all the n-th roots of unity are given by $\alpha_r = e^{\frac{2\pi i r}{n}}$ where $r = 0, 1, \ldots, n-1$ and α_r is primitive if r and n are coprime (in that case the multiplicative order of α_r is n and not a number less than n). Since every n-th root of unity α (primitive or not) must satisfy $\alpha^n = 1$ the order of α divides n and hence for some k dividing n, α is a primitive k-th root of unity while for a given k, the number of k-th roots of unity is equal to $\phi(k)$. We thus have

$$x^n - 1 = \prod_{d|n} \phi_d(x). \tag{12.4}$$

Theorem 12.6.12. *The cyclotomic polynomial $\phi_n(x)$ is given by*

$$\phi_n(x) = \prod_{d|n} (x^d - 1)^{\mu(\frac{n}{d})}$$

Proof Equation (12.4) gives us

$$\log(x^n - 1) = \sum_{d|n} \log(\phi_d(x))$$

So, using the Mobius inversion formula (Theorem 4.3.6), we get

$$\log(\phi_n(x)) = \sum_{d|n} \mu\left(\frac{n}{d}\right) \log(x^d - 1)$$

and hence

$$\phi_n(x) = \prod_{d|n} (x^d - 1)^{\mu(\frac{n}{d})}$$

□

Here are some computations of the cyclotomic polynomials:

$$\begin{aligned}\phi_{12}(x) &= \prod_{d|12}(x^d-1)^{\mu(\frac{12}{d})}\\ &= (x^{12}-1)\times(x^6-1)^{-1}\times(x^4-1)^{-1}(x^2-1)\\ &= \frac{x^{12}-1}{x^6-1}\times\frac{x^2-1}{x^4-1}\\ &= \frac{x^6+1}{x^2+1}\\ &= x^4+x^2+1\end{aligned}$$

$$\begin{aligned}\phi_{18}(x) &= \prod_{d|18}(x^d-1)^{\mu(\frac{18}{d})}\\ &= (x^{18}-1)(x^9-1)^{-1}\times(x^6-1)^{-1}\times(x^3-1)\\ &= \frac{x^{18}-1}{x^9-1}\times\frac{x^3-1}{x^6-1}\\ &= \frac{x^9+1}{x^3+1}\\ &= x^6+x^3+1\end{aligned}$$

It is a fact (from algebra) that the cyclotomic polynomial is an integer (and monic) polynomial which is irreducible over $\mathbb{Z}$. The following factorization of the Riemann zeta function is crucial to many parts of number theory.

Theorem 12.6.13. *The Riemann zeta function has a factorization*

$$\zeta(s) = \prod_p \frac{1}{1-p^{-s}}$$

where the product is taken over all the primes p.

Proof On the L.H.S., the coefficient of n^{-s} is 1. After multiplying out the geometric series, a typical term on the R.H.S. has the form $({p_1}^{a_1}{p_2}^{a_2}\cdots{p_r}^{a_r})^{-s}$. Since the prime power factorization of every natural number n in the form ${p_1}^{a_1}{p_2}^{a_2}\cdots{p_r}^{a_r}$ is unique, the result follows. □

Remark 12.6.14. The factorization of the Riemann zeta function is used in analytic and algebraic number theory. Also, the Riemann zeta function itself is central to all of number theory and gives rise to more involved mathematics in the form of L-series and their analysis.

12.7 Exercises for Chapter 12

12.1 Let $f(x)$ be the g.f. for the sequence (a_n). Show that $f(x)$ has an inverse iff $a_0 \neq 0$.

12.2 Let the g.f. for (a_n) be $f(x)$. What is the g.f. for the following sequences (b_n)?

(a) $(0, 0, 0, 0, a_0, a_1, a_2, \ldots)$

(b) $(a_1, a_0, a_2, a_3, \ldots)$

(c) $(a_0, -a_1, a_2, -a_3, \ldots)$

(d) $b_n = \alpha a_n$ where α is a constant.

(e) $b_n = a_n + c$ where c is a constant.

(f) $b_n = a_n$ if n is odd and $b_n = 0$ if n is even.

(g) $b_n = a_{n+5}$.

(h) $b_n = na_n$ for every n.

12.3 Let

$$H(n) = F(H(n-1), H(n-2), \ldots, H(n-k))$$

be a linear homogeneous recurrence relation with constant coefficients of order k (as in Section 11.3) with its characteristic polynomial $p(x)$. Let $f(x)$ be the g.f. of the sequence $(H(n))$. Show that $f(x) = \frac{q(x)}{p(x)}$ where $q(x)$ is a polynomial that is uniquely determined by the initial conditions.

12.4 Let a_n denote the number of binary words of length n that do not contain consecutive zeros. Find a recurrence relation for a_n's and solve it using g.f.s.

12.5 Solve the following recurrence relations on the sequence (a_n).

(a) $a_n = 4a_{n-1} - 4a_{n-2}$ for $n \geq 2$ subject to $a_0 = 0$ and $a_1 = 1$.

(b) $a_n = 4a_{n-1} - 5a_{n-2}$ for $n \geq 2$ subject to $a_0 = 0$ and $a_1 = 1$.

(c) $a_n = 5a_{n-1} - 6a_{n-2}$ for $n \geq 2$ subject to $a_0 1$ and $a_1 = -2$.

(d) $a_n = 3a_{n-2} - 2a_{n-3}$ for $n \geq 3$ subject to $a_0 = 3, a_1 = 1$ and $a_2 = 8$.

(e) $a_n = a_{n-1} + 9a_{n-2} - 9a_{n-3}$ for $n \geq 3$ subject to $a_0 = 0, a_1 = 1$ and $a_2 = 2$.

(f) $a_n = 5a_{n-1} - 6a_{n-2} - 4a_{n-3} + 8a_{n-4}$ for n $vspace.5cmgeq4$ subject to $a_0 = 0, a_1 = a_2 = 1$ and $a_3 = 2$.

(g) $a_n = 4a_{n-1} - 3$ for $n \geq 1$ subject to $a_0 = 3$.

12.6 Let a recurrence relation be given by

$$a_{n+1} = \begin{cases} 2a_n & \text{if n is odd} \\ 2a_n + 1 & \text{if n is even} \end{cases}$$

with the initial condition $a_1 = 1$. Show that the recurrence relation can be written in the form $a_{n+2} = a_{n+1} + 2a_n + 1$ and then solve it.

12.7 Let a_n denote the shifted Fibonacci number. That is $a_0 = 0$ and $a_n = f_{n-1}$ for all $n \geq 1$. Let $b_n = a_{2n}$ for all n. Find the g.f. of the sequence (b_n).

12.8 We have three different types of chocolates say A, B and C with all the three available in unlimited supply. We have to choose n chocolates under one of the following stipulations and let a_n denote the number of ways of doing this. Find a_n in each one of the cases.

(a) At the most four A's are chosen.

(b) The number of A's is even while the number of B's is odd.

(c) The number of A's is a multiple of 2 and the number of B's is a multiple of 3.

(d) The number of A's is a multiple of 2 or a multiple of 3.

(e) The number of A's is a multiple of 2 or a multiple of 3 *but not both.*

12.9 Show that the infinite sum

$$.1 + .01 + .002 + .0003 + .00005 + .000008 + .00000013 + .000000021 + \cdots$$

where the n-th summand is obtained by writing n zeros followed by the n-th Fibonacci number is a convergent sum that converges to a rational number.

12.10 Show that the g.f. of the sequence with n-th term equal to n^2 is given by $\frac{x^2+x}{(1-x)^4}$.

12.11 Let a_n denote the sum of the first n integer squares. Write a recurrence relation for a_n and then solve it using the g.f. given in Exercise 12.10.

12.12 Let $g(x) = \sum_{n=0}^{\infty} b_n x^n$ and let $c_n = \sum_{j=0}^{n} b_j$. Show that the g.f. of (c_n) is $h(x) = \frac{g(x)}{1-x}$.

12.13 Let $a_{n,k}$ denote the number of permutations on the set $[n]$ that have exactly k inversions (Definition 2.1.1) and let $f_n(x) = \sum_k a_{n,k} x^k$ denote the g.f. of these numbers.

(a) Show that given a permutation α on $[n-1]$, the new symbol n can be inserted in n ways, thus increasing the number of inversions by j where $0 \leq j \leq n-1$.

(b) Show that $f_n(x) = (1 + x + x^2 + \cdots + x^{n-1}) f_{n-1}(x)$.

(c) Prove that $f_n(x) = \prod_{m=1}^{n-1} (1 + x + \cdots + x^m)$.

(d) Prove the result of (c) directly as follows. The number of inversions that end in k is at the most $n-k$ and hence the g.f. for placing k is given by $1 + x + x^2 + \cdots + x^{n-k}$. Thus derive the formula for $f_n(x)$.

12.14 Let $1 \leq k \leq n$. Find the sum of the largest elements of all the k-subsets A of the set $[n]$.

12.15 Let $1 \leq k \leq n$. Find the sum of the second smallest elements of all the k-subsets A of the set $[n]$.

12.16 $\alpha_{n,k}$ the number of k-subsets $A = \{a_1, a_2, \ldots, a_k\}$ of $[n]$ with $1 \leq a_1 < a_2 < \cdots < a_k \leq n$ such that for all $i = 2, 3, \ldots, k$ we have $a_i - a_{i-1} \geq 2$.

(a) Find $\alpha_{n,k}$.

(b) Let $\alpha_{n,0} = 1$ and let $\alpha_n = \sum_k \alpha_{n,k}$. Show that α_n equals f_{n+1}, the $(n+1)$-th Fibonacci number.

12.17 We consider the *Terquem problem* of finding the number of k-subsets $A = \{a_1, a_2, \ldots, a_k\}$ of $[n]$ with $1 \leq a_1 < a_2 < \cdots < a_k \leq n$ such that a_i is even if i is even and a_i is odd if i is odd for all $i = 1, 2, \ldots, k$. Let $c_{n,k}$ denote this number. Show that

$$c_{n,k} = \binom{\lfloor \frac{n+k}{2} \rfloor}{k}$$

12.18 This problem called the *Skolem problem* generalizes the Terquem problem. Let $p \geq 2$ and let $c_{n,k,p}$ denote the number of k-subsets $A = \{a_1, a_2, \ldots, a_k\}$ of $[n]$ with $1 \leq a_1 < a_2 < \cdots < a_k \leq n$ such that $a_i - a_{i-1} \equiv 1 \ (mod\ p)$ for all $i = 1, 2, \ldots, k$ (where $a_0 = 0$). Show that

$$c_{n,k,p} = \binom{\lfloor \frac{n+(p-1)k}{p} \rfloor}{k}$$

12.19 Let $a_1, a_2, \cdots, a_r$ be r natural numbers whose g.c.d. is 1. Let h_n denote the number of ways of expressing n in the form $x_1a_1 + x_2a_2 + \cdots x_ra_r$ where each x_i is non-negative. Let $R(n) = \frac{n^{r-1}}{(r-1)!a_1a_2\cdots a_r}$. Show that

$$\lim_{n\to\infty} \frac{h_n}{R(n)} = 1$$

12.20 Fix a positive integer m and determine the generating function of the sequence whose k-th term is $s(m, k)$ (where $s(m, k)$ denotes the Stirling number of the first kind).

12.21 Using the standard recurrence relation for the Stirling numbers of the second kind, show that

$$\sum_{n=0}^{\infty} S(n,k)x^n = \frac{1}{(1-x)(1-2x)\cdots(1-kx)}$$

where k is a fixed positive integer.

12.22 Use partial fractions in Exercise 12.21 to obtain the inclusion-exclusion formula (Exercise 10.20 in Chapter 10):

$$k!S(n,k) = \sum_{m=1}^{k} (-1)^{k-m} \binom{k}{m} m^n$$

12.23 Let $a = (a_0, a_1, a_2, \ldots)$ be a real sequence. We define generalized Stirling numbers of second kind (associated with the sequence a) by the equation:

$$x^n = \sum_{k=0}^{n} S_a(n,k)(x-a_0)(x-a_1)\cdots(x-a_{k-1})$$

Prove the following.

(a) $S_a(n,k)$ is well-defined for all $n \geq k$.

(b) $S_a(n,k) = S_a(n-1,k-1) + a_k S_a(n-1,k)$.

(c) Let k be fixed. Then

$$\sum_{n=k}^{\infty} S_a(n,k)x^n = \frac{x^k}{(1-a_0x)(1-a_1x)\cdots(1-a_kx)}$$

(d) $S(n,k)$ is a special case of the generalized Stirling numbers of second kind and hence the identity in the previous exercise (Exercise 10.21) holds.

12.24 Let $f(x) = \sum_{n=0}^{\infty} a_n x^n$ be the g.f. of the sequence (a_n).

(a) Prove that the g.f. of the sequence (na_n) is $(xD)f(x)$ (here xD stands for the derivative followed by multiplication by x)

(b) By repeated application of (a) show that the g.f. of the sequence $(n^k a_n)$ is $(xD)^k f(x)$ where $(xD)^k$ refers to $(xD)(xD)\cdots(xD)$ (k times).

(c) Use (b) to show that if $P(n)$ is a polynomial then the g.f. of the sequence $(P(n)a_n)$ is $P(xD)f(x)$.

(d) Use (c) to prove that

$$\sum_{n=0}^{\infty} \frac{n^3 - 8n^2 + 10n + 5}{n!} = 4e$$

12.25 Consider a binary sequence $\mathbf{a} = (a_1 a_2 \cdots a_n)$. A maximal set of consecutive zeros or consecutive ones is called a block of $\mathbf{a}$. For example, with $n = 12$ and $a = 001110010011$, we see that $a_3a_4a_5$ is a block (while a_3a_4 is not) and we have 6 blocks, three blocks of 0's and three blocks of 1's. Prove that the number of binary sequences of length n with exactly k blocks is equal to $2\binom{n-1}{k-1}$. Hence show that the average number of blocks (over the set of all the sequences) is $\frac{n+1}{2}$.

12.26 Consider the set of all the coverings of a $2 \times n$ chessboard by 2×1 (either horizontal or vertical) non-overlapping dominoes. Every horizontal domino contributes one rupee while every vertical domino contributes 4 rupees. The value of a domino arrangement is the sum of all the dominoes in the arrangement. Let γ_m denote the total number of domino covering of all such chessboards (over all n) that fetch us m rupees. Show that γ_m is zero if m is odd and is equal to $f_{m/2}$ if m is even where f_r refers to the r-th Fibonacci number.

12.27 Consider the following two-variable generating function

$$F(x,y) = \sum_{n=0}^{\infty}\sum_{k=0}^{\infty} b_{n,k}x^n y^k$$

where $b_{n,k} = \binom{n}{k}$. Let $\sum_{n=0}^{\infty} b_{n,k}x^n = f_k(x)$ and $\sum_{k=0}^{\infty} b_{n,k}y^k = g_n(y)$. Prove the following.

(a) For a fixed n, we have $g_n(y) = (1+y)^n$.

(b) For a fixed k, we have

$$f_k(x) = \frac{x^k}{k!}(\frac{d}{dx})^k\{(1-x)^{-1}\}$$

(a) Use (b) to show that

$$f_k(x) = \frac{1}{1-x}(\frac{x}{1-x})^k$$

(c) When the expression in (c) is summed over all k we get

$$F(x,y) = \frac{1}{1-x-xy}$$

(d) When the expression in (a) is summed over all n, we get the same expression for $F(x,y)$ as in (d).

12.28 Prove the binomial theorem

$$(x+y)^n = \sum_{k=0}^{n}\binom{n}{k}x^k y^{n-k}$$

by equating the coefficients in the identity: $e^{t(x+y)} = e^{tx}e^{ty}$.

12.29 Find the number of ways of coloring an $n \times 1$ chessboard in three colours red, blue and green with the following stipulations.

(a) The number of blue cells is a multiple of 3 and there are no green cells.

(b) The number of blue cells is not a multiple of 3 and there are no green cells.

(c) The number of red cells is even, and the number of blue cells is odd.

(d) The number of blue cells is even,

(e) The number of blue cells is odd and there is at the most one green cell.

12.30 Write the e.g.f. of the sequence $5^n + (-1)^n 3^n$.

12.31 Solve the recurrence relation $a_n = na_{n-1} + 2n$ with the initial condition $a_0 = 5$.

12.32 Recall that D_n denotes the number of derangements of a set of n elements. Prove the following.

(a) $n! = \sum_{k=0}^{n} \binom{n}{k} D_k$

(b) Let $g(x)$ denote the e.g.f. of (D_n). Show that

$$g(x) = e^{-x}(1-x)^{-1}$$

(c) Use (b) to show that

$$D_n = n! \sum_{j=0}^{n} \frac{(-1)^j}{j!}$$

12.33 A permutation $\pi = a_1 a_2 \cdots, a_n$ on $[n]$ is called *an oscillatory permutation* if the following condition is satisfied.

$$a_1 < a_2 > a_3 < a_4 > \cdots$$

Define u_n to be the number of oscillatory permutations on $[n]$. Define $u_0 = 1$; show that $u_1 = u_2 = 1$. The two oscillatory permutations on $[3]$ are 132 and 231 and hence $u_3 = 2$. Define v_n to be the number of oscillatory permutations on $[n]$ if n is odd and $v_n = 0$ if n is even. Define w_n to be the number of oscillatory permutations on $[n]$ if n is even and $n \geq 2$, $w_0 = 1$ and $w_n = 0$ if n is odd. Verify that $u_n = v_n + w_n$. We now obtain e.g.f. of both v_n and w_n and thus the e.g.f. of u_n. Proceed through the following steps.

(a) Let n be odd and let $\pi = a_1 a_2 \cdots a_k a_{k+1} \cdots a_n$ where $a_{k+1} = n$ and π is an oscillatory permutation. Show that the subsequences $a_1 a_2 \cdots, a_k$ and $a_{k+2} \cdots a_n$ determine two oscillatory permutations on $[k]$ and $[m] = [n-1-k]$ respectively. In particular, both k and m are odd. We, therefore have,

$$v_n = \sum_{k+m=n-1} v_k v_m$$

(b) Let n be even and let $\pi = a_1 a_2 \cdots, a_k a_{k+1} \cdots a_n$ where $a_{k+1} = n$ and π is an oscillatory permutation. Show that the subsequences $a_1 a_2 \cdots, a_k$ and $a_{k+2} \cdots a_n$ determine two oscillatory permutations on $[k]$ and $[m] = [n-1-k]$ respectively. In particular, k is even and m is odd. We, therefore have,

$$w_n = \sum_{k+m=n-1} v_k w_m$$

(c) Let $V(x)$ and $W(x)$ represent the e.g.f.s:

$$V(x) = \sum_{n=0}^{\infty} v_n \frac{x^n}{n!}; \; W(x) = \sum_{n=0}^{\infty} w_n \frac{x^n}{n!}.$$

of the sequence (v_n) and (w_n) respectively. Prove the following.

$$V(x)^2 + 1 = V'(x)$$

and

$$V(x)W(x) = W'(x)$$

(d) Solve the first differential equation to obtain the solution $V(x) = \tan x$ and hence solve the second differential equation to get $W(x) = \sec x$. Conclude that the e.g.f. of the sequence (u_n) is $\tan x + \sec x$.

(e) Use (d) to show that the number of oscillatory permutations on $[6]$ is 61.

12.34 Let X be a set. A function $f : X \to X$ is called an idempotent function if $f(f(x)) = f(x)$ holds for all x. Let $i(n)$ denote the number of idempotent functions on an n-set.

(a) Prove that

$$i(n) = \sum_{k=1}^{n} \binom{n}{k} k^{n-k}.$$

(b) Prove that the e.g.f. of $i(n)$'s is given by e^{xe^x}.

12.35 An involution α on an n-set is a permutation on the set such that α has order 2 or α is identity. Let a_n denote the number of involutions on an n-set. Prove the following.

(a) If α is an involution, then each cycle of α is either a singleton or a 2-cycle and conversely.

(b) $a_n = \sum_{k+2j=n} \frac{n!}{k!j!2^j}$.

(c) $a_n = a_{n-1} + (n-1)a_{n-2} \ \forall\, n \geq 2$

(d) Either (b) or (c) can be used to show that the e.g.f. of (a_n) is $e^{x+x^2/2}$.

(e) a_n is even $\ \forall\ n \geq 2$.

(f) $a_n > \sqrt{n} \ \forall\ n \geq 2$.

12.36 Let $S(n) = \sum_{k=0}^{n} \binom{3n}{3k}$. Show that

$$\lim_{n\to\infty} \sqrt[3n]{S(n)} = 2.$$

(hint: use the multisection of series proved in Chapter 1, Exercise 1.51).

12.37 Let p be a prime number. We wish to find the number of p-subsets A of the set $[2p]$ with the property that $\sum_{a\in A} a$ is a multiple of p. Call this number M and proceed through the following steps to determine M.

(a) Let ω denote a primitive p-th root of 1. Let $f(z) = z^p - 1$ and let $F(z) = (f(z))^2$. Show that

$$F(z) = \prod_{j=1}^{2p} (z - \omega^j) = \sum_{r=0}^{2p} a_r z^r$$

where a_p is the number of p-subsets of $[2p]$.

(b) Show that

$$a_p = \sum_{t=0}^{p-1} c_t\omega^t$$

where c_t denotes the number of p-subsets A of $[2p]$ with the property that $(\sum_{a\in A} a) \equiv t\ (mod\ p)$.

(c) By equating coefficients, show that

$$c_0 + c_1\omega + c_2\omega^2 + \cdots + c_{p-1}\omega^{p-1} = 2$$

and hence ω is the root of the polynomial

$$g(z) = (c_0 - 2) + c_1 z + c_2 z^2 + \cdots + c_{p-1}z^{p-1}$$

(d) Using the fact that ω is an arbitrary primitive p-th root of 1, show that $g(z) = c_{p-1}(1 + z + z^2 + \cdots + z^{p-1})$.

(e) Comparing coefficients, show that $(c_0 - 2) = c_1 = c_2 = \cdots = c_{p-1}$. and hence conclude that $M = c_0$ equals

$$2 + \frac{1}{p}\left(\binom{2p}{p} - 2\right)$$

(a problem in the International Mathematical Olympiad 1995, [21])

12.38 Find the e.g.f. of (b_n) where b_n is the number of all the permutations β of an n-set for which $\beta^3 = 1$.

12.39 Express the sums $1^3 + 2^3 + \cdots + \cdots n^3$ and $1^4 + 2^4 + \cdots + \cdots n^4$ in terms of Bernoulli numbers and then evaluate them.

12.40 Let r be a fixed real number larger than one. A well-known result in analysis tells us that the series

$$H^{(r)} = \sum_{n=1}^{\infty} \frac{1}{n^r}$$

is convergent for all $r > 1$. When $r = m$ is a natural number, we write $\zeta(m) = H^{(m)}$. This exercise is meant to introduce the reader to the connection between $\zeta(m)$ for m even and Bernoulli numbers. To that end, *assume the Euler formula*:

$$z\cot z = 1 - 2\sum_{k=0}^{\infty} \frac{z^2}{k^2\pi^2 - z^2}$$

Prove the following assertions.

(a) Use Euler formula to show that

$$\begin{aligned} z\cot z &= 1 - 2\left(\frac{z^2}{\pi^2}\zeta(2) + \frac{z^4}{\pi^4}\zeta(4) + \frac{z^6}{\pi^6}\zeta(6) + \cdots\right) \\ &= 1 - 2\sum_{m=1}^{\infty} \frac{z^{2m}}{\pi^{2m}}\zeta(2m) \end{aligned}$$

(b) By equating the coefficients and using the table of small Bernoulli numbers, show that

$$\zeta(2) = \frac{\pi^2}{6}$$

(c) Show that

$$\zeta(4) = \frac{\pi^4}{90}$$

(d) In general, show that

$$\zeta(2m) = (-1)^{m-1}\left\{\frac{\pi^{2m}2^{2m-1}B_{2m}}{(2m)!}\right\}$$

The numbers $\zeta(m)$ are very near one as m tends to infinity. The question of closed formula for $\zeta(m)$ for odd values of $m \geq 3$ is a very difficult question. *It has been shown that $\zeta(3)$ is an irrational number but it is still an unproved statement that all $\zeta(m)$ with odd m are transcendental numbers.*

12.41 Define the Bernoulli polynomial of degree m by:

$$B_m(x) = \sum_{k=0}^{m} \binom{m}{k} B_k x^{m-k}$$

(a) Show that the e.g.f. of Bernoulli polynomials is given by:

$$\sum_{m=0}^{\infty} \frac{B_m(x)z^m}{m!} = \frac{ze^{xz}}{e^z - 1}$$

(b) Prove that (Definition 12.5.3):

$$S_{m-1}(n) = \frac{1}{m}(B_m(n) - B_m(0))$$

Both Bernoulli polynomials and Bernoulli numbers are extensively used in the Euler-MacLaurin sum formula which converts a problem of the form of a finite sum (of functions or polynomial functions) into an integral along with an error term.

12.42 For a sequence (a_n) , let $f(x)$ and $g(x)$ respectively denote the ordinary and exponential generating functions. Show that:

$$f(x) = \int_0^{\infty} e^{-s} g(sx)ds$$

12.43 For a positive integer n, we wish to find the number s_n of all the functions f such that $f : [n] \to [n]$ and such that f has the following property: If the range of f contains j then the range of f must also contain every $i \leq j$. Define $s_0 = 1$ and proceed through the following steps to determine s_n by finding first its exponential generating function $s(x) = \sum_{n=0}^{\infty} s_n \frac{x^n}{n!}$.

(a) Show that the number 1 must be in the range of any function f with the desired property.

(b) Let a function f have the desired property and let $|\{f^{-1}(\{1\})| = k$ for some positive integer k (where k is fixed). Then the number of such functions equals $\binom{n}{k}s_{n-k}$ and hence show that we have the recurrence relation

$$s_n = \sum_{k=1}^{n} \binom{n}{k} s_{n-k}$$

and hence

$$2s_n = \sum_{k=0}^{n} \binom{n}{k} s_{n-k}$$

(c) Show that $2s(x) = 1 + s(x)e^x$ and hence we have $s(x) = \frac{1}{2-e^x}$.

(d) Expanding the R.H.S. in (c) as a geometric series in $\frac{e^x}{2}$, prove that

$$s_n = \sum_{k=0}^{\infty} \frac{k^{n+1}}{2^{k+1}}$$

(e) This requires some residue analysis (see Lovasz [37]). Prove the asymptotic result:

$$\lim_{n\to\infty} \frac{s_n(\log 2)^{n+1}}{n!} = 1$$

12.44 We wish to find the g.f. of the (finite) harmonic series

$$h_n = 1 + \frac{1}{2} + \frac{1}{3} + \cdots + \frac{1}{n}$$

by proving the following assertions.

(a) Consider the sequence with n-th term equal to $\frac{1}{n}$ and let the corresponding g.f. be denoted by $g(x)$. Show that $g'(x) = (1-x)^{-1}$ and hence $g(x) = -\log(1-x)$.

(b) Let $H(x)$ denote the g.f. for h_n's. Write an obvious recurrence relation for h_n and then use it to obtain

$$H(x) = \frac{1}{1-x}\log\frac{1}{1-x}$$

(c) Find the sum $\sum_{n=1}^{\infty} \frac{h_n}{10^n}$.

12.45 Show that if f is a multiplicative function, then so is the function g where g is defined by $g(n) = \sum_{d|n} f(d)$.

12.46 Fix a positive integer k and let the number theoretic function H be defined by $H(n) = \sum_{d|n} d^k$. Show that H is a multiplicative function and hence find a formula for $H(n)$ in terms of the prime factorization of n.

12.47 Let ϕ denote the Euler totient function. Then prove the following.

(a) If a divides b then $\phi(a)$ divides $\phi(b)$.

(b) $\phi(mn) = \phi(m)\phi(n) \times \frac{d}{\phi(d)}$, where d is the g.c.d. of a and b.

12.48 Define the Mangoldt function $\Lambda(n)$ by letting $\Lambda(n) = \log p$ if $n = p^a$ (with $a \geq 1$) is a prime power and $\Lambda(n) = 0$ otherwise. Prove the following.

(a) For all $n \geq 1$ we have $\log n = \sum_{d|n} \Lambda(d)$.

(b) For all $n \geq 1$, we have $\Lambda(n) = \sum_{d|n} \mu(d) \log(d)$.

(c) Mangoldt function is not multiplicative.

12.49 With the notations for the number theoretic functions set up in this chapter, prove the following.

(a) $N^{-1} = \mu N$.

(b) $\phi = \mu * N$.

(c) $\phi^{-1} = \mu * (\mu N) = u * \mu N$.

(d) $\phi^{-1}(n) = \sum_{d|n} d\mu(d)$.

(e) $\phi^{-1}(n) = \prod_{p|n}(1-p)$ where the product is over all the primes p dividing n..

12.50 Let λ denote the Liouville function (Definition 12.6.10).

(a) Let $g(n) = \sum_{d|n} \lambda(d)$. Show that g is multiplicative.

(b) Let p be a prime and let a be a positive integer. Show that $g(p^a) = 0$ if a is odd and $g(p^a) = 1$ if a is even.

(c) Let n be a natural number. Show that $\sum_{d|n} \lambda(d)$ equals 1 if n is a perfect square and is zero otherwise.

(d) Show that $\lambda^{-1}(n) = |\mu(n)|$.

12.51 Find the cyclotomic polynomials $\phi_n(x)$ when $n = 23, 27, 20, 30, 36$.

12.52 Let f be a multiplicative function with its Dirichlet g.f. given by $F(s) = \sum_{n=1}^{\infty} f(n)n^{-s}$. Show that

$$F(s) = \prod_p \{1 + F(p)p^{-s} + F(p)p^{-2s} + F(p)p^{-3s} + \cdots\}$$

where the product is over all the primes p.

12.53 Prove the following identity:

$$e^x = \prod_{m=1}^{\infty} (1 - x^m)^{\frac{-\mu(m)}{m}}$$

12.54 Let $\tau(n)$ denote the number of divisors of the natural number n.

(a) Show that $\tau(n)$ is odd iff n is a perfect square.

(b) Show that $\tau(n) < n$ if $n > 2$.

(c) Show that $\tau(n) = 2$ iff n is a prime number.

(d) (*From Erdös to Kiev by Honsberger [32]*): Fix n and consider the sequence $n, \tau(n), \tau(\tau(n)), \tau(\tau(\tau(n))), \ldots$. Show that this sequence is square-free (has no term that is divisible by a square) iff it is of the form $p, 2, 2, 2, \ldots$ where p is a prime number.

12.55 Here is an alternative proof of Theorem 1.2.9 from Chapter 1. *All the congruences are modulo* 2.

(a) Inductively, show that, for all natural numbers k, we have

$$(1+x)^{2^k} \equiv 1 + x^{2^k}$$

(b) Let n be a natural number with the binary representation $n = \sum_{j=1}^{m} 2^{i_j}$ where $i_1 < i_2 < \cdots < i_m$. Show that

$$(1+x)^n \equiv \prod_{j=1}^{m} (1 + x^{2^{i_j}})$$

(c) Conclude that the number of odd binomial coefficients $\binom{n}{r}$ is equal to 2^m.

Chapter 13

Partition theory of integers

13.1 Partitions and Ferrers diagrams

This chapter is aimed at studying generating functions in their application to the theory of integer partitions. Historically, this area marks the beginning of modern combinatorics in the form of a very large number discoveries mainly by Euler and later by many others such as Gauss, Jacobi, Sylvester and Ramanujan.

Definition 13.1.1. Let n and k be positive integers. Then $q(n,k)$ denotes the number of *ordered partitions* of n into k parts. Also, $q(n)$ denotes the total number of ordered partitions of n.

For example,

$$\begin{aligned} 3 &= 1+1+1 \\ &= 1+2 \\ &= 2+1 \\ &= 3 \end{aligned}$$

are the four ordered partitions of 3 (note that we treat $1+2$ and $2+1$ as different ordered partitions).

Theorem 13.1.2. $q(n,k) = \binom{n-1}{k-1}$ *for all* $n, k \geq 1$. *Also,* $q(n) = 2^{n-1}$.

Proof This is a direct consequence of the results in Chapter 1. The second part follows by summing over all k. Alternatively, we may write n in the form $n = 1+1+\cdots+1$ where we have n ones and hence $n-1$ plus signs. Any (ordered) partition of n amounts to choosing a certain number of $+$ signs. For example, in $7 = 3+2+2$, we are actually choosing 2 plus signs out of the available 6 plus signs. This amounts to saying that for every $+$ sign, we have two choices, either we select it or do not select it showing that

$$q(n) = \underbrace{2 \times 2 \times \cdots \times 2}_{n-1\ times} = 2^{n-1}$$

□

Definition 13.1.3. A partition $n = a_1 + a_2 + \cdots + a_k$ (into parts $a_1, a_2, \ldots, a_k \geq 1$) is said to be an *unordered partition* if we do not distinguish between $n = a_1 + a_2 + \cdots + a_k$ and $n = a_{i_1} + a_{i_2} + \cdots + a_{i_k}$ where $\sigma = i_1 i_2 \cdots i_k$ is a permutation of the set $[k]$.

We thus do not distinguish between any two of the following:

$$2+3+4,\ 3+4+2,\ 2+4+3,\ 3+2+4,\ 4+2+3,\ 4+3+2 \tag{13.1}$$

and these six ordered partitions give rise to a single (unordered) partition. Similarly, $5+3$ and $3+5$ are treated as the same (unordered) partitions. Since Theorem 13.1.2 has answered questions relating to ordered partitions, *we make the convention that all the partitions considered from this point on throughout the remainder of the book are unordered partitions and hence will be just referred to as partitions.* It may appear to be the case that if we have a single partition into k parts, then it must give rise to as many as $k!$ ordered partitions (as in the examples above). However, this is not true. The partition $3+3+3$ corresponds to only one ordered partition (since there is no way of distinguishing between the three 3's). *We now follow a strict convention of writing*:

Definition 13.1.4. A partition of n into k parts a_i where $i = 1, 2, \ldots, k$ *will be written in the form* $n = a_1 + a_2 + \cdots + a_k$ where $a_1 \geq a_2 \geq \cdots \geq a_k \geq 1$.

With this exactness in writing a partition, out of the six ordered partitions of 9 given in (13.1), only $9 = 4 + 3 + 2$ is *a partition while others are not.* Similarly, $14 = 2+1+2+1+5+3$ is not a partition but $14 = 5+3+2+2+1+1$ is.

Definition 13.1.5. For a natural number n we denote the *total number of partitions of* n *by* $p(n)$ where $p(0) = 1$ by definition.

For example, $3 = 3,\ 3 = 2+1$ and $3 = 1+1+1$ are the three partitions of 3 and hence $p(3) = 3$. Similarly, the set of all the partitions of 5 is given by

$$\begin{aligned} 5 &= 5 \\ 5 &= 4+1 \\ 5 &= 3+2 \\ 5 &= 3+1+1 \\ 5 &= 2+2+1 \\ 5 &= 2+1+1+1 \\ 5 &= 1+1+1+1+1 \end{aligned}$$

and hence $p(5) = 7$. The following table lists $p(n)$ for values of $n \leq 10$.

n	0	1	2	3	4	5	6	7	8	9	10
$p(n)$	1	1	2	3	5	7	11	15	22	30	?

No closed formula for $p(n)$ is known. In fact, elaborate tables of $p(n)$ for values of n upto 200 were prepared by Hardy and Ramanujan and they also gave estimates (both upper and lower bounds) for $p(n)$. Our first task is to look for a generating function (g.f.) of the numbers $p(n)$:

$$P_{all\ part\ partitions}(x) = \sum_{n=0}^{\infty} p(n)x^n = 1 + x + 2x^2 + 3x^3 + 5x^4 + 7x^5 + \cdots$$

We wish to obtain an alternate expression for this generating function. To that end, first suppose that we wish to find the number of ways in which n can be written as a sum where each summand is j and j is some fixed positive integer. There are only two possibilities here: either we can (uniquely) write n as a sum of j's (and then n must be a multiple of j) or we cannot (and then n is not a multiple of j). The g.f. for this stipulation, is, therefore,

$$\begin{aligned} f_j(x) &= 1 + x^j + x^{2j} + \cdots + x^{mj} + \cdots \\ &= \sum_{m=0}^{\infty} x^{mj} \\ &= \sum_{m=0}^{\infty} (x^j)^m \\ &= (1 - x^j)^{-1} \end{aligned}$$

Theorem 13.1.6. *(all part partition g.f.)*

$$P_{all\ part\ partitions}(x) = \prod_{j=1}^{\infty} (1 - x^j)^{-1}$$

Proof The R.H.S. is

$$(1+x+x^2+\cdots+x^r+\cdots)(1+x^2+x^4+\cdots+x^{2r}+\cdots)(1+x^3+x^6+\cdots+x^{3r}+\cdots)$$
$$\cdots(1 + x^j + x^{2j} + \cdots + x^{mj} + \cdots)\cdots$$

The coefficient of x^n in this expression (for a positive integer n) arises from all the terms of the type

$$(x^{j_1})^{m_1}(x^{j_2})^{m_2}\cdots(x^{j_r})^{m_r} = x^{m_1j_1+m_2j_2+m_rj_r}$$

where $m_1j_1+m_2j_2+m_rj_r = n$ and $j_1 \geq j_2 \geq \cdots j_r \geq 1$ (without loss of generality) and indeed this expression corresponds to the partition

$$n = \underbrace{j_1 + j_1 + \cdots j_1}_{m_1} + \underbrace{j_2 + j_2 + \cdots j_2}_{m_2} + \cdots + \underbrace{j_r + j_r + \cdots j_r}_{m_r}$$

For example,

$$x^{50} = x^{4\times 7}x^{1\times 5}x^{3\times 4}x^{2\times 2}x^1$$

corresponds to

$$50 = 7 + 7 + 7 + 7 + 5 + 4 + 4 + 4 + 2 + 2 + 1.$$

This shows that on the R.H.S. we are actually counting all the partitions of n when we consider the coefficient of x^n. □

As an example consider the coefficient of x^4 on the R.H.S. We should only be looking at $f_1(x)f_2(x)f_3(x)f_4(x)$ because when $j \geq 5$, the non-trivial terms in $f_j(x)$ involve at least 5-th power of x. We begin by choosing x^4 from $f_4(x)$ (and 1 from the rest). This is the only way $f_4(x)$ can contribute to our stipulation. Now consider $f_1(x)f_2(x)f_3(x)$ and we may take x^3 from $f_3(x)$ and then must take x from $f_1(x)$. Again, this is the only way $f_3(x)$ can contribute. Next consider $f_1(x)f_2(x)$. Here we can choose x^4 from $f_2(x)$ (and 1 from $f_1(x)$). We can also choose x^2 from $f_2(x)$ and x^2 from $f_1(x)$. Finally we can choose x^4 from $f_1(x)$ giving a total of 5 partitions. We have not really simplified the problem; in fact these five partitions are

$$\begin{aligned}
\pi_1 : 4 &= 4 \\
\pi_2 : 4 &= 3 + 1 \\
\pi_3 : 4 &= 2 + 2 \\
\pi_4 : 4 &= 2 + 1 + 1 \\
\pi_5 : 4 &= 1 + 1 + 1 + 1
\end{aligned}$$

A slightly different way of doing the same thing, which uses more algebra, but not very useful in this case is the following.

$$\begin{aligned}
&f_1(x)f_2(x)f_3(x)f_4(x) \\
&\qquad = \frac{1}{1-x}\frac{1}{1-x^2}\frac{1}{1-x^3}\frac{1}{1-x^4} \\
&\qquad = \frac{1+x}{(1-x^2)^2}\frac{1}{1-x^4}\frac{1}{1-x^3} \\
&\qquad = \frac{(1+x)(1+x^2)^2}{(1-x^4)^3}\frac{1}{1-x^3} \\
&\qquad = (1+x)(1+x^2)^2 \times \sum_{k=0}^{\infty}\binom{k+2}{k}x^{4k} \times \sum_{r=0}^{\infty}x^{3r} \\
&\qquad = [1 + x + 2x^2 + 2x^3 + x^4 + x^5] \\
&\qquad\qquad \times \sum_{k=0}^{\infty}\binom{k+2}{2}x^{4k} \times \sum_{r=0}^{\infty}x^{3r}
\end{aligned}$$

The coefficient of x^4 can now be read off by making cases $(k, r) = (1, 0)$ when the number is $\binom{3}{2} = 3$ and $(k, r) = (0, 1)$ when the number is 1 and finally when $(k, r) = (0, 0)$ when the number is 1 giving the same answer.

As a convention, we use Greek letters α, β, π etc. to denote partitions. If π is the partition $n = a_1 + a_2 + \cdots + a_k$, then we write $\pi = (a_1, a_2, \ldots, a_k)$ identifying π with the sequence of parts written in a monotone decreasing manner and the integer represented by π is uniquely determined.

Definition 13.1.7. *A Ferrers diagram F* is an array that consists of a certain number of dots, say $n = a_1 + a_2 + \cdots a_k$ arranged in some k rows with i-th row containing exactly a_i dots. Here $a_1 \geq a_2 \geq \cdots \geq a_k \geq 1$ and the dots are arranged regularly with no gaps between two consecutive dots in a row or a column such that the following condition is satisfied: The rows of **F** are *left justified.* That is, the first dot in every row begins at the same column.

In Figure 13.1, the first diagram is an example of a Ferrers diagram while the second is not. The second is not a Ferrers diagram because it fails to be one on all the three counts: the second row is not left justified, the last row is bigger than the one before it and we also have a gap in the second row.

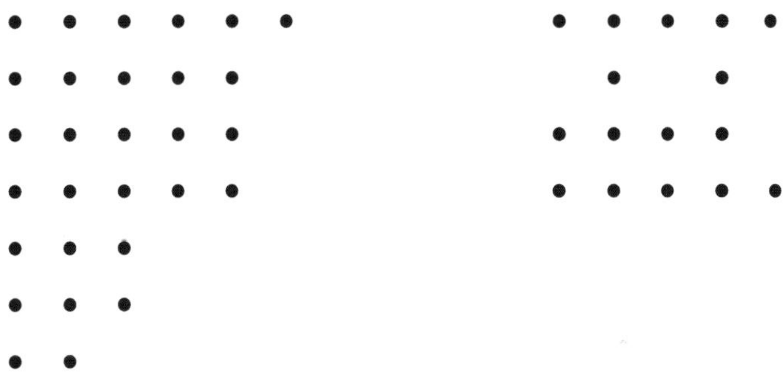

Figure 13.1: A Ferrers diagram and not a Ferrers diagram

It is clear that every partition π gives rise to a Ferrers diagram $F(\pi)$ and conversely. If $\pi = (a_1, a_2, \ldots, a_k)$, then we make $F(\pi)$ by drawing a_i dots in the i-th row. For example, if $\pi = (4, 3, 3, 1, 1)$ is a partition (of the number 12), then its Ferrers diagram is given in Figure 13.2.

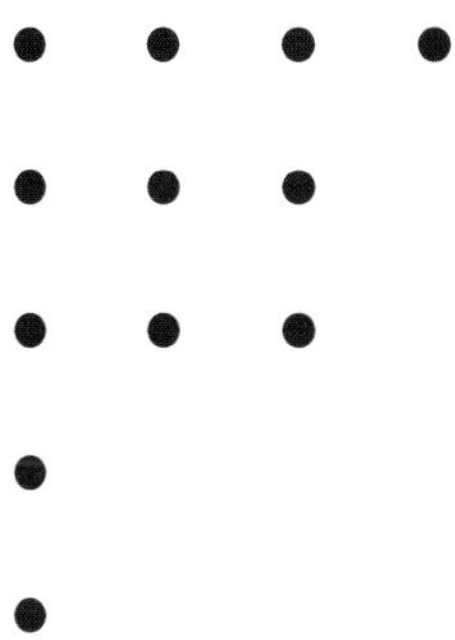

Figure 13.2: Ferrers diagram of the partition $\pi = (4, 3, 3, 1, 1)$

Conversely, every Ferrers diagram corresponds to a unique partition. For example the Ferrers diagram in the left half of Figure 13.1 corresponds to the partition $\sigma = (8, 8, 7, 7, 4, 2, 1)$ of the integer 37. As another example consider the five partitions of 4 given earlier; they have been called $\pi_1, \ldots, \pi_5$ respectively. These partitions have Ferrers diagrams given in Figure 13.3.below (where by F_j we mean the $F(\pi_j)$ and the diagrams are in the same order).

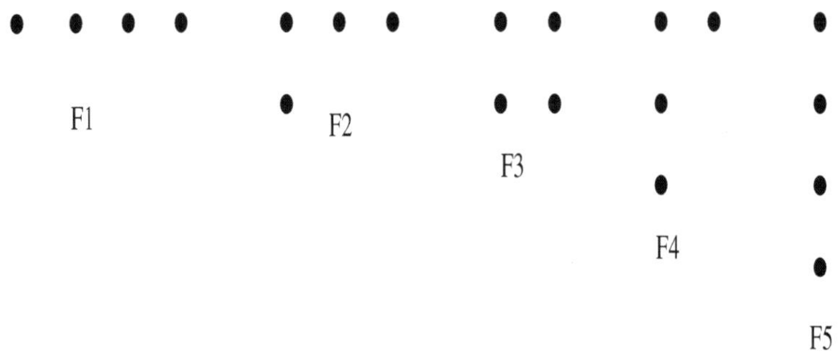

Figure 13.3: Ferrers diagrams of the five partitions of 4

The one-to-one correspondence between partitions and Ferrers diagrams is exploited in various ways. We may treat a Ferrers diagram like a (partial) matrix with dots referring to filled entries and with the additional requirement that if we have a dot at position (i, j) then we must also have dots at all the positions $(i, 1), \ldots, (i, j-1)$ as well as at positions $(1, j), (2, j), \ldots, (i-1, j)$. We can thus look at the transpose. This amounts to reading the diagram columnwise (instead of row-wise). We thus get a new diagram $F' = (b_1, b_2, \ldots)$ where b_1 refers to the number of dots in the first column and b_2, the number of dots in the second column and so on. Because of the stipulation of left justified rows and no gaps in rows (and also columns), it follows that $b_1 \geq b_2 \geq \cdots$. Formally, we have

Definition 13.1.8. Let $\pi = (a_1, a_2, \ldots, a_k)$ be a partition. Then the partition $\pi' = (b_1, b_2, \ldots, b_m)$ is called *the conjugate* of π where π' is defined by:

$$b_i = |\{j : a_j \geq i\}|$$

It is clear that the conjugate π' of π is obtained from its Ferrers diagram F by taking the conjugate F' of F. Also, $(\pi')' = \pi$. As an example, consider the Ferrers diagram given in Figure 13.3. Here F_1 and F_5 are conjugates (hence π_1 and π_5 are conjugates). Likewise, F_2 and F_4 are conjugates. Finally, the conjugate of F_3 is itself. This motivates the following.

Definition 13.1.9. A partition π is called a *self conjugate partition* if it is the same as its conjugate; that is $\pi' = \pi$.

For example, the two partitions $\pi_1 = (3, 1, 1)$ and $\pi_2 = (3, 2, 1)$ are both self conjugate partitions of numbers 5 and 6 respectively and have Ferrers diagrams given in Figure 13.4.

Figure 13.4: Self conjugate partitions of 5 and 6 respectively

Notice that if $\pi = (a_1, a_2, \ldots, a_k)$ and $\sigma = (b_1, b_2, \ldots, b_m)$ are conjugate partitions, then $b_1 = k$ and $m = a_1$. Thus, the number of parts in σ equals the size of the largest part of π and the size of the largest part of σ is the same as the number of parts of π. This correspondence proves the following theorem.

Theorem 13.1.10. *Let n and k be natural numbers. Then the number of partitions of n into k parts is equal to the number of partitions of n in which the largest part is k. Also, the number of partitions of n into at the most k parts is equal to the number of partitions of n in which each part is less than or equal to k.*

Proof Let S denote the set of all the partitions of n. In both the cases, we can construct two subsets of S say T and T' such that $T' = \{\pi' : \pi \in T\}$. For example, in the first case, T consists of those partitions for which the largest part is k. Then the bijection between T and T' proves the required result. For the second part, note that the partitions of n in which each part is $\leq k$ are same as those in which the largest part is $\leq k$. $\square$

As an example, there are three partitions of 6 in which the largest part is 3 and these are:

$$\begin{aligned} 6 &= 3+3 \\ &= 3+2+1 \\ &= 3+1+1+1 \end{aligned}$$

and by conjugation, there are also three partitions of 6 in which we have exactly three parts:

$$\begin{aligned} 6 &= 2+2+2 \\ &= 3+2+1 \\ &= 4+1+1 \end{aligned}$$

Similarly, we have three partitions of 5 in which each part is ≤ 2 and these are:

$$\begin{aligned} 5 &= 2+2+1 \\ &= 2+1+1+1 \\ &= 1+1+1+1+1 \end{aligned}$$

and we also have three partitions of 5 in which the number of parts is at the most 2.

$$\begin{aligned} 5 &= 3+2 \\ &= 4+1 \\ &= 5 \end{aligned}$$

Theorem 13.1.11. *Let k be a fixed positive integer. Let the g.f. $P_{k\ parts}(x)$ count the number of partitions of n into exactly k parts and let g.f.$P_{\leq k\ parts}(x)$ count the number of partitions into at the most k parts. Then*

$$P_{\leq k\ parts}(x) = \prod_{i=1}^{k}(1-x^i)^{-1}$$

and

$$P_{k\ parts}(x) = x^k \prod_{i=1}^{k}(1-x^i)^{-1}$$

Proof Let a_n denote the number of partitions of n with $\leq k$ parts, b_n the number of partitions of n with exactly k parts, c_n the number of partitions of n with each part $\leq k$ and d_n the number of partitions of n with largest part equal to k. Then the bijections we established show that $a_n = c_n$ and $b_n = d_n$. The g.f.s of c_n and d_n are easy to find. $\sum_{n=0}^{\infty} c_n x^n$ equals $f_1(x)f_2(x)\cdots f_k(x)$ since each part is at the most k, where $f_i(x) = (1-x^i)^{-1}$. Hence

$$P_{\leq k\ parts}(x) = \sum_{n=0}^{\infty} a_n x^n = \prod_{i=1}^{k}(1-x^i)^{-1}$$

How do we find $\sum_{n=0}^{\infty} d_n x^n$? Here each part is $\leq k$ and the largest part is k (and hence must occur). So, the g.f. $f_k(x)$ that counts the occurrence of k is modified to

$$x^k + x^{2k} + x^{3k} \cdots = x^k f_k(x) = x^k(1-x^k)^{-1}$$

Therefore,

$$P_{k\ parts}(x) = x^k \prod_{i=1}^{k}(1-x^i)^{-1}$$

□

Theorem 13.1.12. *Let k be a fixed natural number. Let $n \geq k$ be a natural number and write $p_k(n)$ to denote the number of partitions of n into k parts. Let γ_n denote the number $\frac{n^{k-1}}{k!(k-1)!}$. Then*

$$\lim_{n\to\infty} \frac{p_k(n)}{\gamma_n} = 1$$

Proof Each (unordered) partition of n into k parts gives rise to at the most $k!$ ordered partitions and hence $p_k(n)k! \geq q(n,k) = \binom{n-1}{k-1}$. This shows that asymptotically, $p_k(n)$ is lower bounded by γ_n. Now consider a partition $(a_1, a_2, \ldots, a_k)$ of n and let $(b_1, b_2, \ldots, b_k)$ be a rearrangement of $(a_1, a_2, \ldots, a_k)$. If we define $(c_1, c_2, \ldots, c_k)$ by: $c_i = b_i + (k - i)$ for all $i = 1, 2, \ldots, k$, then c_i's are positive integers whose sum is $n + \frac{k(k-1)}{2}$. Using the number of ordered partitions of $n + \frac{k(k-1)}{2}$ into exactly k parts we get

$$k!p_k(n) \leq \binom{n + \frac{(k-1)}{2} - 1}{k-1}$$

We see that $p_k(n)$ is asymptotically bounded above by γ_n. □

Definition 13.1.13. A partition $\pi = (a_1, a_2 \ldots, a_k)$ is called a *distinct part partition* if $a_1 > a_2 > \cdots > a_k \geq 1$. The number of *distinct part partitions of* n *is denoted by* $d(n)$.

For example, the number of distinct part partitions of 6 is four and these are:

$$\begin{aligned} 6 &= 6 \\ &= 5+1 \\ &= 4+2 \\ &= 3+2+1 \end{aligned}$$

Now let $o(n)$ denote the number of partitions of n into parts *each of which is odd.* For example, the number of partitions of 6 into parts with each part odd (odd part partitions) is also four and these are:

$$\begin{aligned} 6 &= 5+1 \\ &= 3+3 \\ &= 3+1+1+1 \\ &= 1+1+1+1+1+1 \end{aligned}$$

We see that $d(6) = o(6)$. Euler proved that this is not a mere coincidence and we have the following.

Theorem 13.1.14. *For every non-negative integer* n, $d(n) = o(n)$ *and thus the number of distinct part partitions of* n *equals the number of odd part partitions of* n.

Proof Let $P_{odd}(x) = \sum_{n=0}^{\infty} o(n)x^n$ and let $P_{distinct}(x) = \sum_{n=0}^{\infty} d(n)x^n$. We show that these two g.f.s are the same.

$$P_{distinct}(x) = \prod_{j=1}^{\infty} g_j(x)$$

where $g_j(x)$ is the g.f. for the occurrence of j as a part. Since j can occur at the most once, $g_j(x) = 1 + x^j$. So

$$P_{distinct}(x) = (1+x)(1+x^2)\cdots(1+x^j)\cdots$$

Since

$$\begin{aligned} 1 + x^j &= \frac{1 - x^{2j}}{1 - x^j} \\ &= (1 - x^{2j})(1 - x^j)^{-1} \end{aligned}$$

it follows that

$$\begin{aligned} P_{distinct}(x) &= \left[(1 - x^2)(1 - x^4) \cdots (1 - x^{2j}) \cdots\right] \times \\ &\quad \left[(1 - x)^{-1}(1 - x^2)^{-1} \cdots (1 - x^j)^{-1} \cdots\right] \\ &= (1 - x)^{-1}(1 - x^3)^{-1} \cdots (1 - x^{2j+1})^{-1} \cdots \\ &= \prod_{k=1}^{\infty} (1 - x^{2k+1})^{-1} \\ &= P_{odd}(x) \end{aligned}$$

□

13.2 Durfee squares and self conjugate partitions

Definition 13.2.1. Given a Ferrers diagram F, *the Durfee square* D_m *of* F is a maximal $m \times m$ square of dots in the given diagram that includes consecutive dots (both from left to right and top to bottom) that begins at the left top corner dot.

The way we defined Durfee square, it should be clear that D_m and hence m are uniquely determined by the Ferrers diagram F. Also a non-empty Ferrers diagram must have a Durfee square D_m where m is a positive integer. For example, the Durfee squares in the following two diagrams (in Figure 13.5) F_1 and F_2 are D_2 and D_3 respectively.
Notice that D_m by itself (that is, the diagram that consists of only the dots in the Durfee square) is a Ferrers diagram and represents the partition:

$$m^2 = \underbrace{m + m + \cdots + m}_{m \ times}$$

Given a Ferrers diagram F, denote by F_R the portion of dots to the right of its Durfee square D_m and by F_B the portion of dots below the Durfee square. Then F_R is a Ferrers diagram in its own right and in fact, corresponds to a partition π_R which is a partition with at the most m parts (else F_R will have some dots protruding below D_m which either contradicts the definition of Ferrers diagram F or the definition of D_m, obtained from F). Also, F_B is a Ferrers diagram in its own right and in fact, corresponds to a partition π_B which is a partition with the largest part $\leq m$ (else the the first part of F_B will protrude beyond the last dot of D_m, a contradiction). So, the given Ferrers diagram F is decomposed into three Ferrers diagrams:

$$F \longrightarrow (D_m, F_R, F_B).$$

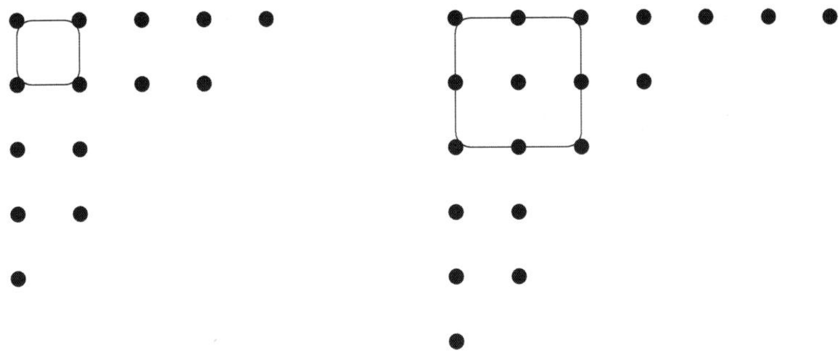

Figure 13.5: Durfee squares in Ferrers diagrams

Consider the two Ferrers diagrams F_1 and F_2 in Figure 13.5. In the case of F_1, we have $m = 2$ and the partition to the right of the Durfee square is $\pi_R = (3, 2)$ while the one below the Durfee square is $\pi_B = (2, 2, 1)$. In the case of F_2, we have $m = 3$ and the partition to the right of the Durfee square is $\pi_R = (4, 1)$ while the one below the Durfee square is $\pi_B = (2, 2, 1)$. *This procedure is reversible.* That is, given $m \geq 1$, given a partition σ with at the most m parts and given a partition τ with the largest part $\leq m$, we can stick σ to the right of D_m and τ under D_m to get a partition π that has D_m as its Durfee square. Check that this works because σ has at the most m parts while the largest part in τ is no larger than m. For example, given $m = 3$, the partition $\sigma = (7, 2, 1)$ (with at most three parts) and the partition $\tau = (3, 3, 3, 3, 2)$ with the largest part at the most 3, we get the partition $\pi = (10, 5, 4, 3, 3, 3, 3, 2)$ by putting σ to the right of D_3 and τ below D_3. This is shown in Figure 13.6.

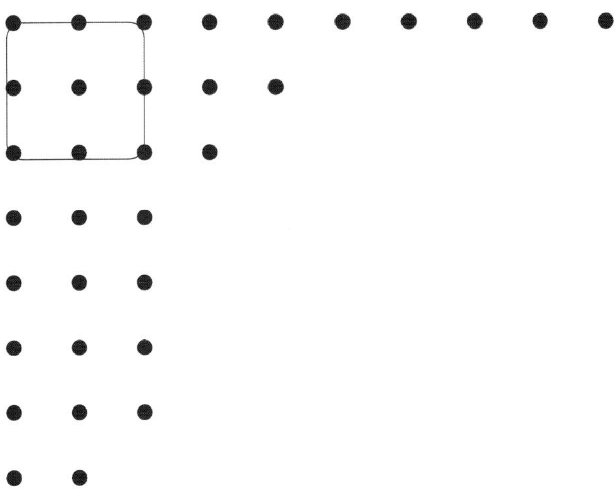

Figure 13.6: Patching the Durfee square D_3 with two partitions

The bijection between π and the triple (D_m, F_R, F_B) allows us to obtain a formula for the g.f. $P_{all\ part\ partitions}(x)$. Using Theorem 13.1.11, the generating functions for F_R and F_B are the same and these are each equal to $\prod_{j=1}^{m}(1-x^j)^{-1}$. This proves the following.

Theorem 13.2.2. *The g.f. for all part partition is given by*

$$P_{all\ part\ partitions}(x) = 1 + \sum_{m=1}^{\infty} x^{m^2} \prod_{j=1}^{m}(1-x^j)^{-2}$$

As an example, consider $p(9)$. We must look at the R.H.S. of the identity in Theorem 13.2.2. We make the following cases depending on the value of m.

(a) $m = 1$: We should look at the coefficient of x^9 in $x(1-x)^{-2}$ and hence the coefficient of x^8 in

$$\begin{aligned}(1-x)^{-2} &= \sum_{r\geq 0}\binom{r+1}{r}x^r \\ &= \sum_{r\geq 0}\binom{r+1}{1}x^r\end{aligned}$$

which is 9.

(b) $m = 2$: We should look at the coefficient of x^9 in $x^4(1-x)^{-2}(1-x^2)^{-2}$ and hence the coefficient of x^5 in

$$\begin{aligned}(1-x)^{-2}(1-x^2)^{-2} &= \left[\sum_{r\geq 0}\binom{r+1}{r}x^{2r}\right] \times \left[\sum_{s\geq 0}\binom{s+1}{s}x^s\right] \\ &= \left[\sum_{r\geq 0}(r+1)x^{2r}\right] \times \left[\sum_{s\geq 0}(s+1)x^s\right] \\ &= \sum_{r,s\geq 0}(r+1)(s+1)x^{2r+s}\end{aligned}$$

When $(r,s) = (0,5)$ we get the coefficient 6. When $(r,s) = (1,3)$ we get the coefficient $2 \times 4 = 8$. Finally when $(r,s) = (2,1)$ we get the coefficient $3 \times 2 = 6$. Hence the total contribution in this case is $6+8+6 = 20$.

(c) $m = 3$: We should look at the coefficient of x^9 in $x^9(1-x)^{-2}(1-x^2)^{-2}(1-x^3)^{-2}$ and hence the constant term in $(1-x)^{-2}(1-x^2)^{-2}(1-x^3)^{-2}$ which is 1.

Therefore, $p(9) = 9 + 20 + 1 = 30$.

Let us now look at the self conjugate partitions (Definition 13.1.9) of 9. If π is such a self conjugate partition, then the Durfee square has size $m \leq 3$ and with $m = 3$, we have exactly one such partition. Now observe that the diagram of a self conjugate partition is symmetric about the Durfee square. This means the portion below the Durfee square must be conjugate of the portion to the right of the Durfee square. It follows at once that the number of dots not covered in the Durfee square must be an even number if π is a self conjugate partition. This shows that in our case, we cannot have Durfee square of size 2 for a self conjugate partition of 9. Hence the only possibility left out is $m = 1$. It is easy to check that in this case, $\pi = (5, 1, 1, 1, 1)$. Hence the number of self conjugate partitions of 9 is 2. If we are dealing with self conjugate partitions, then the Durfee square decomposition obtained in the proof of Theorem 13.2.2 gets specialized. If F is a self conjugate Ferrers diagram then F_R and F_B are conjugate diagrams. Hence, for self conjugate partitions, the decomposition changes into

$$F \longleftrightarrow (D_m, F_R, (F_R)') \longleftrightarrow (D_m, F_R)$$

Conversely, given m and a partition σ that has at the most m parts, let R denote the Ferrers diagram of σ and let B denote the Ferrers diagram of τ, the conjugate of σ (thus τ has largest part $\leq m$), we simply put R to the right of D_m and B under D_m to get a self conjugate partition π. For example, given $m = 3$ and the partition $\sigma = (5, 2, 1)$ that has at the most 3 parts, we get the partition $\pi = (8, 5, 4, 3, 2, 1, 1, 1)$ by putting σ to the right of D_3 and the conjugate of σ below D_3. Evidently, π is a self conjugate partition. This is shown in Figure 13.7.

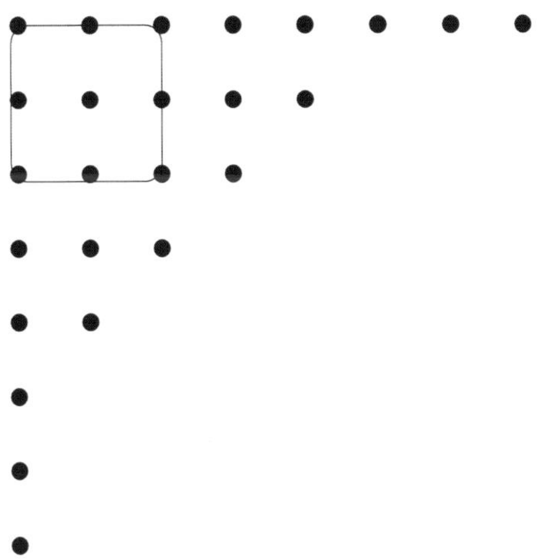

Figure 13.7: Durfee square and self conjugate partitions

Since the g.f. of σ (partition with at most m parts) is $\prod_{j=1}^{m}(1 - x^j)^{-1}$, we have the same g.f. for the conjugate τ. However, the correspondence we have, determines the portion below D_m, (which is $\tau = \sigma'$) uniquely and hence the g.f. for the pair (σ, σ') is obtained by replacing x by x^2 in the g.f. for σ. This proves the following.

Theorem 13.2.3. *Let $sc(n)$ denote the number of self conjugate partitions of n (with the standard convention that $sc(0) = 1$). Let $P_{self\ conjugate}(x) = \sum_{n=0}^{\infty} sc(n)x^n$*

denote the g.f. of $sc(n)$'s. Then

$$P_{self\ conjugate}(x) = 1 + \sum_{m=1}^{\infty} x^{m^2} \prod_{j=1}^{m} (1 - x^{2j})^{-1}$$

As an example, suppose we wish to find the number of self conjugate partitions of 33. Then $m \leq 5$ and the argument we made earlier shows that we need to look only at odd values of m. We use Theorem 13.2.3 and make the following three cases.

1. $m = 1$. Then we must look at coefficient of x^{32} in $(1 - x^2)^{-1}$ and this is clearly 1.

2. $m = 3$. Then we look at the coefficient of x^{24} in $(1 - x^2)^{-1}(1 - x^4)^{-1}$. Here we have

$$\begin{aligned}(1 - x^2)^{-1}(1 - x^4)^{-1} &= (1 + x^2)(1 - x^4)^{-2} \\ &= (1 + x^2) \sum_{r \geq 0} \binom{r+1}{r} x^{4r} \\ &= (1 + x^2) \sum_{r \geq 0} (r + 1)x^{4r}\end{aligned}$$

   and the coefficient is 7.

3. $m = 5$. Then we look at the coefficient of x^8 in $(1-x^2)^{-1}(1-x^4)^{-1}(1-x^6)^{-1}$. Here we have

$$\begin{aligned}(1 - x^2)^{-1}(1 - x^4)^{-1}(1 - x^6)^{-1} &= (1 + x^2)(1 - x^4)^{-2}(1 - x^6)^{-1} \\ &= (1 + x^2) \sum_{r \geq 0} \binom{r+1}{r} x^{4r} \sum_{s \geq 0} x^{6s} \\ &= (1 + x^2) \sum_{r \geq 0} (r + 1)x^{4r} \sum_{s \geq 0} x^{6s}\end{aligned}$$

   The two possibilities are $(r, s) = (0, 1)$ and $(r, s) = (2, 0)$ and the coefficient is $1 + 3 = 4$.

Adding all the numbers, we get $sc(33) = 1 + 7 + 4 = 12$.

Theorem 13.2.4. *Let n be a natural number. Then the number of self conjugate partitions of n is equal to the number of partitions of n into distinct parts each of which is an odd integer.*

Proof Let $SC(n)$ denote the set of all self conjugate partitions of n and let $DO(n)$ denote the set of all distinct part partitions of n such that each part is an odd integer. We wish to show that these two sets have the same cardinality. We work through Ferrers diagrams and set-up an explicit bijection between these two sets. Given $\pi \in SC(n)$, let $\pi = (a_1, a_2, \ldots, a_k)$. Look at the Ferrers diagram F of π. Let b_1 be the number of dots on the first row and the first column of F (avoid duplication of the top left corner dot and count it only once). Then $b_1 = 2a_1 - 1$. Let b_2 equal the number of dots on the second row and second column; avoid duplication again and do not count the dots in the first row and column that have contributed to b_1 as also count the dot at position $(2, 2)$ only once. Then $b_2 = 2a_2 - 3$. Proceeding in this manner, b_i is the total number of dots on the i-th row and the i-th column (without counting the dots that have contributed to $b_1, \ldots, b_{i-1}$). Then $b_i = 2a_i - (2i - 1)$. Continuing in this manner we get $\tau = (b_1, b_2, \ldots, b_j)$. Since we do not lose any dot in the process, τ is also a partition of n such that $b_1 > b_2 > \cdots$ and hence $\tau \in DO(n)$. For the converse, let $\tau = (b_1, b_2, \ldots, b_j) \in DO(n)$ be given with its Ferrers diagram G. We create a new Ferrers diagram F as follows. Distribute the b_1 dots in the first row of G equally on the first row and column of F (this is possible since b_1 is odd; the dot at position $(1, 1)$ contributes to both). Then distribute b_2 dots equally on the second row and the second column (again note that b_2 is odd). This obtains F and hence the partition $\pi = (a_1, a_2, \ldots, a_k)$. Since $b_1 > b_2 > \cdots, > b_j$ and since each b_i is odd we get $a_1 \geq a_2 \geq \cdots \geq a_k$. Also, the procedure ensures that π is a self conjugate partition of the same integer n and hence $\pi \in SC(n)$. □

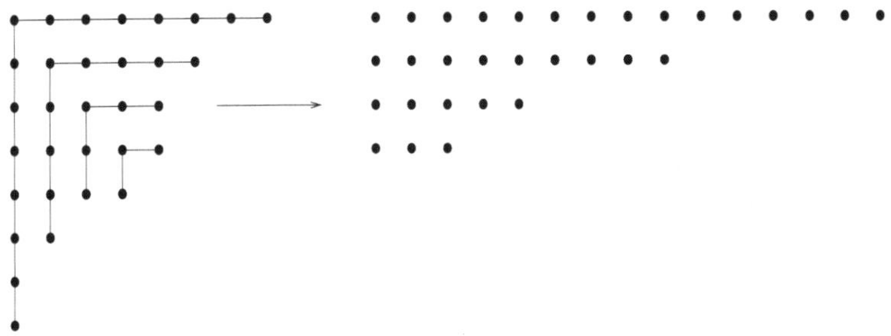

Figure 13.8: Self conjugate to distinct odd part partition

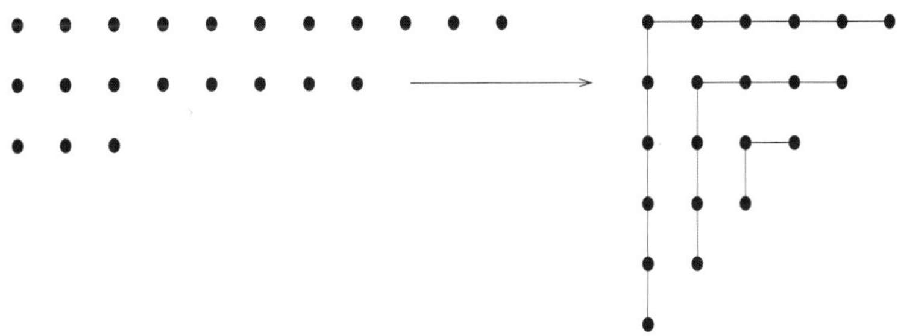

Figure 13.9: Distinct odd part partition to a self conjugate partition

As an example, let $\pi = (8, 6, 5, 5, 4, 2, 1, 1)$ be a self conjugate partition of 32 shown in Figure 13.8. Then the first row and column together contain 15 dots. Excluding these dots, the second row and column contain 9 dots and so on. This gives us $\tau = (15, 9, 5, 3)$ shown as the second Ferrers diagram. As another example, let

$\tau = (11, 7, 3)$ be in $DO(21)$. We break the first row of 11 dots of τ in $1 + 5 + 5$ dots, putting 6 dots on the first row and also 6 dots on the first column of the new Ferrers diagram π. We then look at the second part 5 of τ and break it as $1 + 2 + 2$. This gives 3 additional dots on the second row and also second column. The procedure is symmetric (about rows and columns) and hence ensures that the new partition we get is in $SC(21)$. The new partition $\pi = (6, 5, 4, 3, 2, 1)$. This is shown as the second Ferrers diagram of Figure 13.9.

Corollary 13.2.5. *The following algebraic identity holds.*

$$\prod_{r=0}^{\infty}(1 + x^{2r+1}) = 1 + \sum_{m=1}^{\infty} x^{m^2} \prod_{j=1}^{m}(1 - x^{2j})^{-1}$$

13.3 Euler's pentagonal theorem

We wish to find a closed formula for the g.f.

$$f(x) = (1 - x)(1 - x^2) \cdots (1 - x^k) \cdots = \prod_{k=1}^{\infty}(1 - x^k)$$

This g.f. is related to the g.f.

$$P_{distinct}(x) = (1 + x)(1 + x^2) \cdots (1 + x^k) \cdots = \sum_{n=0}^{\infty} d(n)x^n$$

where $d(n)$ equals the number of distinct part partitions of n as also to the g.f.

$$P_{all\ part\ partitions}(x) = \prod_{j=1}^{\infty}(1 - x^j)^{-1}.$$

In fact $P_{all\ part\ partitions}(x)$ is the multiplicative inverse of $f(x)$. Let $f(x)$ equal $\sum_{n=0}^{\infty} a_n x^n$. What is the relationship between $f(x)$ and $P_{distinct}(x)$? The coefficient a_n in $f(x)$ is obtained as follows. We write n as a sum of distinct positive integers. If the number of parts (summands) is even then we count this as $+1$ and if it is odd, then we count it as -1. Adding all these ± 1 obtains the number a_n. *It thus follows that a_n is the difference between the number of ways of writing n as a sum of even number of distinct parts and the number of ways of writing n as a sum of odd number of distinct parts.* For example, when $n = 6$, we have 4 partitions of 6 in which parts are distinct:

$$\begin{aligned} 6 &= 6 \\ &= 5 + 1 \\ &= 4 + 2 \\ &= 3 + 2 + 1 \end{aligned}$$

Of these, the second and the third have an even number of parts while the first and the fourth have an odd number of parts. So, $a_6 = 0$. Similarly, $a_9 = 0$. Consider a_7. The distinct part partitions are:

$$\begin{aligned} 7 &= 7 \\ &= 6+1 \\ &= 5+2 \\ &= 4+3 \\ &= 4+2+1 \end{aligned}$$

Of these, the second, third and the fourth have an even number of parts while the first and the fifth have an odd number of parts. So $a_7 = 1$. Euler's pentagonal theorem which we now prove shows that a_n is equal to 0 for most values of n and when $a_n \neq 0$, we must have $a_n = \pm 1$.

Theorem 13.3.1. *Euler's pentagonal theorem*

$$\prod_{k=1}^{\infty}(1-x^k) = 1 + \sum_{k=1}^{\infty}(-1)^k \left\{x^{\frac{3k^2+k}{2}} + x^{\frac{3k^2-k}{2}}\right\}$$

Proof Writing

$$\prod_{k=1}^{\infty}(1-x^k) = 1 + \sum_{n=1}^{\infty} a_n x^n$$

we need to show that $a_n = (-1)^k$ if $n = \frac{3k^2+k}{2}$ or $\frac{3k^2-k}{2}$ for some natural number k and is equal to 0, otherwise. Also note that, if k_1 and k_2 are distinct natural numbers, then $\frac{3k_1{}^2 \pm k_1}{2}$ cannot be equal to $\frac{3k_2{}^2 \pm k_2}{2}$. To that end, we now fix the natural number n and let D denote the set of *all distinct part partitions* of n. The set D is divided into two disjoint subsets D_e and D_o that consist of partitions π in D such that π has an even (respectively odd) number of parts (thus when $n = 6$ both the sets D_e and D_o have 2 partitions each). The proof consists of defining an operation on almost all partitions in D that changes the parity of the number of parts. Thus this operation (to be defined) maps D_e and D_o into each other and is actually a bijection when n is not of the form $\frac{3k^2+k}{2}$ or $\frac{3k^2+k}{2}$. The proof is accomplished by first dividing the set D into two disjoint sets in a different manner.

Definition 13.3.2. Let π be a partition in D, say $\pi = (b_1, b_2, \ldots, b_k)$. The *base* $b = b(\pi)$ is the set of dots on the last line of (Ferrers diagram of) the partition π. By a slight abuse of language, we also use b to mean the size of b (that is, the number of dots on the last line of π). The term *slope* s of π refers to the set of dots on the line that makes an angle of 45° with the positive x and y axes such that s begins at the top right hand corner of the Ferrers diagram of π and descends downwards as far as it can (we imagine that the Ferrers diagram is drawn in the first quadrant of the real plane). By a slight abuse of language, we also use s to mean the size of s (that is, the number of dots on the slope of π).

Example 13.3.3. In Figure 13.10, we have $\pi = (11, 10, 9, 8, 3, 2)$ and $(b, s) = (2, 4)$. Note that the slope cannot be extended further (this has to be done if the fifth part was 7).

We now classify various partitions in D according to the relative values of the base and the slope as also whether or not *the base and slope meet*. These are illustrated in Figure 13.11 below. In diagram 1, $(b, s) = (1, 2)$ and the base and slope do not meet. In diagram 2, $(b, s) = (3, 3)$ and the base and slope meet. In diagram 3, $(b, s) = (1, 6)$ and the base and slope meet. In diagram 4, $(b, s) = (2, 1)$ and the base and slope do not meet.
For a partition $\pi \in D$, write $b(\pi)$ and $s(\pi)$ to denote, the base and slope of π respectively. We now form various subsets of D:

$$
\begin{aligned}
D_{\leq} &= \{\pi \in D : b(\pi) \leq s(\pi)\} \\
D_{>} &= \{\sigma \in D : b(\sigma) \geq s(\sigma) + 1\} \\
D_{<} &= \{\pi \in D : b(\pi) \leq s(\pi) - 1\} \\
D_{=} &= \{\pi \in D : b(\pi) = s(\pi) \textit{ and } b \textit{ and } s \textit{ do not meet}\} \\
D_{>>} &= \{\sigma \in D : b(\sigma) \geq s(\sigma) + 2\} \\
D_{>=} &= \{\sigma \in D : b(\sigma) = s(\sigma) + 1 \textit{ and } b \textit{ and } s \textit{ do not meet}\}
\end{aligned}
$$

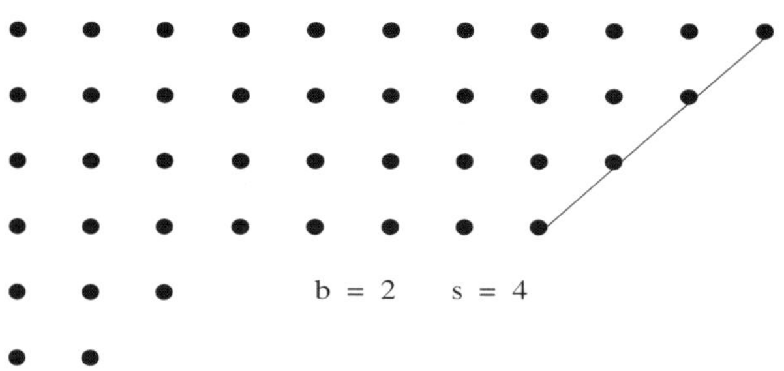

Figure 13.10: A distinct odd part partition with slope not meeting the base

Also, let $G_{\leq} = D_{<}\dot{\cup}D_{=}$ and $G_{>} = D_{>>}\dot{\cup}D_{>=}$. Then $G_{\leq}$ is a subset of $D_{\leq}$ and $G_{>}$ is a subset of $D_{>}$. *The two transformations f and g on the sets $G_{\leq}$ and $G_{>}$ respectively are jointly called Franklin transformations after the American mathematician Franklin who discovered them [29, 57].* Given $\pi \in G_{\leq}$ define $f(\pi) = \sigma$ where σ is obtained by removing the base of π and making it into a new slope as shown in Figure 13.12. In the first case, $(b, s) = (3, 4)$ so that $\pi \in D_{<}$. In the resulting new diagram σ, we have $(b', s') = (5, 3)$ and hence $\sigma \in D_{>>}$. In the second case, $(b, s) = (2, 2)$, but the slope does not meet the base so that $\pi \in D_{=}$. In the resulting new diagram σ, we have $(b', s') = (5, 2)$ and hence $\sigma \in D_{>>}$. In both the diagrams, *it is only the black dots that are part of the diagram; the hollow dots form the newly created slope.* The first diagram is the Ferrers diagram of partition $18 = 6 + 5 + 4 + 3$, which, after application of f becomes $18 = 7 + 6 + 5$. In the second diagram, we have $11 = 5 + 4 + 2$ which, after applying f becomes $11 = 6 + 5$.

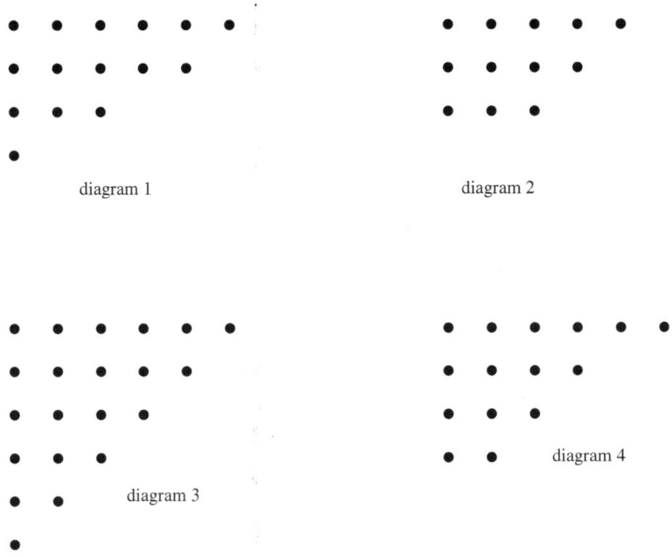

Figure 13.11: Ferrers diagrams of various types

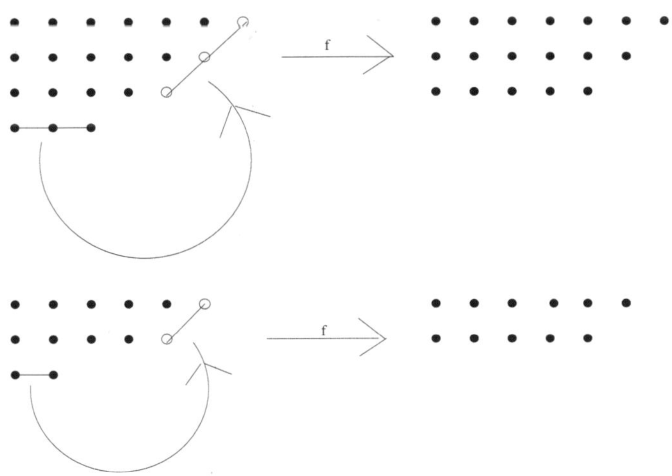

Figure 13.12: Franklin transformation f illustrated

Given $\sigma \in G_>$ define $g(\sigma) = \pi$ where π is obtained from σ by removing the slope of σ and making it into a new base as shown in Figure 13.13. In the first case, $(b, s) = (5, 3)$ so that $\sigma \in D_{>>}$. In the resulting new diagram π, we have $(b', s') = (3, 4)$ and hence $\pi \in D_<$. In the second case, $(b, s) = (3, 2)$, but the slope does not meet the base so that $\sigma \in D_{>=}$. In the resulting new diagram π, we have $(b', s') = (2, 2)$ (with the slope and the base not meeting) and hence $\pi \in D_=$.

The proof will now consist of two major parts, showing that f and g are well-defined maps. First let $\pi \in G_{\leq}$ with $b = b(\pi)$ and $s = s(\pi)$. Let $\sigma = f(\pi)$ and let $b' = b(\sigma)$ and $s' = s(\sigma)$. First suppose that b and s *do not meet.* In that case, when b is lifted and made into a new slope s', no dot of (the old slope) s is lost and since $b \leq s$, the new slope cannot protrude below the last line of σ say $\bar{b}$ and thus f is well-defined. Next, suppose that b and s *meet.* Then $\pi \in D_{<}$ and $b \leq s - 1$. Here, one dot of s is lost when we lift b, but that is okay since $b \leq s - 1$ ensures that the new slope $s' = b$ does not descend below the last line (which is $\bar{b}$ or $\bar{b} + 1$) of the new diagram σ. Hence f is well-defined in this case as well. Consider σ. First suppose that b' and s' *do not meet.* Then $b' = \bar{b} \geq b + 1 = s' + 1$ and hence $\sigma \in D_{<<}$ if $b' \geq s' + 2$ and $\sigma \in D_{<=}$ if $b' = s' + 1$. Therefore, $\sigma \in G_{>}$ as desired. Next suppose that b' and s' *meet.* Then $b' = \bar{b} + 1 \geq (b + 1) + 1 = s' + 2$ and hence $\sigma \in D_{>>}$ which implies $\sigma \in G_{>}$. We have thus shown that $f : G_{\leq} \to G_{>}$ is well defined.

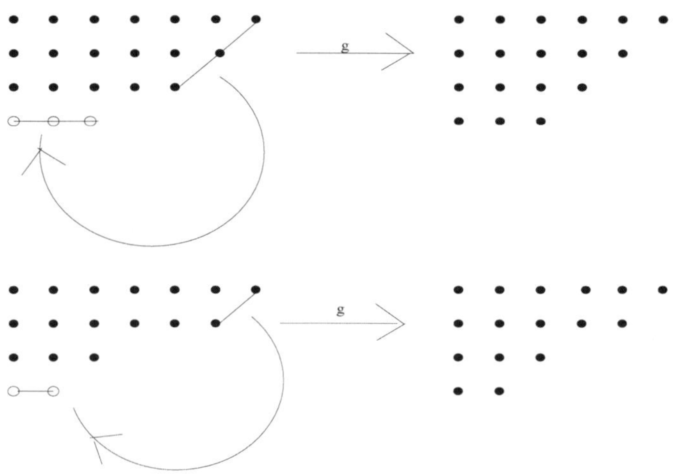

Figure 13.13: Franklin transformation g illustrated

Let $\sigma \in G_{>}$ with $b = b(\sigma)$ and $s = s(\sigma)$. Let $\pi = g(\sigma)$ and let $b' = b(\pi)$ and $s' = s(\pi)$. We are given that $b \geq s + 1$.Write b^* to denote the last but one line of π. Then b^* equals b or $b + 1$. First suppose that b and s *do not meet.* When we remove the slope and make it into a new base, no dot of b is lost. Hence $b' = s \leq b - 1 \leq b^* - 1$. Next suppose that b and s *meet.* Then σ is not in $D_{>=}$ and hence must be in $D_{>>}$ so that $b \geq s + 2$. Here one dot of b is lost when the slope is removed. Then $b' = s \leq b - 2 = (b - 1) - 1 \leq b^* - 1$ and hence in both the cases g is well defined and $\pi \in D$. Consider π. We have $b' = s \leq s'$. Hence if b' and s' do not meet, then $\pi \in D_{<}$ if $b' \leq s' - 1$ and $\pi \in D_{=}$ if $b' = s'$. So, $\pi \in G_{\leq}$. Finally, let b' and s' *meet.* If $s' = s$, then the new slope can reach only upto the last but one line b^* of π which is not true. So, $s' \geq s + 1$. Then $b' = s \leq s' - 1$ and $\pi \in D_{<}$. We have thus shown that $g : G_{>} \to G_{\leq}$ is well defined.

Let $G = G_{\leq} \dot{\cup} G_{>}$. We have established the bijections

$$G_{\leq} \xrightarrow{f} G_{>}; \;\; G_{>} \xrightarrow{g} G_{\leq}$$

and, in fact, g and f are inverses of each other. Hence, if we define $h : G \longrightarrow G$ by $h(\tau) = f(\tau)$ if $\tau \in G_{\leq}$ and $h(\tau) = g(\tau)$ if $\tau \in G_{>}$, then h is a bijection of G to itself such that $h^2 = h \circ h$ is the identity function on G. Notice that the operation h either removes the base of a Ferrers diagram or adds a new base to the Ferrers diagram. Hence the operation h either increases the number of parts of a given partition τ (if $\tau \in G_{\leq}$ i.e. $h = f$) by 1, or decreases the number of parts of a given partition by 1 (if $\tau \in G_{>}$ i.e. $h = g$). Hence if τ is in D and if h can be applied to τ, then $\tau \in D_e$ iff $h(\tau) \in D_o$. Hence if $G = D$, then, $|D_e| = |D_o|$. It remains to see when $G \neq D$. Let n be a number for which $G \neq D$. Then we have a partition τ which has the property that the base meets the slope and either $b = s$ or $b = s + 1$. Write $k = s$ to denote the number of parts of τ. First let $b = s$. Then the partition τ corresponds to the number:

$$n = k + (k+1) + \cdots + (2k-1) = \frac{3k^2 - k}{2} = s_k$$

Next let $b = s + 1$. Then the partition τ corresponds to the number:

$$n = (k+1) + (k+2) + \cdots + (2k) = \frac{3k^2 + k}{2} = t_k$$

It is easily checked that s_k and t_m are distinct for all values of k and m and hence we conclude the following. When $n \neq s_k, t_k$, we must have $G = D$ and therefore $|D_e| = |D_o|$ giving $a_n = 0$. When n is of the form s_k or t_k, there is exactly one partition τ in $G \neq D$ and the number of parts in that partition is k and hence its contribution to a_n is 1 (respectively -1) if k is even (respectively odd). $\square$

Figure 13.14 illustrates pentagonal numbers t_k. Here is a small table of the pentagonal numbers s_k and t_k.

k	1	2	3	4	5	6
s_k	1	5	12	22	35	51
t_k	2	7	15	26	40	57

Corollary 13.3.4. *Let n be a positive integer and let s_k and t_k respectively denote the numbers $\frac{3k^2-k}{2}$ and $\frac{3k^2+k}{2}$. Then the following assertions hold.*

(a) If $\frac{3a^2+a}{2} \leq n < \frac{3(a+1)^2-(a+1)}{2}$, then

$$p(n) = \sum_{k=1}^{a} (-1)^{k-1} \left[p(n - s_k) + p(n - t_k)\right]$$

(b) If $\frac{3a^2-a}{2} \leq n < \frac{3a^2+a}{2}$, then

$$p(n) = \sum_{k=1}^{a-1} (-1)^{k-1} \left[p(n - s_k)\right] + p\left(n - \frac{3a^2 - a}{2}\right)$$

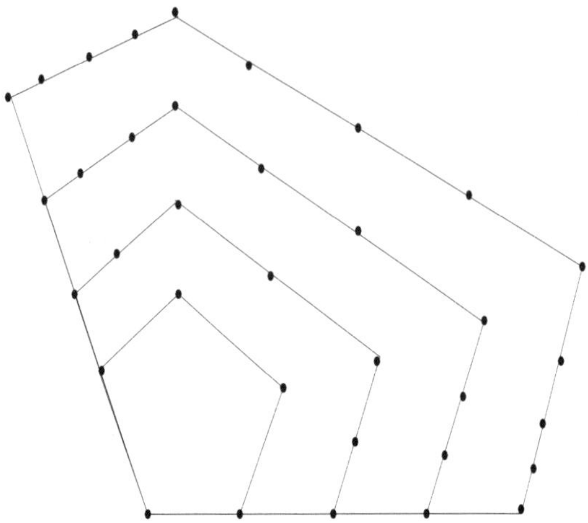

Figure 13.14: Pentagons and pentagonal numbers

Proof As we already noted,

$$P_{all\ part\ partitions}(x) \times f(x) = 1$$

where $f(x)$ is equal to (using Euler's Pentagonal theorem, Theorem 13.3.1)

$$1 + \sum_{k=1}^{\infty} (-1)^k \left\{ x^{\frac{3k^2+k}{2}} + x^{\frac{3k^2-k}{2}} \right\}$$

Look at the first equation and equate coefficient of x^n on both sides which is zero on the R.H.S. On the L.H.S. the the product x^n is obtained by looking at all the terms of the form $(-1)^k p(m) s_k$ or $(-1)^k p(m) t_k$ where $m + s_k = n$ (respectively $m + t_k = n$). This proves the result. □

A restatement of Corollary 13.3.4 is the following.

Corollary 13.3.5. *We have*

$$p(n) = p(n-1) + p(n-2) - p(n-5) + p(n-7) + p(n-12) + p(n-15)$$

$$-p(n-22) - p(n-26) + p(n-35) + p(n-40) - p(n-51) - p(n-57) + \cdots$$

where, by convention we drop those $p(m)$ on the R.H.S. that have $m < 0$ (and thus the sum on the R.H.S. is a finite sum for every n).

Example 13.3.6. Here are a few illustrations.

(a) $p(5) = p(4) + p(3) - p(0) = 5 + 3 - 1 = 8$.

(b) If $n = 9$, then $(s_1, t_1) = (1, 2)$ and $(s_2, t_2) = (5, 7)$ are the only possibilities. So,

$$p(9) = p(8) + p(7) - p(4) - p(2) = (22 + 15) - (5 + 2) = 30$$

(c) When $n = 15$, we have $a = 3$ and hence

$$p(15) = p(14) + p(13) - p(10) - p(8) + p(3) + p(0)$$

13.4 Exercises for Chapter 13

13.1 Show that $q(n, k) = q(n-1, k) + q(n-1, k-1)$ and hence use Pascal identity to prove that $q(n, k) = \binom{n-1}{k-1}$ for all $n, k \geq 1$.

13.2 Obtain a recurrence relation for $q(n)$ and then solve it using generating functions.

13.3 Write conjugate partitions of each one of the following partitions.

(a) $28 = 10 + 5 + 5 + 3 + 2 + 2 + 1$

(b) $17 = 9 + 4 + 4$

(c) $20 = 7 + 5 + 4 + 2 + 1 + 1$

(d) $18 = 3 + 3 + 3 + 3 + 3 + 3$

(e) $29 = \underbrace{3 + 3 + \cdots + 3}_{9\ times} + 2$

13.4 Let $p_k(n)$ denote the number of partitions of n into k parts. Prove the following.

(a) $\sum_{k=1}^{m} p_k(n) = p_m(n + m)$

(b) $p_k(n) = p_k(n - k) + p_{k-1}(n - 1)$

(c) $p_n(2n) = p(n)$

(d) $p(n) \geq \frac{p(n+1)+p(n-1)}{2}$

13.5 Use Theorem 13.2.2 to find $p(10)$.

13.6 Use Theorem 13.2.3 to find the number of self conjugate partitions of 17 and 20.

13.7 Find the expected number of parts in a randomly chosen ordered partition of an integer n.

13.8 Let a_n denote the number of partitions of n into parts such that each part is at least two. Find the g.f. $f(x)$ of a_n's. Show that $(1 - x)f(x) = P_{all\ part\ partitions}(x)$. Use this to prove that the number of partitions of n in which no part equals 1 is equal to $p(n) - p(n - 1)$.

13.9 Find the number $p_2(n)$ of the number of partitions of n into exactly 2 parts.

13.10 We wish to find $p_3(n)$, the number of partitions of n into exactly 3 parts. Proceed through the following steps.

(a) Show that the g.f.

$$h(x) = \sum_{n=0}^{\infty} a_n x^n = (1-x)^{-3}(1+x)^{-1}(1-\omega x)^{-1}(1-\omega^2 x)^{-1}$$

where $\omega = e^{\frac{2\pi i}{3}}$ is a cube root of 1.

(b) Obtain the partial fraction expansion

$$h(x) = \frac{1}{6}(1-x)^{-3} + \frac{1}{4}(1-x)^{-2} + \frac{17}{72}(1-x)^{-1} + \frac{1}{8}(1+x)^{-1}$$
$$+\frac{1}{9}(1-\omega x)^{-1} + \frac{1}{9}(1-\omega^2 x)^{-1}$$

(c) Use the binomial theorem (Theorem 1.2.5) to get

$$\left| a_n - \frac{1}{12}(n+3)^2 \right| < \frac{1}{2}$$

(d) Conclude that $p_3(n) = \left\{ \frac{n^2}{12} \right\}$ where $\{a\}$ denotes the integer nearest to a.

13.11 Show that the number of incongruent triangles T that can be formed by taking the vertices of T to be the vertices of a regular n-gon is same as $p_3(n)$ by exhibiting a bijection between the two sets.

13.12 Prove that the number of partitions of n into exactly k parts is the same as the number of partitions of $n-k$ into at the most k parts.

13.13 Use generating functions to prove the following.

(a) Every natural number can be uniquely expressed as a sum of distinct powers of 2.

(b) Generalize the previous exercise: Given a fixed $m \geq 2$, prove that every n has a unique expression of the form

$$n = \sum_{j \geq 0} \alpha_j m^j$$

with each $0 \leq \alpha_j \leq m-1$ for every j.

13.14 Consider the correspondence between $SC(n)$ and $DO(n)$ set up in the proof of Theorem 13.2.4. Let $\pi = (a_1, a_2, \ldots, a_k) \in SC(n)$ correspond to $\tau = (b_1, b_2, \ldots, b_j) \in DO(n)$.

(a) Show that j is the largest integer i for which $a_i \geq i$.

(b) Show that the Durfee square of π is D_j.

(c) Show that $b_i = 2a_i - (2i-1)$ for all $i = 1, 2, \ldots, j$ while

$$a_i = \frac{b_i + (2i-1)}{2} \quad \forall\ i \leq j$$

and $a_{j+r} = |\{i : a_i \geq j+r\}| \quad \forall\ r \geq 1$

13.15 Use the correspondence given in Theorem 13.2.4 to obtain self conjugate partitions corresponding to each one of the following partitions into distinct odd parts.

(a) $40 = 21 + 11 + 5 + 3$

(b) $88 = 33 + 29 + 15 + 11$

(c) $25 = 9 + 7 + 5 + 3 + 1$

(d) $52 = 29 + 15 + 5 + 3$

(e) $40 = 31 + 5 + 1$

13.16 Show that the following partitions are self conjugate and then use the correspondence given in Theorem 13.2.4 to find distinct odd part partitions corresponding to each one of them.

(a) $18 = 6 + 6 + 2 + 2 + 2$

(b) $30 = 9 + 7 + 4 + 4 + 2 + 2 + 1 + 1$

(c) $42 = 9 + 6 + 6 + 6 + 6 + 6 + 1 + 1 + 1$

13.17 Use Euler's pentagonal theorem (Theorem 13.1.1) to complete the last entry $p(10)$ in the table of $p(n)$ in Section 13.1. Use this to find $p(11)$ and $p(12)$ recursively.

13.18 Find a suitable arithmetic progression that gives the number of dots in Figure 13.14 of the pentagonal numbers t_k.

13.19 We wish to prove that $p(n)$ takes both even and odd values infinitely many times. Proceed through the following steps.

(a) Let a be a sufficiently large positive integer and let $n = \frac{3a^2+a}{2}$. Prove that

$$p(n) = \sum_{k=1}^{a-1}(-1)^{k-1}[p(n - s_k) + p(n - t_k)] + (-1)^{a-1}(p(a) + 1)$$

(b) Show that if we assume that $p(b)$ is even $\forall\, b \geq a$, then reading the equation in the first part modulo 2 gives a contradiction.

(c) Show that we again get a contradiction if we replace even by odd and hence complete the proof.

13.20 Prove that the number of partitions of n in which only the odd parts can repeat any number of times while each even part can occur at the most once is equal to the number of partitions of n in which every part can occur at the most three times.

13.21 Let $m \geq 2$ be a fixed positive integer. Show that the number of partitions of n in which no part is divisible by $m+1$ equals the number of partitions of n in which each part can repeat at the most m times.

13.22 Prove that the number of partitions of n in which consecutive integers do not both appear as parts is equal to the number of partitions of n in which no integer occurs as a part exactly once.

13.23 Prove that the number of partitions of $a-b$ into $c-1$ parts such that no part is greater than b equals the number of partitions of $a-c$ into $b-1$ parts such that no part is greater than c.

13.24 Prove that the number of partitions of n into exactly m parts is equal to the coefficient of $x^n y^m$ in the product $\prod_{i=1}^{\infty}(1-yx^i)^{-1}$.

13.25 Let $e(n)$ (respectively $o(n)$) denote the number of partitions of n into an even (respectively odd) number of parts and let $t(n) = e(n) - o(n)$. Prove the following.

(a) $t(n)$ equals $\sum_{i\geq 1}(u_{2i} - u_{2i-1})$ where u_j is the coefficient of $x^n y^j$ in the product $\prod_{k=1}^{\infty}(1-yx^k)^{-1}$.

(b) The number $t(n)$ is equal to the coefficient of x^n in

$$\prod_{i=1}^{\infty}(1-(-1)x^i)^{-1} = \prod_{k=1}^{\infty}(1+x^k)^{-1}$$

(c) The number $t(n)$ is equal to the coefficient of x^n in $\prod_{k=1}^{\infty}(1-x^{2k-1})$.

(d) The coefficient of x^n in $\prod_{k=1}^{\infty}(1-x^{2k-1})$ and the coefficient of x^n in $\prod_{k=1}^{\infty}(1+x^{2k-1})$ have the same absolute value.

(e) The absolute difference between the number of partitions of n into an even number of parts and the number of partitions of n into an odd number of parts is the same as the number of partitions of n into distinct parts each of which is odd.

13.26 Let a_n denote the number of partitions of n into parts each of which is a power of 2. Prove the following.

(a) $a_{2n+1} = a_{2n}$

(b) $a_{2n} = a_{2n-1} + a_n$

(c) a_n is even $\forall\, n \geq 2$

13.27 We want to look at the partitions $\pi = (a_1, a_2, \ldots, a_m)$ of n with the following property. When we consider the (partial) partitions $a_i + a_{i+1} + \cdots + a_j$ where $1 \leq i \leq j \leq m$, we get all the numbers between 1 and n as sums. For example the partition $n = 1 + 1 + \cdots + 1$ has this property. For which n is it true that $(1, 1, \ldots, 1)$ is the only partition with the stipulated property?

13.28 Prove that the number of partitions of n in which each part appears $2, 3$ or 5 times is equal to the number of partitions of n in which each part is an integer congruent to $2, 3, 6, 9$ or 10 modulo 12.

13.29 Show that the number of partitions of n in which no integer appears as a part exactly once is equal to the number of partitions of n in which each each part is not congruent to 1 or 5 modulo 6.

13.30 Prove that the number of partitions of n in which there is a unique smallest part and the largest part is at most twice the smallest part is equal to the number of partitions of n in which the largest part is odd and the smallest part is larger than one-half of the largest part.

13.31 Apply Franklin transformation to the following partitions (wherever it can be applied).

(a) $23 = 11 + 8 + 4$

(b) $27 = 9 + 8 + 7 + 3$

(c) $23 = 10 + 9 + 8 + 6 + 3$

(d) $100 = 25 + 24 + 20 + 15 + 10 + 6$

13.32 We wish to make an estimate for $p(n)$, the total number of partitions of n. Write $f(x) = P_{all\ part\ partitions}(x)$, and proceed through the following steps.

(a) Show that

$$\log f(x) = \sum_{k=1}^{\infty} \frac{1}{k} \frac{x^k}{1 - x^k}$$

(b) *From this point on fix the real number x in the open interval* $(0, 1)$. Show that

$$\frac{x^{n-1}}{\sum_{j=0}^{n-1} x^j} < \frac{1}{n}$$

(c) Show that

$$\frac{x^n}{1 - x^n} < \frac{1}{n} \frac{x}{1 - x}$$

(d) Show that

$$\log f(x) < \frac{x}{1 - x} \sum_{m=1}^{\infty} \frac{1}{m^2}$$

(e) Using

$$\sum_{m=1}^{\infty} \frac{1}{m^2} < \int_1^{\infty} \frac{dx}{x^2} = 2$$

show that $\log f(x) < \frac{2x}{1-x}$.

(f) Prove that $\log(p(n)) < \frac{2x}{1-x} - n\log x$.

(g) Using $-\log x < \frac{1}{x} - 1 = \frac{1-x}{x}$ show that

$$\log p(n) < \frac{2x}{1 - x} + \frac{n(1 - x)}{x}$$

(h) Make the substitution $x = \frac{\sqrt{n}}{\sqrt{n+1}}$ to get $\log p(n) < 3\sqrt{n}$.

More precise estimate

$$p(n) \approx \frac{1}{4\sqrt{3}n} e^{\pi\sqrt{2/3}\sqrt{n}}$$

was obtained by Hardy-Ramanujan-Rademacher.

Chapter 14

Group action on a set

14.1 Introduction and the class equation

This chapter, combinatorially is intended to be a forerunner to the next chapter 15, where purely combinatorial applications of group action in the form of Polya theory are discussed. However, there appears to be sufficiently large interesting material on group actions with applications that are purely algebraic. That is the basic purpose of separating this chapter from chapter 15. In spite of that, the results as well as proofs have a combinatorial flavour and quite importantly use combinatorial techniques. The material we cover here includes the class equation, group actions on conjugacy classes of subgroups and action on cosets, Sylow theorems, simplicity of the group A_n for all $n \geq 5$, a discussion on outer automorphisms of S_n and the classification of finite subgroups of the rotation groups $SO_2(\mathbb{R})$ and $SO_3(\mathbb{R})$.

Notation 14.1.1. By a permutation group (G, X) is meant a group G acting on a set X. This means that $G \leq S_X$ (the notation $\leq$ is used to denote a subgroup and S_X means the symmetric group with the set X on which it is acting. That is, S_X is the set of all the permutations on the set X). Members of G will be generally denoted by Greek letters α, β, γ etc. and the elements of the (ground) set X on which G is acting will be denoted by lower case letters x, y, z etc. though we use g, h etc. for group elements particularly when the group acts on itself. Unless the group is known to be abelian, it is written multiplicatively with 1 as the multiplicative identity. We also write (by a slight abuse of language) 1 to mean the trivial group $\{1\}$ that consists of the identity element alone. $H \leq G$ means H is a subgroup of G and $H \lhd G$ means that H is a normal subgroup of G.

Definition 14.1.2. Let $x, y \in X$. Then $x \sim y$ if there is some $\alpha \in G$ such that $\alpha(x) = y$.

Lemma 14.1.3. *$\sim$ is an equivalence relation on the set X.*

Proof

- *Reflexivity:* $x \sim x \;\forall\, x \in G$ because $id(x) = x$ where id refers to the identity permutation.
- *Symmetry:* $x \sim y \;\Rightarrow\; \exists\, \alpha \in G$ such that $\alpha(x) = y$. This implies that $\alpha^{-1}(y) = x$. Here $\alpha^{-1} \in G$ since G is a group and we thus have $y \sim x$.
- *Transitivity:* Let $x \sim y$ and $y \sim z$. Then $\exists\, \alpha, \beta$ such that $\alpha(x) = y$ and $\beta(y) = z$. Hence letting $\gamma = \beta \circ \alpha$ we get $\gamma(x) = z$ and $\gamma \in G$ since G is a group. □

Definition 14.1.4. The equivalence classes of X given by $\sim$ (i.e. under the action of G) are called *orbits* of X (under the action of G).

Notation 14.1.5. $x^G = \{y : y \in X \;\; and \;\; x \sim y\}$. Observe that x^G is just the orbit of (or containing) the element x (under the G-action).

Lemma 14.1.6. *Let $x, y \in G$. Then:*

(a) $y \in x^G \Rightarrow y^G \subset x^G$.

(b) $y \in x^G \Rightarrow x \in y^G$.

(c) $y \in x^G \Rightarrow x^G = y^G$.

(d) Either x^G and y^G are equal or are disjoint.

Trivial proof of this Lemma uses the fact that two equivalence classes are the same or are disjoint.

Definition 14.1.7. Let $x \in X$. Then the *stabilizer of x (in G)* is defined by $G_x = \{\sigma \in G : \sigma(x) = x\}$.

Note that G_x is a subgroup of G.

Theorem 14.1.8. *(The Orbit-Stabilizer Theorem) Let $x \in X$. Then, the size of the orbit and the stabilizer are connected by:*

$$|G_x||x^G| = |G|$$

Proof This is equivalent to proving that

$$|x^G| = \frac{|G|}{|G_x|} = [G : G_x]$$

and the last number is the index of $H = G_x$ in G. It is therefore enough to produce a bijection between the set x^G, the orbit of x and the set Γ that consists of all the left cosets of H in G. Define a function $f : x^G \to \Gamma$ by setting $f(y) = \sigma H$ if $\sigma(x) = y$.

- *Claim* f is well-defined.
  Proof Let σ, τ be such that $\sigma(x) = \tau(x) = y$. Then $(\tau^{-1}\sigma)(x) = \tau^{-1}(y) = x$ and hence $\tau^{-1}\sigma \in H$ so that σ in τH and therefore $\sigma H = \tau H$.

- *Claim* f is injective.
  Proof Let $y, z \in x^G$ be such that $f(y) = f(z)$. If $\alpha(x) = y$ and $\beta(x) = z$, then the given condition implies that $\alpha H = \beta H$ so that $\beta^{-1}\alpha \in H$ and therefore $\beta^{-1}\alpha$ fixes x. So , $z = \beta(x) = \alpha(x) = y$.

- *Claim* f is surjective.
  Proof If $\sigma H \in \Gamma$, then let $y = \sigma(x)$. Clearly then $f(y) = \sigma H$. □

Corollary 14.1.9. *The size of any orbit divides the size of G.*

Example 14.1.10. A group of order 40 cannot have an orbit whose size is 12 and a group of prime order p can have orbits of sizes 1 and p and no other size is possible, while the possible orbit sizes in the action of a group of order p^2 are $1, p, p^2$.

Example 14.1.11. Let $X = [12]$ be the set of the first 12 natural numbers and let G be a permutation group acting on X where G equals one of the following.

(a) $G = <\alpha>$ where α denotes the single 12-cycle $(1, 2, 3, \ldots, 11, 12)$ (thus G is a cyclic group of order 12 generated by α). Then we have only one orbit. Note that the stabilizer of any element of X is the trivial group that consists of the identity permutation alone.

(b) $G = <\beta>$ where $\beta = \alpha^2$. Here we get two orbits, the first is $\{1, 3, 5, 7, 9, 11\}$ and the second is $\{2, 4, 6, 8, 10, 12\}$.

(c) $G = <\gamma>$ where $\gamma = \alpha^3$. Then G has 3 orbits.

(d) $G = <\delta>$ where $\delta = \alpha^4$. Here G has 4 orbits on X, one of which is $\{2, 6, 10\}$.

(e) $G = <\omega>$ where $\omega = \alpha^6$. Here G has 6 orbits, one of which is $\{5, 11\}$.

(f) $G = \{1\} = 1$. Then G has 12 orbits on X each one of which is a singleton.

(g) $G = <\mu>$ where $\mu = (1, 2, 3)(4, 5, 6, 7)(8, 9, 10)(11, 12)$. Then G has order 12 and has 4 orbits on X. *Notice the difference between this situation and the one in (a). Both are cyclic groups of order* 12 *but have different combinatorial representations.*

(h) $G = <\pi>$ where $\pi = (1, 2, 3, 4)(5, 6, 7)$. Then G has 7 orbits on X.

Definition 14.1.12. A permutation group G on a set X is called *transitive* if G has only one orbit (on X). Equivalently, given x and y in X there is a permutation $\alpha \in G$ such that $\alpha(x) = y$. A transitive group action is said to be *sharp transitive* if $\forall\, x, y \in X,\ \exists\, !\alpha \in G$ such that $\alpha(x) = y$. Equivalently, a transitive group action of G on X is sharp transitive if $G_x = 1$ for some $x \in X$ (and hence for all $x \in X$).

Here is an example of a sharp transitive group action. Let G be *any group* and let $X = G$. For $g \in G$ and $x \in X = G$, we let $g(x) = gx$ where the product is calculated in the group G. This is clearly a permutation. Further, given x and y, there is a unique $g = yx^{-1}$ that solves $g(x) = y$ showing that G acts sharp transitively on G. This G-action on G is called the *Cayley representation or the left regular representation.* We leave it as an exercise to show that upto an equivalence, this is the only sharp transitive action of G.

In many situations, we do not have G as a permutation group on X but rather *G acts as a permutation group on X*. There is a permutation group (G^*, X) and a surjective homomorphism $\phi : G \to G^*$. Then G acts on X as follows: $g(x) = (\phi(g))(x)$. If K denotes the kernel of the homomorphism ϕ, then K is a normal subgroup of G and $G^* \cong \frac{G}{K}$. Equivalently, G has a quotient (namely G^*) which is a permutation group on X. We can still define the stabilizer $G_x = \{g : g(x) = x\}$. This is indeed a subgroup of G and in fact, we have $\frac{G_x}{K} \cong G^*_x$. Also, the orbits under the G-action are the same as the orbits under the G^* action. Hence in the orbit stabilizer theorem (Theorem 14.1.8) for G^*, replacing G^* by $\frac{G}{K}$ and replacing G^*_x by $\frac{G_x}{K}$ obtains the orbit stabilizer theorem (Theorem 14.1.13) for the more general situation:

Theorem 14.1.13. *(Generalized Orbit Stabilizer theorem) Let G act on X as a permutation group. Then $|G_x| \times |x^G| = |G|$*

We say that G acts faithfully on X if $K = 1$. In that case, we actually have an isomorphism from G to G^* and the group G has an isomorphic copy in S_X which is expressed by saying that G embeds into S_X.

We now describe different actions of a group G on subsets of G. First let $H \leq G$ and let $[G : H] = n$. Let $X = \{H = H_0, H_1, \ldots, H_{n-1}\}$ denote the set of all the n distinct left cosets of H in G. For $g \in G$ and H_i, a left coset, we let $g(H_i) = H_j$ where $gH_i = H_j$; note that this is well-defined since $H_i = xH$ for some $x \in G$ and hence H_j is the left coset $(gx)H$. Further, g gives a permutation since $gH_i = gH_j$ implies $H_i = H_j$ by the standard cancellation in groups. This action is also surjective since given H_k, we have $g(g^{-1}H_K) = H_k$. Thus g gives a permutation on X and hence G acts on X. What is the kernel K of this action? If $g \in K$, then $gH_i = H_i$ for all i and hence in particular, $gH = H$ and therefore $g \in H$. We thus have $K \lhd G$ (this means K is normal in G) and $K \leq H$. A special case of this action is obtained when we take $H = 1$, that is, when H is the trivial group. In that case, cosets can be identified with the group elements and we just have *the left regular action of G on itself.*

Definition 14.1.14. Let $a \in G$. The *centralizer* $Z(a) = \{g \in G : ga = ag\}$. The centralizer of a is indeed a subgroup of G. Given $a, b \in G$, we write $a \sim b$ if there is some $x \in G$ such that $b = x^{-1}ax$ *and call b a conjugate of a.*

Conjugacy is indeed an equivalence relation: If $a \sim b \sim c$ then there are x, y such that $b = x^{-1}ax$ and $c = y^{-1}by$ and hence $c = y^{-1}x^{-1}axy = (xy)^{-1}a(xy)$ and thus $a \sim c$. In fact, here G is acting on itself as follows: given $g \in G$, we have $\phi_g(a) = g^{-1}ag$. Then the orbit of a under this action is called its *conjugacy class* $C(a)$ and the stabilizer of a is clearly

$$\{g \in G : g^{-1}ag = a\} = \{g \in G : ag = ga\} = Z(a)$$

Using the orbit stabilizer theorem (Theorem 14.1.13), we get

$$|Z(a)| \times |C(a)| = |G| \tag{14.1}$$

When is $|C(a)| = 1$? The orbit-stabilizer theorem tells us that in that case $Z(a) = G$ and for such an a, we have $ag = ga \ \forall \ g \in G$.

Definition 14.1.15. *The center $Z = Z(G)$ of a group is defined by:*

$$Z = Z(G) = \{a : ag = ga \ \forall \ g \in G\} = \bigcap_{x \in G} Z(x)$$

The center $Z(G)$ is a normal subgroup of G. Given any $a \in G$, the conjugacy class of $C(a)$ of a is a singleton iff $a \in Z$. *A singleton conjugacy class is called a trivial conjugacy class.* Every non-trivial conjugacy class C has $|C| \geq 2$ and for such a C, every element a of C is not a central element (that is, it is not in the center). Since distinct conjugacy classes are disjoint, we get the class equation of the group G:

Theorem 14.1.16. *(The class equation) Any group G has a unique decomposition into conjugacy classes and we thus have*

$$G = \left(\bigcup_{a \in Z} C_a\right) \bigcup \left(\dot{\bigcup_{j=1}^{r}} C_j\right) = Z(G) \bigcup \left(\dot{\bigcup_{j=1}^{r}} C_j\right)$$

and hence

$$|G| = |Z| + \sum_{j=1}^{r} |C_j| \tag{14.2}$$

Assuming that $|Z| = s$, we have s trivial conjugacy classes and r non-trivial conjugacy classes and *combinatorially most important point is that every non-trivial conjugacy class has a size which divides the order of the group G.* Here is an immediate application.

Theorem 14.1.17. *Let G be a group of order p^n where p is a prime and n is a natural number. Then the center Z is non-trivial.*

Proof Since p is the only prime divisor of $|G|$, every non-trivial conjugacy class C must have size that is of the form p^t where $t \geq 1$. In particular, p divides $|C|$. Since p also divides $|G|$, p must divide $|Z|$ (using equation (14.2)) and hence Z has order p^m where m is some natural number. □

Corollary 14.1.18. *Let p be a prime and let G be a group of order p^2. Then G is abelian.*

Proof Let Z be the center. We have to show that $Z = G$ and Theorem 14.1.17 implies that $Z \neq 1$. Hence we are done if we show that $|Z| \neq p$. Suppose $|Z| = p$. If $a \notin Z$, then $Z(a)$ contains Z as well as a and hence $|Z(a)| \geq p + 1$ and therefore $Z(a) = G$, a contradiction. □

Theorem 14.1.19. *Let β and γ be two permutations in the symmetric group S_n. Then γ is a conjugate of β iff they have the same cycle type.*

Proof Let $\beta \sim \gamma$. Then there is some $\alpha \in S_n$ such that $\gamma = \alpha^{-1}\beta\alpha$. Let $(c_1c_2\cdots c_m)$ be a cycle of β in its cycle decomposition and let $d_i = \alpha(c_i)$ (where all the subscripts are read modulo m). We have:

$$d_i \xrightarrow{\alpha^{-1}} c_i \xrightarrow{\beta} c_{i+1} \xrightarrow{\alpha} d_{i+1}$$

showing that $(d_1d_2\cdots d_m)$ is a cycle in the cycle decomposition of γ. Hence there is a one-to-one correspondence between the m-cycles of β and m-cycles of γ. Converse is in fact proved by finding a suitable α (refer to Chapter 3 and Exercise 14.7). Indeed, a careful counting can also find how many such α's are there. □

Theorem 14.1.20. *Let $\beta \in A_n$ where A_n is the alternating group (subgroup of even permutations), Then $Z_{A_n}(\beta) \leq Z_{S_n}(\beta)$ and $C_{A_n}(\beta) \subset C_{S_n}(\beta) = C$. Further, precisely one of the following is true.*

(a) $Z_{A_n}(\beta) = Z_{S_n}(\beta)$ and then $C_{A_n}(\beta) \subsetneq C_{S_n}(\beta)$. In fact, we then have $C = C' \cup C''$ where C' and C'' are disjoint conjugacy class in A_n and $C_{A_n}(\beta) = C'$. Hence $|C'| = \frac{1}{2}|C|$. In this case, every permutation commuting with β is an even permutation.

(b) $Z_{A_n}(\beta) \lneqq Z_{S_n}(\beta)$ and then $C_{A_n}(\beta) = C_{S_n}(\beta)$. In this case, we have both even and odd permutations commuting with β.

Proof Let H denote the subgroup $Z_{S_n}(\beta)$ of S_n and H' the subgroup $Z_{A_n}(\beta)$ of A_n. Note that β is even implies that all the permutations in C are also even. If two permutations α and γ are in A_n and are conjugates in A_n, then they are conjugates in S_n as well, showing that if C is a conjugacy class in S_n such that C consists of even permutations, then C is a union of conjugacy classes in A_n. Further, any subgroup of S_n either is a subgroup of A_n or has an equal number of even and odd permutations, the former making a subgroup of A_n. The *class equation* (Theorem 14.1.16) implies that $|C'| \times |H'| = \frac{n!}{2}$ while we also have $|C| \times |H| = n!$. Hence if $H' = H$, we are in case (a). Otherwise, $[H : H'] = 2$ and we have $C' = C$ in which case the conjugacy class of β in S_n continues to be conjugacy class of β in A_n. □

As an example, consider $n = 5$ and first let $\beta = (123)$ be a 3-cycle. The number of such 3-cycles in S_5 (or in A_5, since a 3-cycle is an even permutation) is equal to $\binom{5}{3} \times 2! = 20$ and these 3-cycles are conjugates in S_5 using theorem 14.1.19. Therefore, by the orbit stabilizer theorem (Theorem 14.1.13), $|Z_{S_5}(\beta)| = \frac{5!}{20} = 6$. Since the centralizer of β in S_5 must contain the transposition (45), which is an odd permutation (in fact the centralizer is simply a cyclic group of order six generated by $(123)(45)$), we get $[Z_{S_5}(\beta) : Z_{A_5}(\beta)] = 2$. Therefore *all the twenty 3-cycles are in a single conjugacy class in A_5 as well.* Now consider a different situation with $\tau = (12345)$. The number of such 5-cycles is $4! = 24$ and since these are in a single conjugacy class in S_5, we must have (by the orbit-stabilizer theorem, Theorem 14.1.13), $|Z_{S_5}(\tau)| = \frac{5!}{24} = 5$. For three independent reasons, we can now assert that the 24 cycles of length 5 split into two conjugacy classes in A_5. First since $< \tau > \leq Z_{S_5}(\tau)$ and since τ is an even permutation, it follows that $Z_{S_5}(\tau) =< \tau >= Z_{A_5}(\tau)$. Alternatively, $Z_{A_5}(\tau)$

is an index 1 or index 2 subgroup of $Z_{S_5}(\tau)$ which is of order 5. Hence $Z_{A_5}(\tau) = Z_{S_5}(\tau)$. A third way of arguing is as follows. Since $C = C(\tau)$, the conjugacy class of τ in S_5 has order 24 and since 24 does not divide 60, which is the order of A_5, C must split into two classes C' and C'' both of order 12 where $C' = C_{A_5}(\tau)$ in A_5. All the three arguments tell us that the conjugacy class of τ in A_5 consists of twelve 5-cycles. Finally, note that the only other order possible for a non-identity permutation σ in A_5 is 2 (since 4-cycles are odd permutations). σ must be of the form $(ab)(cd)$ where a, b, c, d are all distinct and the number of such σ's is $\binom{5}{1} \times 3 = 15$. These 15 permutations form a single conjugacy class in S_5 and that conjugacy class cannot split into two equal sized conjugacy classes in A_5 since 15 is an odd number. We record these observations in the following lemma.

Lemma 14.1.21. *The conjugacy classes in A_5 have sizes* $1, 20, 15, 12, 12$.

Given a natural number k, one wants to know as to how large the total number of finite groups (upto isomorphism) that have k conjugacy classes, can be. It was shown by Landau that this number is finite *for every k.* We wish to prove this result and the treatment follows Rose [45].

Lemma 14.1.22. *Let k be a natural number and r a real number. Let $t(r, k)$ the total number of solutions $(n_1, n_2, \ldots, n_k)$ to the equation*

$$\sum_{i=1}^{k} \frac{1}{n_i} = r$$

subject to the condition that each n_i is a natural number. Then $t(r, k)$ is finite.

Proof We make an induction on k. When $k = 1$, we are solving $\frac{1}{n} = r$ which either has a single solution if $r = \frac{1}{n}$ (where $n = n_1$) if r has that form and has no solution otherwise. So the assertion is true when $n = 1$. Now let $k \geq 2$ and assume that the assertion is true for $k - 1$. Notice that the given assertion is equivalent to proving that the given equation has a finite number of solutions $(n_1, n_2, \ldots, n_k)$ with the additional requirement $n_1 \geq n_2 \geq \cdots \geq n_k$ (since the difference between the two situations is at the most a factor of $k!$ which is finite). Hence assume w.l.o.g. that $n_1 \geq n_2 \geq \cdots \geq n_k$. Then we have:

$$\begin{aligned} r &= \frac{1}{n_1} + \frac{1}{n_2} + \cdots + \frac{1}{n_k} \\ &\leq \frac{k}{n_k} \end{aligned}$$

Thus $n_k \leq \frac{k}{r}$ and hence $n_k \leq m$ where $m = \lfloor \frac{k}{r} \rfloor$. Letting $j = 1, 2, \ldots, m$, we write $r_j = r - \frac{1}{j}$. Then every solution of the original question corresponds to a solution of

$$r_j = \sum_{i=1}^{k-1} \frac{1}{n_i}$$

for some r_j and hence

$$t(r,k) \leq \sum_{j=1}^{m} t(r_j, k-1)$$

By induction, each summand on the R.H.S. is finite and hence $t(r,k)$ is finite. □

Theorem 14.1.23. *(Landau's theorem) Let k be a natural number. Then the number of finite groups that have k conjugacy classes is a finite number.*

Proof It is enough to show that the order of a finite group G with k conjugacy classes is bounded by some $N(k)$ where $N(k)$ depends only on k. Let the G be such a group with the k conjugacy classes: $C_1, C_2, \ldots, C_k$ where $C_1 = \{1\}$ is the singleton conjugacy class of the identity element. Choose a representative $x_i \in C_i$ and let n_i denote $|Z(x_i)|$. Then using the orbit-stabilizer theorem (Theorem 14.1.13), we have $n = |G| = n_i|C_i|$ for every $i = 1, 2, \ldots, k$ and the class equation (Theorem 14.1.16) gives

$$|G| = \sum_{i=1}^{k} |C_i| = \sum_{i=1}^{k} \frac{|G|}{n_i}$$

from which we get

$$1 = \sum_{i=1}^{k} \frac{1}{n_i}$$

Here $n_1 = n$ and Lemma 14.1.22 shows that we have only a finite number of such sequences $(n_1, n_2, \ldots, n_k)$ satisfying the given equation and hence, in particular, there is a natural number $N(k)$ such that $n \leq N(k)$ showing that the order of G is bounded by $N(k)$. □

14.2 Sylow theorems

Definition 14.2.1. Let G be a group of order $n = k \times p^m$ where p is a prime, m and k are positive integers such that k is coprime to p. Then a subgroup H of order p^m of G is called *a Sylow p-subgroup of G or sometimes just a Sylow subgroup of G* (this is the largest power of p for which a subgroup of that order may exist in G,though the existence of such a subgroup is not obvious).

Definition 14.2.2. Let G be a group and H a subgroup of G. By the *normalizer of H in G* we mean the subgroup $N_G(H) = \{g \in G : g^{-1}Hg = H\}$ (check that $N_G(H)$ is a subgroup).

Definition 14.2.3. Let p be a prime. We use the term *p-element* to mean an element a of the group such that $o(a)$ (the order of a) is a power of p.

Lemma 14.2.4. *Let P be a Sylow p-subgroup of G and let $N = N_G(P)$ be the normalizer of P in G. Then the following assertions hold.*

(a) All the p-elements of N are in P.

(b) P is the only Sylow p-subgroup of N.

Proof Clearly (b) follows from (a), for if Q is a Sylow p-subgroup of N, then all the elements of Q must be in P and hence $Q = P$ since they have the same order. Consider (a) and let x be a p-element of N. In the group $\frac{N}{P}$, we see that xP is a p-element while the order of $\frac{N}{P}$ is not a multiple of p and hence $\frac{N}{P}$ cannot have any p-element, except of course, the identity. Therefore $xP = P$. So $x \in P$. □

Theorem 14.2.5. *(Sylow's first theorem) Let G be a group of order $n = k \times p^m$ where p is a prime, m and k are positive integers such that k is coprime to p. Then G has a Sylow p-subgroup.*

Proof Among the three Sylow theorems, this is the most difficult one to prove because the statement is existential in nature. Let S be the set of all the subsets A of G such that $|A| = p^m$. Then

$$|S| = \binom{p^m k}{p^m} = \frac{(p^m k)(p^m k - 1)\cdots(p^m k - p^m + 1)}{p^m(p^m - 1)\cdots 1}.$$

The R.H.S. has the form $\prod_{j=0}^{p^m-1} a_j$, where a_j is given by

$$a_j = \frac{p^m k - j}{p^m - j} = \frac{N_j}{D_j}$$

Here $N_j = p^m k - j$ and $D_j = p^m - j$. Notice that if j is not a multiple of p then both N_j and D_j do not have p as a factor (and are thus p-free). Let p^i be the highest power of p that divides j. Then both N_j and D_j are divisible by p^i and none is divisible by p^{i+1}. This shows that in the expression for $|S|$, both the numerator and denominators are exactly divisible by the same power of p and hence $|S|$ is p-free.

We now let G act on S in the following manner. If $A \in S$ and $g \in G$, then $g(A) = B$ where $B = gA = \{ga : a \in A\}$. Clearly this defines an action of G on S. Not every orbit under this action can have a size that is a multiple of p since $|S|$ is not a multiple of p. So, there is some orbit, say the orbit A^G of A such that $|A^G|$ is p-free. Let $G_A = U$. This is a subgroup of G and by the orbit stabilizer theorem (Theorem 14.1.13) and Exercise 14.2, $|U|$ is a multiple of p^m. Fix some $y \in A$. Let $g \in U$ Then $y \in A = g(A) = gA$ implies that there is a unique $a \in A$ such that $y = ga$. Define $F : U \to A$ by $F(g) = a$ if $ga = y$. Then $F(g) = F(g') = a$ implies that $ga = y = g'a$ and hence $g' = g$ showing that F is injective. Hence using the Pigeonhole principle, we have $p^m \leq |U| \leq |A| = p^m$ showing that $|U| = p^m$ and U is indeed a Sylow p-subgroup as desired. □

Theorem 14.2.6. *(Sylow's second theorem) Let p be a prime dividing the order of a finite group G. Then all the Sylow p-subgroups of G are conjugates to each other (and hence form a single conjugacy class).*

Theorem 14.2.7. *((Sylow's third theorem) Let p be a prime dividing the order of a finite group G. Let the number of Sylow p-subgroups of G be n_p. Then $n_p \equiv 1 \ (mod\ p)$. Further, the number n_p divides $|G|$.*

Proof *We prove both the second and the third Sylow theorems.* Let P be a Sylow p-subgroup of G and let the conjugacy class of subgroups conjugate to P be

$$X = \{P = P_0, P_1, \ldots, P_r\}$$

We have to prove two things: all the Sylow p-subgroups of G are to be found in X and $r \equiv 1 \ (mod\ p)$. We let G act on X by conjugation: $g(P_i) = g^{-1}P_i g$. Since X is the orbit of P under this G-action, this action is transitive. We now take P (instead of G) and make the elements of P act on X (by conjugation). Since $|P| = p^m$, every orbit under this action has size which is a power of p. Clearly P by itself is a singleton orbit $\{P\}$. If some P_i with $i \geq 1$ is also a single orbit under the conjugation action of P, then $P \leq N_G(P_i) = N$. Then P and P_i are both Sylow p-subgroups of N, which is contrary to the assertion of Lemma 14.2.4. So no orbit except $\{P\}$ is a singleton. Hence every other orbit under this P-action has size some p^t where t is a positive integer. Thus the size of each orbit except $\{P\}$ is a multiple of p and hence $|X| \equiv 1 \ (mod\ p)$. Finally let Q be a Sylow p-subgroup and suppose, for the sake of contradiction, that $Q \notin X$. Again, let Q act on the members of X by conjugation. Exactly for the same reasons as before Q cannot fix any P_i (i.e. P_i cannot form a singleton orbit) and hence each orbit has size that is a multiple of p. But then p divides $|X|$, which is contrary to what we just proved. This contradiction arose because we assumed that Q is not in X. *Hence every Sylow p-subgroup of G is in X and therefore all the Sylow p-subgroups are conjugate to P proving Sylow's second theorem.* Since X is the totality of Sylow p-subgroups of G and $|X| \equiv 1 \ (mod\ p)$ *we already have proved Sylow's third theorem.* □

Lemma 14.2.8. *Let $H \lhd G$. Let $h \in H$ and let $g^{-1}hg$ be a conjugate of h. Then $g^{-1}hg \in H$. Hence H is a union of conjugacy classes of G.*

Corollary 14.2.9. *If H is a unique Sylow p-subgroup of a group G, then H is normal in G.*

Proof This follows from the fact that any conjugate of H is also a Sylow p-subgroup. But since H is unique, H has only one conjugate which is saying the same thing as $H \lhd G$. □

Definition 14.2.10. A group G is *simple* if it has no proper normal subgroups.

We give several applications of Sylow's theorems.

Example 14.2.11. We wish to show that upto isomorphism there is only one group of order 15, which is the cyclic group of order 15. Let G be a group of order $15 = 5 \times 3$. Then the number of Sylow 5-subgroups is $n_5 \equiv 1 \ (mod\ 5)$ (this number is coprime to 5 and also must divide 15 and hence must divide 3). So $n_5 = 1$ and we have a unique Sylow 5-subgroup H which must be normal by the Corollary 14.2.9. Similarly $n_3 = 3k + 1$ and must divide 15 and hence must divide 5. So $n_3 = 1$ showing that we have a unique Sylow 3-subgroup which is also normal in G. Consider $x = hkh^{-1}k^{-1}$ where $h \in H$ and $k \in K$. On one hand, we have $x = (hkh^{-1})k^{-1}$ which shows that $x \in K$ (since $K \lhd G$) but on the other hand, we also have $x = h(kh^{-1}k^{-1})$ which shows that $x \in H$ (because $H \lhd G$). Since the orders of H and K are coprime and $x \in H \cap K$, we get $x = 1$. So $hk = kh$ showing that every element of H commutes with every element of K. Let $h \in H$. Since H is cyclic (and hence abelian) it follows that the centralizer $Z(h)$ contains both H and K and therefore $|Z(h)|$ is a multiple of both 5 and 3. So $Z(h) = G$ and the same argument also holds for any element of K. It then follows that the center Z (of G) contains both H and K and hence $|Z|$ is a multiple of both 3 and 5. Therefore $|Z|$ is a multiple of 15 and hence $G = Z$. So, G is an abelian group. If h and k are respectively generators of H and K, then the order of hk is 15 showing that G cyclic.

Example 14.2.12. Does the same argument work if $|G| = 21 = 7\times 3$? Here n_7 divides 3 and $n_7 \equiv 1 \ (mod\ 7)$ and hence $n_7 = 1$ showing that the unique Sylow 7 subgroup H is normal in G. We do have a Sylow 3-subgroup K (by Sylow's first theorem) and both H and K are generated by h and k respectively. If h and k commute, then $o(hk) = 21$ and G is a cyclic group exactly as in the previous case. But a Sylow 3-subgroup K need not be unique. We have $n_3 \equiv 1 \ (mod\ 3)$ and n_3 divides 7. So $n_3 = 1$ or $n_3 = 7$ where the first possibility has already been considered. If the second possibility holds then we actually have 14 elements of order 3. Such a group exists and is certainly not abelian.

Example 14.2.13. Let $|G| = 63 = 7 \times 3^2$. Then $n_7 \equiv 1 \ (mod\ 7)$ and n_7 divides 9. So, $n_7 = 1$ and hence we have a unique normal Sylow 7-subgroup of G. *So G cannot be simple.*

Example 14.2.14. Let $|G| = 56 = 7\times 2^3$. Then $n_7 \equiv 1 \ (mod\ 7)$ and n_7 divides 8. So, $n_7 = 1$ or $n_7 = 8$. If $n_7 = 1$, then we have a normal Sylow 7-subgroup and G is not simple. Let $n_7 = 8$. Then any two Sylow 7-subgroups have only the identity element in common. Hence the total number of 7-elements (not equal to identity) contained in all the Sylow 7-subgroups, is $8 \times 6 = 48$. This leaves us with only $56 - 48 = 8$ elements. But a Sylow 2-subgroup of G has $2^3 = 8$ elements and hence these elements must constitute the unique Sylow 2-subgroup (there is no space for more Sylow 2-subgroups). Therefore the Sylow 2-subgroup of G is normal. In either case, we see that *a group G of order 56 is not simple.*

Example 14.2.15. Let G be a group of order $n = k \times p^m$ where p is a prime, m and k are positive integers such that k is coprime to p. Let H be a subgroup of order p^r where $r < m$. We look at the conjugacy class X of Sylow p-subgroups given in the proof of Sylow's second and third theorem. Consider the action of H on X (through conjugation of members of X by the elements of H). Since $|H|$ is a power of p, the orbit-stabilizer theorem shows that each orbit of X under this H-action has size either 1 or multiple of p (in fact a power of p). When can the size of an orbit be 1? If H fixes $Q \in X$, then $\forall\ h \in H$, we have $h^{-1}Qh = Q$ and hence $h \in N_G(Q)$. But h is a p-element and by Lemma 14.2.4, $h \in Q$. Hence $H \leq Q$. We thus have two types of orbits. The trivial orbits are singletons and each non-trivial orbit has size which is a multiple of p. The number of non-trivial orbits is $1 + kp$ for some k. We have thus proved that the number of Sylow p-subgroups P containing H is a number which is $\equiv 1 \ (mod\ p)$. In particular, given any subgroup H of order p^r with $r \leq m$, there is at least one Sylow p-subgroup containing H.

Theorem 14.2.16. *Let G be a group of order pqr where p, q, r are three distinct primes. Then G is not a simple group.*

Proof Let, w.l.o.g., $p > q > r$ and let n_i denote the number of Sylow i-subgroups of G where $i = p, q, r$. We are done if we show that one of these three numbers is equal to 1. Suppose, for the sake of contradiction, that n_p, n_q and n_r are all ≥ 2. First consider n_p. By Sylow's theorems, $n_p \equiv 1 \ (mod\ p)$ and n_p divides pqr and hence n_p divides qr. The only other possibilities are n_p is equal to q, r or qr. The first two cannot arise since $q, r < p$ and $n_p \equiv 1 \ (mod\ p)$. So $n_p = qr$. Since these qr Sylow p-subgroups account for exactly $(p-1)qr$ non-identity elements of G (every two Sylow p-subgroups must intersect in identity only), we are left with only qr remaining elements. Next consider n_q. This number must divide pr and exactly for the same reason as before $n_q > q$ and similarly $n_r \geq q$. Thus the number of non-identity elements contained in Sylow q or Sylow r-subgroups is $\geq q(q-1) + q(r-1) = q(q+r-2)$ and hence $q(q+r-2) \leq qr$. We thus get $q \leq 2$, a contradiction. □

Definition 14.2.17. Let G be a group. A *generating set S for G is* a subset of G such that every element of G can be written as a product of some elements in S.

Note that a generating set for a group always exists since G or $G - \{1\}$ certainly are trivial generating sets. One would be interested in knowing the least number of generators (the size of a generating set). For example, any cyclic group can be generated by a single element and conversely, any group with a single generator must be a cyclic group. In Exercise 14.33, some generating sets for the symmetric group are studied.

Theorem 14.2.18. *The group $G = A_n$ is generated by 3-cycles for all $n \geq 3$.*

Proof When $n = 3$, the non-identity even permutations are the 3-cycles and when $n = 4$, we have eight 3-cycles in a group G of order 12 and hence using Lagrange theorem, the subgroup (whose order must divide 12) generated by the eight 3-cycles is G. Let $n \geq 5$. We have $(12)(34) = (12)(23)(23)(34) = (132)(243)$ and hence product of any two disjoint transpositions is a product of two 3-cycles. If the transpositions are not disjoint, then indeed we have $(12)(13) = (123)$. Since every permutation σ in G is a product of an even number of transpositions, it follows that every permutation in G can be written as a product of 3-cycles. □

Theorem 14.2.19. *$G = A_5$ is a simple group.*

Proof Let $H \neq 1$ and $H \lhd G$ and suppose for the sake of contradiction that $H \neq G$. If $\alpha \in H$, then every permutation conjugate to α (in G) must be in H and

hence H is a union of some conjugacy classes of G such that the union includes the conjugacy class of the identity permutation whose size is 1. Since A_5 is generated by 3-cycles (all of which are conjugate in A_5), if H contains one 3-cycle, then it must contain all and hence $H = G$, a contradiction. We may thus assume that H contains no 3-cycle and hence $|H|$ is not a multiple of 3. Let H contain a 5-cycle. Then it must contain the entire conjugacy class (of size 12) of that 5-cycle. It thus follows that H has at least 12 permutations of order 5 and hence at least three Sylow 5-subgroups. Since the number of Sylow 5-subgroups in H is of the form $1 + 5m$ (using Sylow's second and third theorems) and since G has only 6 Sylow 5-subgroups, it follows that H also has 6 Sylow 5-subgroups. Then using Sylow's third theorem, we see that 6 divides $|H|$ forcing H to contain a 3-cycle, a contradiction. Hence H cannot contain any 5-cycle. This forces every non-identity permutation in H to have order two and thus H precisely consists of the identity permutation and 15 permutations of order two, that is H has order 16 which is absurd since 16 does not divide 60.
Perhaps a better and more direct alternative argument is the following: H is a union of some conjugacy classes (Lemma 14.2.8) and hence $|H|$ is a sum from the multi set $\{1, 20, 12, 12, 15\}$ with the proviso that one summand is 1. It is easily checked that none of the numbers $\{20, 12, 12, 15\}$ (along with 1) adds to a sum that divides 60 giving a contradiction. □

Theorem 14.2.20. *Let $n \geq 5$ and let $G = A_n$. Then the following assertions hold.*

(a) 3-cycles generate G.

(b) Any two 3-cycles in G are conjugates (in G).

(c) Let $1 \neq H \lhd G$. If H contains a 3-cycle, then $H = G$.

(d) Let $n = 6$ and let $1 \neq H \lhd G$. Then H contains a 3-cycle.

Proof (a) has been proved earlier in Theorem 14.2.18 and (c) follows from (a) and (b) exactly as in the special case when $n = 5$. It is therefore sufficient to prove (b) (which has already been shown to hold when $n = 5$) and (d). We can thus make an induction on n and let $n \geq 6$. Let $\alpha = (x_1, x_2, x_3)$ and $\beta = (y_1, y_2, y_3)$ be two different 3-cycles. Let B denote the set $\{x_1, x_2, x_3\} \cup \{y_1, y_2, y_3\}$ and first assume that $\{x_1, x_2, x_3\} \cap \{y_1, y_2, y_3\} \neq \emptyset$. Then B has size at the most 5 and since $n \geq 6$, we can obtain a subset C of size 5 of $[n]$ such that C contains B. Let G' denote the subgroup of G such that that $\gamma \in G'$ iff $\gamma(j) = j \ \forall \ j \notin C$. Then α and β are both 3-cycles in G' which is is isomorphic to A_5 (on the set of elements of C). Since 3-cycles in A_5 are in a single conjugacy class, as we have already seen, α and β are conjugates in G' and hence in G as well. We are thus through when there is an element common to the 3-subsets moved by the 3-cycles α and β. Finally assume that $\{x_1, x_2, x_3\}$ and $\{y_1, y_2, y_3\}$ are disjoint. Then the set B as before is a set of order 6. We can now take a suitable 3-cycle say $\gamma = (x_1, x_2, y_3)$. Then using the previous step α and γ are conjugates in G and so are γ and β. Hence α and β are conjugates in G as desired. This proves the correctness of (b).
Finally consider (d). Let H be a non-identity normal subgroup of $G = A_6$. If there is a permutation $\alpha \in H$ such that α fixes some symbol say 6 w.l.o.g. then $H_6 = H \cap G_6$ is

a non-identity normal subgroup of $G_6 \cong A_5$ (*note that* G_6 *means the stabilizer of* 6 *in* G) and since the latter is simple, we see that $H \cap G_6 = G_6$ which, of course contains 3-cycles (because it isomorphic to A_5) and hence H also contains 3-cycles. We are thus reduced to looking at the possibility that each non-identity permutation α of H moves every element and hence is a derangement. Since such an α is an even permutation which moves all the six elements in $[6]$ we see that the cycle decomposition of α must be either a product of a transposition and a 4-cycle or a product of two 3-cycles. In the former case, w.l.o.g. we may assume that $\alpha = (12)(3456)$ and in the latter case $\alpha = (123)(456)$. In fact, in the former case, α^2 is a non-identity permutation in H that fixes both 1 and 2 giving us the case we have already dealt with (namely H has a non-identity permutation fixing some point). So consider $\alpha = (123)(456)$. Let $\beta = (124)$. Then $\beta \in G$ and since H is normal,

$$\gamma = \beta\alpha\beta^{-1}\alpha^{-1} = (\beta\alpha\beta^{-1})\alpha^{-1}$$

is in H. By actual multiplication, γ moves 1 to 2 and fixes 5 showing that γ is a non-identity permutation in H that fixes a point and we are back to the case already discussed. □

Theorem 14.2.21. *Let* $n \geq 5$. *Then* A_n *is a simple group.*

Proof We have shown this to be true for $n = 5$ in Theorem 14.2.19. Let $n = 6$. Then the Theorem 14.2.20(d) shows that any non-identity normal subgroup H of $G = A_6$ contains a 3-cycle and hence by Theorem 14.2.20(c) $H = G$ showing that G is simple. Let $G = A_n$ where $n \geq 7$ and let $H \lhd G$ such that $H \neq 1$. Let $\alpha \in H$ such that $\alpha \neq 1$. Let w.l.o.g $\alpha(1) = 2$. Let β denote the 3-cycle (234). Since $\alpha(1) = 2$ and $\beta(2) = 3$ we see that $\alpha\beta$ takes 1 to 3 while $\beta\alpha$ takes 1 to 2 showing that α and β do not commute. Hence $\gamma = (\beta\alpha\beta^{-1})\alpha^{-1} \in H$ and is not identity. On the other hand, $\gamma = \beta(\alpha\beta^{-1}\alpha^{-1})$ is a product of β and a conjugate of β. Hence γ is a product of two 3-cycles. So γ moves at the most 6 elements say $j_1, j_2, j_3, j_4, j_5, j_6$. If K denotes the subgroup of G that consists precisely of those permutations that are constant on every i not in the set $\{j_1, j_2, j_3, j_4, j_5, j_6\}$, then K is isomorphic to A_6 and $\gamma \in K \cap H$. Hence $K \cap H$ is a non-identity normal subgroup of K. But as we already saw K is simple. So $K \cap H = K$. Since K contains 3-cycles, so does H and then we are done by Theorem 14.2.20(c). □

14.3 Automorphisms of a symmetric group

Definition 14.3.1. *An automorphism* α of a group G is a bijection $\alpha : G \to G$ which is also a group homomorphism. The set of all automorphisms of G is a group and it is denoted by $Aut(G)$ (note that $Aut(G)$ is finite if G is finite).

For an element $g \in G$, we denote, by γ_g the map $\gamma_g : G \to G$ given by $\gamma_g(x) = gxg^{-1}$. It is easy to see that γ_g *is an automorphism:*

$$\gamma_g(xy) = g(xy)g^{-1} = gxg^{-1}gyg^{-1} = \gamma_g(x)\gamma_g(y)$$

Such an automorphism γ_g (with $g \in G$) is called an inner automorphism.

Definition 14.3.2. An automorphism α is an *inner automorphism* if $\alpha = \gamma_g$ for some $g \in G$. If α is not an inner automorphism, then it is called an *outer automorphism*. By the set $Inn(G)$, we mean $\{\gamma_g : g \in G\}$. This is a subgroup of $Aut(G)$ since $\gamma_{gh}(x) = (gh)x(gh)^{-1} = g(hxh^{-1})g^{-1} = (\gamma_g)(\gamma_h)(x)$. We shall use γ to denote an inner automorphism.

Theorem 14.3.3. *Let G be a group. Then the following assertions hold.*

(a) $Inn(G) \lhd Aut(G)$.

(b) Let $f : G \to Inn(G)$ *be defined by* $f(g) = \gamma_g$. *Then* f *is a group homomorphism whose kernel is* $Z(G)$ *and hence* $\frac{G}{Z(G)} \cong Inn(G)$.

(c) If G is abelian then $Inn(G) = 1$ *and hence every proper automorphism of G is an outer automorphism.*

(d) If G is centerless (i.e. $Z(G) = 1$*), then* $G \cong Inn(G)$.

Proof Let α be an automorphism. We verify that $\alpha\gamma_g\alpha^{-1} = \gamma_{\alpha(g)}$. Let $\alpha(g) = h$ and let $\alpha^{-1}(x) = y$. Then we have

$$\begin{aligned} \alpha\gamma_g\alpha^{-1}(x) &= \alpha\gamma_g(y) \\ &= \alpha(gyg^{-1}) \\ &= \alpha(g)\alpha(y)\alpha(g^{-1}) \\ &= hxh^{-1} \\ &= \gamma_{\alpha(g)}(x) \end{aligned}$$

proving (a). For (b), note that we have already verified that f is a homomorphism. $g \in ker(f)$ iff $\gamma_g = 1$ iff $gxg^{-1} = x \; \forall \, x \in G$ iff $g \in Z(G)$ showing that the kernel of f is $Z(G)$. This proves (b). If G is abelian, then $Z(G) = G$ and hence $Inn(G) = 1$. Finally, if G is centerless, then $Z(G) = 1$ and again using (b) we get $G \cong Inn(G)$. □

Lemma 14.3.4. *Let* g_1 *and* g_2 *be conjugate elements in a group G and let* f *be an automorphism of G. Let* $f(g_i) = h_i$. *Then* h_1 *and* h_2 *are also conjugates in G. Thus* f *maps a conjugacy class C in G to a conjugacy class* C' *and if* f *is an inner automorphism then* $C' = C$.

Proof Indeed if $\alpha g_1\alpha^{-1} = g_2$, and if $f(\alpha) = \beta$, then $\beta h_1\beta^{-1} = h_2$. □

Since each non-identity permutation in S_n (for $n \geq 3$) has a conjugacy class whose size is strictly larger than 1, it follows from the orbit-stabilizer theorem (Theorem 14.1.8) that S_n is centerless for all $n \geq 3$. Hence $Inn(S_n) \cong S_n$. For which values of n can S_n have an outer automorphism? Our investigation begins with the following two purely combinatorial results.

Theorem 14.3.5. *Let* $f \in Aut(S_n)$ *such that* f *maps the conjugacy class C of all the transpositions in* S_n *to itself. Then* $f \in Inn(S_n)$.

Proof Let R denote the subgroup consisting of all the automorphisms f of S_n such that f maps C to itself. Clearly R is a subgroup of $Aut(S_n)$ and the theorem is proved if we show that $R = Inn(G)$. Using Lemma 14.3.4 just proved every inner automorphism maps C to itself and hence $Inn(S_n) \leq R$. Since $n \geq 3$, S_n is a centerless group and therefore $Inn(S_n) \cong S_n$. Therefore, $|Inn(S_n)| = n!$. *Thus it suffices to prove that* $|R| \leq n!$, *which is what we will actually prove* (an alternative treatment where f is explicitly obtained as an inner automorphism is given in Exercise 14.41.)

Let $f \in R$. Since S_n is generated by the set
$N = \{(12), (13), \ldots, (1n)\}$ of $n - 1$ transpositions (Exercise 14.33), it is enough to focus on the action of f on N. Consider the complete graph K_n with vertex set $[n]$. We identify every transposition (ij) with the edge (ij) of this complete graph. f can thus be viewed as a permutation of edges of the complete graph K_n. We first claim that if two distinct edges e, e' share vertex in common, then so do the edges $f(e)$ and $f(e')$. Let w.l.o.g. $e = (12)$ and $e' = (13)$. Since f is a permutation (actually an automorphism) on C, $f(e)$ and $f(e')$ are distinct. If $f(e)$ and $f(e')$ do not share a vertex in common, then they are of the form (xy) and (zw) where x, y, z, w are all distinct. But then $(xy)(zw)$ is a permutation of order two while $(12)(13) = (123)$ is a permutation of order 3, which is a contradiction to the fact that f is an automorphism. Next, let $e_2 = (12), e_3 = (13)$ and $e_4 = (14)$ be three edges. The claim just proved shows that $f(e_2) = (x_1x_2)$ and $f(e_3) = (x_1x_3)$, where x_1, x_2, x_3 are all distinct. The same claim also shows that $f(e_4) = (x_1x_4)$ where x_4 is different from x_1, x_2, x_3 or $f(e_4) = (x_2x_3)$. We show that the latter possibility cannot arise: $(12)(13)(14) = (1234)$ has order 4 and hence $f(e_2)f(e_3)f(e_4)$ must also have order 4. But $(x_1x_2)(x_1x_3)(x_2x_3) = (x_1x_3)$ has order two showing that the second case cannot occur.

Definition 14.3.6. *An ordered star* T in K_n is an ordered set of $n - 1$ edges of the form

$$T = \{e_2 = (x_1x_2), e_3 = (x_1x_3), \ldots, e_n = (x_1x_n)\}$$

such that all the e_i's share the unique common vertex x_1 to be called the *head of the star* T and such that $x_1, x_2, \ldots, x_n$ are all distinct and therefore, $\{x_1, x_2, \ldots, x_n\} = [n]$.

Here are two examples, both with $n = 5$. Note that the edges in the ordered star are read from left to right. Thus the second ordered star $T = \{(31), (34), (32), (35)\}$ and the head of this star T is 3.

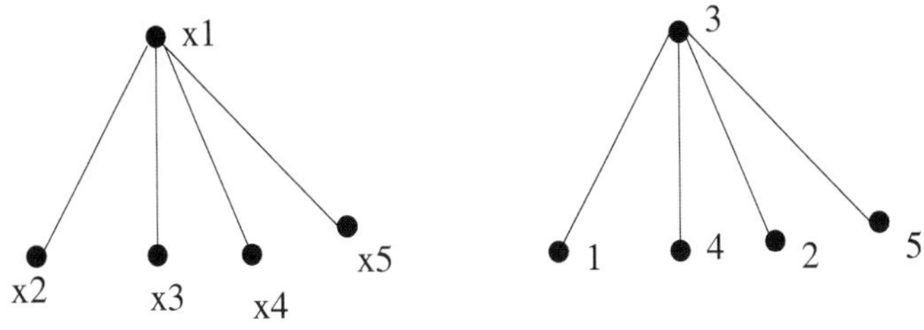

Figure 14.1: Ordered stars

Let Γ denote the set of all the ordered stars. Since the head x_1 can be chosen in n different ways and then $x_2, x_3, \dots, x_n$ can be chosen $n-1, n-2, \dots$ ways respectively, $|\Gamma| = n \times (n-1) \times \cdots \times 1$ which is equal to $n!$. Given an ordered star T as above and $f \in R$, we define $f(T) = T'$ by

$$T' = \{e_2{}', e_3{}', \dots, e_n{}'\}$$

where $e_i{}' = f(e_i)$ for all $i = 2, 3, \dots, n$. From the claim we established earlier, we see at once, that:
Assertion Let

$$T = \{e_2 = (x_1x_2), e_3 = (x_1x_3), \dots, e_n = (x_1x_n)\}$$

and let $f \in R$. Let

$$T' = f(T) = \{e_2{}', e_3{}', \dots, e_n{}'\}$$

Then T' is also an ordered star and hence for some $y_i, i = 1, 2, \dots, n$ we have $e_i{}' = (y_1y_i)$. Thus,

$$T' = \{(y_1y_2), (y_1y_3), \dots, (y_1y_n)\}.$$

Since f is an automorphism, $f(T) = T'$ implies that $f^{-1}(T') = T$ showing that f is a bijection and hence a permutation on the set Γ. We thus see that R acts on Γ with a permutation action. Now let $T = \{(12), (13), \dots, (1n)\}$ denote the standard ordered star. If $f \in R$ and f fixes T, then f must fix each one of the transpositions $(1j)$ where $j = 2, 3, \dots, n$. Since these $n-1$ transpositions generate S_n (Exercise 14.33), f must also fix each permutation in S_n and hence $f = 1$. Therefore, the stabilizer of T in R is the identity automorphism. Using the orbit-stabilizer theorem (Theorem 14.1.13), we get the size of the orbit of T equal to $|R|$. Since $Inn(S_n) \leq R$, we have $n! \leq |R|$ and since the size of the orbit of T cannot be greater than $|\Gamma| = n!$, we get $n! \leq |R| \leq |\Gamma| \leq n!$ showing that equality must hold everywhere and we have $|R| = n!$. This proves that $R = Inn(S_n)$ as desired. $\square$

Theorem 14.3.7. *Let $f \in Aut(S_n)$ be such that f maps the conjugacy class C of transpositions to some other conjugacy class $C' \neq C$. Then $n = 6$ and f maps C to C' where C' consists of 15 permutations that are products of three disjoint transpositions of the form $\alpha = (x_1x_2)(x_3x_4)(x_5x_6)$.*

Proof Since f is an automorphism, f must map C to C' where every permutation in C' also has order 2. The cycle types of all the permutations in C' are the same. Hence there are non-negative integers k and m such that $n = 2k + m$ and $\forall\, \alpha \in C'$ the permutation α is a product of k transpositions and m singleton cycles. Thus α has cycle type $(m, k, 0, \ldots)$ and a representative of C' is $(12)(34)\cdots(2k-1, 2k)$. We must have $k \geq 2$ since $C' \neq C$. Then

$$\begin{aligned} |C'| &= \binom{2k+m}{m} \times \binom{2k}{2, 2\ldots, 2} \times \frac{1}{k!} \\ &= \frac{(2k+m)!}{(2k)!m!} \times \frac{(2k)!}{2^k} \times \frac{1}{k!} \\ &= \frac{(2k+m)!}{m!2^k k!} \end{aligned}$$

Since $|C| = \binom{2k+m}{2} = \frac{(2k+m)(2k+m-1)}{2}$, we see that

$$(2k+m-2)\cdots(m+1) = 2^{k-1}k!$$

If $m \geq 1$, then L.H.S. $\geq (2k-1)\cdots 1$ and hence we get

$$\begin{aligned} (2k-1)! &\leq 2^{k-1}k! \\ \Rightarrow [(2k-1)(2k-3)\cdots 1] \times (k-1)! \times 2^{k-1} &\leq 2^{k-1}k! \\ \Rightarrow \qquad (2k-1)(2k-3)\cdots 1 &\leq k \end{aligned}$$

which is false for all $k \geq 2$. So, $m = 0$. We thus get $(2k-2)! = 2^{k-1}k!$ and working exactly as before we have

$$(2k-3)(2k-5)\cdots 1 = k$$

This does not hold for $k = 2$ and clearly the L.H.S. $> k$ if $k \geq 4$ forcing $k = 3$. Hence we have $n = 2k = 6$ and C' is as described in the statement of the theorem. □

Theorem 14.3.8. *Let $G = S_6$ in its action on the set $X = [6]$. Then G has a subgroup H such that*

(a) $[G : H] = 6$.

(b) H is transitive on X (and hence in particular, H is not one of the stabilizers G_i of some $i \in X$).

(c) H does not contain a transposition and hence H is not one of the stabilizers G_i.

Proof Consider S_5 and a Sylow 5-subgroup P of S_5. Using Sylow's theorem, the number of Sylow 5-subgroups is equal to 6 which is also equal to $[S_5 : K]$ where $K = N_{S_5}(P)$. Since $|S_5| = 120$, we get $|K| = 20$ using the orbit-stabilizer theorem (Theorem 14.1.13). Now let

$$Y = \{K\beta_0, K\beta_1, k\beta_2, K\beta_3, K\beta_4, K\beta_5\}$$

denote the set of all the six right cosets of K in S_5 with $\beta_0 = 1$. S_5 acts on Y by right regular action $g : K\beta_i \to K\beta_i g$. This permutation action F is transitive. We thus have $F : S_5 \to S_Y$, the symmetric group on Y that maps $g \in S_5$ to $F(g)$ with $F(g)(K\beta_i) = K\beta_i g$. Let K^* denote the kernel of F. If $g \in K^*$, then $Kg = K$ implies that $g \in K$ and hence $K^* \leq K$. Thus K^* is a normal subgroup of S_5 contained in K. But the only proper normal subgroup of S_5 is A_5 (with order 60) and hence it follows that $K^* = 1$. Since $S_Y \cong S_6$, the map F gives an embedding of S_5 into S_6 and since we are dealing with a right regular action, $H = F(S_5)$ is a transitive subgroup of S_6 (in its action on the six elements of Y).
To summarize, we have found a subgroup H of order 120 of S_6 which is transitive on the set on which it is acting. It just remains to show that H contains no transposition. Since $|H| = 120$ which is a multiple of 5, H must have a permutation say α of order 5 and we can assume, w.l.o.g. that $\alpha = (12345)$. Suppose $\sigma = (16) \in H$. Then $\alpha^{-1}\sigma\alpha = (26)$ and thus by conjugating σ by various powers $\alpha, \alpha^2, \alpha^3, \alpha^4 = \alpha^{-1}$, we see that all the transpositions $(16), (26), (36), (46), (56)$ are in H and these transpositions generate the group S_6. So, $H = S_6$, a contradiction. If $\sigma = (i6) \in H$, then using the fact that α could be written with i at the front, we get all the five transpositions forming an ordered star with i as head, which also leads to a contradiction (because those five transpositions also generate S_6, using Exercise 14.32). The only case we are left with is $\sigma = (ij)$ where $1 \leq i < j \leq 5$. We can then use transitivity of H to find a permutation $\beta \in H$ such that $\beta(i) = 6$. Let $\tau = \beta^{-1}\sigma\beta$. Then being a conjugate of σ, τ is also a transposition. Further, if $\beta(j) = r$, then τ takes r to 6 and hence $\tau = (r6)$ (where $1 \leq r \leq 5$) and we are back to the previous case. This final contradiction completes the proof. □

Theorem 14.3.9. *S_6 has an outer automorphism F.*

Proof Theorem 14.3.8 provides us with a subgroup H of S_6 such that $[S_6 : H] = 6$ and H contains no transposition. Let

$$Y = \{H\alpha_0, H\alpha_1, H\alpha_2, H\alpha_3, H\alpha_4, H\alpha_5\}$$

be the set of six distinct cosets of H in S_6, where, as in the earlier situations, $\alpha_0 = 1$ and hence $H\alpha_0 = H$. S_6 acts on H by right regular action: If $\beta \in S_6$, then $F(\beta) : H\alpha_i \to H\alpha_i\beta$ for all $i = 0, 1, 2, 3, 4, 5$. Then we already know that $F : S_6 \to S_Y$ (the symmetric group on Y) is a homomorphism whose image is transitive on the 6-set Y. Further, the kernel of F is contained in H. Since the only proper normal subgroup of S_6 is A_6 (Exercise 14.38), whose index is 2, the kernel of F cannot be A_6 (whose order is $360 > 120 = |H|$). It thus follows that the kernel of F is 1 and hence F is in fact, an isomorphism (since S_6 and S_Y have the same order). But Y is 6-set and hence we have an isomorphism, that is an automorphism $F : S_6 \to S_6$. It just remains to show that F cannot be an inner automorphism. Let $\beta = (12)$, a transposition. Suppose $F(\beta)$ in its action on Y is also a transposition (it has to be a permutation of order two since F is an automorphism). A transposition on Y must fix 4 members of Y. Hence for some i from 0 to 5, we must have $H\alpha_i\beta = H\alpha_i$ which implies that $\alpha_i\beta{\alpha_i}^{-1} = \tau \in H$. But τ is a conjugate of β and hence τ is also a transposition showing that H contains some transposition, which is a contradiction to Theorem 14.3.8. Since an inner automorphism must map the conjugacy class of transpositions to itself, it follows that F is indeed an outer automorphism. *The arguments have actually shown more.* $F(\beta)$ *is an order* 2 *permutation without fixed points and hence must be a product of three disjoint transpositions. Thus* F *maps the class of* 15 *transpositions to the class of* 15 *permutations that are products of three transpositions.* □

We cannot resist the temptation of telling the reader about the automorphism tower of a finite centerless group. *In the discussion to follow, we make no distinction between a group G and an isomorphic copy of G.* The basic ingredient of the construction is any finite centerless group G. An elementary result on which we base ourselves is the following.

Theorem 14.3.10. *Let G be a centerless group. Then* $G \lhd Aut(G)$ *and* $Aut(G)$ *is also centerless.*

Proof The first part has already been proved in Theorem 14.3.3(a). Consider the second where we claim the stronger assertion: $C_{Aut(G)}(Inn(G)) = 1$. To that end, suppose $\alpha \in Aut(G)$ commutes with *every inner automorphism* γ_g. From the equation $\alpha\gamma_g\alpha^{-1} = \gamma_h$ where $h = \alpha(g)$ (that was proved in Theorem 14.3.3(a)), it follows that α commutes with γ_g iff $\alpha(g) = g$ and hence α must be identity. □

Begin with a centerless group $G = G_0$ and let $G_1 = Aut(G)$. Using the convention on isomorphic groups, we have $G_0 \lhd G_1$ and G_1 is also a centerless group using Theorem 14.3.10. Therefore $G_1 \lhd G_2$ where $G_2 = Aut(G_1)$ and G_2 is centerless. We thus have the following sequence of groups:

$$G_0 \lhd G_1 \lhd G_2 \lhd \cdots \lhd G_m \lhd \cdots$$

where $G_m = Aut(G_{m-1})$. This sequence is called *the automorphism tower of the centerless group G*. We end the discussion with the following extremely non-trivial assertion of Wielandt (see Rotman [48]).

Theorem 14.3.11. *(Wielandt's theorem) The automorphism tower of a finite centerless group G is stationary (that is, there is some* m *such that* $G_m = G_{m+1}$*).*

14.4 Finite subgroups of the orthogonal group

We close this chapter with a description of finite subgroups of the groups $SO_2(\mathbb{R})$ and $SO_3(\mathbb{R})$, the groups of rotations of $\mathbb{R}^2$ and $\mathbb{R}^3$ respectively. While the proof of the main classification theorem is purely combinatorial which we give in some detail, we shall be brief in developing the prerequisite linear algebra. An $n \times n$ real matrix A (which is the same as a linear transformation g from $\mathbb{R}^n$ to $\mathbb{R}^n$) is called *an orthogonal matrix* if $AA^t = I$. That is, the rows of A have norm 1 and any two rows of A are orthogonal. It is easy to see that the orthogonal matrices form a multiplicative group called the *orthogonal group* $O_n(\mathbb{R})$. Also, the determinant of an orthogonal matrix is ± 1. The index 2 subgroup of $O_n(\mathbb{R})$ that consists of orthogonal matrices with determinant 1 is called the unimodular group (of dimension n) and is denoted by $SO_n(\mathbb{R})$. The interest here is in $SO_2(\mathbb{R})$ and $SO_3(\mathbb{R})$. Let $A \in SO_2(\mathbb{R})$ and let A be equal to

$$\begin{pmatrix} a_{11} & a_{12} \\ a_{21} & a_{22} \end{pmatrix}$$

Then we have the equations:

$$\begin{aligned} (a_{11})^2 + (a_{12})^2 &= 1 \\ (a_{21})^2 + (a_{22})^2 &= 1 \\ a_{11}a_{21} + a_{12}a_{22} &= 0 \end{aligned}$$

From the first two equations, we see that there are real numbers (angles) ϕ and θ such that

$$a_{11} = \cos\phi, \quad a_{12} = \sin\phi, \quad a_{21} = \sin\theta, \quad a_{22} = \cos\theta$$

and then the third equation gives $\sin(\phi + \theta) = 0$ and hence we have $\phi = -\theta$ showing that A has the form

$$\begin{pmatrix} \cos\theta & -\sin\theta \\ \sin\theta & \cos\theta \end{pmatrix}$$

The matrix A is a rotation of the plane $\mathbb{R}^2$ (fixing the origin) through an angle θ. The group $S0_2(\mathbb{R})$ has many interesting properties that are outlined in the Exercise 14.44. In particular, we have

Theorem 14.4.1. *Every finite subgroup of $SO_2(\mathbb{R})$ is a cyclic group of order n for some natural number n.*

Now consider $SO_3(\mathbb{R})$ and let $A \in SO_3(\mathbb{R})$. Let $B = A - I$. Since $\det(A^t) = \det(A) = 1$, we can compute the determinant of A^tB in two ways. On one hand this is equal to $\det(B) = \det(A - I)$. On the other hand, it is also equal to

$$\det(A^tA - A^t) = \det(I - A^t) = \det[I - A]^t = \det(I - A) = -\det(A - I)$$

Hence we get $\det(A - I) = -\det(A - I)$ giving $\det(A - I) = 0$. Therefore we conclude that $A - I$ has a non-zero null-space. We can then choose a unit vector $\underline{v}$ such that $A\underline{v} = \underline{v}$. It then follows that A fixes every point on the line l that joins the origin and $\underline{v}$; *this line l is called the axis of the linear transformation A.* By a similarity transformation, we can bring A to a matrix C (with the same eigenvalues) such that the first row and column have all entries equal to zero except the first. Arguing exactly as before the second and third row (which now correspond to rotations of the plane) have the form that we already discussed. This shows that C is equal to

$$\begin{pmatrix} 1 & 0 & 0 \\ 0 & \cos\theta & -\sin\theta \\ 0 & \sin\theta & \cos\theta \end{pmatrix}$$

To summarize, we have the following theorem.

Theorem 14.4.2. *Every non-identity linear transformation $A \in SO_3(\mathbb{R})$ is a rotation with a unique axis and conversely every rotation of $\mathbb{R}^3$ is an element of $SO_3(\mathbb{R})$. This linear transformation fixes each point on the axis and rotates the entire space (in a cylindrical manner) around this axis.*

Definition 14.4.3. Let A be a non-identity linear transformation in $SO_3(\mathbb{R})$. A unit vector $\underline{v}$ lying on the axis of A is called its *pole*. Given a vector (which is the same as a point) say $\underline{v} = (x, y, z)$ in $\mathbb{R}^3$, the point *antipodal* to $\underline{v}$ is the vector $-\underline{v} = (-x, -y, -z)$.

It is clear that $\underline{v}$ is a pole iff $-\underline{v}$ is a pole of A. Thus poles of A are the points of $\mathbb{R}^3$ at which the axis intersects the unit sphere in $\mathbb{R}^3$. Hence all the poles of all the members of $SO_3(\mathbb{R})$ are on the unit sphere. Let G be a finite subgroup of $SO_3(\mathbb{R})$. If H is a subgroup of G that consists of all the members of G that have a fixed unit vector p as a pole, then H is a finite (cyclic) group; it actually corresponds to a finite subgroup of $SO_2(\mathbb{R})$ and consists of a finite set of planar rotations. We have the following theorem that classifies all the finite subgroups of $SO_3(\mathbb{R})$.

Theorem 14.4.4. *Let $G \leq SO_3(\mathbb{R})$ such that G is of finite order n. Then G is precisely one of the following types.*

(a) G is a finite cyclic group of rotations of a regular n-gon and hence $G \cong \mathbb{Z}_n$ (in its action on a regular n-gon in the plane).

(b) $n = 2m$ and G is the dihedral group D_m of m rotations and m reflections of a regular m-gon in the plane.

(c) G is one of the following three types and G is the rotation group of one the five regular polytopes.

(i) $n = 12$ and G is the group of rotations of a regular tetrahedron; $G \cong A_4$, the alternating group on 4 points.

(ii) $n = 24$ and G is the group of rotations of a cube; $G \cong A_4$, the symmetric group on 4 points.

(iii) $n = 60$ and G is the group of rotations of a regular dodecahedron; $G \cong A_5$, the alternating group on 5 points.

Proof Let $g, h \in G$ with $h \neq 1$. Since h is a non-identity rotation, it has exactly two poles $\pm\underline{v}$. Then $k = g \circ h$ is a rotation in G with poles $\pm\, g\underline{v}$. This shows that if we pool all the poles of all the non-identity rotations in G and call that set $\mathbf{P}$, then G acts as a permutation group on $\mathbf{P}$. Following this crucial part, remaining part of the proof is purely combinatorial. Let $p \in \mathbf{P}$ and let r_p denote the number of rotations in G

that have p as a pole. Let n_p denote the size of the orbit of p. Then the orbit-stabilizer theorem gives $r_p n_p = |G| = n$. Since each non-identity rotation in G has exactly two poles, two-way counting gives

$$\sum_{p \in \mathbf{P}} (r_p - 1) = 2n - 2$$

Let $O_1, O_2, \ldots, O_k$ be the k distinct orbits of poles (under G-action). Then the above equation obtains

$$\sum_{i=1}^{k} \sum_{p \in O_i} (r_p - 1) = 2n - 2$$

Now fix an orbit O_i. Then $\forall\ p \in O_i$, we have $n_p = |O_i|$. So, we get $r_p = r_{p'}$ if $p, p' \in O_i$. Simplifying the notation, we write r_i to denote r_p for $p \in O_i$ and using $|O_i| = n_i$ we obtain

$$\begin{aligned} \sum_{i=1}^{k} |O_i| \times (r_i - 1) &= 2n - 2 \\ \Rightarrow \sum_{i=1}^{k} (r_i n_i - n_i) &= 2n - 2 \\ \Rightarrow \sum_{i=1}^{k} \left(1 - \frac{1}{r_i}\right) &= 2 - \frac{2}{n} \end{aligned}$$

If $k = 1$, then the L.H.S. is < 1 while the R.H.S. is > 1, a contradiction. Hence $k \geq 2$. Since $r_i \geq 2$, we see that each summand on the L.H.S. is $\geq \frac{1}{2}$ and hence the L.H.S. is $\geq \frac{k}{2}$ while the R.H.S. is < 2. Therefore we get $k < 4$ and hence $k = 2$ and $k = 3$ are the only possibilities. We now consider these two cases separately. First let $k = 2$. Then we have

$$\frac{1}{r_1} + \frac{1}{r_2} = \frac{2}{n}$$

and hence $n_1 + n_2 = 2$. So we must have exactly two poles (that are in distinct singleton orbits) and therefore only one axis. Thus all the rotations in G are around this fixed axis. Hence G is isomorphic to the cyclic group $\mathbb{Z}_n$ in its action on a regular n-gon with the axis given by the unique line passing through the center (which is also the origin) and perpendicular to the plane of the n-gon.

Next, let $k = 3$. Rearrange r_i's so that $r_1 \leq r_2 \leq r_3$. Then we have

$$\frac{1}{r_1} + \frac{1}{r_2} + \frac{1}{r_3} = 1 + \frac{2}{n}$$

If $r_1 \geq 3$, then each $r_i \geq 3$ and then L.H.S. ≤ 1 while the R.H.S. > 1, a contradiction. So $r_1 = 2$ and we get

$$\frac{1}{r_2} + \frac{1}{r_3} = \frac{1}{2} + \frac{2}{n}$$

Again, if $r_2 \geq 4$, then $r_3 \geq 4$ and L.H.S $\leq \frac{1}{2}$, a contradiction. So, $r_2 = 2$ and $r_2 = 3$ are the only two possibilities. First let $r_2 = 2$. Then we have $r_3 = \frac{n}{2} = m$ say. In this case, we get $n = 2m$ and $n_1 = m = n_2$ and $n_3 = 2$. So the group has an even order $2m$ with a subgroup H of index 2 that consists of m rotations. These are around an axis perpendicular to the plane of a regular m-gon exactly as in the previous case except that the two special poles are in the same orbit ($n_3 = 2$). The other m rotations have m different axes in the plane of the regular m-gon each passing through a vertex of the regular m-gon. Depending on the parity of m, there are two possibilities. Let m be even. Then one set of $\frac{m}{2}$ axes join antipodal vertices and the other set of $\frac{m}{2}$ axes join mid-points of antipodal pairs of edges. If m is odd then each such axis joins a vertex with the mid-point of an edge opposite to the vertex. *Thus, when m is even, we have two types of rotations, one type that has both the poles as vertices of a regular m-gon* $\mathbf{Q}$ *while the other type has both the poles not on* $\mathbf{Q}$. *If m is odd then each rotation has an axis with one pole, a vertex of* $\mathbf{Q}$ *while the other is not a vertex of* $\mathbf{Q}$. In both the cases, we have two orbits of poles, all in the plane of the regular m-gon. Also, the m rotations we just described are actually planar reflections with each reflection reflecting the plane of the lamina of the regular m-gon in the axis (which is in the plane), and thus 2-dimensional reflections become 3-dimensional rotations. This group is the dihedral group D_m in its action of rotations and reflections of a regular m-gon.

Finally, we are left with the case $r_1 = 2$ and $r_2 = 3$. Then we have

$$\frac{1}{r_3} = \frac{1}{6} + \frac{2}{n}$$

Then $r_3 \leq 5$ and $r_2 \leq r_3$ gives $r_3 = 3, 4, 5$ as the only possibilities. First let $r_3 = 3$. Then $n = 12$. Here G is a group of rotations of a regular tetrahedron. The orbits of poles correspond to the following three distinct categories: mid-points of edge, vertices and centers of opposite (antipodal) faces respectively giving

$$(n_1, r_1) = (6, 2), \;\; (n_2, r_2) = (4, 3), \;\; (n_3, r_3) = (4, 3)$$

respectively. Here the group G is isomorphic to the alternating group A_4. Next let $r_3 = 4$. Then $n = 24$. Here G is a group of rotations of a cube. The orbits of poles correspond to the following three distinct categories: mid-points of edge, vertices and centers of opposite faces respectively giving

$$(n_1, r_1) = (12, 2), \;\; (n_2, r_2) = (8, 3), \;\; (n_3, r_3) = (6, 4)$$

respectively. Here the group G is isomorphic to the symmetric group S_4. This group is also the group of rotations of a regular octahedron. Finally, let $r_3 = 5$. Then we have $n = 60$. Here G is a group of rotations of a regular dodecahedron. The orbits of poles correspond to the following three distinct categories: mid-points of edge, vertices and centers of opposite faces respectively giving

$$(n_1, r_1) = (30, 2), \quad (n_2, r_2) = (20, 3), \quad (n_3, r_3) = (12, 5)$$

respectively. Here the group G is isomorphic to the alternating group A_5 and is the full group of rotations of a regular dodecahedron (or a regular icosahedron). □

We conclude this discussion by noting that the theorem just proved is in the line of the classification of wall paper groups (or planar crystallographic groups) which are 17 in number and their three dimensional analogues called the crystallographic groups (which are 230 in number). A discussion on these is given in Artin [3]. The groups we just described also have connections with reflection groups, the discussion of which is beyond our scope.

14.5 Exercises for Chapter 14

14.1 Show that if H is a subgroup of S_n that consists of both even and odd permutations, then it has an equal number of even and odd permutations.

14.2 Let G be a finite group and let t be a natural number. Let

$$S = \{A : A \subset G \; and \; |A| = t\}$$

Consider that action of G on S given by

$$g(A) = gA = \{ga : a \in A\}$$

For $A \in S$ the orbit of A is $A^G = \{B \in S : g(A) = B \; for \; some \; g \in G\}$. Prove the following.

(a) This defines a proper G-action on S.

(b) Let $G_A = \{g \in G : g(A) = A\}$ where $A \in S$. Then $G_A \leq G$.

(c) *Orbit-stabilizer theorem*: $|A^G| \times |G_A| = |G|$.

14.3 Given a a permutation group (G, X) and (G', X') we say that these two groups are permutation-isomorphic if there is an isomorphism $\phi : G \to G'$ and a bijection $f : X \to X'$ such that $f(\alpha(x)) = \phi(\alpha)(f(x))$ for every $x \in X$ and for every $\alpha \in G$. Prove that given a group G the Cayley representation is the unique sharp transitive group action of G on a set X.

14.4 Let G be a group of order p^n where p is a prime and $n \geq 3$. Show that $|Z(G)| = p^m$ where $m \in \{1, 2, \ldots, n-2, n\}$.

14.5 Consider the symmetric group S_9 and let $\beta = (147)(268)(359)$ and $\gamma = (159)(247)(368)$ be two permutations in their cycle decomposition form.

(a) Find α such that $\gamma = \alpha\beta\alpha^{-1}$.

(b) What is $|Z_{S_9}(\beta)|$?

(c) Show that $[Z_{S_9}(\beta) : Z_{A_9}(\beta)] = 2$.

(d) Show that the conjugacy classes of β in A_9 and S_9 are identical sets.

14.6 Consider S_7 and let β denote the permutation in its cycle decomposition given below. In each case, determine if $Z_{S_7}(\beta) = Z_{A_7}(\beta)$ or $[Z_{S_7}(\beta) : Z_{A_7}(\beta)] = 2$.

(a) $\beta = (1345)(26)$

(b) $\beta = (13)(25)$

(c) $\beta = (1234567)$

14.7 Show that if two permutations β and γ have the same cycle type then they are conjugates in the symmetric group S_n. Given β and γ that have the same cycle type, how many α's do we have with the property that $\gamma = \alpha^{-1}\beta\alpha$? Given β with cycle type $(b_1, b_2, \ldots, b_n)$, how many permutations are there in the conjugacy class of β?

14.8 Let $\beta \in A_n$.

(a) Let the cycle decomposition of β contain a cycle of length $4k + 2$. Then show that $Z_{A_n}(\beta) \neq Z_{S_n}(\beta)$.

(b) Let for some odd number $2m + 1$, the cycle decomposition of β contain two cycles of length $2m + 1$. Then show that $Z_{A_n}(\beta) \neq Z_{S_n}(\beta)$.

(c) Given $\beta \in A_n$, state and prove a necessary and sufficient condition (in terms of the cycle decomposition) for $Z_{A_n}(\beta) = Z_{S_n}(\beta)$ (see Tsuzuku [55]).

14.9 Let P be a Sylow p-subgroup of G and let $H = N_G(P)$ be the normalizer of P. Prove that $N_G(H) = H$.

14.10 Determine all the finite groups that have at the most three conjugacy classes.

14.11 Show that if H is a normal subgroup of G such that $|G|$ is odd and $|H| = 5$ then $H \leq Z(G)$.

14.12 We wish to prove that if H is a subgroup of index p of a finite group G of order n, where p is the smallest prime divisor of n, then H must be a normal subgroup of G. Proceed through the following steps. *Suppose H is not normal.*

(a) Using the fact that there is no subgroup of G properly contained between H and G, conclude that $N_G(H) = H$.

(b) Use the orbit stabilizer theorem (Theorem 14.1.13) to show that H has exactly p conjugates. Let the set of all the conjugates of H be given by $X = \{H = H_0, H_1, \ldots, H_{p-1}\}$.

(c) Show that under the conjugation action of G on X, the stabilizer of H is H.

(d) Now consider the (conjugation) action of H on the set $Y = \{H_1, \ldots, H_{p-1}\}$. Use (a) for H_i to show that no orbit under the H-action on Y has size equal to 1 and hence every orbit under the H-action on Y has size some k where $2 \leq k \leq p-1$. Further k divides $|G|$.

(e) We get a contradiction from (d) because we obtain a proper divisor of n which is strictly less than p.

(f) Show that the assertion of the theorem is not true if p is some prime divisor of n. Also show that we do have groups G of order n with p the smallest prime divisor of n such that G has no subgroup of index p.

For a proof of the assertion that does not use group action, see Luther and Passi [39].

14.13 Let p be a prime and let G be a group of order $p^m k$ where p and k are coprime. Let $H \lhd G$ such that $[G : H]$ is coprime to p. Then show that every Sylow p-subgroup of G is a Sylow p-subgroup of H and vice versa.

14.14 Let p and q be distinct primes with $p < q$ such that p does not divide $q-1$. Let G be a group of order pq. Prove that G is a cyclic group.

14.15 Let G be a group of order 108. Prove that G has a normal subgroup of order 27 or a normal subgroup of order 9.

14.16 Let G be a group of order p^n where p is a prime and let H be a subgroup of order p^m where $m < n$. By considering the (conjugation) action of H on its conjugates, prove that $N_G(H) \neq H$.

14.17 Let G be a group of order p^n where p is a prime. Let H be a subgroup of G such that $|H| = p$. Show that if H is normal, then $H \leq Z(G)$.

14.18 In the following two cases, G is a group of order n. Prove the stipulated assertions.

(a) $n = 231$. Show that the Sylow 11-subgroup is central while the Sylow 7-subgroup is normal.

(b) $n = 385$. Show that the Sylow 7-subgroup is central while the Sylow 11-subgroup is normal.

14.19 Observe that every permutation of odd order in S_n is an even permutation. Hence show that the number of permutations of even order in S_n (with $n \geq 2$) is larger than the number of permutations with odd order.

14.20 Let p be a prime number and let G be the group of all the non-singular $n \times n$ matrices over F, the field with p elements.

(a) Find the order of G.

(b) Let H be a subset of G consisting of those matrices $A = [a_{ij}]$ such that $a_{ii} = 1$ for all i and $a_{ij} = 0$ for all $i > j$. Show that $H \leq G$.

(c) Show that H is a Sylow p-subgroup of G.

14.21 Show that upto isomorphism, there are exactly two groups of order ten.

14.22 Show that $Z(D_n) = 1$ if n is odd and $Z(D_n) \cong \mathbb{Z}_2$ if n is even. Show also that if n is even then $\frac{D_n}{Z(D_n)} \cong D_{n/2}$.

14.23 The quaternion group Q of order 8 has generators a, b, c each of order 4 such that

$$a^2 = b^2 = c^2 = -1,\ ab = c;\ bc = a;\ ca = b:\ ba = -c;\ cb = -a:\ ac = -b$$

Prove that

(a) Q has a unique element of order 2.

(b) $\frac{Q}{Z(Q)} \cong V_4$, the Klein group of order 4.

(c) Q is the only non-abelian group od order 8 with a unique element of order 2.

(d) Upto isomorphism, there are exactly two non-abelian groups of order eight.

Comment: There is an infinite family of finite non-abelian groups (called the family of dicyclic groups) that have a unique element of order 2; refer to Rotman [48].

14.24 Let G be a group of order 12 such that G is not isomorphic to A_4. Prove that G has an element of order 6.

14.25 Let F the field with 4 elements and let G be group of all the 2×2 matrices A over F such that $\det(A) = 1$. Show that $G \cong A_5$.

14.26 A permutation group (G, X) is said to be 2-transitive if given any two ordered pairs (x_1, x_2) and (y_1, y_2), we can find a permutation $g \in G$ such that $g(x_i) = y_i$ for $i = 1, 2$. Show that a 2-transitive permutation group is transitive. Let G be the group $GL(V)$ of all the non-singular linear transformations on an n-dimension vector space V over a field F where $n \geq 2$. Let $V^* = V - \{0\}$.

(a) Show that G is transitive on V^* but not 2-transitive in general.

(b) Let F have only two elements. Show that G is 2-transitive on V^*.

14.27 Let G be a non-abelian group of smallest order n such that G is simple. Show that $n = 60$.

14.28 A subgroup H of a group G is called its characteristic subgroup if $\forall\, \gamma \in Aut(G)$ we have $\gamma(H) = H$. Prove the following.

(a) If H is a normal Sylow p-subgroup of G, then H is a characteristic subgroup of G.

(b) $Z(G)$ is a characteristic subgroup of G.

(c) The Klein group consisting of the identity and three permutations that are products of two disjoint transpositions, is a characteristic subgroup of S_4.

(d) $\forall\, n \geq 5,\ A_n$ is a characteristic subgroup of S_n.

14.29 Let G be a group and let $\Phi(G)$ denote the intersection of all the maximal subgroups of G. Show that $\Phi(G)$ is a characteristic subgroup of G (*this subgroup is called the Frattini subgroup of* G).

14.30 Let G be a group of order $n = k \times p^m$ where p is a prime, m and k are positive integers such that k is coprime to p. Let H be a subgroup of order p^r where $r < m$. Here is an alternative (and direct) proof to show that H is contained in some Sylow p-subgroup P of G. Follow the proof of Sylow's first theorem and argue in the same manner with G acting on all the p^m-subsets (containing H) of the n-set G (containing H) under the action of G.

14.31 Let G be a finite group and let $K \lhd G$. Let P be a Sylow p-subgroup of K. Then show that $G = KN_G(P)$ (this assertion is called the *Frattini argument*) .

14.32 In an automorphism tower of a centerless group G, suppose we have $G_m = G_{m+1}$ for some m. Show that $G_n = G_m\ \forall\, n \geq m$.

14.33 Let $G = S_n$. Prove the following assertions.

(a) G is generated by the set of all the transpositions.

(b) G is generated by the set S of transpositions where

$$S = \{(12), (13), \ldots, (1n)\}$$

(hint: use induction on n).

(c) Let S be a subset of G such that every member of S is a transposition. Draw a graph Γ with vertex set $[n]$ and with edge-set consisting of transpositions in S. Show that S is generating set for G iff the edge-subgraph on S is a connected graph.

(d) Let S be a subset of G such that every member of S is a transposition and S is a set of generators for G. Show that S is a minimal generating set iff the edge-subgraph on S is a spanning tree of Γ.

(e) Show that G is also generated by the transposition $\alpha = (12)$ and the full n-cycle $(1, 2, \ldots, n)$.

14.34 Prove Theorem 14.2.20(b) using the fact that there is a transposition commuting with a 3-cycle disjoint from it.

14.35 Let G be a finite group with a proper subgroup H. Show that the the union of all the conjugates of H is a proper subset of G.

14.36 Let G be a finite group with center Z such that the quotient group $\frac{G}{Z}$ is cyclic. Prove that G must be abelian.

14.37 Show that $\forall\, n \geq 5$, A_n is the only proper normal subgroup of S_n. What happens when $n = 4$?

14.38 Complete the following alternative treatment of the proof of simplicity of A_n for all $n \geq 6$ (Theorem 14.2.21). Assume that A_5 is simple and work inductively. Let $n \geq 6$ and let H denote a non-identity normal subgroup of $G = A_n$. We wish to show that $H = G$.

(a) Show that if $h \in H$ is a non-identity permutation fixing a point, then $H = G$.

(b) Suppose every non-identity permutation σ of H is a derangement. Let w.l.o.g. $\sigma(1) = 2$ and $\sigma(3) = 4$ (this can be done since $n \geq 6$). Let $\tau = (12)(3456)$. Let γ denote the permutation $\sigma\tau^{-1}\sigma\tau$. Show that γ fixes 1 and moves 3 to 5 which gives the required contradiction.

14.39 Show that if G is a finite cyclic group, then $Aut(G)$ is abelian. Show further that if G has order p where p is a prime, then $|Aut(G)| = p - 1$.

14.40 Let f be an automorphism of $G = S_n$ that takes every transposition to a transposition. We wish to show that$f \in Inn(G)$. Work through the following steps which construct an inner automorphism equal to f. We write γ to denote an inner automorphism and thus γ_β is an inner automorphism given by $\gamma_\beta(\alpha) = \beta\alpha\beta^{-1}$ for a permutation $\alpha \in S_n$.

(a) Let $f((12)) = (ij)$ and let $\beta_2 = (1i)(2j)$ with the convention that $(1i) = 1$ if $i = 1$. Show that $f((12)) = \gamma_\beta(12)$ and hence the automorphism $f_2 = (\gamma_{\beta_2})^{-1}f$ fixes the transposition (12) and takes transpositions to transpositions. *To simplify the notation, write* γ_i *to mean* γ_{β_i}.

(b) Show that $f_2((13))$ equals $(1j)$ or $(2j)$ for some $j \geq 3$. Hence find a suitable a β_3 such that γ_3 fixes (12) and takes (13) to $(1j)$ (or $(2j)$ as the case may be.) Therefore show that $f_3 = (\gamma_3)^{-1}f_2$ fixes both (12) and (13) and takes transpositions to transpositions.

(c) Inductively assume that for $k \geq 3$, we have constructed $\beta_2, \beta_3, \ldots, \beta_k$ and hence $\gamma_i = \gamma_{\beta_i}$ such that $f_i = (\gamma_i)^{-1}f_{i-1}$ where $i = 2, 3, \ldots, k$ (with $f_{-1} = f$) satisfying the following: f_k fixes the transpositions $(12), (13), \ldots, (1k)$ and takes transpositions to transpositions. Then show that $f_k((1, k+1)) = (1j)$ where $j \geq k+1$ and for a suitable β_{k+1}, we have $f_{k+1} = (\gamma_{k+1})^{-1}f_k$ where f_{k+1} is an automorphism fixing each one of the transpositions $(12), (13), \ldots, (1k), (1, k+1)$ and takes transpositions to transpositions.

(d) Finally, let γ equal $\gamma_2\gamma_3 \cdots \gamma_n$. Show that $f_n = \gamma^{-1}f$ fixes each one of $(12), (13), \ldots, (1n)$ and since transpositions generate the symmetric group (Exercise 14.33), f_n is the identity permutation and hence $f = \gamma$ where $\gamma \in Inn(G)$

14.41 Consider S_6. Prove the following assertions.

(a) S_6 has exactly 11 conjugacy classes of which three classes consists of permutations of order 2. Two of these classes say C_1 and C_2 have 15 permutations each and the third has 45 permutations. C_1 is the class of transpositions and C_2 the class of permutations that are products of three disjoint transpositions (with no fixed point).

(b) Any outer automorphism F must interchange C_1 and C_2.

(c) If F_1 and F_2 denote two outer automorphisms, then ${F_1}^{-1}F_2$ takes transpositions to transpositions and hence is an inner automorphism.

(d) $[Aut(S_6) : Inn(S_6)] = 2$.

The curious existence of an outer automorphism of S_6 (which exists only when $n = 6$) has been exploited in many intricate constructions in finite geometries and finite groups. For example, this object occurs in the construction of the sporadic simple group called the Mathieu group M_{12} and a related Steiner system (refer to Rotman [48]).

14.42 Show that $Aut(A_n) = S_n$ for all $n \neq 6$. What happens when $n = 6$?

14.43 For a natural number m, let $\overline{S_m}$ denote the set of all the bijections α on $\mathbb{N}$ such that $\alpha(j) = j \; \forall \; j \geq m + 1$.

(a) Show that $\overline{S_m} \cong S_m$.

(b) Let $\overline{A_m}$ denote the subgroup of $\overline{S_m}$ that corresponds to even permutations in S_m (i.e. corresponds to A_m).

(c) Now let $A = \cup_{m=1}^{\infty} \overline{A_m}$ and let H denote a normal subgroup of A such that $H \neq 1$. Show that for some $m \geq 5$, we have $H_m = H \cap \overline{A_m}$ normal in A_m and hence $H_m = \overline{A_m}$.

(d) Use the previous assertion to conclude that A is a countably infinite simple group.

14.44 A reflection of $\mathbb{R}^2$ is a linear transformation which fixes a line through the origin and reflects the entire plane using this fixed line as a mirror. Prove the following.

(a) Every orthogonal transformation is either a rotation or a reflection.

(b) If A is a matrix representing a reflection, show that A has the form

$$\begin{pmatrix} \cos\theta & \sin\theta \\ \sin\theta & -\cos\theta \end{pmatrix}$$

(c) Show that if A is a proper rotation of $\mathbb{R}^2$, then the eigenvalues of A are not real while the eigenvalues of a reflection are real.

(d) Show that product of two reflections is a rotation through an angle which is twice the angle between the two mirrors.

(e) Show that a regular m-gon has $2m$ symmetries of which m are rotations through an angle which is a multiple of $\theta = \frac{2\pi}{m}$ and m are reflections.

(f) Argue both geometrically as well as algebraically to assert that a 2-dimensional reflection can also be viewed as a 3-dimensional rotation (refer to the proof of Theorem 14.4.4).

14.45 Prove the following statements on the group $SO_2(\mathbb{R})$.

(a) Prove that any member of $SO_2(\mathbb{R})$ with a finite order is a rotation.

(b) Prove that every finite subgroup of $SO_2(\mathbb{R})$ is a cyclic group (Theorem 14.4.1).

(c) Let T denote the multiplicative group of complex numbers z with $|z| = 1$ (and hence $z = e^{i\theta}$ for a suitable θ). Show that $SO_2(\mathbb{R}) \cong T$.

(d) Show that for every natural number n, T has a unique subgroup of order n and this group is also cyclic.

(e) Let p be a prime and let H denote the subgroup of T that consists of all the p^m-th roots of unity for $m = 0, 1, 2, \ldots$. Show that H is not cyclic but every proper subgroup of H is cyclic. In fact, every proper subgroup of H is finitely generated but H has no finite set of generators. (*H is called a Prufer group.*)

Chapter 15

Polya theory of enumeration

15.1 Introduction

This chapter deals exclusively with group action on combinatorial objects. It has become imperative at this stage after having developed all the necessary algebraic tools through earlier chapters mainly in Chapter 12 and in Chapter 14. Unlike the earlier chapters, configurations, and therefore related geometric transformations form the main objects of the study and the chapter concerns itself as much with the qualitative aspects of counting as with the quantitative aspects.

A major part of combinatorics is concerned with counting of objects: arrangements or possibilities or sets of a given type, to just name a few. We have so far acquired a number of techniques of counting in this book. These include the basic counting parameters such as the permutations and combinations, the binomial and multinomial theorems, Stirling, Catalan and various other special numbers and their interconnections. We studied theories of partitions, both partitioning numbers as well as partitioning sets. We also studied recurrence relations and were able to solve linear homogeneous recurrence relations with constant coefficients. This considerably eases out difficulties in terms of making many counting techniques available to us. A very useful tool in this regard is the idea of a *generating function*. This counts not just one number but an infinite class of numbers. Frequently, the technical difficulty of finding a general formula is overcome by the method of generating functions. However, many difficulties of counting related questions in combinatorics are of a *conceptual* rather than of a technical nature. The difficulty here is not that of lack of counting techniques but the difficulty lies in not knowing as to what one is counting. This happens when the counting becomes confused and blurred because under different rules, seemingly different objects have to be considered as identical. In the modern 'abstract' language, we are not counting objects as such but are counting equivalence classes of objects. These classes are a result of an equivalence relation. This is the second difficulty, where we need to know, instead of the number of objects, the number of equivalence classes of objects. This equivalence is caused by the group action on a set. The third difficulty in counting is that we do not always count all the objects with the same weightage. For example in a class with 50 students, the statement of the teacher that there are 50 heads puts weight one on every head (except the teacher's own head) while a statement that there are 70 (though there are only 50 students) heads implies that there are some students in the class that are much brighter and merely putting weight one on each student is not a correct idea. All these three aspects of counting: generating functions, equivalence under a group action on a set and the idea of weights are very elegantly blended in a remarkable theorem of Polya which is our topic of discussion in this chapter.

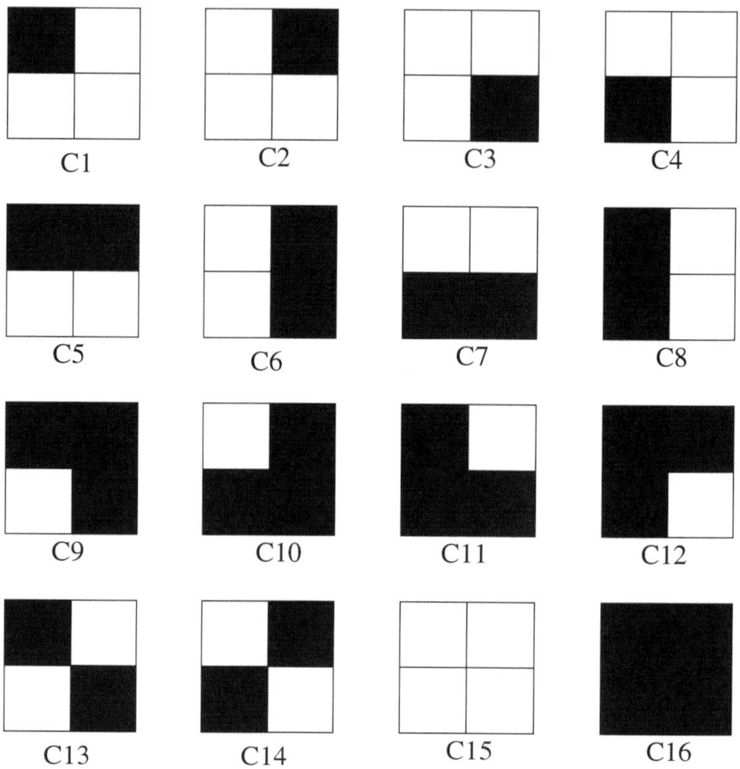

Figure 15.1: All 2×2 chessboards

Consider the question of colouring a 2×2 chessboard using two colours white and black. Since each cell can be coloured by any one of the two colours, (independent of how other cells are coloured) and there are 4 cells to be coloured, there $2 \times 2 \times 2 \times 2 = 16$ ways. This the number of functions from a 4-set to a 2-set and hence the required number is $2^4 = 16$. These 16 possibilities are shown in Figure 15.1. Notice that each one of $C1, C2, C3$ and $C4$ are 'same' in some sense. They all have exactly one black cell. Most people would not consider these four chessboards as different. What one has in mind here is the fact that any of these chessboards can be obtained from the other by a (suitable) rotation. These four chessboards are thus considered 'equivalent' because any one can be obtained from the other by a rotation through multiples of 90°. A clockwise rotation takes $C1$ to $C2$, $C2$ to $C3$, $C3$ to $C4$ and $C4$ back to $C1$. Under this idea of equivalence, we get only 6 chessboards (6 equivalence classes or patterns) and these are listed in Figure 15.2. Notice that there are two different colourings with exactly two black cells. These are $A2$ and $A3$ and neither can be obtained from the other by a rotation.

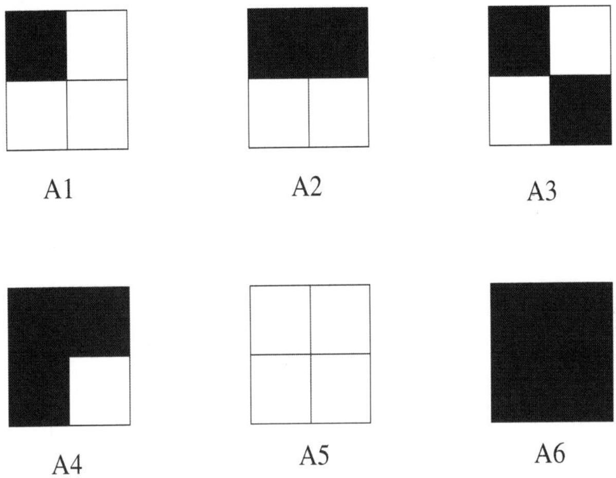

Figure 15.2: Representatives of Equivalence classes of 2×2 chessboards

Polya theory should be able to provide us with the answer 6 to the above question without actually having to compute using such elementary (but most of the times quite tedious) methods. Let the cells of the chessboard be numbered $1, 2, 3, 4$ as the four quadrants of the real plane. Here we have the following group in mind. G is a cyclic group of all the four rotations of the square. The full group is described as follows. Let σ denote a rotation (of the lamina of the chessboard in the plane) through an angle of 90° in the anticlockwise direction. Then σ takes 1 to 2, 2 to 3, 3 to 4 and 4 back to 1. σ^2 is a 180° rotation, σ^3 is a 270° rotation (this is same as rotation through an angle of -90° and thus equals σ^{-1}) and finally σ^4 is the identity permutation (that fixes each one of $1, 2, 3$ and 4). In terms of the two line notations, these permutations are respectively:

$$id = \begin{pmatrix} 1\,2\,3\,4 \\ 1\,2\,3\,4 \end{pmatrix}, \ \sigma = \begin{pmatrix} 1\,2\,3\,4 \\ 2\,3\,4\,1 \end{pmatrix}, \ \sigma^2 = \begin{pmatrix} 1\,2\,3\,4 \\ 3\,4\,1\,2 \end{pmatrix}, \ \sigma^3 = \begin{pmatrix} 1\,2\,3\,4 \\ 4\,1\,2\,3 \end{pmatrix}$$

Thus the group G in the mind is the cyclic group of order four generated by σ and acting on the set of all the 16 colored chessboards $C1$ through $C16$. The interest is thus not in G acting on a set of order four but in G acting on the set of order 16 and we have to look at things differently. Here, σ written as a product of disjoint cycles reads:

$$\sigma = (C1C2C3C4)(C5C6C7C8)(C9C10C11C12)C13C14)(C15)(C16)$$

The group has 6 orbits on the set of 16 coloured chessboards and these 6 orbits (actually their representatives) are given in Figure 15.2. Thus the 16 colored chessboards get partitioned into 6 orbits.
As a different example, think of a 3×3 chessboard with cells numbered 1 through 9 as shown in Figure 15.3.

1	2	3
4	5	6
7	8	9

Figure 15.3: 3×3 chessboard

We still have the same four rotations and hence the same group in an abstract algebraic sense. *But their representations are different: the rotations are now acting on a set of order 9 and hence the same group has a different combinatorial description. For example, the permutation σ will now look like the following.*

$$\begin{pmatrix} 1\,2\,3\,4\,5\,6\,7\,8\,9 \\ 7\,4\,1\,8\,5\,2\,9\,6\,3 \end{pmatrix}$$

This example shows that the same (meaning isomorphic) group can act as a permutation group on two sets of different sizes and hence can have different combinatorial representations. In this question it takes quite some effort to see that the number of distinct ways of colouring the 9 cells in two colours (with two colourings considered the same if one can be obtained from the other by a rotation) is equal to 140.

In this chapter, we have to break the convention (that we have stuck to, in all the earlier part of this book) and multiply permutations (like functions) from right to left. We are thus going to treat multiplication of permutations as a composition of functions. Thus, if α and β are two permutations on X, then

$$(\beta\alpha)(x) = \beta \circ \alpha(x) = \beta(\alpha(x))$$

Note that this is different from the earlier stand where we first computed $\beta(x)$ and then applied α to it. Our framework in this chapter is in the line of composite of two functions: $g \circ f(x) = g(f(x))$. As an example, let $n = 5$ and let

$$\alpha = \begin{pmatrix} 1 & 2 & 3 & 4 & 5 \\ 2 & 4 & 1 & 3 & 5 \end{pmatrix}; \ \beta = \begin{pmatrix} 1 & 2 & 3 & 4 & 5 \\ 4 & 2 & 1 & 5 & 3 \end{pmatrix}$$

be given in the two-line notation form. Then $\beta\alpha(1) = \beta \circ \alpha(1) = \beta(\alpha(1)) = 2$ while $\alpha\beta(1) = 3$. In fact, $\beta\alpha$ and $\alpha\beta$ are respectively given by

$$\begin{pmatrix} 1 & 2 & 3 & 4 & 5 \\ 2 & 5 & 4 & 1 & 3 \end{pmatrix}; \quad \begin{pmatrix} 1 & 2 & 3 & 4 & 5 \\ 3 & 4 & 2 & 5 & 1 \end{pmatrix}$$

Counting the number of orbits under a group action is one of the key questions answered by Polya theory and this is given in the following.

Lemma 15.1.1. *(Burnside's Lemma) Let (G, X) be a permutation group. Then the number of orbits of X under the action of G is given by*

$$\frac{1}{|G|} \sum_{\pi \in G} f(\pi)$$

where $f(\pi) = |\{x \in X : \pi(x) = x\}|$.

Proof Let the set of all the orbits of G on X be given by $\{X_1, X_2, \ldots, X_k\}$. Then X is a disjoint union $\cup_{i=1}^{i=k} X_i$. Let S denote the set,

$$S = \{(\pi, x) : \pi \in G, x \in X, \ \pi(x) = x\}$$

Then the contribution to $|S|$, from a given $\pi \in G$, is $f(\pi)$ and hence $|S| = \sum_{\pi \in G} f(\pi)$. On the other hand, for a fixed $x \in X$ the contribution to $|S|$ is $|G_x|$. Using a two-way counting, we get

$$\sum_{\pi \in G} f(\pi) = \sum_{x \in X} |G_x|$$

The second sum can be written as $\sum_{i=1}^{i=k} \sum_{x \in X_i} |G_x|$. Fix $i = 1, 2, \ldots, k$ and let $x \in X_i$. Then the orbit-stabilizer theorem (Theorem 14.1.8 tells us that $|G_x|$ equals $|G|/|x^G|$. But $x \in X_i$ implies $x^G = X_i$ and this is true for all $x \in X_i$. Hence

$$\sum_{x \in X_i} |G_x| = |G| \frac{|X_i|}{|X_i|} = |G|$$

Substitution of this in the original equation gives $|S| = k|G|$ and hence

$$k = \frac{1}{|G|} |S| = \frac{1}{|G|} \sum_{\pi \in G} f(\pi)$$

as desired. □

A situation that we come across most of the times is a bit more involved than the set-up of the orbit-stabilizer theorem and Burnside Lemma that we just discussed. We have a permutation group (G^*, X) and a surjective group homomorphism $\phi : G \to G^*$. In this situation we say that G *acts as a permutation group* on X. Thus for $g \in G$ and $x \in X$, we define the action of g on x by letting $g(x) = \alpha(x)$ where $\phi(g) = \alpha$. In view of Theorem 14.1.13, all the machinery we describe goes through. We record this in the following theorem.

Theorem 15.1.2. *(Burnside's lemma; general form) Let G act as a permutation group on X. Then the number of orbits of G on X is given by $\frac{1}{|G|}\sum_{g\in G} f(g)$ where $f(g)$ equals $|\{x \in X : g(x) = x\}|$.*

Example 15.1.3. We now have a re-look at the colouring problem of a 2×2 chessboard in two colours white and black. Here we have the group action of the cyclic group G (consisting of four rotations of the chessboard) that is generated by the permutation σ (where σ stands for a rotation through 90°). The set X on which G acts is a set of order 16 whose members are shown in Figure 15.1. Consider $f(\sigma)$. Since σ takes $C1$ to $C2$, $C2$ to $C3$ etc. it follows that there σ fixes only two members of X and these are the boards $C15$ and $C16$. Thus $f(\sigma) = 2$. The rotation through 180° given by σ^2 can fix only the chessboards $C13$ and $C14$ besides $C15$ and $C16$. Clearly $\sigma^3 = \sigma^{-1}$ has the same cycle type as σ and hence $f(\sigma^3) = 2$. Finally $f(id) = 16$ and therefore the number of orbits, using Burnside Lemma (Theorem 15.1.1) is equal to

$$\frac{1}{4}(2+4+2+16) = 6$$

which we already saw in Figure 15.2.

Example 15.1.4. Now consider the question of colouring a 3×3 chessboard in two colours under the action of the same group G of previous example (this is also the group of rotations of the new chessboard). Refer to Figure 15.3. The set X on which the group G is acting has order 2^9. For a chessboard to be fixed by σ, we must have the same colour on the cells $1, 7, 9, 3$ as also same colour on the cells $2, 4, 8, 6$ (and it does not matter which colour is given to the central cell 5). It follows that $f(\sigma) = 2^3$ and by symmetry $f(\sigma^{-1})$ also equals 2^3. The same argument gives $f(\sigma^2) = 2^5$ and finally $f(id) = 2^9$. Therefore, using Burnside Lemma, the number of orbits is equal to to

$$\frac{1}{4}(2 \times 2^3 + 2^5 + 2^9) = 140$$

In order to be able to apply Burnside Lemma, we need to have a fair idea as to to how the given group G in its action on the set X looks (and, in particular, which set X we have in mind). We begin with a couple of important examples of group actions. These are both groups of *symmetries* of a regular polygon in the plane.

Example 15.1.5. Let $\mathbf{P}$ be a regular n-gon (with $n \geq 3$) and let G be the group of all the rotations of $\mathbf{P}$. Here, we fix the centre of $\mathbf{P}$ and rotate $\mathbf{P}$ around the centre through angles that are multiples of $\theta = \frac{2\pi}{n}$. This has been described in Theorem 14.4.3. The group G is thus a cyclic group of order n and is generated by a rotation α where α is rotation of $\mathbf{P}$ through θ in the anticlockwise direction. Let the vertices of $\mathbf{P}$ be ordered $1, 2, \ldots, n$ in an anticlockwise manner. Then, in its action on the vertices of $\mathbf{P}$, α as a permutation is a single n-cycle $(1, 2, 3, \ldots, n-1, n)$. Take the particular case $n = 10$ and suppose the vertices of $\mathbf{P}$ are coloured in two colours red and blue. What is the total number of distinct equivalence classes of colourings? This problem is similar to the 2×2 chessboard coluring question (there, we had $n = 4$). So, we see that $f(\alpha) = f(\alpha^{-1}) = f(\alpha^9) = 2$ as also $f(\alpha^3) = f(\alpha^7) = 2$. To find $f(\alpha^2)$, notice that $1, 3, 5, 7, 9$ must all receive the same colour as also $2, 4, 6, 8, 10$ must receive the same colour. Hence $f(\alpha^2) = 2^2$. Similarly, $f(\alpha^8) = (\alpha^4) = f(\alpha^6) = 2^2$ Finally $f(\alpha^5) = 2^5$. We thus have the number of equivalence classes of colourings equal to

$$\frac{1}{10}\left(4 \times 2 + 4 \times 4 + 2^5 + 2^{10}\right) = 108$$

In general, if $\mathbf{P}$ is an n-gon and $\beta = \alpha^k$, what can we say about $f(\beta)$?

Example 15.1.6. Let G now denote the group of all the rotations as well as reflections of $\mathbf{P}$, a regular n-gon. A reflection of the real plane, pointwise fixes some line in the plane and reflects every (other) point of the plane through this fixed line as a mirror, as we already saw in Section 14.4. For convenience, label the vertices of $\mathbf{P}$ as $0, 1, 2, \ldots, n-1$ in an anticlockwise manner. We wish to describe reflections of $\mathbf{P}$ in some more details. Here, we need to make two cases. First, let n be odd, say $n = 2m+1$. Holding the vertex 0 and the mid-point of the side $(m, m+1)$ fixed (and thus the straight line joining these two points is the mirror), we reflect $\mathbf{P}$. Read modulo $2m+1$, observe that a vertex i is reflected to the vertex $-i$ and vice versa. This fixes the vertex 0 and pairs the other vertices in m pairs $(i, -i)$. What we did with vertex 0 can also be done with any other vertex and hence we get $2m + 1 = n$ reflections in all. Note that each reflection has exactly one fixed vertex. Now let $n = 2m$ be an even number (with $m \geq 2$). Here we have *two types of reflections*. We can fix a pair of antipodal vertices say 0 and m (and hence the mirror passes through these two points) and reflect $\mathbf{P}$. This reflects i to $-i$ and vice versa. Clearly this reflection (in its action on vertices of $\mathbf{P}$) has only two fixed points. There are clearly m permutations of this kind since there are m pairs of antipodal vertices. The other types of reflections are obtained as follows. Take a line joining the mid-points of antipodal edges such as the mid-point of the edge $(0, 1)$ and the mid-point of the edge $(m, m+1)$ and reflect the polygon $\mathbf{P}$ in this line. This refection pairs all the $2m$ vertices in m pairs. Clearly such a reflection has no fixed point and since there are m pairs of antipodal edges, we see that there are m such reflections. *To sum up, irrespective of whether n is odd or even, we get n reflections in all.*

Here are some more examples of the use of Burnside Lemma (Theorem 15.1.2).

Example 15.1.7. Consider the problem of colouring the four cells of a 2×2 chessboard where two colourings are considered equivalent if one can be obtained from the other under of a suitable member of the dihedral group D_4. We are now allowing both rotations and reflections and the set on which the group acts is the same set of 16 chessboards shown in Figure 15.1 while the group now has order 8. Thus $C1$ and $C3$ are equivalent (also) under a reflection. In this case, the number of equivalence classes is

$$\frac{1}{8}\left[2^4 + 2 \times 2 + 2^2 + 2 \times 2^3 + 2 \times 2^2\right] = 6$$

These equivalence classes have already been shown in Figure 15.2 (though the group here is larger).

Example 15.1.8. Consider the problem of colouring in two colours red and blue, the beads of a necklace with 6 beads where the beads are arranged regularly on the vertices of a regular hexagon. First assume that the symmetries we allow are only rotations. In this case, arguing as before, we see that the number of equivalence classes is

$$\frac{1}{6}\left[(2 \times 2) + (2 \times 2^2) + 2^3 + 2^6\right] = 14$$

Next assume that the group of symmetries is the dihedral group D_6. In this case, the number of equivalence classes is equal to

$$\frac{1}{12}\left[(2 \times 2) + (2 \times 2^2) + 2^3 + 2^6 + (3 \times 2^3) + (3 \times 2^4)\right] = 13$$

It is evident, that generally the number of equivalence classes is smaller if the group is larger. We thus have two coloured necklaces with 6 beads that are not equivalent under the action of the rotation group but are equivalent under the dihedral group action. As an exercise (Exercise 15.9), the reader is asked to identify these two necklaces.

15.2 Group action on functions and cycle index of a group

Burnside's lemma (Theorem 15.1.2) has a shortcoming in that we have to compute $f(\sigma)$, the number of fixed points of σ for each σ in the group G. This can be a cumbersome task. Another drawback of Burnside Lemma is that it does not give us any qualitative idea of the nature of the equivalence classes. For example, in the 2×2 chessboard colouring problem, the formula given by Burnside lemma does not tell us that there are two equivalence classes that with exactly two cells black. *Both these shortcomings are overcome by Polya theory, the discussion of which we now begin.* Following is the set-up of Polya theory. D and R are finite sets and G a permutation group acting on D. The notation R^D stands for the set of all the functions from D to R (the domain to the range); $R^D = \{f | f : D \to R\}$ and using multiplication principle, we already know that $|R^D| = |R|^{|D|}$.

Definition 15.2.1. Let $\sigma \in G$. Then we define $a_\sigma : R^D \to R^D$ *by:*

$$a_\sigma(f) = g \;\; if \;\; \forall \; d \in D \;\; f(d) = g(\sigma(d))$$

We proceed through the following three claims.
Claim a_σ is well-defined.
Proof If $e \in D$, then let $g(e) = f(\sigma^{-1}(e))$. Then setting $e = \sigma(d)$ (which can be done since σ is a permutation), we see that $f(d) = g(\sigma(d))$ holds $\forall\, d \in D$.
Claim a_σ is one-one.
Proof If both $a_\sigma(f) = g$ and $a_\sigma(h) = g$, then we see that $f(d) = h(d) \;\; \forall \;\; d \in D$ (using the fact again that σ is a bijection).
Claim a_σ is onto.
Proof Given $g \in R^D$, define f by $f(d) = g(\sigma(d)) \; \forall d \in D$. Then clearly $a_\sigma(f) = g$. Alternatively, observe that $a_\sigma : R^D \to R^D$ is injective and hence must be surjective as well.

Lemma 15.2.2. *The following assertions hold.*

(a) a_σ *is a permutation on* R^D.

(b) $a_\sigma a_\tau = a_{\sigma \circ \tau}$

Proof It suffices to prove only (b) since (a) has already been proved. Let $a_\tau(f) = g$ and $a_\sigma(g) = h$. Defining $e = \tau(d) \ \forall \ d \in D$, we see that $f(d) = g(\tau(d)) \ \forall \ d \in D$ and $g(e) = h(\sigma(e)) \ \forall \ e \in D$. So,

$$f(d) = g(\tau(d)) = g(e) = h(\sigma(e)) = h(\sigma(\tau(d))) = h(\sigma \circ \tau(d))$$

This holds for all $d \in D$ and hence

$$a_{\sigma \circ \tau}(f) = h = a_\sigma a_\tau(f)$$

□

We have now translated the G-action on D to the G action on R^D (in the sense that G has a homomorphic copy inside the symmetric group on R^D).

Definition 15.2.3. A *pattern* is an equivalence class (or orbit) of functions in R^D under G-action on R^D (as defined in Lemma 15.2.2).

Counting the number of patterns is a by product of Polya theory. Consider the original example of the 2×2 chessboard once again. Here the set D is the set of the 4 cells of a 2×2 chessboard shown in Figure 15.3. The set R has two elements white and black and hence R^D consists of 16 functions that are shown in Figure 15.1 as coloured chessboards. The permutation group G on D is the cyclic group G that consists of 4 rotations (generated by a single rotation through an angle of 90°). G-action on R^D partitions R^D into *patterns*. These (actually representatives of patterns) are shown in Figure 15.2. Thus the pattern $A1$ actually consists of four functions in R^D given by the colourings $C1, C2, C3, C4$ of Figure 15.1. Notice that not all the patterns consist of the same number of functions.

One of the ingredients of Polya theory is the idea of cycle index of a permutation group which we now describe.

Definition 15.2.4. Let (G, X) be a permutation group with $|X| = n$. Let $\alpha \in G$ and let α have type $(b_1, b_2, \ldots, b_n)$ (recall that this means that in the cycle decomposition of α, we have b_k cycles of length k for all k). Then *a (formal) monomial associated with* α *(or monomial of* α*)* is ${x_1}^{b_1} {x_2}^{b_2} \cdots {x_n}^{b_n}$ where we treat the monomial to be an element of a polynomial ring over a field (we can thus add and subtract monomials and also divide by non-zero field elements) in the variables $x_1, x_2, \ldots, x_n$.

Definition 15.2.5. Let (G, X) be a permutation group with $|X| = n$. Then *the cycle index of G (in its action on* X*)* is given by

$$P_G(x_1, x_2, \ldots, x_n) = \frac{1}{|G|} \sum_{\sigma \in G} (monomial\ of\ \sigma)$$

Example 15.2.6. Look at the cyclic group $G \cong \mathbb{Z}_n$ (group of integers modulo n) as a permutation group in its action on the set $[n]$ where $G =< \alpha >$ and $\alpha = (1, 2, \ldots, n)$ and thus α is a single n-cycle. For an arbitrary $\beta \in G$ with $\beta = \alpha^j$, β has order k where $k = \frac{n}{(n,j)}$ where (n, j) denotes the g.c.d. of n and j and all the cycles in β are of the same length k and thus, in its cycle decomposition, β is a product of $\frac{n}{k}$ cycles each of length k. Given a k that divides n how many permutations in G have order k? All such permutations β are generators of the unique cyclic subgroup of order k of G and hence we have exactly $\phi(k)$ permutations with this cycle type $(0, 0, \ldots, \frac{n}{k}, 0, \ldots)$ where the positive entry is at the k-th place. Thus, there are $\phi(k)$ monomials each equal to ${x_k}^{n/k}$. Therefore, the cycle index of G is

$$P_{Z_n}(x_1, x_2, \ldots) = \frac{1}{n} \sum_{k|n} \phi(k){x_k}^{n/k}$$

Example 15.2.7. Now let X denote the set of vertices of a regular polygon $\mathbf{P}$ with n vertices and consider the group action by the dihedral group D_n. We just need to compute the monomials associated with reflections since the those associated with rotations have already been computed. First let n be odd. Then every reflection of $\mathbf{P}$ fixes exactly one vertex and pairs the others into $\frac{n-1}{2}$ pairs. So the monomial of a reflection is $x_1{x_2}^{\frac{n-1}{2}}$ (and there are n such monomials). Let n be even. Then there are $\frac{n}{2}$ reflections each of which fixes two antipodal vertices and pairs the others in $\frac{n-2}{2}$ pairs. The remaining $\frac{n}{2}$ reflections have no fixed points and pairs all the vertices in $\frac{n}{2}$ pairs. Thus the monomial of a reflection of the first type (respectively second type) is ${x_1}^2{x_2}^{\frac{n-2}{2}}$ (respectively ${x_2}^{\frac{n}{2}}$). Thus, the cycle index of D_n, in its action on vertices of $\mathbf{P}$ is

$$P_{D_n}(x_1, x_2, \ldots) = \frac{1}{2n}\left\{\left(\sum_{k|n} \phi(k){x_k}^{n/k}\right) + nx_1{x_2}^{\frac{n-1}{2}}\right\} \quad if\ n\ is\ odd$$

$$P_{D_n}(x_1, x_2, \ldots) = \frac{1}{2n}\left\{\left(\sum_{k|n} \phi(k){x_k}^{n/k}) + \frac{n}{2}({x_1}^2{x_2}^{\frac{n-2}{2}} + {x_2}^{\frac{n}{2}}\right)\right\} \quad if\ n\ is\ even$$

Example 15.2.8. Let G be the group of rotations of a regular n-gon $\mathbf{P}$ in its action on the set $X = X_v \cup X_e$ consisting of the disjoint union of the sets X_v of all the n vertices and the set X_e consisting of all the n edges. G has two orbits X_v and X_e on X and G acts sharp transitively on these two orbits. Modifying the argument in Example 15.2.6, we see that if $\beta \in G$ is such that $o(\beta) = k$, then the β has cycle type $(0, \ldots, 0, \frac{2n}{k})$ (with the last non-zero entry at the k-th place). Hence, the cycle index of G on X (in this changed situation) is:

$$P_G(x_1, x_2, \ldots) = \frac{1}{n} \sum_{k|n} \phi(k){x_k}^{2n/k}$$

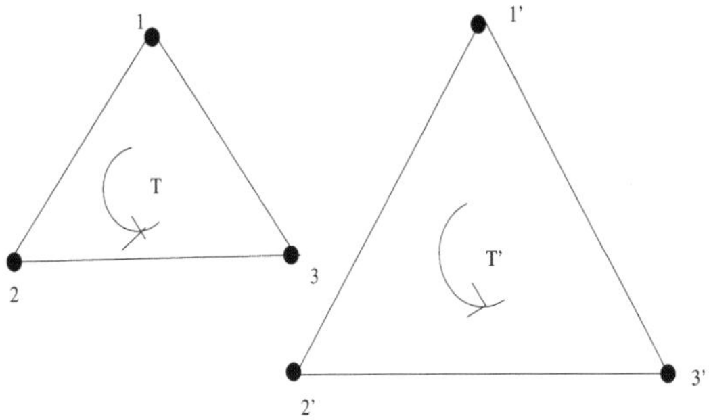

Figure 15.4: Rotation group of two incongruent triangles

Example 15.2.9. Let X be a set of order 6 consisting of two disjoint equilateral triangles say T and T' with vertex sets $\{1,2,3\}$ and $\{1',2',3'\}$ (written anticlockwise as shown in Figure 15.4 such that these two triangles are not congruent; this just means that the length of their sides are different and hence w.l.o.g. T is smaller than T'. The group consists of rotations of both T and T'. Since we can independently rotate T and T' each with a rotation group of order three, we see that the group of G of rotational symmetries of X has order $3 \times 3 = 9$ and in fact, we have two orbits $\{1,2,3\}$ and $\{1',2',3'\}$. It is then clear that

$$P_G(x_1, x_2, \dots,) = \frac{1}{9}\left((x_1)^6 + 4(x_1)^3 x_3 + 4(x_3)^2\right)$$

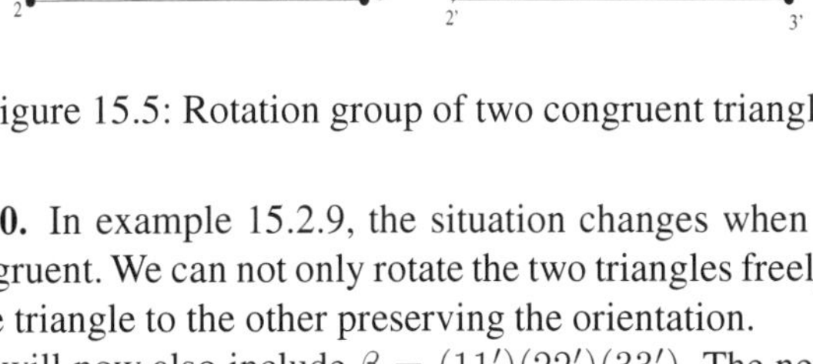

Figure 15.5: Rotation group of two congruent triangles

Example 15.2.10. In example 15.2.9, the situation changes when the two triangles $T$and T' are congruent. We can not only rotate the two triangles freely (independently) but also map one triangle to the other preserving the orientation.
The symmetries will now also include $\beta = (11')(22')(33')$. The new group $G* \geq G$ is thus generated by β, α, α' where $\alpha = (123)$ and $\alpha' = (1'2'3')$ and has order $2 \times 9 = 18$. Then

$$P_G(x_1, x_2, \dots,) = \frac{1}{18}\left\{\left((x_1)^6 + 4(x_1)^3 x_3 + 4(x_3)^2\right) + \left(6x_6 + 3(x_2)^3\right)\right\}$$

Example 15.2.11. Consider the group G of all the rotations of $\mathbf{P}$, a regular tetrahedron with vertex set $X = [4]$. As was shown in Chapter 14 (Theorem 14.4.4) this group has order 12. We consider the G-action on X.

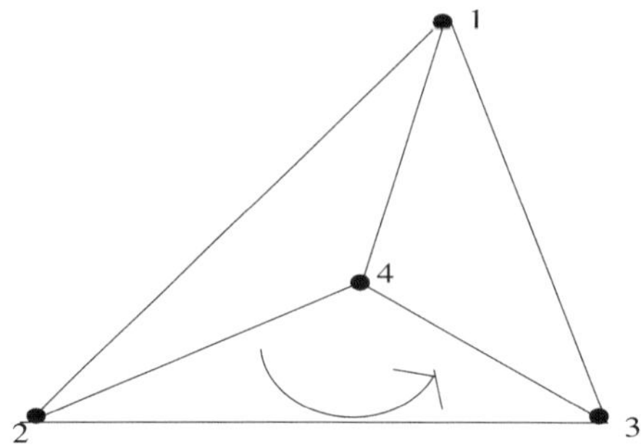

Figure 15.6: A regular tetrahedron and its vertex based rotations

The identity permutation evidently has cycle type $(4, 0)$ and the associated monomial ${x_1}^4$. The non-identity rotations are of two types: Vertex based rotations and edge based rotations. For the first type, fix a vertex say 1 and the center of the antipodal face taking the straight line joining them as the axis. Now rotate the space $\mathbb{R}^3$ and in particular $\mathbf{P}$ around this axis first through an angle of 120° and then 240° giving the two 3-cycles (234) and (243) respectively. Rotations fixing a vertex give us a Sylow 3-subgroup of G. The two non-identity permutations we get have cycle type $(0, 0, 1)$ and hence the associated monomial x_3. We have $4 \times 2 = 8$ rotations of this kind that fix a vertex. Next consider two antipodal (skew) edges such as (14) and (23) and take an axis that joins the mid-points of these edges. With this as the axis rotate $\mathbf{P}$ through an angle of 180°. Let α denote this rotation of $\mathbf{P}$. Then $\alpha(1) = 4$ and $\alpha(4) = 1$. Similarly $\alpha(2) = 3$ and $\alpha(3) = 2$ giving $\alpha = (14)(23)$ and hence the cycle type of α is $(0, 2)$ with the corresponding monomial ${x_2}^2$. Since we have three pairs of skew edges, we get three such rotations with monomial ${x_2}^2$. The cycle index is thus

$$P_G(x_1, x_2, \ldots) = \frac{1}{12}\left((x_1)^4 + 8x_1x_3 + 3(x_2)^2\right)$$

Example 15.2.12. Let G denote the rotation group of a cube $\mathbf{C}$ in its action on the six faces of the cube.

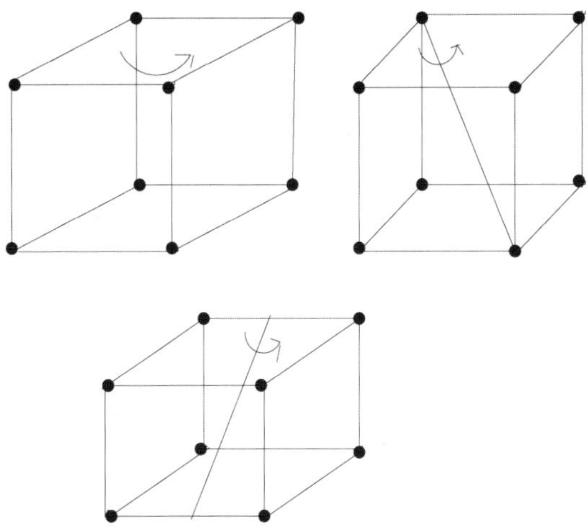

Figure 15.7: Face, Vertex and Edge based rotations of a cube

G has order 24 (Theorem 14.4.4). Let F, B, L, R, U, D respectively denote the front, back, left wall, right wall, top (up), and bottom (down) faces of $\mathbf{C}$. Besides identity whose cycle type is $(6,0)$ and with associated monomial ${x_1}^6$, we have 23 rotations classified below.

(a) *Vertex based rotations*: Fix an axis passing through a pair of antipodal vertices such as the meeting point of F, L and D the meeting point of B, R and U. Rotate the cube $\mathbf{C}$ around this axis through an angle of 120° and then through 240° (or -120°). This gives us three rotations with the given axis (in fact,these constitute a Sylow 3-subgroup of G).Two of these three rotations are proper and are given by (in terms of cycle decomposition): $(FLD)(BRU)$ and $(FDL)(BUR)$. These have the cycle type $(0,0,2)$ with associated monomial ${x_3}^2$. Since there are four pairs of antipodal vertices, we get $4 \times 2 = 8$ such rotations.

(b) *Edge based rotations*: Fix an axis passing through a pair of antipodal edges with an axis joining the mid-points of these edges such as the edge common to F and D and the edge common to B and U; this can be done by actually passing a rod joining these mid-points and passing through $\mathbf{C}$. Now rotate $\mathbf{C}$ through an angle of 180° around this axis. We get the permutation whose cycle decomposition is $(FD)(BU)(RL)$ and hence has the cycle type $(0,3)$ with the associated monomial $(x_2)^3$. The number of such rotations is 6 since we have 6 pairs of antipodal edges.

(c) *Face based rotations*: Fix an axis passing through the centers of a pair of two opposite faces such as say U and D. With the line joining these two points as the axis (thus we have a vertical axis in this case), rotate $\mathbf{C}$ through an angle $\theta = 90^o$ and multiples of θ. That gives us four rotations of which one is the identity and the other three are: $(FRBL)(U)(D)$ and its inverse both with the same cycle type $(2,0,0,1)$ and the square of this rotation (i.e. 180° rotation) which has the cycle decomposition $(FB)(RL)$ and with the cycle type $(2,2,0)$.

The axis can be chosen in three ways and we have 6 monomials each equal to $(x_1)^2 x_4$ corresponding to a 90° (or -90°) rotation and 3 monomials equal to $(x_1)^2 (x_2)^2$ corresponding to a 180° rotation.

We also note that a regular octahedron is the dual of a cube in the following sense. If we take the centers of faces of a cube and take the convex hull of these six vertices, the regular polytope we get is the regular octahedron; this also works in the other direction. Thus the faces of a cube correspond to the vertices of a regular octahedron and the faces of a regular octahedron correspond the vertices of a cube. Hence they both have the same (that is, isomorphic) automorphism group. We see that the cycle index of the group of rotations of a cube in its action on the faces of the cube (or the cycle index of the group of rotations of a regular octahedron in its action on the vertices of the regular octahedron) has the cycle index given by

$$P_G(x_1, x_2, \ldots) = \frac{1}{24}\left((x_1)^6 + 8(x_3)^2 + 6(x_2)^3 + 6(x_1)^2 x_4 + 3(x_1)^2(x_2)^2\right)$$

Example 15.2.13. Consider a regular dodecahedron $\mathbf{D}$ with G, the group of all the sixty rotations of $\mathbf{D}$ in its action on the 12 faces of $\mathbf{D}$. As seen in Chapter 14 (Theorem 14.4.4), there is a one-to-one correspondence between faces (respectively vertices) of a regular dodecahedron and vertices (respectively faces) of a regular icosahedron. To be more precise, the convex hull of the 12 points that are centers of the 12 faces of $\mathbf{D}$ is a regular icosahedron. Hence the cycle index of G in its action on the vertices of a regular icosahedron is the same as the cycle index we intend to compute. Since the number of faces is 12, the cycle type of identity is $(12, 0, 0)$ and the associated monomial is $(x_1)^{12}$. The other 59 rotations of $\mathbf{D}$ are of three different types exactly as in the case of a cube.

(a) *Vertex based rotations*: Fix a pair of antipodal vertices with an axis passing though these two vertices (this will be a body diagonal passing through the center of $\mathbf{D}$). Rotate $\mathbf{D}$ though an angle of 120° and 240° around this axis. Since each vertex is incident with three faces, the cycle type of such a rotation is $(0, 0, 4)$ with associated monomial $(x_3)^4$. We have 10 pairs of antipodal vertices and hence the number of such rotations is equal to $10 \times 2 = 20$.

(b) *Edge based rotations*: Take a pair of antipodal edges and fix the mid-points of these edges. With an axis through these two mid-points, rotate $\mathbf{D}$ through an angle of 180°. This rotation pairs the 12 faces into 6 pairs and hence has cycle type $(0, 6)$ with the associated monomial $(x_2)^6$. Since $\mathbf{D}$ has 30 edges, we see that there are 15 such rotations.

(c) *Face based rotations*: Take a pair of opposite faces and fix the centers of these faces. With an axis through the centers of these faces, rotate $\mathbf{D}$ through an angle of 72° and then through multiples of 72°. This creates 4 proper rotations. Each such rotation fixes two faces and rotates the other ten in two 5-cycles (since each face is surrounded by 5 faces) giving the cycle type $(2, 0, 0, 0, 2)$. Since we have 12 faces and hence 6 pairs of opposite faces we get $6 \times 4 = 24$ monomials each equal to $(x_1)^2(x_5)^2$.

To conclude, the cycle index of G in it action on the faces of $\mathbf{D}$ is given by

$$P_G(x_1, x_2, \ldots) = \frac{1}{60}\left((x_1)^{12} + 20(x_3)^4 + 15(x_2)^6 + 24(x_1)^2(x_5)^2\right)$$

Example 15.2.14. Now consider the symmetric group S_n in its natural action on the set $[n]$. The number of types of permutations in S_n is $p(n)$, the partition number and each permutation has type $(b_1, b_2, \ldots)$ where $\sum_i ib_i = n$. Given such a sequence $(b_1, b_2, \ldots)$, the number of permutations of type $(b_1, b_2, \ldots)$ is given by (refer to Chapter 3, Theorem 3.1.5)

$$\frac{n!}{\prod_i i^{b_i} b_i!}$$

and therefore we get

$$P_{S_n}(x_1, x_2, \ldots) = \sum_{\sum ib_i = n} \prod_i \frac{x_i^{b_i}}{i^{b_i} b_i!}$$

Besides the cycle index, the other basic ingredient of Polya theory is the idea the weight assignment to the elements of the range. Thus w, the weight is a mapping from R to some suitable set. What this suitable set is depends on what we want to do or what we want to extract from Polya's theorem. The total weight of R, which the range or the store is given by

$$store\ enumerator\ = w(R) = \sum_{r \in R} w(r)$$

We are tempted to give the following example borrowed from de Bruijn [4]. Treat the range (the store) as a shop. The shopkeeper has 7 packs of which he has two packs of coffee, three packs of tea and two packs of sugar. If the coffee packs are denoted by c_1, c_2, the tea packs by t_1, t_2, t_3 and the sugar packs by s_1, s_2, then the store enumerator is the formal sum

$$w(R) = c_1 + c_2 + t_1 + t_2 + t_3 + s_1 + s_2$$

If one coffee pack say c_1 is of a higher quality (and hence has a higher price) we can distinguish between c_1 and c_2 by calling them c and c' respectively. Suppose the three tea packs are all identical (except for being physically different) and suppose that so are the two sugar packs. We can then give a common weight say t to each t_i and a common weight say s to each s_j and the store enumerator now is

$$w(R) = c + c' + t + t + t + s + s = c + c' + 3t + 2s$$

Now let the price of the finer brand of coffee (i.e. c) be Rupees 100 and that of the lower brand be Rupees 60. Assume that the tea packs have price Rupees 50 each and the sugar packs have price Rupees 30 each. Then the store enumerator is

$$w(R) = (100 + 60) + (3 \times 50) + (2 \times 30) = 370$$

This store enumerator tells us how much amount the items will fetch the shopkeeper. Finally, assume that it is not the money that interests the shopkeeper, but what interests him is the total number of items (in a stock checking sense) he has. Then each pack can be given weight 1 (thus $c = c' = t = s = 1$) and then the new store enumerator $w(R) = 7$ giving us the total number of packs in the store. *Formally, we have the following definition.*

Definition 15.2.15. Let $\overline{R}$ be a commutative ring with 1 and let $w : R \to \overline{R}$ be any function. Then w is called a *weight* (assignment) for R.

Which $\overline{R}$ we should choose for a given problem depends on the situation under consideration, the most popular being Z, the ring of integers with weight assignment 1 for every element of R. To sum up, weight is assigned in such a way that we are able to add and multiply weights as we would normally expect. In that case, it does not really matter which $\overline{R}$ one is talking about. *From this point on, assume that w is a weight assignment on R.*

Definition 15.2.16. Let $f \in R^D$. Then *the weight of f* is defined by

$$w(f) = \prod_{d \in D} w(f(d))$$

We give some examples.

Example 15.2.17. Let the faces of a cube be colored in three colors say green, blue and yellow where we make the weight assignment $w(green) = g, w(blue) = b$ and $w(yellow) = y$. If f is a function (from the set of six faces to the set of three colors) that paints three faces green, two faces blue and one face yellow, then

$$w(f) = g \times g \times g \times b \times b \times y = g^3 b^2 y$$

Example 15.2.18. Now suppose f is a function (color assignment) for which we have the same number of faces of each color, then $w(f) = g^2 b^2 y^2$. If f is a function which colors all the six faces yellow, then $w(f) = y^6$. Finally, if we change the weight assignment and let $w(any\ color) = 1$ then clearly, $w(f) = 1$.

Example 15.2.19. Let the faces of the cube be coloured in two colours say green and yellow with $w(green) = g$ and $w(yellow) = y$. If f is a colouring with three faces say top, bottom and right coloured green and the other three faces left, back and front coloured yellow, then $w(f) = (g)^3 (y)^3 = g^3 y^3$. Now suppose f' is another assignment of colours in which the top, left and the front are coloured green (and thus the green faces are incident at the top left corner vertex) and the yellow faces are bottom, right and back (that are incident at the corner vertex at the back), then no rotation (and actually no symmetry) can map f to f' (since in the latter case, we have a vertex common to the three faces that are coloured by the same colour while there is no such vertex in the former case). However, $w(f')$ also equals $g^3 y^3$ *and the example clearly shows that inequivalent functions can have the same weight.* If the green colour costs Rupees 5 per face and the yellow colour costs Rupees 2 per face, then

$$w(f) = w(f') = (5 \times 5 \times 5) \times (2 \times 2 \times 2) = 1000$$

Lemma 15.2.20. *Let $f \sim g$ where $f, g \in R^D$. Then $w(g) = w(f)$.*

Proof Since $f \sim g$, there is some $\alpha \in G$ such that $a_\alpha(f) = g$ and thus we have $f(d) = g(\alpha(d))\ \forall\ d \in D$. Therefore,

$$w(f) = \prod_{d \in D} w(f(d)) = \prod_{d \in D} w(g(\alpha(d)))$$

In the expression on the R.H.S. let $e := \alpha(d)\ \forall\ d \in D$. Then the fact that α is a permutation on D, allows us to write

$$w(f) = \prod_{d \in D} w(g(\alpha(d))) = \prod_{e \in D} w(g(e)) = w(g)$$

completing the proof. □

Definition 15.2.21. Let F be a pattern (refer to Definition 15.2.3). Then *the weight of F, written* $w(F)$ is $w(f)$ where f is some function in F.

Lemma 15.2.20 just proved ensures that $w(F)$ is well-defined since equivalent functions have the same weight.

Definition 15.2.22. *The pattern inventory written P.I.* is defined by $P.I. = \sum_F w(F)$ where the sum is over all the patterns F (in the action of a group G acting on the domain D).

The pattern inventory gives us all the information we are looking for about patterns as the following examples illustrate.

Example 15.2.23. Consider the problem of colouring a 2×2 chessboard in two colours white and black with respective weights w and b. Any function then has a weight of the form $w^i b^j$ such that $i + j = 4$ and $0 \leq i \leq 4$. For each i between 0 and 4 we have exactly one pattern with weight $w^i b^j$ except when $i = 2$. In that case, we have two patterns with the same weight w^2b^2 (the two black cells may either be placed diagonally or they may be placed in two adjacent cells giving rise to two different patterns). Hence the pattern inventory is

$$P.I. = b^4 + wb^3 + 2w^2b^2 + w^3b + w^4$$

Example 15.2.24. Consider the colouring of the six faces of a cube in two colours say red and blue. Assuming that we have the group G of rotations of the cube, what is the number of patterns? Evidently, there is one way of colouring the cube with no face red and exactly one way of colouring the cube with one face red (the other faces blue) since G is transitive on the set of faces. If we wish to colour two faces red, then there are two possibilities: either the two faces could be opposite to each other (like colouring both top and the bottom red and the remaining four faces blue) or we could colour two adjacent faces red (like colouring the top and the front red and the remaining faces blue). What happens, when we colour the cube in which exactly three faces are red and three are blue? As was already seen in Example 15.2.19, we have two patterns, the one in which the three red faces share a vertex in common and the other in which the three red faces have no vertex in common (and hence contain a pair of opposite faces). The patterns when $4, 5$ and 6 faces are red, are, by symmetry similar to the cases when $2, 1$ or no face is red respectively. We thus get 10 patterns in all. This answer, that we have 10 patterns, could have been obtained using Burnside's Lemma. What Burnside's Lemma cannot answer is the nature of these patterns. We have

$$P.I. = b^6 + rb^5 + 2r^2b^4 + 2r^3b^3 + 2r^4b^2 + r^5b + r^6$$

Polya' theorem (Theorem 15.3.1) finds an expression for the pattern inventory without going through such computations, in terms of the cycle index of the group acting on the domain.

Definition 15.2.25. Let $S \subset R^D$. Then *the weight of S* is given by

$$w(S) = \sum_{f \in S} w(f)$$

Thus the weight of a subset of functions is obtained in a natural way, by summing the weights of all the functions in that set. Consider the following situation. A travel agent gives a choice of three tourist destinations say r_1, r_2, r_3 (with the weight $w(r_i) = r_i$) to his 11 customers. Among the customers is a family of four consisting of a couple and their two children, a second family of three members consisting of a couple and a small child, a couple that is newly married and two bachelors. All the members in the same family must naturally choose to go to the same destination. Let S denote the set of all the travel arrangements for the 11 people under this restriction. Thus the first family of four must choose some destination while the second family may choose the same or some other destination and so on. For example, one such arrangement (which is the same as a function from the 11-set to the 3-set of three destinations) could map all the four members in the first family to r_2, all the members in the second family to r_3, the newly married couple to r_1 and the bachelors to r_1 and r_2 respectively giving the weight of this function equal to $(r_2)^4(r_3)^3(r_1)^2(r_1)(r_2) = (r_1)^3(r_2)^5(r_3)^3$. To find $w(S)$, we may look at *any function* $f \in S$ in the following manner. Let E_1 denote the set of the four members in the first family, E_2, the set of three members in the second family, E_3 the set of two people who are newly married, E_4, E_5 respectively the singleton sets of bachelors. If $f \in S$, then f is constant on E_1 and hence the part of $w(f)$ coming from E_1 has the form $(r_1)^4 + (r_2)^4 + (r_3)^4$. This also holds for each $i = 2, 3, 4, 5$. In fact, $f \in S$ iff f is constant on each E_i. This gives us $w(S)$ equal to

$$\{(r_1)^4 + (r_2)^4 + (r_3)^4\} \times \{(r_1)^3 + (r_2)^3 + (r_3)^3\} \times \\ \{(r_1)^2 + (r_2)^2 + (r_3)^2\} \times \{r_1 + r_2 + r_3\}^2$$

Lemma 15.2.26. *Let $D = D_1 \cup D_2 \cup \cdots \cup D_k$ denote a partition (disjoint union) of D into k non-empty subsets D_i where $i = 1, 2, \ldots, k$. Let S be a subset of functions defined by*

$$S = \{f : f(d) = f(d') \ \forall \, d, d' \in D_i, \ \forall \, i = 1, 2, \ldots, k\}$$

Then weight of S is given by

$$w(S) = \prod_{i=1}^{k} \left\{ \sum_{r \in R} w(r)^{|D_i|} \right\} \tag{15.1}$$

Proof Let $f \in S$. Then f is constant on each D_i (and conversely). If $f(d) = r_i$ for every $d \in D_i$ (and for every $i = 1, 2, \ldots, k$), then $w(f)$ equals

$$(w(r_1))^{|D_1|}(w(r_2))^{|D_2|} \cdots (w(r_k))^{|D_k|}$$

Thus, in order to find $w(S)$, we may simply choose k elements $r_1, r_2, \ldots, r_k$ in R (that are not necessarily distinct) and let f map D_i to r_i giving the expression for the weight of f as obtained above. It follows that

$$w(S) = \sum_{r_1, r_2, \ldots, r_k \in R} (w(r_1))^{|D_1|}(w(r_2))^{|D_2|} \cdots (w(r_k))^{|D_k|} \tag{15.2}$$

We have to prove that the right hand sides of (15.1) and (15.2) are the same. A typical summand on the R.H.S. of (1) has the form $(w(r_1))^{|D_1|}(w(r_2))^{|D_2|} \cdots (w(r_k))^{|D_k|}$ where we choose $r_1, r_2, \ldots, r_k$ in any manner from R. Since the typical summands on the right hand sides of both (15.1) and (15.2) are the same, we are done. □

15.3 Polya's theorem and applications

Theorem 15.3.1. *(Polya's fundamental theorem) Let G be a permutation group acting on the domain set D and let w be a weight assignment on the range R. Then the pattern inventory is given by*

$$\begin{aligned} P.I. &= P_G\left(\sum_{r \in R} w(r), \sum_{r \in R} w(r)^2, \ldots, \sum_{r \in R} w(r)^j, \ldots\right) \\ &= P_G(x_1, x_2, \ldots, x_j, \ldots)\Big| \left\{ x_j = \sum_{r \in R} w(r)^j \right\} \ \forall\, j \end{aligned}$$

That is, the pattern inventory is obtained by replacing each variable x_j in the cycle index of the group G by the sum $\sum_{r \in R} w(r)^j$ (where each summand is the j-th power of the weight).

Proof Let $W_1, W_2, \ldots, W_i, \ldots$ be the distinct entities that occur as weights of (all the) patterns. Let m_i denote the number of patterns whose weight is W_i. Then, by definition, we have $P.I. = \sum_i m_i W_i$. Now fix an i and consider the set T of all the functions f whose weight W_i. If $F_1, F_2, \ldots, F_{m_i}$ denotes the m_i patterns whose weight is W_i, then $T = F_1 \cup F_2 \cup \cdots \cup F_{m_i}$. Let $f \in T$ and let $\alpha \in G$. Then $a_\alpha(f) \in T$ and hence $a_\alpha | T$ is a permutation on T. Therefore, we may restrict the action of the permutation group G to T and that gives us m_i orbits of G on T. Using Burnside's lemma (Theorem 15.1.2), we have

$$m_i = |G|^{-1} \sum_{\alpha \in G} |(Fix)_{\alpha,i}|$$

where

$$(Fix)_{\alpha,i} = \{f : f \in T \; and \; a_\alpha(f) = f\}$$

and we let

$$(Fix)_\alpha = \{f : f \in R^D \; and \; a_\alpha(f) = f\}$$

Hence the pattern inventory is given by summing over all i,

$$\begin{aligned} P.I. &= \sum_i W_i m_i \\ &= \sum_i W_i |G|^{-1} \left\{ \sum_{\alpha \in G} |(Fix)_{\alpha,i}| \right\} \\ &= |G|^{-1} \sum_{\alpha \in G} \left\{ \sum_i |(Fix)_{\alpha,i}| W_i \right\} \end{aligned}$$

We now interpret the expression in the parentheses. Fix an $\alpha \in G$. Then we are summing over all i. Thus $\sum_i |(Fix)_{\alpha,i}| W_i$ is the sum of weights of all the functions (irrespective of their weights) that are fixed by α. Let

$$S_\alpha = \{f : a_\alpha(f) = f\}$$

Then the expression for the pattern inventory (using Definition 15.2.25) simplifies to

$$P.I. = |G|^{-1} \sum_{\alpha \in G} w(S_\alpha)$$

To simplify the notation, let $S = S_\alpha$. Let the cycle type of α be $(b_1, b_2, \ldots, b_j, \ldots)$. Let $D_{j,t}$ denote the t-th j-cycle in the cycle decomposition of α. Thus $D = \cup_j \cup_{t=1}^{b_j} D_{j,t}$ is a partition of D. Let $D_{j,t} = (a_1 a_2 \cdots a_j)$. Then α fixes f iff $f(a_i) = f(\alpha(a_i)) = a_{i+1}$ where we read subscripts modulo j. Thus $f \in S$ iff f is constant on each $D_{j,t}$. Lemma 15.2.26 gives the weight of S to be

$$\begin{aligned} w(S) &= \prod_j \prod_{t=1}^{b_j} \sum_{r \in R} w(r)^{|D_{j,t}|} \\ &= \prod_j \left\{ \sum_{r \in R} w(r)^j \right\}^{b_j} \end{aligned}$$

because $|D_{j,t}| = j$ and we have b_j cycles of length j. It just remains to interpret the right hand side of the equation just obtained. Since α has cycle type $(b_1, b_2, \ldots, b_j, \ldots)$, the monomial of α is

$$(x_1)^{b_1}(x_2)^{b_2}\cdots(x_j)^{b_j}\cdots = \prod_j (x_j)^{b_j}$$

Thus $w(S_\alpha) = w(S) = \prod_j (x_j)^{b_j}$ where the variable x_j is replaced by the expression $\sum_{r\in R}\{w(r)^j\}$. Thus $w(S_\alpha)$ is nothing but the monomial of α with the j-th variable x_j replaced by $\sum_{r\in R}\{w(r)^j\}$. Since *P.I.* is obtained by summing all $w(s_\alpha)$ (over all $\alpha \in G$) and dividing by $|G|$, we see, using the definition of cycle index, that

$$P.I. = |G|^{-1}\sum_{\alpha\in G}(monomial\ of\ \alpha)$$

where, in the monomial of α, we replace the j-th variable x_j by $\sum_{r\in R}\{w(r)^j\}$. It thus follows that the *P.I.* equals the polynomial given by the cycle index of the group G in which x_j is replaced by $\sum_{r\in R}\{w(r)^j\}$. This completes the proof of Polya's theorem. □

Corollary 15.3.2. *The total number of patterns is equal to*

$$P_G(x_1, x_2, \ldots, x_j, \ldots)|[x_j = |R|\ \forall\, j] \ = P_G(|R|, |R|, \ldots, |R|, \ldots)$$

Proof. Make the special weight assignment $w(r) = 1\ \forall\ r \in R$. Then weight of every function is 1 and hence weight of every pattern is also equal to 1. So the pattern inventory merely computes the total number of patterns. Now use Polya's theorem. Under the special weight assignment, $\sum_{r\in R}\{w(r)^j\}$ is equal to $|R|$, the size of the range proving the assertion. □

We now give a number of examples that illustrate the use of Polya's theorem.

Example 15.3.3. We wish to find out how many distinct organic molecules can be formed under the following stipulation. The geometry of a typical organic molecule has a carbon atom at the center of a regular tetrahedron with four valencies going towards the four corners of the regular tetrahedron.

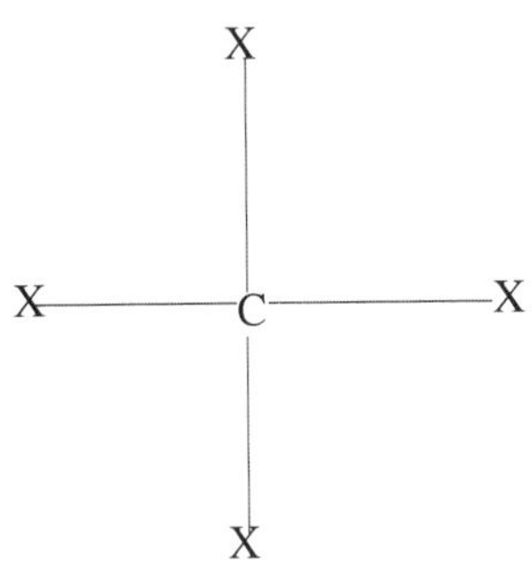

Figure 15.8: An organic molecule

So, we may attach four radicals at these four corners of the regular tetrahedron. Assume that these four radicals are one of the following: H, Cl, CH_3, C_2H_5 representing hydrogen, chlorine (chloride), methyl and ethyl radicals respectively. As an example, we may construct CH_4 the methane molecule by attaching H at all the four vertices of the regular tetrahedron. We may attach H at three vertices and Cl at the fourth vertex and that obtains CH_3Cl, called methyl chloride. Similarly CCl_4 is called carbon tetrachloride and $CHCl_3$ is the chloroform. Since we have four vertices of the regular tetrahedron forming the domain and the range consisting of the four radicals, the total number of molecules that can be formed is $4^4 = 256$. However, the actual number of molecules that can be formed is much less since two molecules are the same when one can be rotated into the other. Thus, what we are looking for is not the set of functions from a 4-set to a 4-set but patterns of functions under the group action of G of the rotation group of the regular tetrahedron. We can thus ask various questions.

(a) We wish to find out as to how many distinct compounds (molecules) can be made? Using Corollary 15.3.2, we see that this number is

$$P_G(4, 4, \ldots) = \frac{1}{12}[4^4 + (8 \times 4^2) + (3 \times 4^2)] = 36$$

(b) Now suppose we wish to know as to how many of these 36 compounds have hydrogen atom (as a radical) in them. Since the total is 36, we may as well find the number of compounds *that do not have H at all.* This amounts to having a range that consists of only the remaining three radicals and hence the required answer is

$$P_G(3, 3, \ldots) = \frac{1}{12}\left[3^4 + (11 \times 3^2)\right] = 15$$

and we conclude that there are $36 - 15 = 21$ compounds that use at least one H as a radical.

(c) Finally we want to find out *the nature of all the compounds that contain at least on H*. Give weight h to H and give weight 1 to every other radical. The pattern inventory then becomes

$$\begin{aligned} P.I. &= P_G(h+3, h^2+3, h^3+3, \ldots) \\ &= \frac{1}{12}\left[(h+3)^4 + 8(h+3)(h^2+3) + 3(h^2+3)^2\right] \\ &= h^4 + 3h^3 + 6h^2 + 11h + 15 \end{aligned}$$

This pattern inventory gives us the full information. There is one molecule that has all the four H's. There are three that have exactly three H's. We have 6 molecules that use two H's and there are 11 molecules that use only one H. Finally, there are 15 molecules that have no H in them.

Example 15.3.4. We take a re-look at the problem of colouring the six faces of a cube in two colours red and blue where the group under consideration is the group G of rotations. Let the weights of red and blue be r and b respectively. Since the cycle index of G is already computed (Example 15.2.12), Polya's theorem (Theorem 15.3.1) gives the pattern inventory when we substitute $r^j + b^j$ in place of x_j in the cycle index. This obtains

$$\begin{aligned} P.I. &= \frac{1}{24}\Big[(r+b)^6 + 8(r^3+b^3)^2 + 6(r^2+b^2)^3 + 6(r+b)^2(r^4+b^4) \\ &+ \quad 3(r+b)^2(r^2+b^2)^2\Big] \\ &= r^6 + r^5 b + 2r^4b^2 + 3r^3b^3 + 2r^2b^4 + rb^5 + b^6 \end{aligned}$$

This has been already obtained in a direct manner (Example 15.2.24). However, the advantage now is that we can handle any number of colours (see the exercises)!

Example 15.3.5. Consider the problem of colouring a necklace with 6 beads coloured in k colours where the group G we have in mind is that of rotations (thus the necklace can be rotated but not flipped over). The cycle index is given by (Example 15.2.7)

$$P_G(x_1, x_2, \ldots) = \frac{1}{6}\left[(x_1)^6 + 2x_6 + 2(x_3)^2 + (x_2)^3\right]$$

and hence the number of patterns is given by

$$P_G(k, k, \ldots) = \frac{1}{6}\left[k^6 + k^3 + 2k^2 + 2k\right]$$

which is a degree 6 polynomial in k. When $k = 2$, we get 14 patterns and the reader can list all of them.

Example 15.3.6. Consider the problem of colouring the 12 faces of a regular dodecahedron in two colours red and blue with the weights of red and blue r and b respectively. We have earlier computed the cycle index of the rotation (Example 15.2.13) group of a regular dodecahedron. Using Polya's theorem (Theorem 15.3.1), the pattern inventory is given by

$$P.I. = \frac{1}{60}\left[(r+b)^{12} + 20(r^3+b^3)^4 + 15(r^2+b^2)^6 + 24(r+b)^2(r^5+b^5)^2\right]$$

We can now ask several questions.

(a) How many patterns have 4 faces red (and 8 faces blue)? This is the coefficient of r^4b^8 in the $P.I.$ and is equal to

$$\frac{1}{60}\left\{\binom{12}{4} + 15 \times \binom{6}{2}\right\} = 12$$

(b) How many patterns have 3 faces red (and 9 faces blue)? This is the coefficient of r^3b^9 in the $P.I.$ and is equal to

$$\frac{1}{60}\left\{\binom{12}{3} + 20 \times 4\right\} = 5$$

(c) The total number of patterns is

$$\frac{1}{60}\left\{2^{12} + 20 \times 2^4 + 15 \times 2^6 + 24 \times 2^4\right\} = 96$$

15.4 de Bruijn's generalization of Polya's theorem

Again consider the coloured 2×2 chessboards coloured in one of the two colours white or black. If a person is merely interested in finding the number of colour contrast patterns, then it is as good as saying that the negative and positive of a black and white picture are the same for her. Referring to the 6 patterns we already obtained (Figure 15.2), we see that $A1$ and $A4$ merge into a single pattern and so do $A5$ and $A6$. However, $A2$ and $A3$ will continue to be different since $A3$ cannot be obtained from $A2$ by rotation and/or changing black colour into white and white colour into black. Note that we are not in a framework where there is no distinction between white and black, which can be the case if the observer is colour blind. What it means is that when white is taken to be black, simultaneously, black is taken to be white. We thus have only four patterns. Note also that merely counting the number of black cells is not enough (there are still two patterns with two black cells namely $A2$ and $A3$ which cannot be mapped into each other). We thus have a more general set up in which we not only have a group G acting on the domain D but also have a group H acting on the range R. *The question of pattern inventory and the number of patterns under this general set up was considered by de Bruijn's deep generalization of Polya's theorem which we now discuss.*

Definition 15.4.1. Let (G, D) and (R, H) be permutation groups. Given the pair $(\sigma, \tau) \in G \times H$, we define $a_{\sigma,\tau} : R^D \to R^D$ as follows. $a_{\sigma,\tau}(f) = g$ if $\tau f(d) = g(\sigma(d))$ for every $d \in D$.

The routine checking to show that $a_{\sigma,\tau}$ is a permutation on R^D as also the assertion that the map $(\sigma, \tau) \to a_{\sigma,\tau}$ is a homomorphism is left to the reader as an exercise (Exercise 15.39). The equivalence classes induced by $G \times H$ on R^D are *the patterns.* The special case $H = 1$ corresponds to the classical Polya theory which was discussed in Section 15.3. Unfortunately, one cannot talk of weight of a pattern under this general set up as the following example shows. Let $D = \{a, b\}$ with the trivial group $G = 1$ on D and let $R = \{x, y\}$ with the group $H = \{1, \tau\}$ acting on R where τ is the transposition (xy). If we now take two functions f and g with $f(a) = f(b) = x$ and $g(a) = g(b) = y$, then $w(f) = x^2 \neq y^2 = w(g)$ but f and g are in the same pattern since $a_{1,\tau}(f) = g$. *To avoid all the resulting complications a strong assumption is made: the weights are assigned in such a manner that equivalent functions have equal weights. We go even a step further and confine ourselves to computing the number of patterns, all through the remaining part of this chapter.*

Definition 15.4.2. Let $\sigma \in G$ and $\tau \in H$ where (G, D) and (H, R) are permutation groups. Let $f \in R^D$. Let $A = (\alpha_1\alpha_2 \cdots \alpha_k)$ be a cycle of σ and $B = (\beta_1\beta_2 \cdots \beta_j)$ be a cycle of τ. Then we say that f *maps A cyclically onto B* if (after cyclically permuting, if required), we have:

$$f(\alpha_1) = \beta_1, f(\alpha_2) = \beta_2, \ldots, f(\alpha_i) = \beta_i, \ldots$$

where the subscripts on the R.H.S. are to be read modulo j.

Theorem 15.4.3. *Let* $(\sigma, \tau) \in G \times H$ *and let* σ *and* τ *have types* $(b_1, b_2, \ldots, b_k, \ldots)$ *and* $(c_1, c_2, \ldots, c_j, \ldots)$ *respectively. Let* $a_{\sigma,\tau}(f) = f$. *Let* $f(d) = r$ *where* d *is in some* k*-cycle* $A = (\alpha_0, \alpha_1, \ldots, \alpha_{k-1})$ *and* r *is in some* j*-cycle* $B = (\beta_0, \beta_1, \ldots, \beta_{j-1})$ *respectively (in the respective sets* D *and* R*) where, w.l.o.g. we assume that* $d = \alpha_0$ *and* $r = \beta_0$. *Then the following assertions hold.*

(a) $f(d)$ *determines* f *on the cycle* A *uniquely.*

(b) *The restriction of* f *to* A *maps* A *onto* B.

(c) *Read the subscripts modulo* j. *If* $0 \leq u \leq k-1$ *and if* $u = mj + v$, *with* $0 \leq v \leq j-1$, *then* $f(\alpha_u) = (\beta_v)$.

(d) j *divides* k.

Proof Since $f(\alpha_0) = \beta_0$, we assume that $f(\alpha_u) = \beta_v$. Then

$$\tau f(\alpha_u) = \tau(\beta_v) = \beta_{v+1} = f(\sigma\alpha_u) = f(\alpha_{u+1})$$

proving the first three assertions. Let $k = qj + t$ where $0 \leq t \leq j-1$. Then inductively we have $f(\alpha_{qj-1}) = \beta_{j-1}$. Here $f(\alpha_{k-1}) = \beta_t$ for some t between 0 and $j-1$. If $t \neq j-1$, then we get $f(\alpha_0) = \tau(\beta_t) = (\beta_{t+1})$ which is a contradiction if $t \neq j-1$. Hence $t = j-1$ showing that j divides k. □

Theorem 15.4.4. $a_{\sigma,\tau}(f) = f$ *iff for every* k, f *maps each* k*-cycle of* α *to some* j*-cycle of* τ *in a cyclic manner where* j *divides* k. *For* $(\sigma, \tau) \in G \times H$, *let* $(Fix)_{\sigma,\tau}$ *denote the set of functions fixed by* (σ, τ) *and let the types of* σ *and* τ *be respectively* $(b_1, b_2, \ldots, b_k, \ldots)$ *and* $(c_1, c_2, \ldots, c_j, \ldots)$. *Then*

$$|(Fix)_{(\sigma,\tau)}| = \prod_k \left[\sum_{j|k} jc_j\right]^{b_k} \tag{15.3}$$

Proof The proof of the first statement follows from Theorem 15.4.3. Consider the second statement and let $f \in (Fix)_{(\sigma,\tau)}$. Let k be fixed and let A be some k-cycle in σ. Using Theorem 15.4.3, f must map A onto some j-cycle where j divides k, say B. We can now fix some element $d \in A$ and choose the image $f(d) = r \in B$ in j different ways. Since there are c_j such j-cycles, the image of d can be chosen in jc_j ways for a fixed j and hence in $\sum_{j|k} jc_j$ ways in all. This independently holds for every k-cycle A and the number of such k-cycles is b_k showing that we could choose f in $\prod_k (\sum_{j|k} jc_j)^{b_k}$ ways. □

The following theorem, generalizing Polya's fundamental theorem, now follows as an immediate consequence of Burnside's lemma.

Theorem 15.4.5. *Let* (G, D) *and* (H, R) *be permutation groups. Then the number of patterns of* $G \times H$ *on* R^D *is given by*

$$\frac{1}{|G| \times |H|} \sum_{\sigma \in G,\, \tau \in H} |(Fix)_{(\sigma,\tau)}|$$

where $|(Fix)_{(\sigma,\tau)}|$ *is as given in equation* (15.3).

Here are some examples.

Example 15.4.6. Consider the 2×2 chessboard colouring problem in two colours white and black where the group acting on the domain is the rotation group G generated by the rotation α through an angle of 90^0. The group on the range is $H = \{1, \beta\}$ where β interchanges the colours and hence has the cycle type $(0, 1)$ while identity has the cycle type $(2, 0)$. We thus have a group $G \times H$ of order 8 and denoting a generic element in that group by (σ, τ), the following table gives the number of chessboards fixed by each one of them.

σ	τ	*type of* σ	*type of* τ	$(Fix)_{\sigma,\tau}$
1	1	$(4,0)$	$(2,0)$	2^4
1	β	$(4,0)$	$(0,1)$	0
α	1	$(0,0,0,1)$	$(2,0)$	2
α	β	$(0,0,0,1)$	$(0,1)$	2
α^2	1	$(0,2)$	$(2,0)$	4
α^2	β	$(0,2)$	$(0,1)$	4
α^3	1	$(0,0,0,1)$	$(2,0)$	2
α^3	β	$(0,0,0,1)$	$(0,1)$	2

So, the number of patterns is equal to $\frac{32}{8} = 4$.

Example 15.4.7. Six books are to be distributed among four children, two girls and two boys. The girls form a pair of identical twins and the boys also form a pair of identical twins. In how many ways can we distribute the books? Here G is the trivial group with cycle index $(x_1)^6$ while H has 4 permutations (this is the Klein group) with cycle types $(4, 0), (2, 1), (2, 1), (0, 2)$. Doing the same computation as before, we get the required number of patterns to be

$$\frac{1}{4}[4^6 + (2 \times 2^6) + 0^6] = 4^5 + 2^5 = 1056$$

Example 15.4.8. Look at the same problem with four books instead of six. The answer then changes to

$$\frac{1}{4}\left[4^4 + (2 \times 2^4) + 0^4\right] = 64 + 8 = 72$$

A closed expression for $|(Fix)_{\sigma,\tau)}|$ can be obtained where instead of the cycle type numbers b_k's and c_j's the partial derivatives and polynomials appear. Here we replace the variable x_k by $\frac{\partial}{\partial z_k}$ and the variable y_j (in the cycle index of τ) by an exponential. This is elaborated in the exercises (Exercises 15.46 through 15.49). This does not quite simplify the computations but merely obtains a concise and sophisticated expression. We now ask the question as to how many patterns of injective functions are there. First note that if $a_{\sigma,\tau}(f) = g$ and if f is an injective function, then so is g and therefore, it makes sense to ask such a question. Let then $(Fix)^*_{\sigma,\tau}$ denote the set of all the injective functions f that are fixed by the pair (σ, τ). Arguing exactly as before, we have

Theorem 15.4.9. *Let $\sigma \in G$ and $\tau \in H$. Then the number of injective functions fixed by (σ, τ) is given by*

$$(Fix)^*_{\sigma,\tau} = \prod_k k^{b_k} c_k(c_k - 1) \cdots (c_k - b_k + 1)$$

Proof Let A be a k-cycle in σ. Then $f \in (Fix)^*{}_{\sigma,\tau}$ must map A onto some k-cycle B of τ (this, of course, requires that τ has at least as many k-cycles as σ). Thus the b_k cycles of length k in σ have to be paired with distinct k-cycles in τ and this can be done in $[c_k]_{b_k}$ ways. Having done that we take up a pair of k-cycles A in σ and B in τ. If $d \in A$, then the image of d can be chosen to be any one of the k elements of B (and that decides the images of all the elements of A uniquely) and this can be done in k ways. Since we have b_k cycles of length k in σ, the formula is proved. □

Theorem 15.4.10. *Let $|D| = |R|$. Then*

$$(Fix)^*_{\sigma,\tau} = \prod_k k^{b_k} (b_k)!$$

if σ and τ have the same cycle type $(b_1, b_2, \ldots, b_k, \ldots)$ and is zero otherwise.

Here are some examples.

Example 15.4.11. Let the faces of a regular dodecahedron be coloured in 12 distinct colours with each face receiving a different colour. We then have the rotation group G of 60 rotations on the dodecahedron and the identity (trivial) group $H = 1$ on the range of 12 colours. Hence, for all $(\sigma, \tau) \in G \times H$, we have $(Fix)_{(\sigma,\tau)} = \emptyset$ except when $(\sigma, \tau) = (1, 1)$ when we have $(Fix)_{(\sigma,\tau)} = (12)!$ and therefore, the number of patterns is equal to

$$\frac{(12)!}{60} = 12 \times 11 \times 10 \times 9 \times 8 \times 7 \times 6 \times 2$$

Example 15.4.12. Consider the problem of distributing 6 distinct chocolates to 6 children consisting of three pairs of identical twins. Here G is the trivial group while H is a group of order 8 with $H \cong S_2 \times S_2 \times S_2$. Exactly as before, $(Fix)_{(\sigma,\tau)} = \emptyset$ unless $(\sigma, \tau) = (1, 1)$ when we have $(Fix)_{(\sigma,\tau)} = (6)!$ and therefore, the number of patterns is equal to $\frac{6!}{8} = 72$. This result could have been derived independently without using Polya's theory.

Example 15.4.13. Let a necklace with 6 beads admit the full group of symmetries, i.e., the dihedral group. The beads are coloured in 6 different colours with each bead receiving a different colour. Since $|G| = 12$ and H is the trivial group, the number of patterns is equal to $\frac{6!}{12} = 60$.

Example 15.4.14. Let the faces of a regular tetrahedron be coloured in three colours red, magenta and crimson. A somewhat colour blind person makes the following mistake. When he takes red to be magenta, he also takes magenta to be crimson as also crimson to be red. Similarly, when he takes red to be crimson, he also takes crimson to be magenta as also magenta to be red. What this means in plain terms is that we have a cyclic group H of order 3 acting on the range; we assume that the tetrahedron can be rotated so that the group G acting on the domain is (isomorphic to) the alternating group A_4. *We wish to find the total number of patterns.* Let $\sigma \in G$ and let $\tau \in H$. Then we have $Fix(\sigma, \tau) = \emptyset$ except when τ is the identity permutation. We also have $|Fix(id, id)| = 3^4$ and $|Fix(\sigma, \tau)| = 3^2$ if σ is a non-identity permutation. Hence the number of patterns is given by

$$\frac{1}{12 \times 3}\left(3^4 + 11 \times 3^2\right) = 5$$

It is a good idea to find these five patterns explicitly.

Conclusion Original work of Polya appeared in German around 1940s; a part of Polya theory was also discovered by Redfield about twenty years before Polya. Incidentally, Burnside's Lemma was actually discovered by Frobenius and not by Burnside. An extremely lucid exposition of Polya theory was given (in English) by de Bruijn [4] in the 1960s. In this very well-written leisurely exposition, among many other things, de Bruijn generalized Polya's original theorem by permitting a permutation group action on the range. This article also discusses the wreath product and Kranz group but we will not discuss them here. We just inform the reader of a couple of nice and curious things done in the de Bruijn exposition. Consider 10 identical cubes with faces coloured in two colours where the group action is both the rotations of the individual cubes as well as permutations of the cubes (since the cubes are identical). How many patterns do we get? Look at both a regular simplex as well as the hypercube (or a regular hyperoctahedron) (with vertices or faces coloured) in the n-dimensional space. How many patterns do we get?

15.5 Exercises for Chapter 15

15.1 Prove that the full group of symmetries of a regular m-gon is the dihedral group D_m.

15.2 Prove that the full group of symmetries of a regular tetrahedron is the group consisting of 12 rotations and 12 reflections making a total of 24 symmetries and this group is isomorphic to the symmetric group S_4.

15.3 Consider the group G of all the 24 rotations of a cube in its action on the set $\mathbf{D}$ of four body diagonals of the cube. Show that this action is faithful and hence $G \cong S_4$.

15.4 Find the number of ways of coloring the corners of a triangle in m distinct colors.

15.5 Consider a regular 12-gon with vertices colored in five distinct colors. Two 12-gons are considered the same if one can be obtained from the other by rotations through an angle which is a multiple of 30°. Find the number of colourings.

15.6 Show that the map that reflects the cube about its center (antipodal reflection) is a symmetry of the cube. Use this to prove that the full group G of symmetries of a cube is the group consisting of 24 rotations and 24 reflections making a total of 48 symmetries and this is the full group of symmetries of a cube.

15.7 An $n \times n$ signed permutation matrix $A = [a_{ij}]$ is a matrix with each entry 0 or ± 1 such that each row and each column of A has precisely one non-zero entry, Let G' denote the group of all the 3×3 signed permutation matrices. Show that $G \cong G'$ where G refers to the full group of symmetries of the cube given in Exercise 15.6.

15.8 Find the number of ways of colouring the six faces of a cube when two colours are used and the group is the full group of 48 symmetries of the cube using Corollary 15.3.2. Then show that this answer could have been obtained by looking at the ten patterns when the group under consideration is that of rotations of the cube (since no two of the 10 patterns are equivalent under the full group of 48 symmetries).

15.9 Consider the problem of colouring the 6 beads of a necklace in two colours when the group we have in mind is the full group of symmetries (that is, the dihedral group D_6) and show that the number of patterns is 13. Which two patterns, when we have the group consisting only that of rotations (Example 15.1.8) merge to give a single pattern when the permutation group is larger?

15.10 Find the number of ways of colouring the faces of a cube in three colours first under the action of the rotation group and then under the action of the full group of 48 symmetries.

15.11 Use Burnside's lemma to find the total number of ways of colouring a 3×3 chessboard in m when the group of symmetries is the dihedral group D_m.

15.12 We are given a cube and a set of m distinct colors. In each one of the following cases, find out the number of ways of performing the given task.

(a) Six faces are painted with the stipulation that antipodal faces must receive the same colour.

(b) All the 12 edges are coloured with the stipulation that antipodal edges must receive the same color.

(c) All the 8 vertices are colored with the stipulation that antipodal vertices must receive the same color.

15.13 Find the number of ways to color the five faces of a regular pyramid with a square base using the colors red, white blue and yellow.

15.14 Consider a regular pentagon with sides colored in three colors red, blue and yellow. If the symmetries consist of all the rotations, find the number of patterns. Does this number change if the symmetries also include reflections?

15.15 At the corners of a square are placed $0, 1$ or 2 identical jelly beans. Find the number of patterns (under the group of rotations) in which we have five jelly beans in all.

15.16 In how many ways can the seven horses on a merry-go-round (a merry-go-round can be rotated but not flipped) be colored so that we have three blue, three yellow and one red horse? What is the answer if we wish to have three blue, two yellow and two red horses?

15.17 A stick, 8 feet long is to be painted with colors given to each of the 8 one-foot segment chosen from among the colors blue, red, green and yellow. In how many ways can it be painted so that three segments are blue, three are red and two are green? What is the answer if three segments are to be blue, two are to be green, two are yellow and one is red?

15.18 The edges of a cube are to be painted in three colors red, blue and green. What is the number of patterns?

15.19 Consider the following situation when $\mathbf{P}$ is a regular polytopes which is one of the three types: cube, a regular dodecahedron or a regular icosahedron. Assume that $\mathbf{P}$ is transparent and all the body diagonals of $\mathbf{P}$ are painted in two colours red and blue with two colourings considered equivalent if one can be obtained from the other by a rotation. Find the number of patterns in each one of the three cases.

15.20 Let the 6 vertices of a regular octahedron be coloured in m colours. How many patterns do we obtain (when the group under consideration is the rotation group)?

15.21 Show that the number of patterns when the eight vertices of a cube are coloured in two colours red and blue is 23. Prove that if we also allow reflections, then this number reduces by 1.

15.22 Find the number of distinct colourings (under the action of the rotation group) of the faces of a regular dodecahedron with equal number of faces of each colour.

15.23 Consider a regular dodecahedron **P**. Show that the full group of symmetries of **P** is obtained by taking all the 60 rotations and 60 reflections obtained by taking a product of these with the central reflection (antipodal map). Find the cycle index of this group of order 120 in its action on the faces of **P**. Use this to find the number of distinct colourings of the faces of **P** when two colours are used and the group is the full group of symmetries of **P**.

15.24 Suppose we color the four faces of a tetrahedron in n colors both from inside and outside. What is the number of patterns?

15.25 Find the number of ways of coloring the six faces of a cube in six different colors, with each face receiving a different color if the group under consideration is the rotation group of the cube, first using Polya's theorem (Theorem 15.3.1) and then using Burnside's lemma (Theorem 15.1.1).

15.26 What is the number of ways of distributing 10 identical jellybeans to four children among whom there is a pair of identical twin females and identical twin males?

15.27 Consider the cycle index of A_n (in its natural action). Show that

$$P_{A_n}(x_1, x_2, \ldots) = \frac{P_{S_n}(x_1, x_2, \ldots) + P_{S_n}(x_1, -x_2, x_3, -x_4, \ldots)}{2}$$

15.28 Prove that the cycle index of the group of all the rotations and reflections of a regular tetrahedron is given by

$$P(x_1, x_2, \ldots) = \frac{1}{24}\left[x_1{}^4 + 8x_1{}^4x_3 + 3x_2{}^2 + 6x_4\right].$$

15.29 All the twenty faces of a regular icosahedron are coloured in two colours red and blue. Find the number of patterns (under the action of the group of rotations).

15.30 Suppose a 2×2 chess board is colored in 6 colors. These 6 colors consist of three shades of red say r_1, r_2, r_3 and three shades of blue b_1, b_2, b_3. A person who is not very good at distinguishing the colors may take r_i to be r_{i+1} (modulo 3), as also b_j to be b_{j+1} (modulo 3). If the group acting on the chess board is that of rotations, what is the number of patterns?

15.31 A Boolean function of degree n is a function from the set $D = [n]$ to the set $\{0, 1\}$.

(a) If the group acting on the domain is the symmetric group S_n, how many patterns of Boolean functions does one obtain?

(b) If the group acting on D is trivial but the group acting on the range is the symmetric group S_2, then how many patterns does one obtain?

(c) If G is S_n and H is S_2 then how many patterns does one obtain?

15.32 Let the vertices of a regular 17-gon be coloured in three colours red, blue and yellow with two colourings considered the same if one can be obtained from the other by rotation of the polygon. What is the number of colourings in which the number of vertices of any two colours differ my at the most one?

15.33 An aromatic chemical molecule has a central hexagonal Benzene ring with radicals Cl, Br and I (for chlorine bromine and iodine respectively) attached to it). Thus the basic molecule Benzene (formula C_6H_6) has the structure of the following form shown in the first diagram of Figure 15.9 while Benzyl chloride has the form shown in the second diagram of Figure 15.9.

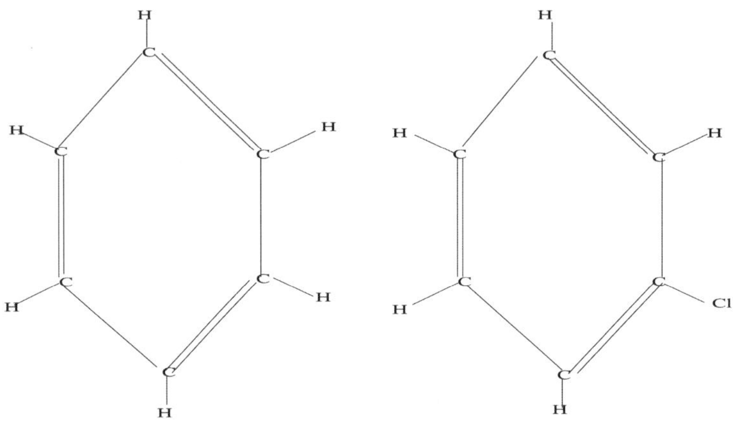

Benzene   Benzyl Chloride

Figure 15.9: Benzene and Benzyl Chloride

We have the hexagonal Benzene ring in which 6 carbon atoms are at the vertices of a regular hexagon with radicals H, Cl, Br and I attached. Two molecules are considered identical if one could be obtained from the other by a rotation or reflection of the regular hexagon. How many molecules have the following formulas?

(a) C_6H_4ClBr

(b) $C_6H_2Cl_2Br_2$

(c) $C_6H_2IClBr_2$

15.34 Let

$$f(z) = 1 + \sum_{n=1}^{\infty} \frac{P_{S_n}(x_1, x_2, \ldots, x_n)}{n!} z^n$$

denote the exponential generating function of the sequence whose n-th term is the cycle index of the symmetric group S_n. Prove that

$$f(z) = \exp\left[zx_1 + \frac{z^2x_2}{2} + \frac{z^3x_3}{3} + \cdots + \frac{z^jx_j}{j} + \cdots\right]$$

15.35 Find the number of patterns on the set of necklaces with 6 beads in three colours red, blue and green where the group on the set of necklaces is the rotation group and we are merely interested in colour contrasts (that is a colours can be permuted among each other).

15.36 Find a formula for the number of patterns when we colour a necklace with m beads in two colours when the two necklaces are considered equivalent if one can be obtained from the other by rotation and/or interchange of colours.

15.37 (de Bruijn [4]) Here is an alternative proof to the expression of the e.g.f. of the cycle indices of the symmetric groups S_n obtained in Exercise 15.34. Let $|D| = |R| = n$ and let $R = \{u_1, u_2, \ldots, u_n\}$ with the weights given by $w(u_i) = u_i \;\; \forall \; i$. Let $G = S_n$.

(a) Show that the pattern inventory

$$P.I. = P_G\left(\sum_{i=1}^{n} u_i, \sum_{i=1}^{n}(u_i)^2, \sum_{i=1}^{n}(u_i)^3, \ldots\right)$$

(b) Show that the weight of any function and therefore the weight of any pattern is of the form $(u_1)^{t_1}(u_2)^{t_2}\cdots(u_n)^{t_n}$ where t_i's are non-negative integers with sum equal to n and every such monomial is weight of some pattern.

(c) Show that two functions are equivalent iff they have the same weight.

(d) Show that $P.I.$ is equal to the coefficient of z^n in the expansion of

$$\prod_{i=1}^{n}\left(1 + zu_i + z^2(u_i)^2 + z^3(u_i)^3 + \cdots\right)$$

(e) Writing $zu_i + z^2(u_i)^2 + z^3(u_i)^3 + \cdots$ in terms of a log function, show that the $P.I.$ is equal to the coefficient of z^n in the expansion of

$$\prod_{i=1}^{n}\exp\left(zu_i + \frac{z^2(u_i)^2}{2} + \frac{z^3(u_i)^3}{3} + \cdots\right)$$

which is equal to

$$\exp\left[z\left(\sum_{i=1}^{n} u_i\right) + \frac{z^2}{2}\left(\sum_{i=1}^{n}(u_i)^2\right) + \frac{z^3}{3}\left(\sum_{i=1}^{n}(u_i)^3\right) + \cdots\right]$$

which is nothing but the result of the substitution

$$x_j = (u_1)^j + (u_2)^j + \cdots + (u_n)^j$$

in

$$\exp\left[zx_1 + \frac{z^2x_2}{2} + \frac{z^3x_3}{3} + \cdots + \frac{z^jx_j}{j} + \cdots\right]$$

(f) Interpret (e) as an identity and hence conclude the result of Exercise 15.34.

15.38 Show that the number of coloured regular dodecahedrons with faces coloured in three colours red, blue and yellow and two colourings considered the same if one can be obtained from the other by rotation is equal to 9099.

15.39 Given permutation groups (G, D) and (H, R) consider $a_{\sigma,\tau}$ defined in Section 15.4. Prove the following.

(a) $a_{\sigma,\tau}$ is a permutation on R^D.

(b) The homomorphy condition is satisfied:

$$a_{\sigma_2,\tau_2} \circ a_{\sigma_1,\tau_1} = a_{\sigma_2 \circ \sigma_1, \tau_2 \circ \tau_1}$$

(c) If $a_{\sigma,\tau}(f) = g$, and f is injective then so is g.

15.41 Prove the assertion of Example 15.4.12 without use of Polya theory.

15.42 Find the total number of ways of painting 20 vertices of a regular dodecahedron under the action of the group of rotations.

15.43 Derive the following general formula when we have a necklace with n beads in two colours red and blue with the group G of rotations acting on the set of coloured necklaces. Show that the number of necklaces with k red beads is equal to the coefficient of x^k in

$$P_G(1 + x, 1 + x^2, \ldots)$$

and hence the number of such necklaces is

$$\sum_j \frac{\phi(j)}{j} \binom{\frac{n}{j}}{\frac{k}{j}}$$

where the sum is over all divisors j of the $gcd(n, k)$.

15.44 We wish to distribute 3 identical red balls and 2 identical blue balls into 4 boxes of which three are identical square boxes and one is a round box. In how many ways can this be done?

15.45 A die is constructed by writing different labels from 1 to 6 on the six faces of the cube. Show that the number of inequivalent dice is equal to 30. Show that this number is 15 if we also allow reflections of a die besides rotations.

15.46 Six labels $d_1, d_2, d_3, d_4, d_5, d_6$ are to be pasted on the six faces of a cube with each face receiving a different label. We are allowed to rotate the cube and two such labeled cubes are considered equivalent. Among the set of six labels, d_1 is yellow, d_2 is black while d_3, d_4 are both violet and d_5, d_6 are both purple. Two labels of the same colour are indistinguishable. In addition, the person dealing with putting these labels is not very colour sensitive so he may confuse between purple and violet. Show that the group G acting on the labels is a group of order 8 with cycle index given by

$$P_G(x_1, x_2, \ldots) = \frac{1}{8}\left((x_1)^6 + 2(x_1)^4 x_2 + 3(x_1)^2(x_2)^2 + 2(x_1)^2 x_4\right)$$

Then use this to show that the total number of patterns is 5.

15.47 Let $(\sigma, \tau) \in G \times H$ where (G, D) and (H, R) are permutation groups. Let the cycle index of σ be $(b_1, b_2, \ldots, b_k, \ldots)$ and $(c_1, c_2, \ldots, c_j, \ldots)$ respectively. Prove that

$$|(Fix)_{\sigma,\tau}| = \left\{\prod_k (\frac{\partial}{\partial z_k})^{b_k}\right\}\left[\prod_j \{\exp j\{z_j + z_{2j} + \cdots\}\}^{c_j}\right]$$

evaluated at $z_i = 0$ for every i. Hence show that the number of patterns is given by

$$P_G\left(\frac{\partial}{\partial z_1}, \frac{\partial}{\partial z_2}, \ldots, \frac{\partial}{\partial z_k} \cdots\right) \times$$

$$P_H(\exp(z_1 + z_2 + \cdots), \exp 2(z_2 + z_4 + \cdots), \cdots, \exp j(z_j + z_{2j} + \cdots), \cdots)$$

evaluated at $z_i = 0$ for every i.

15.48 (continuation of Exercise 15.47) Show that

$$|(Fix)^*_{\sigma,\tau}| = \left\{\prod_k (\frac{\partial}{\partial z_k})^{b_k}\right\}\left\{\prod_j (1 + j z_j)^{c_j}\right\}$$

evaluated at $z_i = 0$ for every i. Hence prove that the total number of patterns of injective functions is equal to

$$P_G(\frac{\partial}{\partial z_1}, \frac{\partial}{\partial z_2}, \ldots, \frac{\partial}{\partial z_k} \cdots) \times$$

$$P_H(1 + z_1, 1 + 2z_2, \cdots, 1 + j z_j, \cdots)$$

evaluated at $z_i = 0$ for every i.

15.49 (continuation of Exercise 15.48) Show that when $|R| = |D|$ the number of patterns of one-to-one functions is equal to

$$P_G\left(\frac{\partial}{\partial z_1}, \frac{\partial}{\partial z_2}, \cdots, \frac{\partial}{\partial z_k} \cdots\right) \times$$

$$P_H(z_1, 2z_2, \cdots, j z_j, \cdots)$$

evaluated at $z_i = 0$ for every i.

15.50 Consider the following situation. The domain consists of n distinct labels (that cannot be permuted into each other) and hence has the trivial automorphism group G. The range also consists of $m \geq n$ elements and the group acting on the range is H. Then a one-to-one function is just a labeled set and we wish to find the number of patterns of such labeled sets. Show that this number is equal to

$$\left(\frac{d}{dz}\right)^n P_H(1+z,1,1,\ldots)$$

evaluated at $z = 0$.

Chapter 16

Systems of distinct representatives

16.1 System of distinct representatives and P. Hall's theorem

In this last chapter, we deal with a question which is very much combinatorial in nature but is qualitatively very different from the kind of questions we dealt with and the combinatorial tools we developed in all the earlier chapters. The problem is the following. A (professional) matchmaker has a list of some boys and some girls who have approached him. Looking at their likings and disliking, he has formed a compatibility or suitability graph that declares a boy-girl pair suitable (for matching) or not suitable. Compatibility is always to be taken on a two-way basis (giving rise to a bipartite graph). The matchmaker wishes to marry off as many boys as he can. Suppose we have m boys and n girls. Then an obvious condition, for all the boys to get married, is that $n \geq m$ since there must be at least m girls available to marry all the m boys. Moreover, every boy must be suitable for at least one girl (else he cannot be married off). The list of girls suitable for at least one of the two boys must have at least two girls (else there would be a tie and only one of the two boys can get married). In general, given any subset of k boys, the list of all the girls suitable to at least one of the k boys must have at least k girls. What turns out to be true is that this simple looking necessary condition is also sufficient to marry off all the m boys (Theorem 16.1.3). This chapter is organized as follows. In this first Section, we prove the standard form of P. Hall's theorem on the existence of a system of distinct representatives along with a defect version of this theorem. In Section 16.2, we translate these results in the convenient set up of bipartite graphs. In Section 16.3, we deal with some algorithmic aspects of this problem. Applications of matching theorem are given in Section 16.4. In Section 16.5, we deal with posets and prove Dilworth's theorem, which is equivalent to Hall's theorem. In Section 6, we look at the question of the validity of similar assertions for the infinite situations.

Definition 16.1.1. Let $\mathbf{A} = \{A_1, A_2, \ldots, A_n\}$ be a family of finite sets. A *System of Distinct Representatives (abbreviated SDR)* for the (ordered) family $\mathbf{A}$ is an ordered n-tuple $(a_1, a_2, \ldots, a_n)$ such that

(a) $a_i \in A_i \; \forall\, i = 1, 2, \ldots, n.$

(b) $a_i \neq a_j \; \forall \, i \neq j$.

We say that a_i *represents* A_i (in the SDR $(a_1, a_2, \ldots, a_n)$).

Example 16.1.2. Let $\mathbf{A} = \{A_1, A_2, \ldots, A_5\}$ where $A_1 = \{a, b, c\}, A_2 = \{b, c\}, A_3 = \{c, d, e\}, A_4 = \{c, d, e\}$ and $A_5 = \{d, e\}$. Then $(a_1, a_2, a_3, a_4, a_5) = (a, b, c, d, e)$ is an SDR for $\mathbf{A}$. A different SDR for $\mathbf{A}$ is (a, b, c, e, d). It may appear that arbitrarily picking up elements from the sets A_i's will produce an SDR but that is no true. For example, if we begin by picking c as a representative for A_1 and b to represent A_2, then d and e are the only elements left to represent all the three of A_3, A_4 and A_5 showing that this initial choice cannot be completed to an SDR.

In order for an SDR to exist, observe that any k sets must contain at least k elements in their union (so that each set gets represented by a different element). Thus, a necessary condition for $\mathbf{A}$ to have an SDR is:

$$|A_{i_1} \cup A_{i_2} \cup \cdots \cup A_{i_k}| \geq k \tag{16.1}$$

must hold for every k with $1 \leq k \leq n$ and $\forall \, 1 \leq i_1 < i_2 < \cdots < i_k \leq n$.

Theorem 16.1.3. *(Marriage Theorem or P.Hall's Theorem)* (16.1) *is both a necessary and sufficient condition for the family* $\mathbf{A} = \{A_1, A_2, \ldots, A_n\}$ *to have an SDR.*

Proof As we already observed it, (16.1) is necessary. We prove the sufficiency by making an induction on n. The condition is clearly sufficient when $n = 1$ since any element of the non-empty set A_1 may be chosen to represent it. Let $n \geq 2$ and assume that the condition (16.1) is sufficient for all families with n' sets where $n' < n$.
Case 1 $\forall \, m < n$ and $\forall \, 1 \leq i_1 < i_2 < \cdots < i_m \leq n$, we have

$$|A_{i_1} \cup A_{i_2} \cup \cdots \cup A_{i_m}| \geq m + 1 \tag{16.2}$$

The given condition (16.1) implies that $A_n \neq \emptyset$. Let $y = y_n \in A_n$. Consider the family $\mathbf{B} = \{B_1, B_2, \cdots, B_{n-1}\}$ where $B_i = A_i - \{y\}$. Then $\forall \, 1 \leq i_1 < i_2 < \cdots < i_m \leq n - 1$, we have

$$\begin{aligned} |B_{i_1} \cup B_{i_2} \cup \cdots \cup B_{i_m}| &= |(A_{i_1} - \{y\}) \cup (A_{i_2} - \{y\}) \cup \cdots (\cup A_{i_m} - \{y\})| \\ &= |(A_{i_1} \cup A_{i_2} \cup \cdots \cup A_{i_m}) - \{y\}| \\ &\geq (m + 1) - 1 = m \end{aligned}$$

Hence $\mathbf{B}$ satisfies (16.1) and by induction, it has an SDR say $(y_1, y_2, \cdots, y_{n-1})$. Then $(y_1, y_2, \ldots, y_{n-1}, y_n)$ is an SDR for $\mathbf{A}$ since $y_n = y \in A_n$ and $\forall \, i = 1, 2, \ldots, n - 1$ we have $y_i \in B_i \subset A_i$ and B_i does not contain y.

Case 2 Negation of case 1: $\exists \;\; m$ such that $1 \leq m \leq n - 1$ and some $1 \leq i_1 < i_2 < \cdots < i_m \leq n$ such that

$$|A_{i_1} \cup A_{i_2} \cup \cdots \cup A_{i_m}| = m \tag{16.3}$$

To simplify the notation, assume, w.l.o.g. that $|A_1 \cup A_2 \cup \cdots \cup A_m| = m$ and let $A = A_1 \cup A_2 \cup \cdots \cup A_m$. Let $\mathbf{A}' = \{A_1, A_2, \ldots, A_m\}$ and let $\mathbf{A}'' = \{B_j : j =$

$m+1,\ldots,n\}$ where $B_j = A_j - A$ for all $j = m+1,\ldots,n$. By induction, $\mathbf{A}'$ has an SDR say $(a_1, a_2, \ldots, a_m)$. Then $a_i \in A_i$ and indeed $A = \{a_1, a_2, \ldots, a_m\}$. Now consider $B_{j_1} \cup B_{j_2} \cup \cdots \cup B_{j_r}$ where $m+1 \leq j_1 < j_2 < \cdots < j_r \leq n$. We have

$$\begin{aligned}
|B_{j_1} \cup Bj_2 \cup \cdots \cup B_{j_r}| &= |(A_{j_1} - A) \cup (A_{j_2} - A) \cup \cdots \cup (A_{j_r} - A)| \\
&= |[(A_{j_1} \cup A) - A] \cup \cdots \cup [(A_{j_r} \cup A) - A]| \\
&= |[A \cup (A_{j_1} \cup A_{j_2} \cup A_{j_r})] - A| \\
&= |(A_1 \cup A_2 \cup \cdots \cup A_m) \cup (A_{j_1} \cup A_{j_2} \cup A_{j_r})| - |A| \\
&\geq (m+r) - m \\
&= r
\end{aligned}$$

showing that $|B_{j_1} \cup Bj_2 \cup \cdots \cup B_{j_r}| \geq r$ and hence condition (16.1) is also satisfied by the family $\mathbf{A}''$. By induction, $\mathbf{A}''$ has an SDR $(a_{m+1}, a_{m+2}, \ldots, a_n)$. Then for all $j \geq m+1$ we have $a_j \notin A$ since $a_j \in B_j = A_j - A$. Thus $(a_1, a_2, \ldots, a_n)$ is an SDR for $\mathbf{A}$ as desired. □

In view of the brevity of the argument, we are tempted to give the following alternative proof of Theorem 16.1.3 from Mirsky [42]. *If $I \subset [n]$, we write A_I to mean $\cup_{i \in I} A_i$.* Let $\mathbf{A} = \{A_1, A_2, \ldots, A_n\}$ be a given family of sets satisfying condition (16.1). What happens if each A_i is a singleton set? Then condition (16.1) implies that no two sets have an element in common. Since the sets are disjoint and non-empty, we, in fact get a unique SDR for the family $\mathbf{A} = \{A_1, A_2, \ldots, A_n\}$. Now let the set A_i contain two distinct elements say b and c. Let $B_i = A_i - \{b\}$ and let $C_i = A_i - \{c\}$ and let $\forall\, j \neq i, B_j = C_j = A_j$. Let $\mathbf{B} = \{B_1, B_2, \ldots, B_n\}$ and $\mathbf{C} = \{C_1, C_2, \ldots, C_n\}$. *We claim that one of* $\mathbf{B}$ *or* $\mathbf{C}$ *satisfies* (16.1). Clearly an SDR for $\mathbf{B}$ (or $\mathbf{C}$) is also an SDR for $\mathbf{A}$ since B_j (or C_j) is contained in A_j for all j. Recursively, we can delete an element from a set with size at least 2 in the family of sets under consideration, making certain that at every stage, the new family of smaller sets also satisfies (16.1) where we will finally be left with n sets each of which is a singleton and since the n elements in these n sets are all different we are done. To prove the claim, suppose it is false. Then for some indexing sets I and J, we must have

$$|B_I| = |\cup_{j \in I} B_j| \leq |I| - 1$$

and

$$|C_J| = |\cup_{j \in J} B_j| \leq |J| - 1$$

Clearly both I and J contain i (since $\mathbf{A}$ satisfies (16.1)). Let $I' = I - \{i\}$ and let $J' = J - \{i\}$. Then we have

$$\begin{aligned}
|A_{I \cup J}| &= |\cup_{r \in I \cup J} A_r| \\
&= |(\cup_{j \in I} B_j) \cup (\cup_{j \in J} C_j)| \\
&= |B_I \cup C_J|
\end{aligned}$$

Since $B_I \cap C_J \supset A_j \;\forall\, j \in I' \cap J'$, we have $B_I \cap C_J \supset A_{I' \cap J'}$ and hence

$$|B_I \cap C_J| \geq |A_{I' \cap J'}| \geq |I' \cap J'|$$

using the given condition (16.1). Therefore, we have

$$\begin{aligned}|A_{I\cup J}| &= |B_I \cup C_J| \\ &= |B_I| + |C_J| - |B_I \cap C_J| \\ &\leq |B_I| + |C_J| - |A_{I'\cap J'}| \\ &\leq |B_I| + |C_J| - |I' \cap J'| \\ &\leq (|I| - 1) + (|J| - 1) - |I' \cap J'| \\ &= |I'| + |J'| - |I' \cap J'| \\ &= |I' \cup J'| \\ &= |I \cup J| - 1\end{aligned}$$

which is a contradiction to the fact that $\mathbf{A}$ satisfies (16.1) and this completes proof of the claim. □

Example 16.1.4. Let $\mathbf{A} = \{A_1, A_2, A_3, A_4\}$ where $A_1 = \{1, 2\}$, $A_2 = \{1, 3\}$, $A_3 = \{2, 3\}$ and $A_4 = \{1, 2, 3\}$. Then $\mathbf{A}$ does not have an SDR but *any* subfamily with 3 sets has an SDR.

Given a family $\mathbf{A}$ of finite sets, we want to find out the largest number r such that we have some subfamily consisting of r sets that has an SDR. This is given in the following two theorems.

Theorem 16.1.5. *Let $\mathbf{A} = \{A_1, A_2, \ldots, A_n\}$ be a family of finite sets. Let $r \leq n$. Then there is a subfamily of $\mathbf{A}$ consisting of r sets that has an SDR iff $\forall\ k$ and $\forall\, 1 \leq i_1 < i_2 < \cdots < i_k \leq n$,*

$$|A_{i_1} \cup A_{i_2} \cup \cdots \cup A_{i_k}| \geq k - (n - r) \tag{16.4}$$

Note that the Theorem 16.1.5 implies the marriage theorem (Theorem 16.1.3) by taking $r = n$. In contrast we use the marriage theorem (Theorem 16.1.3) to prove Theorem 16.1.5.

Proof Let F be a set of order $n - r$ such that F is disjoint from $\cup_{i=1}^n A_i$. Let $\mathbf{D} = \{D_1, D_2, \ldots, D_n\}$ where $D_i = A_i \cup F\ \forall\ i = 1, 2, \ldots, n$. If $\mathbf{D}$ has an SDR say $(d_1, d_2, \ldots, d_n)$, then at the most $n - r$ of the d_i's can come from F and hence at least r of the d_i's that represent $D_i = A_i \cup F$ actually come from A_i and therefore at least r of the A_i's have an SDR. Conversely let $F = \{f_1, f_2, \ldots, f_{n-r}\}$ and assume w.l.o.g. that that the subfamily $\{A_1, A_2, \ldots A_r\}$ has an SDR say $(a_1, a_2, \ldots, a_r)$. Then $(a_1, a_2, \ldots, a_r, f_1, \ldots, f_{n-r})$ is an SDR for the family $\mathbf{D}$ since $a_i \in A_i \subset D_i$ and $f_j \in D_{r+j}\ \forall\ j = 1, 2, \ldots, n - r$.

In turn, using the marriage theorem (Theorem 16.1.3), the family $\mathbf{D}$ has an SDR iff $\forall\, k$ and $\forall\, 1 \le i_1 < i_2 < \cdots < i_k \le n$ we have

$$\begin{aligned} |D_{i_1} \cup \cdots \cup D_{i_k}| &\ge k \\ \Leftrightarrow |(A_{i_1} \cup F) \cup \cdots \cup (A_{i_k} \cup F)| &\ge k \\ \Leftrightarrow |A_{i_1} \cup \cdots \cup A_{i_k}| + |F| &\ge k \\ \Leftrightarrow |A_{i_1} \cup \cdots \cup A_{i_k}| + (n-r) &\ge k \\ \Leftrightarrow |A_{i_1} \cup \cdots \cup A_{i_k}| &\ge \quad k-(n-r) \end{aligned}$$

as desired. □

Theorem 16.1.6. *Let* $\mathbf{A} = \{A_1, A_2, \ldots, A_n\}$ *be a family of sets. Then the largest number of sets that have an SDR equals the minimum value of the expression*

$$|A_{i_1} \cup A_{i_2} \cdots \cup A_{i_k}| + (n-k) \tag{16.5}$$

where the minimum is taken over all all k *with* $1 \le k \le n$ *and over all* $1 \le i_1 < i_2 < \cdots < i_k \le n$.

Proof Let the largest number of sets in $\mathbf{A}$ that have an SDR be α and let the minimum value of the expression $|A_{i_1} \cup A_{i_2} \cdots \cup A_{i_k}| + (n-k)$ be β. Theorem 16.1.5 applies with $r = \beta$ to give $|A_{i_1} \cup A_{i_2} \cdots \cup A_{i_k}| \ge k - (n-\alpha)$ and hence $|A_{i_1} \cup A_{i_2} \cdots \cup A_{i_k}| + (n-k) \ge \alpha$ for every choice of k and for every $(i_1, i_2, \cdots, i_n)$ where $1 \le i_1 < i_2 < \cdots < i_k \le n$ showing that $\alpha \le \beta$. On the other hand, $|A_{i_1} \cup A_{i_2} \cdots \cup A_{i_k}| + (n-k) \ge \beta$ for all choices of $1 \le i_1 < i_2 < \cdots < i_k \le n$ and hence $|A_{i_1} \cup A_{i_2} \cdots \cup A_{i_k}| \ge k + (n-\beta)$ for all such choices. So, by Theorem 16.1.5, β sets in $\mathbf{A}$ have an SDR showing that $\beta \le \alpha$. □

16.2 Bipartite graphs and matchings

A setting in which the problem of a system of distinct representatives may be studied and understood in a better way is that of a bipartite graph. Recall that a bipartite graph $G = (X, Y, \Delta)$ is a triple where X and Y are disjoint vertex subsets whose union is the set of all the vertices, with the set of edges $\Delta \subset X \times Y$. Thus no two vertices in X are adjacent (joined) nor are any two vertices in Y joined and all the adjacencies are between X and Y. If $\mathbf{A} = \{A_1, A_2, \ldots, A_m\}$ is the given family of sets, we take $X = \{x_1, x_2, \ldots, x_m\}$ and $Y = \cup_{i=1}^{m} A_i$ where for $y = y_j \in Y$ we have $(x_i y_j) \in \Delta$ iff $y_j \in A_i$. Clearly this gives us a one-to-one correspondence. That is, given the adjacency relations in the bipartite graph G (that is, given Δ), we can obtain a family $\mathbf{A}$ of sets and conversely. We illustrate this first on the family $\mathbf{A}$ of sets given in Example 16.1.2. Here $\mathbf{A}$ has five sets and hence $X = \{x_1, x_2, x_3, x_4, x_5\}$ while $Y = \cup_{i=1}^{5} A_i$ is given by the set $\{a, b, c, d, e\}$. The edge-set Δ is given by

$$\Delta = \{(x_1a), (x_1b), (x_1c), (x_2b), (x_2c), (x_3c), (x_3d), (x_3e), (x_4c), (x_4d), (x_4e), (x_5d), (x_5e)\}$$

The graph $G = (X, Y, \Delta)$ is drawn in Figure 16.1. Similarly the bipartite graph G corresponding to the family of sets in Example 16.1.4 is drawn in Figure 16.2.

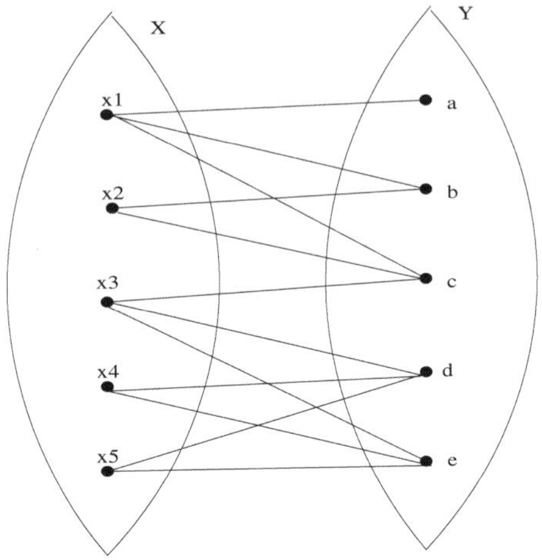

Figure 16.1: Family of Sets in Example 16.1.2 converted to a bipartite graph

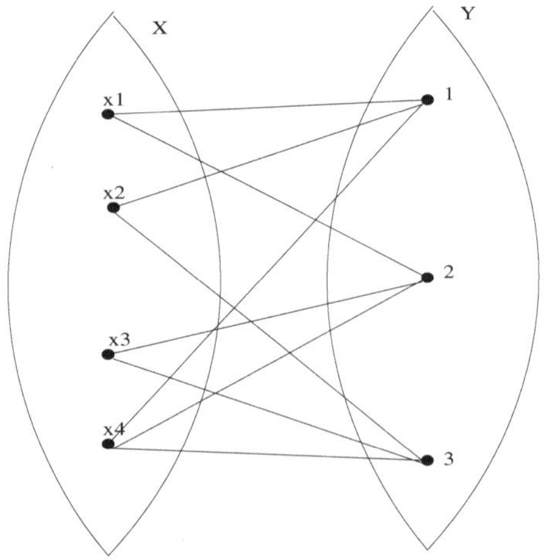

Figure 16.2: Bipartite graph for the family of sets in Example 16.1.4

Example 16.2.1. Consider the bipartite graph given in Figure 16.3.
Then we have the edge-set given by:

$$\Delta = \{(x_1y_2), (x_1y_3), (x_2y_1), (x_2y_2), (x_3y_3), (x_3y_4), (x_4y_7),$$

$$(x_4y_4), (x_5y_6), (x_5y_1), (x_5y_2)\}$$

and hence $\mathbf{A} = \{A_1, A_2, A_3, A_4, A_5\}$ where

$$A_1 = \{y_2, y_3\}, A_2 = \{y_1, y_2\}, A_3 = \{y_3, y_4\}, A_4 = \{y_4, y_7\}, A_5 = \{y_1, y_2, y_6\}$$

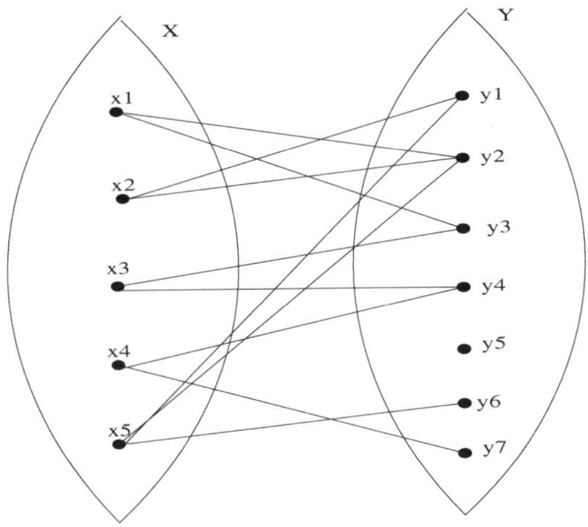

Figure 16.3: A Bipartite graph converted to a family of sets

Definition 16.2.2. A *matching* M in G is a set of edges $M = \{e_1, e_2, \ldots, e_r\}$ where e_i and e_j do not share a common vertex if $i \neq j$.

Thus, if $e_i = (x_iy_i)$, then all the x_i's and all the y_i's are distinct and under the one-to-one correspondence we described, we have an SDR $(y_1, y_2, \ldots, y_r)$ of the subfamily $\mathbf{A}' = \{A_1, A_2, \ldots, A_r\}$.

Definition 16.2.3. A matching M is said to be a *maximal (respectively a maximum) matching* if no edge can be added to it to make it a larger matching (respectively if it has the largest number of edges among the set of all the matchings in G).

Evidently, a a maximum matching is a maximal matching but not conversely. The following one-to-one correspondence is then obvious.

Theorem 16.2.4. *Let $\mathbf{A} = \{A_1, A_2, \ldots, A_m\}$ be a family of sets and let $G = (X, Y, \Delta)$ be the corresponding bipartite graph. Then the following assertions hold.*

(a) $M = \{(x_{i_1}, y_{i_1}), (x_{i_2}, y_{i_2}), \ldots, (x_{i_k}, y_{i_k})\}$ is a matching in G iff $(y_{i_1}, y_{i_2}, \ldots, y_{i_k})$ is an SDR for the subfamily $\mathbf{A}' = \{A_{i_1}, A_{i_2}, \ldots, A_{i_k}\}$.

(b) $M = \{(x_1, y_1), (x_2, y_2), \ldots, (x_m, y_m)\}$ is a (maximum) matching in G iff $(y_1, y_2, \ldots, y_m)$ is an SDR for $\mathbf{A} = \{A_1, A_2, \ldots, A_m\}$.

(c) The largest number of sets in $\mathbf{A}$ that have an SDR is equal to the number of edges in a maximum matching M of G.

Consider Example 16.2.1. The SDR of **A** given by $(y_3, y_1, y_4, y_7, y_6)$ corresponds to the matching

$$M = \{(x_1y_3), (x_2y_1), (x_3y_4), (x_4y_7), (x_5y_6)\}$$

In Example 16.1.2, $M = \{(x_1b), (x_3c), (x_4d), (x_5e)\}$ is a matching of the bipartite graph G. This matching is *a maximal matching* since none of the edges at x_2, viz (x_2b) or (x_2c) can be used to enlarge M. However,

$$M' = \{(x_1a), (x_2b), (x_3c), (x_4d), (x_5e)\}$$

is a larger matching since $|M| = 4 < 5 = |M'|$ and hence M is *not a maximum matching.* This shows that *a maximal matching need not be a maximum matching.*

Definition 16.2.5. Let $G = (X, Y, \Delta)$ be a bipartite graph. A subset C of the vertex set $V(G)$ of G is called *a cover* if $\forall\, e = (xy) \in \Delta$, we have $x \in C$ or $y \in C$. A cover C is called *a minimum cover* if $|C|$ has the smallest size.

Notice that covers exist: X and Y are themselves covers and any superset of these sets is also a cover (though there can be covers other than these).

Example 16.2.6. Let $G = (X, Y, \Delta)$ where

$$\Delta = \{(x_1y_1), (x_1y_2), (x_2y_1), (x_2y_2), (x_3y_1), (x_3y_2), (x_4y_4), (x_5y_1)\}$$

as shown in Figure 16.4.

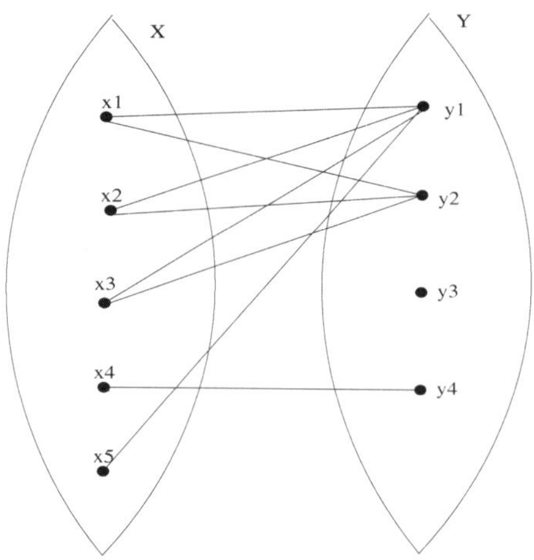

Figure 16.4: A cover of a bipartite graph

Then X is a cover and so is Y. However, the cover $C = \{y_1, y_2, x_4\}$ is a smaller cover. As such C is a minimum cover and the size of a maximum matching in G is 3 as the following theorem shows.

Theorem 16.2.7. *(Minimax Theorem) Let $G = (X, Y, \Delta)$ be a bipartite graph. Then the number of edges in a maximum matching of G is equal to the minimum size of a cover.*

Proof Let α denote the size (the number of edges) of a maximum matching of G and let β denote the minimum size of a cover. Let $M = \{e_1, e_2, \ldots, e_r\}$ be any matching and let C denote any cover. If $e_i = (x_i y_i)$, then either $x_i \in C$ or $y_i \in C$ and since M is a matching, no vertex of C can cover two edges of the matching M. So $|M| \le |C|$ *and hence* $\alpha \le \beta$ *by taking maximum over all* M *and minimum over all* C. Let C be a cover with β vertices. From the graph G, we construct the family of sets $\mathbf{A} = \{A_1, A_2, \ldots, A_m\}$ as has been already outlined: if $X = \{x_1, x_2, \ldots, x_m\}$ and if $Y = \{y_1, y_2, \ldots, y_n\}$, then $y_j \in A_i$ iff $(x_i y_j) \in \Delta$. Then a matching M in G corresponds to a subfamily of $\mathbf{A}$ with an SDR and conversely as asserted in Theorem 16.2.4. If $M = \{(x_{i_1} y_{i_1}), (x_{i_2} y_{i_2}), \ldots, (x_{i_k} y_{i_k})\}$ then $\mathbf{A}' = \{A_{i_1}, A_{i_2}, \ldots, A_{i_k}\}$ is the subfamily with the SDR $(y_{i_1}, y_{i_2}, \ldots, y_{i_k})$. So the largest size of a subfamily with an SDR is equal to the largest size of matching M in G which is equal to α. By Theorem 16.1.6, equation (16.5) gives: α equals minimum over $|A_{i_1} \cup A_{i_2} \cdots \cup A_{i_k}| + (m-k)$ and hence for some k and some $1 \le i_1 < \cdots < i_k \le m$, we have

$$\alpha = |A_{i_1} \cup A_{i_2} \cdots \cup A_{i_k}| + (m-k)$$

Let

$$C = X - \{x_{i_1}, x_{i_2}, \ldots, x_{i_k}\} \cup (A_{i_1} \cup A_{i_2} \cup \cdots \cup A_{i_k})$$

Then C is a vertex subset that has $(m-k) + |A_{i_1} \cup A_{i_2} \cup \cdots \cup A_{i_k})| = \alpha$ vertices. We claim that C is a cover. Let $e = (xy) \in \Delta$. If $x = x_j$ for some $j = i_1, i_2, \ldots, i_k$, then $y \in A_{i_1} \cup A_{i_2} \cup \cdots \cup A_{i_k}$ and hence C covers e. If not then $x \in X - \{x_{i_1}, x_{i_2}, \ldots, x_{i_k}\}$ and by construction of C, it covers $e = (xy)$. So C is a cover of size α showing that $\beta \le \alpha$ completing the proof. □

16.3 Hungarian algorithm

We now make a detour and discuss algorithmic aspects of Theorem 16.2.7. *For the entire discussion to follow,* $G = (X, Y, \Delta)$ *is a (given) bipartite graph* and the attempt is to find a maximum matching M of G.

Definition 16.3.1. Given a matching M of G a vertex z is said to be M*-saturated (or just saturated)* if it is on some edge e of the matching; otherwise z is M*-unsaturated (or just unsaturated)*. A matching M that saturates every vertex is called *a perfect* matching (evidently, this requires $|X| = |Y|$).

Definition 16.3.2. Given a matching M, *an* M*-alternating chain (or just an alternating chain when the matching is clear by context)* is a path γ from a vertex u in X to a vertex v in Y of the form

$$\gamma = x_0 y_0 x_1 y_1 x_2 y_2 \cdots x_r y_r$$

where $x_0 = u$ and $y_r = v$ such that all the $x_i \in X$ and all the $y_i \in Y$ satisfying all of the following conditions:

(a) γ has an odd number of edges: the $r+1$ edges $(x_0y_0), (x_1y_1), \ldots, (x_ry_r)$ which we call *forward edges* and the r edges $(y_0x_1), (y_1x_2), \ldots, (y_{r-1}x_r)$ which we call *backward edges*.

(b) Both the start vertex u and the end vertex v are M-unsaturated.

(c) All the forward edges (that is, the first, third, fifth etc.) *are not in* M.

(d) All the backward edges (that is, the second, fourth, sixth etc.) *are in* M.

Lemma 16.3.3. *Let γ be an alternating chain w.r.t. the matching M. Then there exists a matching M' such that $|M'| = |M| + 1$ and consequently, M is not a maximum matching.*

Proof Since γ is an alternating chain w.r.t. M, we have two disjoint sets N_0 and M_0 of edges of γ such that N_0 consists of the first, third, $\ldots$, $(2r+1)-th$ edges of γ while M_0 consists of the second, fourth, $\ldots$, $(2r)-th$ edges of γ and the edges in N_0 are not in M while those in M_0 are in M. Since the initial vertex u and the end vertex v of γ are unsaturated, it follows that by replacing all the edges of M_0 in M by the edges of N_0, we get a new matching: $M' = (M - M_0) \cup N_0$ and since $|N_0| = |M_0| + 1$, it is also clear that $|M'| = |M| + 1$ showing that M' has more edges than M. □

Lemma 16.3.4. *Let M be a matching such that is not maximum. Then there exists an alternating chain γ w.r.t. M.*

Proof Since M is not a maximum matching, $\exists$ M' such that $|M'| > |M|$ and M' is a matching. Consider the edge subgraph $H = M \triangle M'$ where $\triangle$ is the symmetric difference (thus $A \triangle B$ means those elements that are in A but not in B and those that are in B but not in A; this is equal to $A \cup B - A \cap B$). *We have to make do with the use of the same symbol $\triangle$ to mean both the edge-set of G and symmetric difference, which could not be avoided for the lack of a better symbol.* In H, each vertex has degree 1 or 2 (since the edges in H come only from M or M'). Let C be a non-empty component (that is, with at least one edge) of H. Then being a connected graph, C is either a cycle or a path. Let C be a cycle. Since the edges in C must alternate between M and M' and since G is a bipartite graph, C must have an even number of edges with equal number from both M and M'. But $|M'| > |M|$ and hence there is some component of H say C such that C is not a cycle and hence is a path with an odd number of edges with edges alternating between M' and M (with the first and the last edge in M'). Thus the edges in the path C have the form $e_1, f_1, e_2, f_2, \ldots, e_k, f_k, e_{k+1}$ where $e_i = (x_iy_i) \in M'$ for all $i = 1, 2, \ldots, k+1$ and $f_j = (y_jx_{j+1}) \in M$. Clearly then x_1 and y_{k+1} are both M-unsaturated and the forward edges e_j in C that come from M' are not in M while the backward edges f_j are all in M. It thus follows that C gives us an alternating chain γ w.r.t. M. □

Example 16.3.5. Consider the bipartite graph of Figure 16.5 along with the matchings $M = \{(x_2y_2), (x_3y_3), (x_5y_5)\}$ and $M' = \{(x_1y_2), (x_2y_3), (x_3y_4), (x_5y_5)\}$.

Then $M \triangle M' = \{(x_1y_2), (y_2x_2), (x_2y_3), (y_3x_3), (x_3y_4)\}$ and we have an alternating chain $\gamma = x_1y_2x_2y_3x_3y_4$. Here x_1 and y_4 are M-unsaturated. Hence replacing the backward arcs $(y_2x_2), (y_3x_3)$ in M by the three forward arcs $(x_1y_2), (x_2y_3), (x_3y_4)$, we obtain the larger matching $\{(x_1y_2), (x_2y_3), (x_3y_4), (x_5y_5)\}$.

The two Lemmas 16.3.3 and 16.3.4 together prove the following.

Theorem 16.3.6. *Let M be a matching. Then M is a maximum matching iff $\nexists$ an alternating chain γ w.r.t. M.*

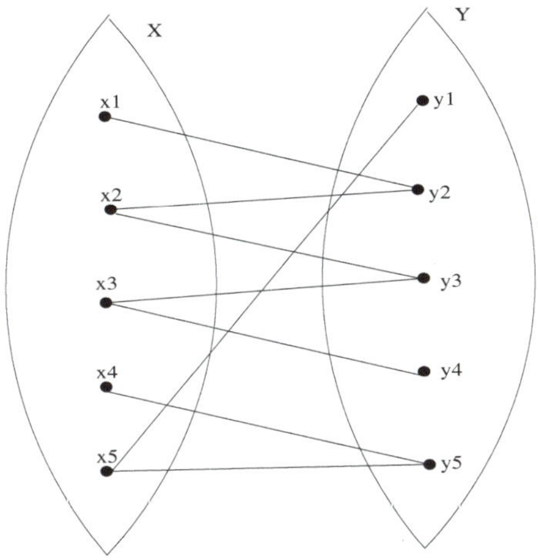

Figure 16.5: Forward and backward arcs

Having gone this far, we are tempted to describe a nice procedure to find a maximum matching M. The procedure we describe here is called the *Hungarian algorithm or the labeling algorithm* which is actually a special case of the Ford-Fulkerson algorithm for finding a maximum flow in a capacity network. The labeling algorithm has, as its input, a given (initial) matching M of the bipartite graph G. After applying it, it should do one of the two things. Either it should declare M to be a maximum matching (and then the algorithm stops at that stage) or it should construct an alternating chain γ w.r.t. M in which case as Lemma 16.3.3 has already proved it, we get a better matching M' with $|M'| > |M|$. These two mutually exclusive cases will be called *non-breakthrough* and *breakthrough* respectively.

Labeling algorithm: We are given a bipartite graph $G = (X, Y, \Delta)$ and a matching M of G. In step 1, label every M-unsaturated vertex x in X by $(*)$ (if there is no such vertex, then the matching M saturates every vertex of X and hence is a maximum matching). In step 2, look at all the vertices $y \in Y$ that are on an edge $e = (xy)$ where x has been labeled in step 1 (such an edge cannot be in M). We then label y by (x) (note that y itself may or may not be saturated). If there is a tie, that is, if y could be labeled both x and x' (both of which have been labeled in step 1), then we resolve the tie arbitrarily and label y by (x) or (x') (but not both). At the end of step

2, some vertices in Y have been (newly) labeled. If any one of these vertices say y is unsaturated and is labeled (x), then the single edge (xy) forms an alternating chain and could be added to M to get a larger matching. In step 2, we label every unlabeled vertex in Y following the given stipulation and do not leave out any vertex of Y that could be labeled under the stipulation. Assume then that each vertex y labeled in step 2 is saturated. We then proceed to step 3 where we attempt to label new vertices in X (we do not relabel any vertex that has been labeled earlier). If x' is a vertex that is not yet labeled and $e' = (x'y) = (yx')$ is an edge in M such that y has been labeled in step 2 (and hence e' is to be called a backward edge) then label x' by (y). Do this for as many new vertices in X as possible. At the end of step 3, some new vertices in X have been labeled (with labels coming from Y). Then in step 4, we look at the newly labeled vertices x in X (in step 3); if $y \in Y$ is unlabeled and is on an edge $e = (xy)$ such that $e \notin M$, then we label y by x. Observe that at the end of step 4, we have constructed a putative alternating chain: If y is labeled (x) in step 4 and x is labeled (y') in step 3 and y' is labeled (x') in step 2, then x' must have been labeled $(*)$ in step 1 and hence must be unsaturated. We thus have $(x'y') \notin M$ while $(y'x) \in M$ (this is a backward arc) and $(xy) \notin M$. If it so happens, that y is unsaturated then indeed we have found an alternating chain $x'y'xy$ and if it happens for any vertex y which is newly labeled then we have a breakthrough. If not, we proceed to step 5 and label new vertices in X using the same rule. In general then, an odd step labels vertices in X labeling a so far unlabeled vertex x using a backward arc $e = (yx)$ (where y has been labeled in the previous step and $e \in M$). An even step labels vertices in Y labeling a so far unlabeled vertex z using a forward arc $f = (xz)$ (where x has been labeled in the previous step and $f \notin M$). A breakthrough can occur only at the end of an even step where a newly labeled vertex $z \in Y$ is unsaturated and we have thus found an alternating chain γ.

A breakthrough thus constructs an alternating chain γ by backward tracing of labels and then the given matching is not maximum and one can then obtain a larger matching M'. We may not be so lucky and may end up in a situation where no more vertices could be labeled and no (newly) labeled vertex in Y is unsaturated. Note that this must eventually happen since a breakthrough obtains a new matching that has a larger number of edges. When no more vertices can be labeled nor do we have a breakthrough, We call that situation a non-breakthrough. Correctness of the labeling algorithm is now established by the following.

Theorem 16.3.7. *Let the labeling algorithm produce a non-breakthrough. Let A be the set of all the unlabeled vertices in X and let B be the set of all the labeled vertices in Y. Let $C = A \cup B$. Then $|M| = |C|$ and hence C is a minimum cover and M a maximum matching. Thus, the labeling algorithm terminates after a finite number of steps and produces a maximum matching.*

Proof Suppose $e = (xy)$ is not covered by C. Then $x \notin A$ and $y \notin B$. Suppose first that $e \in M$. Then x is saturated and hence does not get labeled in step 1. Since x is labeled, it must have been labeled (y) using the backward arc $e = (yx) \in M$. But then y itself must be labeled before x and thus $y \in B$, a contradiction. So $e \notin M$. Then $x \notin A$ implies that x is labeled and the algorithm then can label y by (x)

showing that y must be labeled and hence $y \in B$, a contradiction. This proves that C is a cover. For every $x \in A$, x is not labeled implies that x is M-saturated and hence each such x is on a unique edge e of M. Let $y \in B$. Then y is labeled and we are in a situation of non-breakthrough. So there is a unique edge $e' = (x'y) \in M$. Then x' must be labeled (y) and hence $x' \notin A$. It thus follows that each vertex in $C = A \cup B$ is on a unique edge of M and the $|C|$ edges thus determined are all distinct. So $|C| \leq |M|$ and therefore, equality must hold to give $|C| = |M|$ and thus C is a minimum cover and M a maximum matching as desired. □

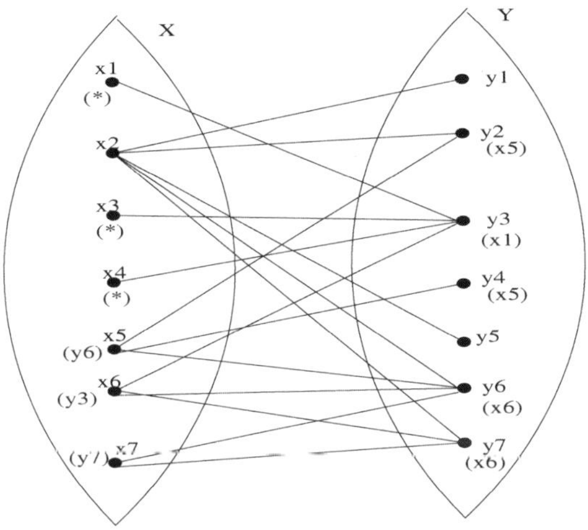

Figure 16.6: Labeling algorithm

Example 16.3.8. Consider the bipartite graph G in Figure 16.6 where we begin with the matching $M = \{(x_2y_2), (x_6y_3), (x_7y_7), (x_5y_6)\}$.
Since x_1, x_3, x_4 are M-unsaturated, we label them $(*)$. Next, we have a forward arc (x_1y_3) which is not in M and hence y_3 is labeled (x_1). Since y_3 is M-saturated, we trace the backward arc (y_3x_6) (which is in the matching M) and label the so far unlabeled vertex x_6 by y_3. From x_6, we look for a forward arc which is not in M and we have the arc (x_6y_7) which is not in M and y_7 is not yet labeled. We now label y_7 by x_6. For the same reason, y_6 also gets label x_6. Then using the backward arcs (y_6x_5) and (y_7x_7), both of which are in M, we label x_5 by y_6 and x_7 by y_7. Next using the forward arcs (x_5y_2) and (x_5y_4) both of which are not in M, we label y_2 and y_4 both by x_5. At this stage, observe that y_4 is M-unsaturated and hence we are in a breakthrough situation: y_4 has label x_5, which has label y_6, which has label x_6 which has label y_3, which has label x_1 and finally x_1 has label $(*)$ showing that we have an alternating chain

$$\gamma = x_1y_3x_6y_6x_5y_4$$

and hence replacing the backward edges $(y_3x_6), (y_6x_5)$ (which are in M) on this chain by three forward edges $(x_1y_3), (x_6y_6), (x_5y_4)$, we obtain a new larger matching $M' = \{(x_1y_3), (x_2y_2), (x_5y_4), (x_6y_6), (x_7y_7)\}$.

Repeat the labeling algorithm with the new matching M' (instead of M). Then the vertices x_3 and x_4 both of which are unsaturated, get the label $(*)$. Next, the forward arc (x_3y_3) is not in the matching and we may give label x_3 to y_3. Finally the backward arc (y_3x_1) is in the matching and hence x_1 gets the label y_3. At this stage we observe that no new labels can be given and no unsaturated vertex in Y has been labeled. We thus have a non-breakthrough. Since x_3, x_4, x_1 are the labeled vertices in X, the set consisting of all the unlabeled vertices in X along with those that are labeled in Y is the set $\{x_2, x_5, x_6, x_7, y_3\}$ and this forms a cover C such that $|C| = 5 = |M'|$ and hence we have a maximum matching M' (consisting of 5 edges).

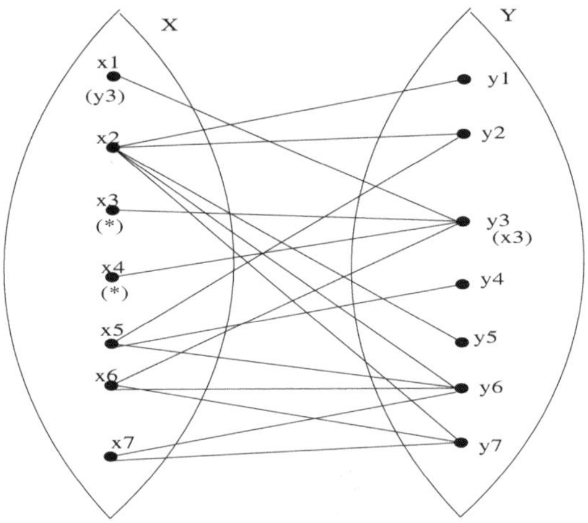

Figure 16.7: A non-breakthrough

16.4 An application of the SDR theorem: doubly stochastic matrices

Recall that a bipartite graph $G = (X, Y, \Delta)$ is said to be *k-regular* if each vertex of G has degree equal to k. Let $k \geq 1$. Then making a two-way counting of the set of edges of G obtains $k \times |X| = k \times |Y|$ and hence $|X| = |Y| = n$, say. It is also clear that the number of edges in G is nk. Since each vertex can cover no more than k edges, it follows that the minimum size of a cover is $\frac{nk}{k} = n$ and hence (using Theorem 16.2.7) G has a matching with n edges, that is, a perfect matching. An equivalent approach to this is through matrices which is what we now discuss.

Definition 16.4.1. Let $A = [a_{i,j}]$ denote an $n \times n$ matrix. A is called *a doubly stochastic matrix* if the following conditions are satisfied.

1. $a_{i,j}$ is a real number and $a_{i,j} \geq 0 \; \forall \, i, j$.
2. $\sum_{j=1}^{n} a_{i,j} = 1 \; \forall \, i = 1, 2, \ldots, n$, that is all the row sums in A are equal to 1.
3. $\sum_{i=1}^{n} a_{i,j} = 1 \; \forall \, j = 1, 2, \ldots, n$, that is all the column sums in A are equal to 1.

Here is an example of a doubly stochastic (d.s.) matrix of order 3.

$$\begin{pmatrix} 0.2 & 0.7 & 0.1 \\ 0.4 & 0.3 & 0.3 \\ 0.4 & 0 & 0.6 \end{pmatrix}$$

Doubly stochastic matrices occur naturally in many different settings in mathematics. In statistics, they appear as probabilities when we deal with Markov chains. An important class of doubly stochastic (d.s.) matrices is the class of permutation matrices whose number is finite for any given n.

Definition 16.4.2. *A convex combination* of matrices $M_1, M_2, \ldots, M_r$ is a matrix M of the form $M = \sum_{j=1}^{r} \alpha_j M_j$ where $0 \leq \alpha_j \leq 1$ for each j and $\sum_{j=1}^{r} \alpha_j = 1$. Let S be a set of matrices. Then *the convex hull H of S* is the set of all those matrices M that consist of convex combinations of (some of the) matrices in S.

The idea of a convex hull H is somewhat similar to that of a 'linear span'. Every matrix in H is a convex combination of matrices in S (and these are the only members of H). S is thus like a basis (for the span H). It is clear that if we take S to be the set of all the permutation matrices, then any convex combination of the members of S must be a doubly stochastic matrix. The following theorem shows that the converse is also true.

Theorem 16.4.3. *(Birkhoff and von Neumann theorem) Every doubly stochastic matrix is a convex combination of the set of permutation matrices hence the set of all the doubly stochastic matrices is the convex hull of the set of permutation matrices.*

Proof Let $A = [a_{i,j}], i, j = 1, 2, \ldots, n$ denote a doubly stochastic matrix. We have to show that we can find permutation matrices $P_1, P_2, \ldots, P_k$ such that $A = \sum_{j=1}^{k} r_j P_j$ is a convex combination. Let the content of A, denoted by $c(A)$ stand for the number

$$|\{(i, j) : a_{i,j} > 0, \ \ i, j = 1, 2, \ldots, n\}|$$

Note that this definition is made for any matrix with non-negative entries. Since each row (column) of A (where A is a d.s. matrix) must have at least one positive entry, it follows that $c(A) \geq n$ with equality iff each row and each column contains exactly one positive entry, which must be 1 and hence $c(A) = n$ iff A is itself a permutation matrix. In that case, we do not have to prove anything. So let $c(A) \geq n + 1$ and we make induction on $c(A)$ (with $c(A) = n$ forming a basis). Thus assume that the assertion of the theorem is true for a d.s. matrix C with $c(C) < c(A)$. Draw a bipartite graph $G = (X, Y, \Delta)$ where $X = \{x_1, x_2, \ldots, x_n\}$ and $Y = \{y_1, y_2, \ldots, y_n\}$ and $(x_i y_j) \in \Delta$ if $a_{i,j} > 0$. We wish to show that G has a perfect matching. By a line of A, we mean either a row or a column of A. Since an edge in G corresponds to a cell in A with a positive entry, it follows that a (vertex) cover in G actually is a collection of lines of A that cover all the positive entries in A among them. Hence G has a cover of size $n - 1$ or less iff we have a cover of all the positive entries in A by $n - 1$ or fewer lines. This is not possible since each line sum is 1 and the sum of all the (positive) entries in A is n. Thus no set of $n-1$ lines can cover all the positive entries in A and hence the minimum size of a cover in G is n. This shows (using Theorem 16.2.7) that G has a matching with n edges. Thus G has a perfect matching say M. The perfect matching M is nothing but a pairing of the sets X and Y (with x_i paired with y_j if the edge $(x_i y_j)$ is in M and hence is the same thing as a permutation α on $[n]$ with the property that $a_{i,\alpha(i)} > 0$ for every $i = 1, 2, \ldots, n$. Let

$$r = \min\{a_{1,\alpha(1)}, a_{2,\alpha(2)}, \ldots, a_{n,\alpha(n)}\}$$

Then $r > 0$ and let $r = a_{m,\alpha(m)}$. Let $P = [p_{i,j}]$ be the permutation matrix corresponding to the permutation α (thus $p_{i,\alpha(i)} = 1$ for every i and $p_{i,j} = 0$ for any other (i,j)). Let $B = A - rP$. Since A is not a permutation matrix, $r < 1$. Hence B is a matrix with non-negative entries. Also,

$$\sum_{j=1}^{n} b_{i,j} = \sum_{j=1}^{n} a_{i,j} - r = 1 - r$$

and similarly, $\sum_{i=1}^{n} b_{i,j} = 1 - r$ showing that the row and column sums in B are equal to $1 - r > 0$. Further the content of B omits the cell $(m, \alpha(m))$ and thus $c(B) < c(A)$. Let $D = (1-r)^{-1}B$. Then the row and column sums in D are all equal to 1 and $c(D) = c(B) < c(A)$ and hence induction applies to get

$$D = t_2 P_2 + t_3 P_3 + \cdots + t_k P_k$$

where each P_i is a permutation matrix and $t_i \geq 0$ with $t_2 + \cdots + t_k = 1$. We thus have

$$\begin{aligned} A &= rP + (1-r)D \\ &= rP + (1-r)[t_2 P_2 + t_3 P_3 + \cdots + t_k P_k] \\ &= r_1 P_1 + r_2 P_2 + + \cdots + r_k P_k \end{aligned}$$

where $r_1 = r$ and $r_j = (1-r)t_j$ for $j = 2, \ldots, k$. Hence $r_1 + r_2 + \cdots + r_k$ equals $r + (1-r)[\sum_j t_j] = r + (1-r) = 1$ and we have obtained A is obtained as a convex combination as desired. □

16.5 Posets and Dilworth's theorem

The matching theorem or the SDR theorem is a statement in optimization which has the following form. The problem of finding a maximum in a certain set up can be solved (and is in fact equivalent to) by solving the problem of finding a minimum of in some other set up. The class of such kinds of problems is studied in the general theory of duality in combinatorial optimization. In this area, the most versatile theorem is the max flow min cut theorem (or the flow theorem) which we do not intend to study in this book. The flow theorem, in fact, obtains the matching theorem as a special case. We content ourselves by discussing a nice result called Dilworth's theorem in the theory of partially ordered sets. Apart from being a very interesting result in the theory of partially ordered sets, Dilworth's theorem will also give an alternative short proof of the matching theorem. We begin with the following set up.

Definition 16.5.1. Let S be a non-empty set and let $\leq$ be a relation on S. That is, $\leq$ is a subset of the Cartesian product $S \times S$ and we conveniently write $a \leq b$ to mean the pair (a, b) is in the relation. A relation $\leq$ on S is called *a partial order* on S if the following axioms are satisfied.

(a) *Reflexivity* $\forall\ a \in S$, we must have $a \leq a$.

(b) *Antisymmetry* $a \leq b$ and $b \leq a$ together imply $a = b$.

(c) *Transitivity* $a \leq b$ and $b \leq c$ together imply $a \leq c$.

A set S equipped with a partial order $\leq$ is called a *a partially ordered set or a poset* and we denote it by $(S, \leq)$ or just by S when the relation $\leq$ is clear. For two elements a and b of S, we write $a < b$ to mean $a \leq b$ and $a \neq b$.

Partial order is a natural concept and myriad examples of posets occur in mathematics. Here are some important ones.

Example 16.5.2. Let X be a set and let $\mathbf{P}$ denote the set of all the subsets of X ordered by inclusion. Thus, if A and B are subsets of X, then we define $A \leq B$ if $A \subseteq B$. Then $\mathbf{P}$ is called *the Boolean poset* and we already know that $\mathbf{P}$ has 2^n members if $|X| = n$.

Example 16.5.3. Let V denote a finite dimensional vector space over a field K. The members of the poset $\mathbf{P}$ are the subspaces of V ordered by inclusion.

Example 16.5.4. Let N denote a natural number and let $\mathbf{P}$ be the set of all the divisors of N where two members a, b of $\mathbf{P}$ have $a \leq b$ if a divides b. If the prime factorization of N is ${p_1}^{a_1}{p_2}^{a_2}\cdots{p_r}^{a_r}$, then $\mathbf{P}$ has $\prod_i a_1 + 1$ members.

Example 16.5.5. Let n denote a fixed natural number and let $\mathbf{P}$ denote the set of all the unordered partitions α of $[n]$. Order the partitions by refinement: $\alpha \leq \beta$ if β is obtained from α by breaking (some) parts in α.

The topic of combinatorics of posets is too vast and deep to be contained in our brief treatment. We need the following definitions for our purpose.

Definition 16.5.6. Let $(\mathbf{P}, \leq)$ denote a poset and let $x, y \in \mathbf{P}$. Then x and y are said to be *comparable* if either $x \leq y$ or $y \leq x$ holds. A subset C of $\mathbf{P}$ is called a *chain* if every two elements in C are comparable. In contrast, a subset A of $\mathbf{P}$ is called an *antichain* if every two elements in C are incomparable (not comparable).

Example 16.5.7. Let $\mathbf{P}$ be the Boolean poset of all the subsets of the set $[9]$ ordered by inclusion. Then

$$\begin{aligned} C &= \{\{2,3\},\{2,3,4,7,9\},\{2,3,4,6,7,9\},\{2,3,4\}\} \\ &= \{\{2,3\},\{2,3,4\},\{2,3,4,7,9\},\{2,3,4,6,7,9\}\} \end{aligned}$$

is a chain. In general, if C is a finite chain with k elements, we can arrange the elements of C in the form $\{x_1, x_2, \ldots, x_k\}$ where $x_1 \leq x_2 \leq \cdots \leq x_k$. Similarly,

$$A = \{\{2,3\},\{3,4\},\{4,7,8\},\{4,6,8,9\}\}$$

is an antichain in $\mathbf{P}$ since no two subsets in A are comparable. Finally,

$$R = \{\{1,3\},\{1,3,7\},\{2,3,7\}\}$$

is neither a chain nor an antichain since $\{1,3\} \subsetneqq \{1,3,7\}$ but $\{1,3\}$ and $\{2,3,7\}$ are not comparable.

Definition 16.5.8. Let $\mathbf{P}$ be a poset and let C be a chain in $\mathbf{P}$. Then C is called a *maximal chain* if we cannot enlarge C to a larger chain. That is, C is maximal, if there is no z such that $C \cup \{z\}$ is also a chain and $z \notin C$. A *chain decomposition of* $\mathbf{P}$ *(into* r *chains)* is an expression of the form

$$\mathbf{P} = \cup_{i=1}^{r} C_i$$

where each C_i is a chain and C_i and C_j have no common element if $i \neq j, i, j = 1, 2, \ldots, r$. A chain decomposition is called a *minimum chain decomposition* if it uses the smallest number of chains.

Example 16.5.9. Let $\mathbf{P}$ denote the Boolean poset of all the subsets of $[7]$. Then

$$C = \{\emptyset, \{3\}, \{3,4\}\{3,4,7\}, \{1,3,4,7\}\{1,3,4,6,7\},$$
$$\{1,3,4,5,6,7\}, \{1,2,3,4,5,6,7\}\}$$

is a maximal chain while

$$C' = \{\emptyset, \{2,3\}, \{2,3,7\}\{1,2,3,5,6,7\}, \{1,2,3,4,5,6,7\}\}$$

is not since any one of the subsets $\{2\}$ or $\{1,2,3,7\}$ could be added to C' to obtain a larger chain. Observe that the set A of all the 3-subsets of $[7]$ is an antichain in $\mathbf{P}$ since no two different sets of the same size are comparable.

Let A be an antichain in a poset $\mathbf{P}$. If we have any chain decomposition of $\mathbf{P}$, then each chain C in such a decomposition can pick up at the most one member of A and hence any chain decomposition of $\mathbf{P}$ must use at least $|A|$ chains. We formulate this in the following obvious lemma.

Lemma 16.5.10. *Let α denote the largest size of an antichain in a poset $\mathbf{P}$ and let β denote the number of chains in a minimum chain decomposition of $\mathbf{P}$. Then $\alpha \leq \beta$.*

Proof Let $|A| = \alpha$ where A is an antichain. Then each element of A must belong to a unique chain in the chain decomposition while no chain in the chain decomposition can contain two elements of A (else, they will become comparable, a contradiction). This two-way counting shows that we have at least α chains in the chain decomposition. □

For the Boolean poset $\mathbf{P}$ consisting of all the subsets of the set $[n]$ we have already seen that all the subset of the same size $m = \lfloor \frac{n}{2} \rfloor$ constitute an antichain A. Since the binomial coefficient $\binom{n}{k}$ takes its largest value when $k = m$ we intuitively feel that this is a maximum sized antichain. Consider also the following question. What is a minimum chain decomposition of $\mathbf{P}$? Both the questions get solved by connecting them with each other in the following result.

Theorem 16.5.11. *(Dilworth's theorem) Let $\mathbf{P}$ be a finite poset and let the largest size of an antichain in $\mathbf{P}$ be k. Then the minimum number of chains in a chain decomposition of $\mathbf{P}$ is also equal to k.*

Proof Let the minimum number of chains in a chain decomposition of $\mathbf{P}$ be k'. Then we have already seen (Lemma 16.5.10)that $k \leq k'$ and hence it suffices to prove that $k' \leq k$. To that end, let the largest size of an antichain in $\mathbf{P}$ be k. Using induction on $|\mathbf{P}|$, we prove the existence of a chain decomposition of $\mathbf{P}$ into at the most k disjoint chains. First, as a basis of induction, let $|\mathbf{P}| = 1$. Then $k = k' = 1$. So, let $|\mathbf{P}| \geq 2$ and we assume that the induction hypothesis holds for all posets P' for which the number of elements strictly less than $|P|$.
Let C be a maximal chain in $\mathbf{P}$. Let $\mathbf{Q} = \mathbf{P} - C$. Then $\mathbf{Q}$ is also a poset. Let the largest size of an antichain in $\mathbf{Q}$ be m. Since $\mathbf{Q} \subset \mathbf{P}$, we must have $m \leq k$. If $m \leq k-1$, then by induction (since $|\mathbf{Q}| < |\mathbf{P}|$), $\mathbf{Q}$ has a chain decomposition that uses no more than m chains say $\mathbf{Q} = \cup_{i=1}^{m} C_i$ and hence $\mathbf{P} = C \cup (\cup_{i=1}^{m} C_i)$ is a chain decomposition of $\mathbf{P}$ into $m + 1 \leq k$ chains as desired. Hence we are forced to assume that $m = k$. Let $A = \{a_1, a_2, \ldots, a_k\}$ be a (maximum) antichain in $\mathbf{Q}$. We now form two subsets U and L (called the upper and lower shadow of A) as follows.

$$U = \{x \in \mathbf{P} : x \geq a_i \; for\; some \; i = 1, 2, \ldots, k\}$$

$$L = \{y \in \mathbf{P} : y \leq a_j \; for\; some \; j = 1, 2, \ldots, k\}$$

Since A is also a maximum antichain in $\mathbf{P}$, each element of $\mathbf{P}$ belongs to U or L. Further, by the definition of U and L, we have $A \subset U \cap L$. Let $z \in U \cap L$. Then for some i and j we have $z \geq a_i$ and $z \leq a_j$. Therefore, $a_i \leq z \leq a_j$. This implies $a_i = a_j$ which forces $i = j$ since A is an antichain showing that $a_i = z = a_j$. Hence $z \in A$. We have thus proved that $U \cap L$ equals A. Now let w be the smallest element of C. If $w \in U$, then for some j, we have $w \geq a_j$. Since $A \subset P - C$, clearly $a_j \notin C$. Therefore, if $w > a_j$, then we get a larger chain $C \cup \{a_j\}$ contradicting the maximality of the chain C. Hence we get $w = a_j$. But then $a_j \in C$, which is also a contradiction. We thus see that the minimal element w of C cannot be in U and hence $|U| < |\mathbf{P}|$. A similar argument shows that the largest element of C is not in L. Therefore both U and L are proper subsets of $\mathbf{P}$. By induction hypothesis, (since neither U nor L can have antichains of size larger than k) we get two chain decompositions:

$$U = \cup_{i=1}^{k} D_i \text{ and } L = \cup_{i=1}^{k} E_i$$

Notice that $A \subset U$ and $|A| = k$ implies that each D_i contains a unique element of A and similarly each E_i also has a unique element of A. By relabeling the chains in the chain decompositions, we may assume that $a_i \in D_i$ and $a_i \in E_i$. Then a_i is the smallest element of D_i and is also the largest element of E_i for every i. We now let $C_i = D_i \cup E_i$. Then each C_i is a chain and $\mathbf{P} = \cup_{i=1}^{k} C_i$ is a chain decomposition of $\mathbf{P} = L \cup U$ as desired. □

As an immediate application of Dilworth's theorem (Theorem 16.5.11), we give *a different proof of the matching theorem (Theorem 16.1.3)*. The set up is as follows. We are given a bipartite graph $G = (X, Y, \triangle)$ with $|X| = m$ and $|Y| = n$. For a subset $B = \{x_{i_1}, x_{i_2}, \ldots, x_{i_k}\}$ of X we write

$$N(B) = N(\{x_{i_1}, x_{i_2}, \ldots, x_{i_k}\}) = \cup_{r=1}^{k} N(x_{i_r})$$

where $N(x) = \{y \in Y : (xy) \in \triangle\}$. Then the marriage condition (16.1) implies

$$|N(B)| \geq |B| \;\; \forall\, B \subset X \tag{16.6}$$

Construct a poset $\mathbf{P}$ as follows. $\mathbf{P} = X \cup Y$ and thus $\mathbf{P}$ has $m + n$ elements. The relation $\leq$ is defined as follows: $\forall\, a \in P$, we have $a \leq a$. Further, given $x \in X$ and $y \in Y$ we have $x \leq y$ iff $(xy) \in \triangle$. Note that these are the only relations. In particular, all the elements of m elements of X are incomparable and so are all the n elements of Y. Hence Y is an antichain in the poset $\mathbf{P}$. How does a (non-empty) chain in $\mathbf{P}$ look like? Either it is a singleton $\{a\}$ or consists of two elements $\{x, y\}$ where $(xy) \in \triangle$. In particular, we have no chains with three or more elements. By taking $B = X$in the marriage condition (16.6), we see that $|Y| \geq |N(X)| \geq |X|$ and hence $n \geq m$ and therefore we expect Y to be a maximum antichain in $\mathbf{P}$. To check the validity of that statement, let A be an antichain with k elements from X and r elements from Y. Thus A has the form

$$A = \{x_{i_1}, x_{i_2}, \ldots, x_{i_k}\} \cup \{y_{j_1}, y_{j_2}, \ldots, y_{j_r}\}$$

where x's are from X and y's are from Y. Since A is an antichain, there is no relation between $X \cap A$ and $Y \cap A$. Hence we see that

$$N(\{x_{i_1}, x_{i_2}, \ldots, x_{i_k}\}) \subset Y - \{y_{j_1}, y_{j_2}, \ldots, y_{j_r}\}$$

The crucial marriage condition (16.6) implies that

$$k \leq |N(\{x_{i_1}, x_{i_2}, \ldots, x_{i_k}\})| \leq n - r$$

and hence $k \leq n - r$ showing that $|A| = k + r \leq n$. Therefore the largest antichain in $\mathbf{P}$ has size n and no bigger. Using Dilworth's theorem (Theorem 16.5.11), we have a chain decomposition of $\mathbf{P}$ into n chains. How would that look like? We have some

k chains C_i in this chain decomposition each of which is a doubleton, some t singleton chains D_j with elements coming from X and some s singleton chains E_j with elements from Y:

$$P = (C_1 \cup \cdots \cup C_k) \cup (D_1 \cup \cdots \cup D_t) \cup (E_1 \cup \cdots \cup E_s)$$

Since the total number of chains is n and since $|P| = m + n$, we get:

$$k + t + s = n \ \textit{and} \ \ 2k + t + s = m + n$$

and hence we must have $k = m$ (which implies that $t = 0$). Therefore, w.l.o.g. the chain decomposition is

$$\{\{x_{i_1}, y_{j_1}\}, \{x_{i_2}, y_{j_2}\}, \ldots, \{x_{i_m}, y_{j_m}\}, \{y_{j_{m+1}}\}, \ldots, \{y_{j_n}\}\}$$

Since this is a chain decomposition, each $\{x_{i_p}, y_{j_p}\}$ represents an edge $(x_{i_p} y_{j_p})$. Further all the the y_{j_p} are distinct and so are all the x_{i_p} and hence

$$\{x_{i_1}, x_{i_2}, \ldots, x_{i_m}\} = X$$

showing that

$$M = \{(x_{i_1} y_{j_1}), (x_{i_2} y_{j_2}), \ldots, (x_{i_m} y_{j_m})\}$$

is a matching saturating each vertex of X as desired. □

We now look at a related question, though we do not apply Dilworth's theorem to answer that question. Given a finite poset **P**, what is the size of its largest antichain? The following theorem called Sperner's theorem (Theorem 16.5.13), gives the answer to this question when **P** is the Boolean poset of all the subsets of the set $[n]$. Apart from being an important theorem, the proof technique of this theorem is based on a general inequality called *the LYM inequality*. This inequality also holds for the posets given by subspaces of a finite dimensional vector space over a finite field. Hence what has prompted us to include Sperner's theorem (Theorem 16.5.13)in the book is its proof based on the important LYM inequality.

Theorem 16.5.12. *(LYM inequality for the Boolean poset) Let* **A** *be an antichain in the poset of all the subsets of* $[n]$. *Let* α_k *denote the number of subsets of size* k *in* **A**. *Then*

$$\sum_{k=0}^{n} \frac{\alpha_k}{\beta_k} \leq 1$$

where $\beta_k = \binom{n}{k}$.

Proof Let C be a maximal chain in **P**. Since we are free to insert or delete elements of X at our will, it follows that C must have a subset of every possible k where $0 \leq k \leq n$. That is, C has the form

$$C = \{\emptyset, \{i_1\}, \{i_1, i_2\}, \ldots, \{i_1, i_2, \ldots, i_k\}, \ldots, \{i_1, i_2, \ldots, i_n\}\}$$

where $i_1, i_2, \ldots, i_n$ are all distinct elements of $X = [n]$. It thus follows that in the one-line notation of a permutation, C determines the unique permutation $\sigma = i_1 i_2 \cdots i_n$ where we write the elements in the order in which they get added to the sets in C. Conversely, given a permutation $\sigma \in S_n$, we determine a unique maximal chain C and hence the number of maximal chains is equal to the number of permutations, which is equal to $n!$. Let $\mathbf{C}$ denote the set of all the maximal chains. We then make a two-way counting of the set

$$S = \{(Y, C) : Y \in \mathbf{A},\ \ C \in \mathbf{C} \text{ and } Y \in C\}$$

On the one hand, each $C \in \mathbf{C}$ determines at the most one member Y of $\mathbf{A}$ and hence, $|S| \leq n!$. On the other hand fix a $Y \in \mathbf{A}$ and let $|Y| = k$. If $Y = \{i_1, i_2, \ldots, i_k\}$, then to construct a maximal chain C, we must place Y at the k-th level in C. Thus a typical C containing Y has the form

$$C = \{\emptyset, \{i_1\}, \{i_1, i_2\}, \ldots, \{i_1, i_2, \ldots, i_k\} = Y, \ldots, \{i_1, i_2, \ldots, i_n\}\}$$

It follows that C needs to be chosen in a two-stage process. We choose a permutation of the set $\{i_1, i_2, \ldots, i_k\}$ followed by a permutation of the set $X - \{i_1, i_2, \ldots, i_k\}$. Clearly the first permutation can be chosen in $k!$ ways and the second in $(n-k)!$ ways. Therefore any given Y in $\mathbf{A}$ of size k determines exactly $k!(n-k)!$ members of $\mathbf{C}$ that contain it and hence the number of elements of S determined by Y is $k!(n-k)!$. Therefore,

$$|S| = \sum_k \alpha_k k!(n-k)!$$

We therefore have

$$\sum_k k!(n-k)!\alpha_k \leq n!$$

and hence

$$\sum_k \frac{\alpha_k}{\binom{n}{k}} \leq 1$$

□

Theorem 16.5.13. *(Sperner's theorem) The largest size of an antichain in the in the Boolean poset* $\mathbf{P}$ *consisting of all the subsets of the set* $[n]$ *is* $\binom{n}{\lfloor \frac{n}{2} \rfloor}$.

Proof Let $X = [n]$ and let $\mathbf{A}$ denote an antichain with the largest number of subsets. Since all the subsets of order $\lfloor \frac{n}{2} \rfloor$ form an antichain of $\mathbf{P}$, it follows that $|\mathbf{A}| \geq \binom{n}{\lfloor \frac{n}{2} \rfloor}$. Hence, it will suffice to show that $\mathbf{A}$ cannot have more than $\binom{n}{\lfloor \frac{n}{2} \rfloor}$ members. Let α_k denote the number of subsets of order k in the antichain $\mathbf{A}$ so that $\sum_{k=0}^{n} \alpha_k = |\mathbf{A}|$. Let $\beta_k = \binom{n}{k}$ denote the number of subsets of order k in the poset $\mathbf{P}$. Then the largest value of β_k is $\beta = \binom{n}{\lfloor \frac{n}{2} \rfloor}$. The remainder of the proof is a direct consequence of the LYM inequality (Theorem 16.5.12). We note that the LYM inequality proves the assertion of the theorem: Since $\beta^{-1} \leq \beta_k^{-1}$, we have $|\mathbf{A}| = \sum_{k=0}^{n} \alpha_k \leq \beta$ which is the assertion of the theorem. □

16.6 From finite to the infinite

Is there a version of the matching theorem if $G = (X, Y, \triangle)$ with X an infinite set? Unfortunately, this is not true if we allow vertices of X to have infinite degrees as the following Example shows.

Example 16.6.1. Let $\mathbf{A} = \{A_1, A_2, \ldots, A_n, \ldots\}$ be a family of subsets indexed by the set of natural numbers $\mathbb{N}$ and let $A_1 = \{1, 2, \ldots, n, \ldots\} = \mathbb{N}$ and let

$$A_2 = \{1\}, A_3 = \{1, 2\}, \ldots, A_n = \{1, 2, \ldots, n-1\}, \ldots$$

Clearly the union of any k sets has at least k elements and hence the marriage condition (16.1) is satisfied. However, $\mathbf{A}$ does not have an SDR: In any such SDR, A_2 gets represented by 1, and then A_3 must be represented by 2, and continuing in this manner, A_n must be represented by $n-1$ and so on. Since this exhausts all the elements of $\mathbb{N}$, we have no element left to represent A_1. Thus $\mathbf{A}$ has no SDR, though the marriage condition is satisfied. The correct version of the Hall's SDR theorem for infinite sets is given by the following.

Theorem 16.6.2. *(SDR theorem for an infinite family of sets) Let $\mathbf{A} = \{A_i : i \in I\}$ denote a family of finite sets. Let, for every natural number k and for every subfamily J of I with $|J| = k$, the marriage condition (16.7) given below hold:*

$$|\cup_{i \in J} A_i| \geq k \tag{16.7}$$

Then $\mathbf{A}$ has an SDR.

Notice that if we have to formulate the condition (16.7) in terms of a bipartite graph $G = (X, Y, \triangle)$, then we require, besides (7) which requires that in the neighborhood of k vertices in X has size at least k, as also the requirement that the degree of each vertex of X be finite (this is called local finiteness). We have to postpone the proof of Theorem 16.6.2 since we need to first develop the necessary background. The proof of the infinite avatar of the matching theorem, of course, uses the finite form which we have already proved. But more important than that is the device that makes the transition from finite to infinite possible. This device called *the Rado selection principle* (or in its equivalent form called the Tychnoff's theorem in topology) is logically equivalent to the foundational statements in mathematical logic called the Zorn's Lemma and the axiom of choice. In what follows we plan to briefly describe Tychnoff's theorem and the derivation of the Rado selection principle from Tychnoff's theorem. We expect the reader to be familiar with the basic notions in topology including that of a compact topological space. The following statement is assumed without proof.

Theorem 16.6.3. *(Tychnoff's theorem) Let $\{X_i : i \in I\}$ be a family of compact topological spaces. Then the space $\prod_{i \in I} X_i$ is a compact topological space in the product topology.*

Note that the elements of the product space $\prod_{i \in I} X_i$ are the choice functions θ; thus θ maps each set X_i to an element of X_i. A typical θ is $(\theta(i) : i \in I)$ where $\theta(i) \in X_i$ for every $i \in I$.

Definition 16.6.4. Let $\mathbf{A} = \{A_r : r \in R\}$ be a family of non-empty sets. A *choice function* θ *on* $\mathbf{A}$ is a function $\theta : R \longrightarrow \cup_{r\in R} A_r$ such that for every $r \in R$, we have $\theta(r) \in A_r$.

In many combinatorial applications of Tychnoff's theorem, the spaces X_i are finite and hence are trivially compact. The following theorem provides us with a satisfactory answer to the question of a universal choice function.

Theorem 16.6.5. *(Rado Selection Principle) Let* $\mathbf{A} = \{A_i : i \in I\}$ *be a family of non-empty finite sets* A_i. *Let* $\mathbf{J}$ *denote the set of all the finite subsets of* I. *Assume that for every* $J \subset I$ *such that* $J \in \mathbf{J}$, *we are given a choice function* θ_J *on the family of sets* $\{A_i : i \in J\}$. *Then there is a universal choice function* θ *on* $\mathbf{A}$ *with the property that for every* $J \in \mathbf{J}$, *there exists a* $K \in \mathbf{J}$ *such that* $K \supset J$ *and* $\theta|J = \theta_K|J$.

Proof Fix $J \in \mathbf{J}$. *Let us call a universal choice function* α *on* I, *good for* J *if there is some finite subset* K *of* I *such that* $\alpha|J = \theta_K|J$. We have to prove the existence of a universal choice function α which is *good for every member of* $\mathbf{J}$. Let F_J denote the set of all the choice functions α on I with the property that $\alpha|J = \theta_K|J$ for some $K \supset J$ with $K \in \mathbf{J}$. Thus F_J consists of all the universal choice functions that are good for J. Let β be a choice function on I and let $S_J(\beta)$ denote the subset of $X = \prod_{i\in I} A_i$ given by $\prod_{i\in I} B_i$ where $B_i = \{\beta(i)\}$ if $i \in J$ and $B_i = A_i$ if $i \notin J$. Thus we have

$$S_J(\beta) = \{\phi \in X : \phi|J = \beta|J\}$$

Note that $\beta \in S_J(\beta)$. Further, given choice functions α and β, we have $\alpha \in S_J(\beta)$ iff $\beta \in S_J(\alpha)$ and then $S_J(\alpha) = S_J(\beta)$. Let $\alpha \in S_J(\beta)$. Then $\alpha \in F_J$ iff $\beta \in F_J$. We claim that

$$X - F_J = \sum_{\beta \in X - F_J} S_J(\beta)$$

If β is on the L.H.S. then $\beta \in S_J(\beta)$ implies that β is on the R.H.S. Conversely let α be on the R.H.S. Then $\alpha \in S_J(\beta)$ for some $\beta \in X - F_J$ and since $\beta \notin F_J$, we have $\alpha \notin F_J$ so that α is on the L.H.S. proving the claim. Since $S_J(\beta)$ is an open set in the product topology X, so is $X - F_J$.
Suppose for the sake of contradiction that $\cup_{J\in\mathbf{J}} X - F_J = X$. Then $\{X - F_J : J \in \mathbf{J}\}$ is an open cover of X which must have a finite subcover. Hence we can find some $J_1, J_2, \ldots, J_m \in \mathbf{J}$ such that

$$\begin{aligned} X &= (X - F_{J_1}) \cup \cdots \cup (X - F_{J_m}) \\ &= X - (F_{J_1} \cap F_{J_2} \cap \cdots \cap F_{J_m}) \end{aligned}$$

and hence $F_{J_1} \cap \cdots \cap F_{J_m} = \emptyset$. Let $J = J_1 \cup \cdots \cup J_m$. Then $J \in \mathbf{J}$ and $F_J \subset F_{J_1} \cap \cdots \cap F_{J_m}$. But $F_J \neq \emptyset$ and this gives the required contradiction. Thus we have shown $\cup_{J\in\mathbf{J}} X - F_J$ is a proper subset of X and hence there is a choice function γ such that γ does not belong to any $X - F_J$ and therefore γ is in every F_J as desired. □

Proof of the infinite avatar of the SDR theorem Let $\mathbf{J}$ denote the set of all the finite subsets of the indexing set I. For each $J \in \mathbf{J}$, let $C_J = \{A_i : i \in J\}$ denote the finite family of sets indexed by J. Then condition (16.1) holds for C_J and hence using the finite version of the marriage theorem (Theorem 16.1.3), we see that C_J has an SDR say θ_J. If $J = \{i_1, i_2, \ldots, i_k\}$ then θ_J is a mapping from the set $\{A_{i_1}, A_{i_2}, \ldots, A_{i_k}\}$ to $A_{i_1} \cup A_{i_2} \ldots \cup A_{i_k}$ such that $\theta_J(A_{i_r}) \in A_{i_r}$ and if $r \neq s$, then $\theta_J(A_{i_r}) \neq \theta_J(A_{i_s})$. Using the Rado selection principle, we have a choice function

$$\theta : \{A_i : i \in I\} \longrightarrow \cup_{i \in I} A_i$$

such that $\forall\ J \in \mathbf{J}$ there exists a $K \in \mathbf{J}$ with the property that $K \supset J$ and $\theta | J = \theta_K | J$. We claim that θ gives the required SDR. Since for each $i \in I$, we have $\theta(A_i) \in A_i$, it enough to verify that for $r \neq s$ we must have $\theta(A_r) \neq \theta(A_s)$. Let $J = \{r, s\}$. Then we have a $K \supset J$ such that $K \in \mathbf{J}$ and $\theta | J = \theta_K | J$. Since θ_K is an SDR, we must have $\theta(A_r) \neq \theta(A_s)$ as desired. □

Conclusion The matching theorem as well as the Hungarian algorithm are special cases of the max flow min cut theorem (or the flow theorem for short) and the Ford-Fulkerson augmenting path algorithm respectively. We do not intend to cover the flow theorem but refer the reader to West [60]. The complexity of the Ford-Fulkerson algorithm is polynomial and because of the importance of the problem, much research work is directed towards finding better algorithms to obtain an optimal matching (optimal network flow). In short, this is a major part of combinatorial optimizations. The proof of Rado selection principle is from Mirsky [42]. Rado selection principle is among the most powerful tool in the passage from finite combinatorics to infinite combinatorics. We have illustrated this in the proof of the infinite avatar of the matching theorem (or the SDR theorem). As an example, the infinite version of the famous four colour theorem (every planar map can be coloured in 4 colours) is true because the finite version of the 4-colour theorem (due to Appel and Haken [60]) is true. Rado selection principle (as well as Tychnoff's theorem) can be used to prove an infinite version of Dilworth's theorem. Finally a word on the combinatorics of finite sets. Sperner's theorem is just the beginning of this area of research that has many interesting results such as the Erdös-Ko-Rado theorem and the Kruskal-Katona theorem. It is possible to use Dilworth's theorem to prove Sperner's theorem. This is done by actually decomposing the Boolean poset into

$$\binom{n}{\lfloor \frac{n}{2} \rfloor}$$

symmetric chains that begin with a set of size r and end at a set of size $n - r$. This was done by de Bruijn and others and we have illustrated this in Exercise 16.37.

16.7 Exercises for Chapter 16

16.1 Find the total number of SDRs for the family $\mathbf{A}$ of Example 16.1.1.

16.2 Given a pruned chessboard C, we wish to cover it by non-overlapping dominoes so that each cell is covered by some domino (and there is no overlap of dominoes on any cell). Such a cover of C by dominoes is called a perfect cover. Let A denote a subset of the set B of black cells of C and let W denote the set of white cells of C. If a is a black cell, then we let $w(a)$ denote the set of white cells that are adjacent to a (adjacency here means either horizontally or vertically adjacent and thus $|w(a)| \leq 4$). Let $w(A) \cup_{a\in A} w(a)$. Prove that C has perfect cover by dominoes iff $|B| = |W|$ and forall subsets A of B, we have $|w(A)| \geq |A|$.

16.3 Let $\mathbf{A} = \{A_1, A_2, \ldots, A_n\}$. For each one of the following, find the total number of SDRs of the family $\mathbf{A}$

(a) $A_i = [n] - \{i\}$ for every i.

(b) $A_1 = \{1, 2\}, A_2 = \{2, 3\}, A_3 = \{3, 4\}, \ldots, A_{n-1} = \{n-1, n\}$ and $A_n = \{n, 1\}$.

(c) $A_1 = \{1, 2\}, A_2 = \{2, 3\}, A_3 = \{3, 4\}, \ldots, A_{n-1} = \{n-1, n\}$ and $A_n = \{n, n+1\}$.

(d) $A_i = \{1, 2, \ldots, i\}$ for all $i = 1, 2, \ldots, n$.

(e) $A_i = \{1, 2, \ldots, i, i+1\}$ for all $i = 1, 2, \ldots, n$.

(f) The family with

$$A_1 = \{1, 2, 3\}, A_2 = \{2, 3, 4\}, \ldots, A_{n-2} = \{n-2, n-1, n\}$$

$$A_{n-1} = \{n-1, n, 1\}, A_n = \{n, 1, 2\}$$

16.4 Let $\mathbf{A} = \{A_1, A_2, \ldots, A_n\}$ be a family of sets. Let for every k and $\forall\ 1 \leq i_1 < i_2 < \ldots < i_k \leq n$, the following inequality hold:

$$|A_{i_1} \cup A_{i_2} \cup \cdots \cup A_{i_k}| \geq k+1$$

Let $y \in A_n$. Show that $\mathbf{A}$ has an SDR of the form $(x_1, x_2, \ldots, x_n)$ where $x_n = y$ represents A_n.

16.5 Let $\mathbf{A} = \{A_1, A_2, \ldots, A_n\}$ be a family of sets and let $(b_1, b_2, \ldots, b_n)$ be a sequence of positive integers. Consider the question of finding an SDR for family of sets

$$\mathbf{C} = \{A_1, A_1, \ldots, A_1, A_2, A_2, \ldots, A_2, \ldots, A_n, A_n, \ldots, A_n\}$$

where the set A_i is repeated b_i times (and thus counting multiplicities, we have $\sum_{i=1}^{n} b_i$ sets in all). Show that

(a) $\exists$ a sequence of subsets $(B_1, B_2, \ldots, B_n)$ satisfying the following properties

(i) $|B_i| = b_i \; \forall \, i = 1, 2, \ldots, n.$

(ii) $B_i \subset A_i \; \forall \, i = 1, 2, \ldots, n.$

(iii) $B_i \cap B_j = \emptyset \; \forall \, i \neq j.$

iff the family **C** has an SDR.

(b) $\exists$ a sequence of subsets $(B_1, B_2, \ldots, B_n)$ satisfying all the properties in (a) iff the following generalized marriage condition is satisfied: $\forall \, m$ and $\forall \, 1 \leq i_1 < i_2 < \cdots < i_m \leq n$ we have

$$|A_{i_1} \cup A_{i_2} \cup \cdots \cup A_{i_m}| \geq b_{i_1} + b_{i_2} + \cdots + b_{i_m}$$

(c) Show that the assertion in (b) implies P.Hall's theorem (Theorem 16.1.3).

16.6 Consider the problem of domino covering of a chessboard.

(a) Let the standard 8×8 chessboard be pruned by removing one white and one black square *anywhere on the board.* Is it possible to cover such a chessboard with 31 dominoes?

(b) Let an $m \times n$ chessboard C with both m and n odd have more white squares than black. Suppose a white square has been removed from anywhere on C. Is it possible to cover this pruned chessboard by dominoes?

16.7 For the following families of sets $\mathbf{A} = \{A_1, A_2, \ldots, A_n\}$ find a largest family of subsets that has an SDR.

(a) $n = 5$ and

$$A_1 = \{a, d\}, A_2 = \{b, d\}, A_3 = \{a, c, d, e\}, A_4 = \{a, d\}, A_5 = \{a, b\}$$

(b) $n = 6$ and

$$A_1 = \{x, y\}, A_2 = \{x\}, A_3 = \{u, v, w\}, A_4 = \{x, y, z\}$$

$$A_5 = \{u, v\}, A_6 = \{z, w\}$$

(c) $n = 4$ and

$$A_1 = \{1, 5\}, A_2 = \{1, 2, 3\}, A_3 = \{3, 4, 5\}, A_4 = \{3, 4, 5\}$$

(d) $n = 4$ and

$$A_1 = \{2, 3\}, A_2 = \{1, 4, 5\}, A_3 = \{1, 2, 4\}, A_4 = \{2, 5\}$$

(e) $n = 4$ and

$$A_1 = \{c\}, A_2 = \{a, c\}, A_3 = \{a, c\}, A_4 = \{a, b, c, d, e\}$$

(f) $n = 7$ and

$$A_1 = \{a, b, f\}, A_2 = \{a\}, A_3 = \{a, b, d, f\}, A_4 = \{a, b\},$$

$$A_5 = \{b, f\}, A_6 = \{d, e, f\}, A_7 = \{a, f\}$$

16.8 Consider the following subsets of the set $\{1, 2, 3, 4, 5\}$.

$$A_1 = \{1, 2\}, A_2 = \{3, 4\}, A_3 = \{5\}, B_1 = \{1, 3\}, B_2 = \{5\}, B_3 = \{2, 4\}$$

Find a permutation σ of $\{1, 2, 3\}$ such that $A_i \cap B_{\sigma(i)}$ are all distinct.

16.9 A 1-factor in a graph G is a set F of edges in G such that no two edges in F share a common vertex and every vertex of G is incident with a unique edge of F. Let k be a positive integer and let G be a bipartite graph which is also k-regular. Prove that the edge set of G can partitioned into k edge disjoint 1-factors.

16.10 $\mathbf{A} = \{A_1, A_2, \ldots, A_n\}$ and let $x \in \cup_{i=1}^{n} A_i$. Let $\mathbf{A}$ have an SDR. Show that there is an SDR in which x occurs as a representative of some A_i. If $x \in A_j$, then is it possible to choose an SDR in which x represents A_j?

16.11 A farmer has divided his 600 acres of farmland property equally among his six sons. The crops grown on this land of 600 acres are wheat, rice, cotton, sugar, corn and sunflower with each one of the six crops being cultivated on an area of 100 acres. Unfortunately, the farm each son inherits does not necessarily have a crop of only one kind but may have crops of two or more kinds and hence the same crop may be grown on the share of two or more sons. The farmer wishes that each son should become a specialist in a particular crop that his farm cultivates. Can this be done so that each son specializes in one crop but no two sons specialize in the same crop? (see [38])

16.12 Consider the following ladder graph G with $2r$ vertices as shown in Figure 16.8.

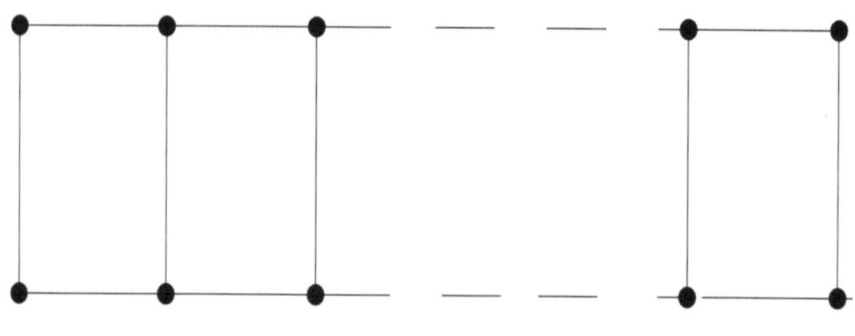

Figure 16.8: The Ladder Graph on $2r$ vertices

where, in general, each vertex has three neighbors except the end vertices that have two neighbors each.

(a) Show that G is a bipartite graph.

(b) Let a_r denote the number of perfect matchings of G (with $2r$ vertices). Obtain a recurrence relation for a_r and solve it.

16.13 Find the largest number of sets in the family $\{A_i | i = 1, 2, \ldots, 10\}$ that have a system of distinct representatives, where

$$\begin{aligned} A_1 &= \{1, 8, 10, 13\} & A_6 &= \{1, 4, 5, 7, 11\} \\ A_2 &= \{5, 8\} & A_7 &= \{8, 13\} \\ A_3 &= \{2, 3, 4, 11, 12\} & A_8 &= \{5, 6, 10, 13\} \\ A_4 &= \{10, 13\} & A_9 &= \{5, 8, 10, 13\} \\ A_5 &= \{1, 5, 8\} & A_{10} &= \{1, 5, 8, 10, 13\} \end{aligned}$$

16.14 What is the maximum number of perfect matchings a bipartite n-vertex graph G can have, *where the maximum is taken over all the n-vertex bipartite graphs*?

16.15 A deck of 52 cards is shuffled and cards are distributed in 26 piles of two cards each. Is it possible to to select a black ace from one pile, a red ace from a second pile, a black two from a third pile, a red two from a fourth pile and so on?

16.16 For the following doubly stochastic matrix A, write A as a convex combination of permutation matrices.

$$\begin{pmatrix} 0 & 0.5 & 0.5 \\ 0.8 & 0.2 & 0 \\ 0.2 & 0.3 & 0.5 \end{pmatrix}$$

16.17 Write the matrix

$$\begin{pmatrix} 1 & 2 & 3 & 4 \\ 2 & 1 & 4 & 3 \\ 3 & 4 & 1 & 2 \\ 4 & 3 & 2 & 1 \end{pmatrix}$$

as a positive integral combination of permutation matrices.

16.18 Let $G = (X, Y, \triangle)$ be a bipartite graph and let for all $A \subset X$, $|N(A)| > |A|$. Prove that every edge of G is in some matching M such that M saturates each vertex of X.

16.19 In this exercise, we wish to obtain a proof of the de Bruijn-Erdös theorem (Theorem 1.4.1). Proceed through the following steps. In the set up of Theorem 1.4.1, let $\mathbf{I} = (\mathbf{P}, \mathbf{L})$ be a linear space in which the number of lines $|\mathbf{L}| = b \leq v = |\mathbf{P}|$. Let x be a (fixed) point not on the (fixed) line X. Consider first the family

$$\overline{\mathbf{L}} = \{\overline{L} = \mathbf{P} - L : L \in \mathbf{L}\}$$

We wish to show that the family $\overline{\mathbf{L}}$ has an SDR in which x represents $\overline{X}$.

(a) Show that if $L_1, L_2, \ldots, L_r$ are r lines and if $r \leq b - 1$, then

$$|\overline{L_1} \cup \overline{L_2} \cup \cdots \cup \overline{L_r}| \geq r + 1$$

if $r \leq b - 1$ and is $\geq b$ if $r = b$.

(b) Show that there is an injective function $\phi : \overline{\mathbf{L}} \longrightarrow \mathbf{P}$ such that $\phi(\overline{Z}) \notin Z$ for all the lines Z.

(c) Use the previous assertion and

$$\sum_{p \in \mathbf{P}} r_p \geq \sum_{L \in \mathbf{L}} r_{\phi(\overline{L})} \geq \sum_{L \in \mathbf{L}} k_L = \sum_{p \in \mathbf{P}} r_p$$

to show that $r_x = k_X$ for all $x \notin X$, $b = v$ and every two lines intersect each other.

(d) Complete an alternative proof of Theorem 1.4.1.

16.20 Let $G = (X, Y, \triangle)$ be a bipartite graph. A subset B of $V = X \cup Y$ is said to be independent if no two vertices in B are adjacent to each other. Thus X is independent and so is Y. Show that G has a perfect matching iff the maximum size of an independent set in G is $\frac{|V|}{2}$.

16.21 Let $G = (X, Y, \triangle)$ be a bipartite graph and let G have a matching that saturates every vertex of X. Let A and B be subsets of X such that $|N(A)| = |A|$ and $|N(B)| = |B|$. Prove that $|N(A \cap B)| = |A \cap B|$.

16.22 Let $G = (X, Y, \triangle)$ be a bipartite graph. For a subset A of X, define $\sigma(A) = |A| - |N(A)|$, called the deficiency of A. Let

$$\sigma(G) = \max_{A \subset X} \{\sigma(A)\}$$

which we call the deficiency of the bipartite graph G.

(a) Let A and B be subsets of X. Then prove that

$$\sigma(A \cup B) + \sigma(A \cap B) \geq \sigma(A) + \sigma(B)$$

(b) Let A and B be subsets of X such that $\sigma(A) = \sigma(B) = \sigma(G)$. Then prove that

$$\sigma(A \cup B) = \sigma(A \cap B) = \sigma(G)$$

(c) Let $\sigma(G) > 0$ and let $A_1, A_2, \ldots, A_r$ be the set of all the subsets of X whose deficiency is equal to $\sigma(G)$. Then show that the set $S = \cap_{i=1}^{r} A_i$ satisfies the property that $\sigma(S) = \sigma(G)$ and S is the smallest subset of X with this property.

(d) Let $x \in S$ and let $X' = X - \{x\}$ and $S' = S - \{x\}$. Let $G' = (X', Y, \triangle')$ denote the bipartite subgraph of G on the vertex set that contains all the vertices except x. Show that every subset of X' has deficiency less than $\sigma(G)$ and hence $\sigma(G') < \sigma(G)$.

(e) Further, show that $\sigma(S') = \sigma(S) - 1$ and hence we have $\sigma(G') = \sigma(G) - 1$.

(f) Repeat the procedure to find a set $X^* \subset X$ and the corresponding bipartite subgraph G^* of G such that $\sigma(G^*) = 0$.

(g) Use the matching theorem (Theorem 16.1.3) to prove the defect version of the matching theorem: the largest number of vertices that can be matched into Y by a matching of G is equal to $|X| - \sigma(G)$.

16.23 Let $N \geq 2$ be a positive integer and let $N = \{1, 2, \ldots, n\}$. Let X be the set of all the k-subsets of N where $k \leq \frac{n-1}{2}$ is a fixed non-negative integer and let Y denote the set of all the $(k+1)$-subsets of N. We define the bipartite graph $G = (X, Y, \triangle)$ as follows. $A \in X$ is adjacent to $A' \in Y$ if $A \subset A'$. Prove the following.

(a) Each vertex of X has degree $n - k$ and each vertex of Y has degree $k + 1$.

(b) G has a matching that saturates every vertex of X and hence, in particular, $|X| \leq |Y|$.

(c) A sequence $(\alpha_0, \alpha_1, \ldots, \alpha_k, \ldots, \alpha_n)$ is called *a unimodal sequence* if for some $m \leq n$, we have:

$$\alpha_0 \leq \alpha_1 \leq \cdots \leq \alpha_m \geq \alpha_{m+1} \geq \cdots \geq \alpha_{n-1} \geq \alpha_n$$

Show that the sequence $(\alpha_0, \alpha_1, \ldots, \alpha_k, \ldots, \alpha_n)$ with $\alpha_k = \binom{n}{k}$ is a unimodal sequence.

16.24 Let $G = (X, Y, \triangle)$ be a bipartite graph. Let $d(z)$ denote the degree of z (that is the number of vertices adjacent to z). Let

$$\min_{x \in X}\{d(x)\} \geq \max_{y \in Y}\{d(y)\}$$

Show that G has a matching that saturates every vertex of X.

16.25 Let M_1 and M_2 be two edge-disjoint matchings of the bipartite graph $G = (X, Y, \triangle)$ such that $|M_1| > |M_2|$. Show that there exist edge disjoint matchings $M_1{}'$ and $M_2{}'$ such that $|M_1{}'| = |M_1| - 1$ and $|M_2{}'| = |M_2| + 1$ and $M_1{}' \cup M_2{}' = M_1 \cup M_2$.

16.26 Let $G = (X, Y, \triangle)$ be a bipartite graph. Prove the following.

(a) Let $X_i \subset X$ and $Y_i \subset Y$ for $i = 1, 2$. Let M_i be a matching that matches X_i onto Y_i for $i = 1, 2$. Then show that there is matching $M' \subset M_1 \cup M_2$ such that M' saturates both X_1 and Y_2.

(b) Let X' be the set of all the vertices of maximum degree in X and let Y' consists of all the vertices of maximum degree in Y. Show that G has a matching that saturates both X' and Y'.

16.27 Let G be a bipartite graph such that the degree of each vertex is $\leq k$. Show that the edge set of G can be expressed as a union of edge disjoint matchings $M_1, M_2, \ldots, M_k$ such that for all i from 1 to k we have

$$\left\lfloor \frac{m}{k} \right\rfloor \leq |M_i| \leq \left\lceil \frac{m}{k} \right\rceil$$

where m is the number of edges of G.

16.28 Let $A = [a_{i,j}]$ denote an $n \times n$ real matrix. Prove that the least number of lines covering all the non-zero entries in A is equal to n iff there do not exist permutation matrices P and Q such that $PAQ = B$ where B has the form

$$\begin{pmatrix} B_{11} & B_{12} \\ 0 & B_{21} \end{pmatrix}$$

such that the zero matrix is an $r \times s$ matrix with $r + s > n$.

16.29 Let the standard pack of 52 cards be arbitrarily arranged in a 4×13 array with 13 columns, each containing 4 cards.

(a) Show that we can find 13 cards in the thirteen columns with one card in each column such that the denominations (values) of all the chosen 13 cards are distinct.

(b) Use (a) or otherwise to prove that repeatedly exchanging cards of the same denominations, we can arrange the pack in such a way that each column has exactly one card of each suit.

16.30 Let r and n be natural numbers. An $r \times n$ *Latin rectangle* $A = a_{ij}$ is an $r \times n$ array with entries that are numbers from the set $[n]$ such that the following properties are satisfied.

(a) Every row of A is a permutation of the set $[n]$.

(b) Each column of A has distinct entries.

Note that the second stipulation forces $r \leq n$ and when $r = n$ we have a square array in which each column is also a permutation of the set $[n]$. This situation is called a *Latin Square* (of order n). Latin Squares are well-studied objects in the combinatorics of configurations and block designs and are extensively used the statistical theory of design of experiments. Let A be an $r \times n$ Latin rectangle with $r < n$. An $(r + 1) \times n$ Latin rectangle C is called an extension of A if the first r rows of C are the same as those of A (and hence C is constructed by adding an extra row to A). A completion of A is a Latin Square of order n whose first r rows are the same as those of A. Let A_j denote the set of numbers that appear on the j-th column of A (with $j = 1, 2, \ldots, n$). Thus $|A_j| = r$ and let $B_j = [n] - A_j$ denote the complement of A_j. Prove the following.

(a) The size of each B_j is $n - r$.

(b) Every i is in exactly r A_j's and hence in exactly $n - r$ of the B_j's.

(c) The family of sets $\{B_j : j = 1, 2, \ldots, n\}$ satisfies the marriage condition and hence has an SDR.

(d) A has an extension to an $(r + 1) \times n$ Latin Square.

(e) Inductively, every Latin rectangle can be completed to a Latin square.

(f) Show that the total number of Latin squares is at least

$$n!(n-1)! \cdots (n-r)! \cdots 2!1!$$

16.31 Let $G = (X, Y, \triangle)$ be a bipartite graph. Show that the following statements are equivalent.

(a) G is connected and every edge e of G is contained in a perfect matching of G.

(b) $|X| = |Y|$ and for all proper subsets A of X, we have $|N(A)| > |A|$.

(c) $\forall\ x \in X$ and $\forall\ y \in Y$, the subgraph G' with vertex set $V - \{x, y\}$ has a perfect matching.

Bipartite graphs G satisfying the condition stated in the assertion are called elementary bipartite graphs. See Lovasz [37].

16.32 Let $\{A_1, A_2, \ldots, A_n\}$ and $\{B_1, B_2, \ldots, B_n\}$ be two partitions of a set S.

(a) Show that there is a permutation τ of $[n]$ such that

$$A_1 \cap B_{\tau(1)} \neq \emptyset, A_2 \cap B_{\tau(2)} \neq \emptyset, \ldots, A_n \cap B_{\tau(n)} \neq \emptyset$$

iff for every k from 1 to n, no k of the sets $A_1, A_2, \ldots, A_n$ are contained in the union of $k - 1$ of the sets $B_1, B_2, \ldots, B_n$.

(b) Show that if $|A_i| = |B_j| = r$ (where r is a fixed positive integer) for $i, j = 1, 2, \ldots, n$, then a permutation satisfying the assertion in (a) holds.

(c) There are mn couple at a dance. The men are divided into m groups of n men each according to their ages and the women are divided into m groups of n women each according to their heights. Show that it it is possible to select m couples so that every age group of men and every height group of women is represented.

(d) Let H be a subgroup of finite index n of a group G. Show that we can find n elements $g_1, g_2, \ldots, g_n$ of G such that

$$G = g_1 H \cup g_2 H \cup \cdots \cup g_n H;$$

is a left coset decomposition and

$$G = Hg_1 \cup Hg_2 \cup \cdots \cup Hg_n$$

is a right coset decomposition of G.

16.33 Let $A = (a_0, a_1, \ldots, a_{n-1})$ be a sequence of natural numbers and let $M = (m_0, m_1, \ldots, m_{n-1})$ be a monotone increasing rearrangement of A. That is, $m_0 \leq m_1 \leq \ldots \leq m_{n-1}$ and the sequence M is a permutation of the sequence A. For a monotone increasing sequence $M = (m_0, m_1, \ldots, m_{n-1})$, define

$$F_n(m_0, m_1, \ldots, m_{n-1}) = \prod_{r=0}^{n-1} (m_r - r)_*$$

where $(m_r - r)_*$ means $\max\{m_r - r, 1\}$. For a sequence $A = (a_0, a_1, \ldots, a_{n-1})$, define

$$f_n(a_0, a_1, \ldots, a_{n-1}) = F_n(m_0, m_1, \ldots, m_{n-1})$$

where $(m_0, m_1, \ldots, m_{n-1})$ is a monotone increasing rearrangement of $(a_0, a_1, \ldots, a_{n-1})$.

(a) Fix $i \in \{0, 1, \ldots, n-1\}$ and let $a_i{}' > a_i$ and let

$$m_0 \leq m_1 \leq \ldots \leq m_{j-1} \leq m_j = a_i \leq m_{j+1} \leq m_{n-1}$$

and

$$m_0 \leq m_1 \leq \cdots \leq m_{j-1} \leq m_{j+1} \leq m_k \leq a_i{}' \leq m_{k+1} \leq m_{n-1}$$

Prove that

$$\frac{f_n(a_0, \ldots, a_i{}', \ldots, a_{n-1})}{f_n(a_0, \ldots, a_i, \ldots, a_{n-1})} = \frac{(m_{j+1} - j)_*}{(m_j - j)_*} \times \frac{(a_i{}' - k)_*}{(m_k - k)_*} \times \prod_{r=j+1}^{k-1} \frac{(m_{r+1} - r)_*}{(m_r - r)_*}$$

and hence f_n is a monotone increasing function.

(b) (Harder: see van Lint and Wilson [57]) Use induction on n and monotonicity of the function f_n to prove the following.
Let $\mathbf{A} = \{A_0, A_1, \ldots, A_{n-1}\}$ be a family of sets that satisfied the condition (16.1). That is, assume that in the union of any r sets we have at least r elements. Let $|A_i| = m_i$ where we also assume (w.l.o.g) that $m_0 \leq m_1 \leq \cdots \leq m_{n-1}$. Then the total number of SDRs of the family $\mathbf{A}$ is at least $F_n(m_0, m_1, \ldots, m_{n-1})$.

16.34 Show that in a set of 64 cats, we must have a set of 10 cats that can be arranged in such a way that the i-th cat is an ancestor of the $(i+1)$-th cat for $i = 1, 2, \ldots, 9$ or there is a set of 8 cats so that any two cats in this set are pairwise unrelated (that is, no one is an ancestor of the other). Also, show that the assertion of the previous sentence can be derived using techniques other than those of this chapter.

16.35 Let r be a fixed real number and let V_r denote the set of all the $n \times n$ real matrices $A = [a_{i,j}]$ with the following two properties:

$$\sum_{j=1}^{n} a_{i,j} = r \; \forall \, i = 1, 2, \ldots, n$$

$$\sum_{i=1}^{n} a_{i,j} = r \; \forall \, j = 1, 2, \ldots, n$$

Let $V = \bigcup_{r \in \mathbb{R}} V_r$.

(a) Show that V is vector space over $\mathbb{R}$ with the operations of matrix additions and scalar multiplication by real numbers.

(b) Find a set of matrices in V whose linear span equals V.

(c) Show that the dimension of V is $(n-1)^2+1$.

16.36 Given a real $n \times n$ matrix $A = [a_{ij}]$ the *permanent* of A is defined by:

$$per(A) = \sum_{\sigma \in S_n} \prod_{i=1}^{n} a_{i,\sigma(i)}$$

(a) What is the difference between the permanent of A and the determinant of A?

(b) Show that if A is a doubly stochastic matrix, then $per(A)$ is a positive real number.

(c) Show that if all the entries of A are $\frac{1}{n}$, then $per(A) = \frac{n!}{n^n}$.

The permanent conjecture (solved in the last thirty years) states that if A is a doubly stochastic matrix of order n, then $per(A) \geq \frac{n!}{n^n}$ with equality iff A is matrix with all the entries equal to $\frac{1}{n}$ as described above. For a proof, see van Lint and Wilson [57].

16.37 Let n be a natural number and let $\mathbf{P}$ denote the Boolean poset consisting of all the subsets of $[n]$. A symmetric chain C is a chain in $\mathbf{P}$ such that for some $k \leq \frac{n}{2}$, C consists of subsets of all the orders from k to $n-k$. For example, with $n = 9$ and $k = 3$

$$C = \{\{3,5,6\},\{2,3,5,6\},\{1,2,3,5,6\},\{1,2,3,5,6,8\}\}$$

is a symmetric chain. Notice that any symmetric chain contains exactly one subset of order $\lfloor \frac{n}{2} \rfloor$. Show that if we can decompose $\mathbf{P}$ into symmetric chains, then the largest size of an antichain in $\mathbf{P}$ is equal to $\binom{n}{\lfloor \frac{n}{2} \rfloor}$. *A systematic procedure for obtaining a symmetric chain decomposition of the Boolean poset was given by de Bruijn and others*; see Rota [47].

16.38 Let $X = \mathbb{N} \times \mathbb{N}$ and define the relation $\leq$ on X by $(a,b) \leq (c,d)$ if $a \leq c$ and $b \leq d$. Prove the following.

(a) $\leq$ is a partial order on X.

(b) $\forall\, m \in \mathbb{N}$ there exists an antichain of size m in X.

(c) X cannot be covered by a finite number of chains.

(d) X does not have an infinite antichain.

(e) Compare the results of the previous statements with the statement of Dilworth's theorem (Theorem 16.5.11).

References

[1] M. Aigner, Combinatorial Theory, Springer-Verlag (1997).

[2] M. Aigner, Moving into the Desert with Fibonacci, Math. Magazine, Volume 70, No.1 (Feb 1997), pages 11-20.

[3] M. Artin, Algebra, Prentice Hall (1994).

[4] E.F. Beckenbach, Applied Combinatorial Mathematics, John Wiley and Sons (1964).

[5] E.A. Bender and S.G. Williamson, A Short course in Discrete Mathematics, Dover (2005).

[6] G. Berman and K.D.Fryer, Introduction to Combinatorics, Academic Press (1972).

[7] P.B. Bhattacharya, S.K. Jain and S.R. Nagpal, Basic Abstract Algebra, Cambridge University Press (1986).

[8] N.L. Biggs, Discrete Mathematics, Oxford University Press (2003).

[9] P. Billingsley, Probability and Measure, John Wiley and Sons (2012)

[10] B. Bollobas, Modern Graph Theory, Springer-Verlag (1998).

[11] M. Bona, Combinatorics of Permutations, CRC Press, Chapman-Hall (2004).

[12] M. Bona, Introduction to Enumerative Combinatorics, McGraw-Hill (2007).

[13] R.A. Brualdi, Introductory Combinatorics, North-Holland (1977).

[14] D. Burton, Elementary Number Theory, McGraw Hill (2002).

[15] P.J. Cameron, Combinatorics: Topics, Techniques and Algorithms, Cambridge University Press (1998).

[16] M. Capinski and T. Zastawniak, Probability Through Problems, Springer-Verlag (2001).

[17] K.L.Chung, A course in Probability Theory, Academic Press (2001).

[18] D.I.A. Cohen, Basic Techniques of Combinatorial Theory, John Wiley and Sons (1978).

[19] H.S.M. Coxeter, Introduction to Geometry, Wiley Classics Library (1969).

[20] R. Diestel, Graph Theory, Springer-Verlag (2000).

[21] D.Djukic, V. Jankovic, I. Matic and N. Petrovic, The IMO Compendium, A collection of Problems for the International Mathematical Olympiads: 1959-2004, Springer-Verlag (2006).

[22] A. Engel, Problem-Solving Strategies, Problem Books in Mathematics, Springer International Edition (1998).

[23] P. Erdös, R.L. Graham, P. Montgomery, B.L. Rothschild, J.H. Spencer and E.G. Straus, Euclidean Ramsey Theorem, J. Combin. Theory, Series A, Vol. 14, pages 341-363 (1973).

[24] W. Feller, An Introduction to Probability Theory and its Applications, Volumes I and II, John Wiley and Sons (1968).

[25] R.A.Gangoli and D.Ylvisaker, Discrete Probability, Harbrage College Mathematics Series, Harcourt Brace and World Inc.(1967).

[26] I.P.Goulden and D.M.Jackson, Combinatorial Enumerations, Wiley Interscience (2003).

[27] R.L. Graham, D. Knuth and O. Patashnik, Concrete Mathematics, Addison-Wesley Publishing Company (1989).

[28] R.L. Graham, B.L. Rothschild and J.H. Spencer, Ramsey Theory, Springer-Verlag (1998).

[29] M. Hall, Jr., Combinatorial Theory, Wiley Interscience (1983).

[30] R. Honsberger, Mathematical Gems I, Mathematical Association of America (1973).

[31] R. Honsberger, Mathematical Gems II, Mathematical Association of America (1976).

[32] R. Honsberger, From Erdös to Kiev, Problems of Olympiad Caliber, Mathematical Association of America (1996).

[33] M. Kac, G.C. Rota and J. Schwartz, Discrete Thoughts, Essays on Mathematics, Science and Philosophy, Birkhauser (1985).

[34] D. E. Knuth, Art of Computer programming, Volume 1: Fundamental algorithms, and Volume 3: Sorting and Searching, Addison-Wesley (1998).

[35] V. Krishnamurty, Combinatorics: Theory and Applications, Affiliated East-West Press (1985).

[36] C.L. Liu, Introduction to Combinatorial Mathematics, McGraw Hill (1968).

[37] L. Lovász, Combinatorial Problems and Exercises, North-Holland (1991).

[38] L. Lovász, J. Pellikan and K. Vesztergombi, Discrete Mathematics, Elementary and beyond, Springer Undergraduate texts in mathematics (2003).

[39] I.S. Luthar and I.B.S. Passi, Algebra, Volume 1: Groups, Narosa Publications (2009).

[40] F.S. Roberts, Applied Combinatorics, Prentice-Hall (1984).

[41] G. Metafune, D. Pallaro and C. Sempi, Euler Convergence: Probabilistic considerations, Mathematics Magazine Volume 72/4 (1999), 314-316.

[42] L. Mirsky, Transversal Theory, Mathematics in Science and Engineering, Volume 75, Academic Press (1971).

[43] D. Mumford, The dawning of the age of stochasticity, in Mathematics: Frontiers and Perspectives, edited by V. Arnold, M. Atiyah, P. Lax and B. Mazur, American Mathematical Society (2000).

[44] I. Niven and H.S. Zuckerman, An Introduction to Number Theory, Wiley Eastern Limited (1989).

[45] J.S. Rose, A Course in Group Theory, Dover (1994).

[46] K.E. Rosen, Discrete Mathematics and its Applications, McGraw-Hill (2007).

[47] G.C.Rota, Studies in Combinatorics, edited by G.C. Rota, Mathematical Association of America (1978).

[48] J.J. Rotman, Introduction to the Theory of Groups, Allyn and Bacon (1994).

[49] R. Stanley, Enumerative Combinatorics, Volume 1, Cambridge University Press (1997).

[50] D. Stanton and D. White, Constructive Combinatorics, Springer Graduate Texts in Mathematics (1986).

[51] M.N.S.Swamy and K. Thulsiraman, Graphs, Networks and Algorithms, Wiley Interscience (1981).

[52] J. Tanton, Solve this: Mathematics Activities for Students and Clubs, Classroom Source Material, Mathematical Association of America (2001).

[53] Thomas, R., A Combinatorial Construction of a Nonmeasurable Set, American Math. Monthly, 92 (1985), no. 6, 421-422.

[54] I. Tomescu and R.A. Melter, Problems in Combinatorics and Graph Theory, John Wiley and Sons (1985).

[55] T. Tsuzuku, Finite Groups and Finite Geometries, Cambridge University Press (1982).

[56] A. Tucker, Applied Combinatorics, John Wiley and Sons (1987).

[57] J.H. van Lint and R.M. Wilson, A Course in Combinatorics, Cambridge University Press (1998).

[58] J.J.Watkins, Across the Board; The Mathematics of Chessboard Problems, Universities Press (2005).

[59] website: www.cuttheknot.org: Interactive Mathematical Miscellany.

[60] D.B. West, Introduction to Graph Theory, Prentice-Hall (2001).

[61] H.S. Wilf, Generatingfuctionology, Academic Press (1994).

[62] J. Wimp, Review of "Selected Problems in Real Analysis, Volume 107, AMS series" in: Mathematical Intelligencer, Volume 16, no. 4 (1994), pages 68-72.

Index

Texts and Readings in Mathematics

1. R. B. Bapat: Linear Algebra and Linear Models (3/E)
2. Rajendra Bhatia: Fourier Series (2/E)
3. C.Musili: Representations of Finite Groups
4. Henry Helson: Linear Algebra (2/E)
5. Donald Sarason: Complex Function Theory (2/E)
6. M. G. Nadkarni: Basic Ergodic Theory (3/E)
7. Henry Helson: Harmonic Analysis (2/E)
8. K. Chandrasekharan: A Course on Integration Theory
9. K. Chandrasekharan: A Course on Topological Groups
10. Rajendra Bhatia(ed.): Analysis, Geometry and Probability
11. K. R. Davidson: C* -Algebras by Example
12. Meenaxi Bhattacharjee et al.: Notes on Infinite Permutation Groups
13. V. S. Sunder: Functional Analysis - Spectral Theory
14. V. S. Varadarajan: Algebra in Ancient and Modern Times
15. M. G. Nadkarni: Spectral Theory of Dynamical Systems (2/E)
16. A. Borel: Semi-Simple Groups and Symmetric Spaces
17. Matilde Marcoli: Seiberg Witten Gauge Theory
18. Albrecht Bottcher:Toeplitz Matrices, Asymptotic Linear Algebra and Functional Analysis
19. A. Ramachandra Rao and P Bhimasankaram: Linear Algebra (2/E)
20. C. Musili: Algebraic Geomtery for Beginners
21. A. R. Rajwade: Convex Polyhedra with Regularity Conditions and Hilbert's Third Problem
22. S. Kumaresen: A Course in Differential Geometry and Lie Groups
23. Stef Tijs: Introduction to Game Theory
24. B. Sury: The Congruence Subgroup Problem - An Elementary Approach Aimed at Applications
25. Rajendra Bhatia (ed.): Connected at Infinity - A Selection of Mathematics by Indians
26. Kalyan Mukherjea: Differential Calculas in Normed Linear Spaces (2/E)
27. Satya Deo: Algebraic Topology - A Primer (2/E)
28. S. Kesavan: Nonlinear Functional Analysis - A First Course (2/E)
29. Sandor Szabo: Topics in Factorization of Abelian Groups
30. S. Kumaresan and G.Santhanam: An Expedition to Geometry
31. David Mumford: Lectures on Curves on an Algebraic Surface (Reprint)
32. John. W Milnor and James D Stasheff: Characteristic Classes(Reprint)
33. K.R. Parthasarathy: Introduction to Probability and Measure
34. Amiya Mukherjee: Topics in Differential Topology
35. K.R. Parthasarathy: Mathematical Foundation of Quantum Mechanics (Corrected Reprint)
36. K. B. Athreya and S.N.Lahiri: Measure Theory
37. Terence Tao: Analysis - I (4E)
38. Terence Tao: Analysis - II (4E)
39. Wolfram Decker and Christoph Lossen: Computing in Algebraic Geometry
40. A. Goswami and B.V.Rao: A Course in Applied Stochastic Processes
41. K. B. Athreya and S.N.Lahiri: Probability Theory
42. A. R. Rajwade and A.K. Bhandari: Surprises and Counterexamples in Real Function Theory
43. Gene H. Golub and Charles F. Van Loan: Matrix Computations (Reprint of the 4/E)

44. Rajendra Bhatia: Positive Definite Matrices
45. K.R. Parthasarathy: Coding Theorems of Classical and Quantum Information Theory (2/E)
46. C.S. Seshadri: Introduction to the Theory of Standard Monomials (2/E)
47. Alain Connes and Matilde Marcolli: Noncommutative Geometry, Quantum Fields and Motives
48. Vivek S. Borkar: Stochastic Approximation - A Dynamical Systems Viewpoint (2/E)
49. B.J. Venkatachala: Inequalities - An Approach Through Problems (2/E)
50. Rajendra Bhatia: Notes on Functional Analysis (2/E)
51. A. Clebsch: Jacobi's Lectures on Dynamics (2/E)
52. S. Kesavan: Functional Analysis (2/E)
53. V.Lakshmibai and Justin Brown: Flag Varieties - An Interplay of Geometry, Combinatorics and Representation Theory (2/E)
54. S. Ramasubramanian: Lectures on Insurance Models (2/E)
55. Sebastian M. Cioaba and M. Ram Murty: A First Course in Graph Theory and Combinatorics (2/E)
56. Bamdad R. Yahaghi: Iranian Mathematics Competitions 1973-2007
57. Aloke Dey: Incomplete Block Designs
58. R.B. Bapat: Graphs and Matrices (2/E)
59. Hermann Weyl: Algebraic Theory of Numbers(Reprint)
60. C L Siegel: Transcendental Numbers(Reprint)
61. Steven J. Miller and RaminTakloo-Bighash: An Invitation to Modern Number Theory (Reprint)
62. John Milnor: Dynamics in One Complex Variable (3/E)
63. R. P. Pakshirajan: Probability Theory: A Foundational Course
64. Sharad S. Sane: Combinatorial Techniques
65. Hermann Weyl: The Classical Groups-Their Invariants and Representations (Reprint)
66. John Milnor: Morse Theory (Reprint)
67. Rajendra Bhatia(Ed.): Connected at Infinity II- A Selection of Mathematics by Indians
68. Donald Passman: A Course in Ring Theory (Reprint)
69. Amiya Mukherjee: Atiyah-Singer Index Theorem- An Introduction
70. Fumio Hiai and Denes Petz: Introduction to Matrix Analysis and Applications
71. V. S. Sunder: Operators on Hilbert Space
72. Amiya Mukherjee: Differential Topology
73. David Mumford and Tadao Oda: Algebraic Geometry II
74. Kalyan B. Sinha and Sachi Srivastava: Theory of Semigroups and Applications
75. Arup Bose and Snigdhansu Chatterjee: *U*-Statistics, M_m-Estimators and Resampling
76. Rajeeva L. Karandikar and B. V. Rao: Introduction to Stochastic Calculus
77. S. Kesavan: Measure and Integration
78. Avrim Blum, John Hopcroft and Ravi Kannan: Foundations of Data Science
79. Anima Nagar, Riddhi Shah and Shrihari Sridharan (Eds.): Elements of Dynamical Systems: Lecture Notes from NCM School
80. Hervé Queffélec and Martine Queffélec: Diophantine Approximation and Dirichlet Series (2/E)
81. M. Vidyasagar: An Introduction to Compressed Sensing
82. Apoorva Khare: Matrix Analysis and Entrywise Positivity Preservers
83. Vivek Borkar and Mallikarjuna Rao: Elementary Convexity with Optimization
84. Rajaraman Bhat & Tirthankar Bhattacharyya: Dilations, Completely, Positive Maps and G
85. M S Raghunathan: Lie Groups and Lie Algebras